# Nonlinear Vibration of Mechanical Systems

# Nonlinear Vibration of Mechanical Systems

Guest Editors

**Francesco Pellicano**
**Yuri Mikhlin**
**Konstantin V. Avramov**
**Antonio Zippo**

MDPI

Basel • Beijing • Wuhan • Barcelona • Belgrade • Novi Sad • Cluj • Manchester

*Guest Editors*
Francesco Pellicano
Department of Engineering
Enzo Ferrari
University of Modena and
Reggio Emilia
Modena
Italy

Yuri Mikhlin
Department of Applied
Mathematics
National Technical University
"Kharkiv Polytechnic
Institute"
Kharkiv
Ukraine

Konstantin V. Avramov
Anatolii Pidhornyi Institute of
Mechanical Engineering
Problems of the National
Academy of Sciences of
Ukraine
Kharkiv
Ukraine

Antonio Zippo
Department of Engineering
Enzo Ferrari
University of Modena and
Reggio Emilia
Modena
Italy

*Editorial Office*
MDPI AG
Grosspeteranlage 5
4052 Basel, Switzerland

This is a reprint of the Special Issue, published open access by the journal *Vibration* (ISSN 2571-631X), freely accessible at: https://www.mdpi.com/journal/vibration/special_issues/PEC39T1TWW.

For citation purposes, cite each article independently as indicated on the article page online and as indicated below:

Lastname, A.A.; Lastname, B.B. Article Title. *Journal Name* **Year**, *Volume Number*, Page Range.

**ISBN 978-3-7258-7811-6 (Hbk)**
**ISBN 978-3-7258-7812-3 (PDF)**
**https://doi.org/10.3390/books978-3-7258-7812-3**

# Contents

# About the Editors

**Francesco Pellicano**

Francesco Pellicano is an Aeronautical Engineer and holds a Ph.D. in Theoretical and Applied Mechanics. He is a Full Professor at the University of Modena and Reggio Emilia, a member of the European Academy of Sciences and Arts, and a member of the Executive Committee of the ASME Applied Mechanics Division. He serves as General Chair of the International Symposium on Vibrations of Continuous Systems (ISVCS) and Track Chair at the ASME IMECE Conference. He has been President of two BSc and two MSc programs and Head of the Intermech MoRe Centre.

He has coordinated several national and international research projects (as PI or local coordinator), including COMETA, a NATO project on composite metamaterials; THEORETIC (electric powertrain); DIADEM (diagnostics and innovative sensors); REFIMAN (industrial retrofitting); METaGEAR POR-FESR (gears, materials, robotics); INDGEAR; EU-FP7 projects on condition monitoring; and HPGA Fortissimo on high-performance computing applications.

He is the author of 3 books, about 100 journal papers, and more than 100 conference papers. His bibliometric record includes over 200 Scopus-indexed papers, a h-index of 39, and roughly 4,500 citations.

His research focuses on nonlinear dynamics and complexity in mechanical and biomechanical systems; fluid–structure interaction involving shells with incompressible, compressible, and supersonic fluids, including experiments with non-Newtonian fluids; gear stress and vibration modeling, non-smooth and chaotic dynamics, optimization via genetic algorithms, diagnostics and prognostics; active and passive vibration control (piezoelectric actuators, variable stiffness suspensions, nonlinear absorbers, quasi-zero stiffness systems, origami isolators); shell dynamics and stability (modal interactions, nonlinear random response, thermal effects); and vibration modeling of single- and multi-walled carbon nanotubes, including size effects and van der Waals interactions.

**Yuri Mikhlin**

Yuri Mikhlin was born on September 21, 1947, in Dnipropetrovsk (now Dnipro, Ukraine). He received his MS in 1970 and his PhD in 1974 in Dynamics and Strength of Machines from Dnepropetrovsk State University. In 1988, he earned his Dr. Sci. in Theoretical Mechanics from the Institute for Problems in Mechanics of the Russian Academy of Sciences. He became a Professor in 1991 and, from 1995, held a position at the National Technical University "Kharkiv Polytechnic Institute," where he worked for the rest of his life.

His scientific career began in 1967 under the influence of Prof. Leonid I. Manevitch, whose school he later followed. Mikhlin devoted his research to nonlinear normal modes, contributing fundamental results on stability of nonlinear vibrations, asymptotic methods (including Padé approximants), resonances in dissipative systems, vibration suppression, rotor dynamics, chaotic and vibro-impact oscillations, and strongly nonlinear systems.

He authored over 300 publications, including 7 monographs, in leading journals such as *Journal of Sound and Vibration*, *Nonlinear Dynamics*, and *International Journal of Non-Linear Mechanics*. He also served as a guest editor for several journals and contributed influential review papers.

A dedicated teacher, he developed courses in nonlinear dynamics, asymptotic methods, and mathematical modeling, supervising numerous students, including two Doctors of Science and eight PhDs.

Mikhlin was active internationally, serving on editorial boards and participating in conferences worldwide. He organized ENOC minisymposia and chaired major conferences on nonlinear dynamics in Kharkiv. He was a member of AMS, GAMM, EUROMECH, and NODYS.

Remembered as an outstanding scientist and a warm, generous person, he leaves a lasting legacy among colleagues and friends.

**Konstantin V. Avramov**

Konstantin V. Avramov, Dr. Tech. Sci., is a Ukrainian mechanical engineer and Professor of Systems Engineering at Kharkiv National University of Radio Electronics. He graduated from NTU Kharkiv Polytechnic Institute in 1991 with a degree in Dynamics and Strength of Machines, received his Candidate of Technical Sciences degree in 1993, and his Doctor of Technical Sciences in Deformable Solid Mechanics in 2005. He became a Full Professor in 2006.

His teaching covers system design and advanced engineering topics. His research spans structural mechanics, stress–strain analysis with defects, condition assessment, and residual life of long-lived structures, fluid–structure interaction, launch vehicle dynamics, and vibration protection systems. He has authored more than 150 publications and presented over 80 conference papers.

Avramov is widely recognized for his contributions to nonlinear dynamics and structural vibration, including nonlinear normal modes and asymptotic methods for nonlinear systems. His work addresses dynamic instability, forced vibrations of shells, flexural torsional vibrations of rotating beams, traveling wave and parametric vibrations in cylindrical shells, and geometrical nonlinear effects in plates and shells. He has co-authored influential studies with Yuri V. Mikhlin on nonlinear normal modes and their applications.

His papers have appeared in leading journals such as the *Journal of Vibration and Control*, *Nonlinear Dynamics*, *Communications in Nonlinear Science and Numerical Simulation*, *Meccanica*, *Acta Mechanica*, the *Journal of Sound and Vibration*, and *International Applied Mechanics*. His research integrates advanced analytical methods with practical engineering applications in aerospace and structural systems.

**Antonio Zippo**

Antonio Zippo, Ph.D., is an Associate Professor at the Department of Engineering Enzo Ferrari, University of Modena and Reggio Emilia. He teaches applied mechanics, mechanical vibrations, multibody dynamics, vehicle mechanics and testing, predictive maintenance, and biomechanics. He received his Ph.D. in High Mechanics and Automotive Design and Technology with a dissertation on nonlinear dynamics and active vibration control of thin-walled structures, and he is a qualified Professional Industrial Engineer.

His research integrates theory, computation, and experiments to investigate complex mechanical dynamics, including nonlinear responses, coupling effects, and stability transitions. A central focus is on the nonlinear dynamics of mechanical systems, with emphasis on stability, bifurcations, and reduced order modeling. These approaches support vibration analysis and control, including both active and passive strategies for complex structures under realistic operating conditions.

A major application area is the structural dynamics of thin-walled structures, especially shells and plates, where geometric nonlinearities and modal interactions dominate system behavior. His work develops control-oriented and design-ready models for mitigation, optimization, and monitoring. He also studies powertrain and transmission dynamics, including gear systems and coupled driveline behavior, with attention to how misalignment, wear, and stiffness variations affect vibration signatures and enable fault detection.

His research further addresses NVH reduction in industrial and automotive systems, including advanced design concepts and metamaterial-inspired solutions. This perspective connects with predictive maintenance and digital twins, combining physics-based models and data-driven methods for diagnostics and prognostics. He also contributes to fluid structure interaction and bioengineering applications, including the modeling of human motion and tremor dynamics.

# Preface

This Special Issue Reprint addresses nonlinear vibrations in engineering systems, from macro- to nanoscale devices/structures. Motivated by the need to model governing nonlinearities in modern systems, it aims to present recent theoretical, computational, experimental, and data-driven advances. It is intended for researchers, engineers, and advanced students in applied mechanics and dynamics.

**Francesco Pellicano, Yuri Mikhlin, Konstantin V. Avramov, and Antonio Zippo**
*Guest Editors*

vibration

MDPI

*Editorial*

# Editorial for the Special Issue of *Vibration*: Nonlinear Vibration of Mechanical Systems

**Francesco Pellicano [1,*], Yuri V. Mikhlin [2], Konstantin V. Avramov [3] and Antonio Zippo [1]**

1 Department of Engineering Enzo Ferrari, Centre InterMech MoRe, University of Modena and Reggio Emilia, 41125 Modena, Italy; antonio.zippo@unimore.it
2 Department of Applied Mathematics, National Technical University "Kharkiv Polytechnic Institute", 61000 Kharkiv, Ukraine; muv@kpi.kharkov.ua
3 Anatolii Pidhornyi Institute of Mechanical Engineering Problems of the National Academy of Sciences of Ukraine, 02000 Kharkiv, Ukraine; kvavr@kharkov.ua
* Correspondence: francesco.pellicano@unimore.it

Nonlinear vibration phenomena play a central role in modern engineering, spanning applications from large-scale civil infrastructure to microscale and nanoscale systems. The recent contributions published in the journal *Vibration* highlight both the richness of nonlinear dynamic behaviours and the growing complexity of analytical, numerical, experimental, and data-driven tools.

In terms of fundamentals, nonlinear oscillations arising from contact mechanics remain a topic of great relevance. The recent work on Hertzian-type restoring forces provides new analytical approximations for strongly nonlinear oscillators, where classical closed-form solutions are unavailable (Contribution 1). By preserving the dominant nonlinear scaling, while incorporating higher-order corrections, such solutions offer accurate and computationally efficient alternatives to purely numerical approaches, supporting theoretical insight and parametric studies of contact-driven vibrations.

Nonlinearities also play a decisive role in vibration isolation and mitigation technologies. Simplified yet physically consistent modelling strategies, applied to wire-rope isolators, demonstrate how the geometric nonlinearities and the hysteretic damping can be modelled using reduced finite-element formulations (Contributions 2,3). Similarly, studies on high-static–low-dynamic stiffness and quasi-zero-stiffness mechanisms confirm the effectiveness of nonlinear kinematics of the suspensions in achieving broadband vibration reduction under harmonic, seismic, and real-world excitations, while maintaining acceptable static load-bearing capacity (Contribution 4).

In smart and adaptive structures, nonlinear electromechanical coupling and hysteresis present both challenges and great perspectives. The advanced modelling methods for piezoelectric composite beams, including hysteretic operators, reveal the critical influence of memory-dependent nonlinearities on dynamic response and limit-cycle behaviours. Regarding mechanical transmissions, investigations on the motor–gearbox coupling underline the importance of considering bidirectional electromechanical feedback, showing that neglecting the coupling may lead to inaccurate predictions of the stability and performance of electric drive systems (Contribution 5).

At nanoscales, nonlinear vibration phenomena are increasingly studied in conjunction with multiphysics effects and data-driven methods. The integration of machine learning techniques with physical models, including fluid interactions, of carbon nanotubes shows a promising direction for handling high-dimensional parameter spaces and complex environmental interactions (Contribution 6). These hybrid approaches enhance predictive capability and open new perspectives for applications in nanotechnology, biomedical devices, and energy harvesting.

Beyond isolation and control, nonlinear vibrations are also central to flow-induced dynamics, frictional systems, and asymmetric excitation mechanisms. Recent studies focused on stability and bifurcation of systems with friction and geometric nonlinearities, as well as unidirectional flow generation in compliant tubes with time-varying geometry, show how asymmetry and nonlinearity can be used to achieve targeted dynamic functionalities. Moreover, methodologies for designing vibration devices with prescribed asymmetry prove the practical relevance of nonlinear techniques in industrial applications (Contributions 7–9).

Finally, applications to infrastructure and energy systems show the importance of research on nonlinear dynamics. Finite-element investigations of damaged pipelines, with reinforcement of composite linings, show how dynamic analyses complete the static strength assessments, contributing to safer and more resilient designs under actual seismic loading (Contribution 10,11). Advances in condition monitoring and reduced-order modelling support the performance-based design and structural health monitoring strategies (Contribution 12,13).

In general, the contributions of this Special Issue reflect the multidisciplinary and multiscale nature of modern research on vibration. They show a clear trend in combining nonlinear mechanics with efficient modelling, experimental validation, and data-driven techniques. As engineering systems become more efficient, lighter, and smarter, the role of studies on nonlinear vibrations will become more and more important, making the journal *Vibration* a central tool to disseminate both fundamental advances and applications.

**Conflicts of Interest:** The authors declare no conflicts of interest.

**List of Contributions:**


1. Kontomaris, S.; Ismail, G.; Malamou, A.; Stylianou, A. Mechanical Nonlinear Oscillations Using a Hertzian-Type Restoring Force. *Vibration* **2025**, *8*, 74. https://doi.org/10.3390/vibration8040074.
2. Marin-Artieda, C. Balancing Accuracy and Efficiency in Wire-Rope Isolator Modeling: A Simplified Beam-Element Framework. *Vibration* **2025**, *8*, 55. https://doi.org/10.3390/vibration8030055.
3. Alazemi, A.; Kurdila, A. Modeling Hysteretically Nonlinear Piezoelectric Composite Beams. *Vibration* **2025**, *8*, 37. https://doi.org/10.3390/vibration8030037.
4. Iarriccio, G.; Zippo, A.; Eskandary-Malayery, F.; Ilanko, S.; Mochida, Y.; Mace, B.; Pellicano, F. Tunable High-Static-Low-Dynamic Stiffness Isolator under Harmonic and Seismic Loads. *Vibration* **2024**, *7*, 829–843. https://doi.org/10.3390/vibration7030044.
5. Zippo, A.; Molaie, M.; Pellicano, F. Nonlinear Dynamics of a Coupled Electromechanical Transmission. *Vibration* **2025**, *8*, 34. https://doi.org/10.3390/vibration8030034.
6. Yinusa, A.; Amokun, R.; Eke, J.; Sobamowo, G.; Oguntala, G.; Ehinmowo, A.; Salami, F.; Osigwe, O.; Adelaja, A.; Ojolo, S.; Usman, M. Machine Learning Approach to Nonlinear Fluid-Induced Vibration of Pronged Nanotubes in a Thermal–Magnetic Environment. *Vibration* **2025**, *8*, 35. https://doi.org/10.3390/vibration8030035.
7. Gerasimov, M.; Lubimyi, N.; Polshin, A.; Chetverikov, B.; Chetverikova, A. Methodology for Designing Vibration Devices with Asymmetric Oscillations and a Given Value of the Asymmetry of the Driving Force. *Vibration* **2025**, *8*, 3. https://doi.org/10.3390/vibration8010003.
8. Hur, D.; Hong, S. Analysis of Displacement Transmissibility and Bifurcation Behavior in Nonlinear Systems with Friction and Nonlinear Spring. *Vibration* **2024**, *7*, 1210–1225. https://doi.org/10.3390/vibration7040062.
9. Manopoulos, C.; Tsangaris, S.; Georgantopoulou, C.; Mathioulakis, D. Unidirectional Flow Through Time-Dependent Cross-Sectional Areas of a Compliant Tube and a Valve: A Nonlinear Model. *Vibration* **2024**, *7*, 987–998. https://doi.org/10.3390/vibration7040052.


https://doi.org/10.3390/vibration9010010

10. Moldagaliyev, A.; Zhangabay, N.; Bonopera, M.; Ristavletov, R.; Ibraimova, U.; Yeshimbetov, S. Dynamics of Steel Gas Pipelines: Finite Element Simulation of Damaged Sections Reinforced with Composite Linings. *Vibration* **2025**, *8*, 8. https://doi.org/10.3390/vibration8010008.
11. Landar, S.; Velychkovych, A.; Ropyak, L.; Andrusyak, A. A Method for Applying the Use of a Smart 4 Controller for the Assessment of Drill String Bottom-Part Vibrations and Shock Loads. *Vibration* **2024**, *7*, 802–828. https://doi.org/10.3390/vibration7030043.
12. Hamedi, B.; Shrikanthan, S.; Taheri, S. The Reduced-Order Modeling Approach for a Double-Damper Concept: A Comparison with a Single Damper for Comfort Analysis. *Vibration* **2024**, *7*, 644–661. https://doi.org/10.3390/vibration7030034.
13. Dolbachian, L.; Harizi, W.; Aboura, Z. Experimental Linear and Nonlinear Vibration Methods for the Structural Health Monitoring (SHM) of Polymer-Matrix Composites (PMCs): A Literature Review. *Vibration* **2024**, *7*, 281–325. https://doi.org/10.3390/vibration7010015.



https://doi.org/10.3390/vibration9010010

vibration

MDPI

*Article*

# Mechanical Nonlinear Oscillations Using a Hertzian-Type Restoring Force

**Stylianos Vasileios Kontomaris [1,*], Gamal M. Ismail [2], Anna Malamou [3] and Andreas Stylianou [1,*]**

[1] School of Sciences, European University Cyprus, Nicosia 2404, Cyprus
[2] Department of Mathematics, Faculty of Science, Islamic University of Madinah, Madinah 42351, Saudi Arabia; gamalm2010@yahoo.com
[3] BioNanoTec Ltd., Nicosia 2043, Cyprus
* Correspondence: s.kontomaris@research.euc.ac.cy (S.V.K.); an.stylianou@euc.ac.cy (A.S.)

**Abstract:** This paper examines the generic case of nonlinear mechanical oscillation under the influence of Hertzian-type restoring forces, a model relevant to phenomena involving elastic contact. The study addresses the complexity of strongly nonlinear systems by focusing on the differential equation governing the oscillation of a rigid sphere interacting with an elastic half-space, which includes a full series expansion to account for large deformations. Since no closed-form solution exists for the amplitude-dependent oscillation period, a new approximate analytical approach is introduced. This method preserves the system's dominant Hertzian scaling while incorporating higher-order corrections through an averaged factor. For amplitudes where the deformation is less than or equal to the sphere's radius, this approximation is nearly identical to the numerical solution. For larger amplitudes, the accuracy is further enhanced by introducing a semi-empirical linear adjustment to the relative error. This framework provides a reliable analytical description of the system's behavior, offering a useful tool for theoretical studies and comparison with numerical results.

**Keywords:** nonlinear oscillations; hertzian contact; elastic half-space; analytical approximation; semi-empirical correction; amplitude-dependent period; rigid sphere; large deformations

## 1. Introduction

Nonlinear oscillations arise across many areas of science, with significant applications in mechanics and engineering [1–6], physics [7–11], and mathematics [12–15]. They are typically described by nonlinear differential equations (NDEs), whose strong nonlinearities make their analysis particularly challenging. This complexity has driven researchers to develop a variety of approximation techniques, and over the past few decades many solution methods have been proposed. Classical perturbation methods, however, fall short when dealing with strongly nonlinear oscillators, as they fail to provide accurate results. To overcome these limitations, several alternative analytical approaches have been introduced. A wide spectrum of methods has been applied to nonlinear oscillators, such as the harmonic balance method [16,17], variational iteration method [18,19], homotopy analysis method [20,21], homotopy perturbation method [22,23], Li-He's modified homotopy perturbation method [24], enhanced homotopy perturbation method [25], asymptotic techniques [26,27], energy balance method [28], differential transformation method [29], parameter expansion schemes [30,31], variational principles [32], frequency–amplitude formulations [33], among others.

Recent advancements in the evaluation of mechanical nonlinearities have introduced refined experimental and analytical methodologies for capturing complex dynamic behaviors. Bonisoli et al. investigated the nonlinear dynamics of cantilever beams subjected to magnetic and contact-induced nonlinearities, employing a laser profilometer for precise displacement measurements [34]. Their Multi-Phi linearization framework effectively demonstrated how nonlinearities influence the dynamic response of mechanical structures. Similarly, other studies have focused on identifying instantaneous modal parameters, such as natural frequencies and damping coefficients, and their dependence on vibration amplitude and excitation conditions [35]. Using signal processing techniques, including the Hilbert transform, these works have provided efficient schemes for detecting nonlinear effects in single-degree-of-freedom systems exhibiting hardening behavior. Collectively, these approaches emphasize the importance of integrating experimental observations with analytical models to characterize nonlinear mechanical responses and to complement theoretical formulations. Additionally, the wavelet transform provides a time–frequency representation for system identification, enabling accurate estimation of natural frequencies and viscous damping ratios from free vibration responses. Its advantages in analyzing free decay are highlighted through comparisons with previous techniques, and its accuracy is confirmed via numerical simulations and real-world measurements from the Queensborough Bridge in Vancouver, demonstrating agreement with established dynamic behavior [36].

Extensive research has also aimed at developing general models that can describe a wide range of nonlinear oscillations. For example, approximate solutions have been proposed for systems with cubic and harmonic restoring forces [37,38], a problem of particular importance, as it captures phenomena that appear across mathematics, physics, and engineering [38,39]. This formulation encompasses well-known cases such as the Duffing oscillator [40], the simple pendulum [41], the cubic–quintic Duffing system [42], and the capillary oscillator [43,44]. Consequently, obtaining reliable approximations for such general cases is of fundamental importance in nonlinear science. In addition, special attention has been given to cases with polynomial type restoring forces [45]:

$$F = \lambda_1 x + \lambda_2 x^2 + \lambda_3 x^3 + \ldots \tag{1}$$

where $\lambda_1$, $\lambda_2$, $\lambda_3, \ldots$ are arbitrary constants. This is a very important case, since Equation (1) can be used to describe many cases of one-dimensional force fields through Maclauren series expansions [46]. The differential equation describing such cases is providing below [47]:

$$\ddot{x} + \sum_{i=1}^{j} \lambda_i |x|^{i-1} x = 0 \tag{2}$$

Another interesting case is the one involving elastic forces [47]. In this case, the force magnitude is usually described by a power law of the form $F = \lambda x^m$, where $\lambda$, $m$ are positive constants that depend on the material properties and the contact geometry of the interacting bodies [48,49]. The most well-known nonlinear example of such forces is the classical Hertzian model, which describes the contact between a sphere and an elastic half-space, with the corresponding force magnitude given as follows [50]:

$$F = \frac{4}{3} E^* R^{1/2} x^{3/2} \tag{3}$$

where

$$\frac{1}{E^*} = \frac{1 - v_{sph}^2}{E_{sph}} + \frac{1 - v^2}{E} \tag{4}$$

In Equations (3) and (4), $E_{sph}$, $E$, $v_{sph}$, $v$ represent the Young's moduli and Poisson's ratios of the contacting elastic materials (the sphere and the elastic half space), $R$ is the sphere's radius, $F$ is the applied force and $x$ is the displacement of the sphere within the elastic material. If considering that the sphere is much stiffer than the elastic half space (i.e., $E_{sph} \gg E$), then $E^* = E/\left(1 - v^2\right)$. A typical oscillatory motion under the conservative Hertz force field in this case is described by the equation below:

$$\ddot{x} + \frac{4}{3}\frac{E}{m(1-v^2)}R^{1/2}|x|^{1/2}x = 0 \quad (5)$$

where $m$ is the mass of the oscillating sphere. It is interesting to note that the half-period of the oscillatory motion described by Equation (5) can also be used to determine the duration of interaction between a rigid sphere and an elastic wall during a collision [51]. It is also important to note that the presented in this paper nonlinear oscillator behavior, is relevant to Atomic Force Microscopy (AFM) technology. Specifically, the undamped oscillations of a rigid sphere interacting with a linearly elastic surface are of particular interest. In many AFM force modes used to determine the Young's modulus of a sample, the spherical tip experiences forced oscillations while in contact with the surface [52]. Determining the frequency of the undamped free oscillation provides useful insight into the magnitude of the resonance frequency, since, in nonlinear systems, the oscillation frequency depends on amplitude. Although the resonance and free oscillation frequencies are not identical, they exhibit similar amplitude-dependent shifts, allowing the latter to serve as a practical approximation. Equation (5) belongs to the generic form below [47]:

$$\ddot{x} + c_\lambda x|x|^{a-1} = 0 \quad (6)$$

where $c_\lambda = \frac{4ER^{1/2}}{3m(1-v^2)}$ and $a = 3/2$. The accurate equation for the oscillation period in this case has been previously presented [47,53] and is provided below:

$$T = \frac{2\sqrt{2\pi}}{\sqrt{c_\lambda}A^{\frac{a-1}{2}}\sqrt{\alpha+1}}\left[\frac{\Gamma\left(\frac{1}{\alpha+1}\right)}{\Gamma\left(\frac{3+\alpha}{2\alpha+2}\right)}\right] \quad (7)$$

where $\Gamma$ is the gamma function and $A$ is the oscillation's amplitude. For example, for $a = 3/2$:

$$T = \frac{6.58}{\sqrt{c_\lambda}}A^{-\frac{1}{4}} \quad (8)$$

While the case of oscillatory motion in a classic Hertz field (Equation (3)) has already been solved, there are many other cases that still require further investigation. More specifically, Equation (3) is valid only in the limit $x << R$. For large deformations, the Hertz equation provides significant errors in force calculations [54,55]. The accurate equations of the force between a rigid sphere and an elastic half space has been given by Sneddon's equations below [56,57]:

$$F = \frac{E}{2(1-v^2)}\left[\left(r_c^2 + R^2\right)ln\left(\frac{R+r_c}{R-r_c}\right) - 2r_cR\right] \quad (9)$$

where $r_c$ is the contact radius between the indenter and the sample. The deformation $x$ is connected to the contact radius and the indenter radius through the following relation [56,57]:

$$ln\left(\frac{R+r_c}{R-r_c}\right) = \frac{2x}{r_c}. \quad (10)$$

There is no analytical solution of the system of Equations (9) and (10). Müller et al. presented an approximate force equation using a truncated power series expansion; according to the paper, this expression was obtained through personal communication with Wolfgang Dobler (JPK Instruments, Berlin) [58]:

$$F = \frac{4}{3}\frac{E}{1-v^2}R^{1/2}x^{3/2}\left[1 - \frac{1}{10}\left(\frac{x}{R}\right) - \frac{1}{840}\left(\frac{x}{R}\right)^2 + \frac{11}{15,120}\left(\frac{x}{R}\right)^3 + \frac{1357}{6,652,800}\left(\frac{x}{R}\right)^4\right] \quad (11)$$

Since Sneddon's equations (Equations (9) and (10)) cannot be analytically solved, two approximate expressions have been proposed in the literature [57,58]. The present study adopts the formulation presented in [58] (i.e., Equation (11)), as it ensures high accuracy for $x \leq 3R$ and smoothly converges to the Hertzian solution at small indentation depths whereas the equation presented in [57] provides better results for very large deformations but introduces a small error in the Hertzian limit.

Equation (11) accounts for large deformations of the elastic material when interacting with the rigid sphere. In other words, the accurate equation relating the force applied between a rigid sphere and a linearly elastic material is given by Equation (11). Therefore, the differential equation involving oscillations under the force field described by Equation (11) is provided below:

$$\ddot{x} + Q\left(R^{\frac{1}{2}}|x|^{\frac{1}{2}}x - \frac{1}{10}R^{-\frac{1}{2}}|x|^{\frac{3}{2}}x - \frac{1}{840}R^{-\frac{3}{2}}|x|^{\frac{5}{2}}x + \frac{11}{15,120}R^{-\frac{5}{2}}|x|^{\frac{7}{2}}x + \frac{1357}{6,652,800}R^{-\frac{7}{2}}|x|^{\frac{9}{2}}x\right) = 0 \quad (12)$$

where

$$Q = \frac{4}{3}\frac{E}{m(1-v^2)} \quad (13)$$

This paper is organized as follows. In Section 2 we present the numerical method used to identify the oscillation period of the differential Equation (12). In Section 3, we present an approximate analytical approach for determining the oscillation period as a function of amplitude. The results are compared with numerical solutions, and the relative errors are also reported. Additionally, a simple sinusoidal solution, based on the analytically determined period, is presented and compared with the numerical results. In Section 4, the accuracy and limitations of the proposed method are also presented and discussed. Comparisons with the Harmonic Balance Method [45] are also presented.

## 2. Numerical Solution

Since the magnitude of interest involving a large deformation is $\xi = x/R$, $\ddot{x} = R\ddot{\xi}$ and Equation (12) can be expressed as follows:

$$\ddot{\xi} + QR\left(|\xi|^{\frac{1}{2}}\xi - \frac{1}{10}|\xi|^{\frac{3}{2}}\xi - \frac{1}{840}|\xi|^{\frac{5}{2}}\xi + \frac{11}{15,120}|\xi|^{\frac{7}{2}}\xi + \frac{1357}{6,652,800}|\xi|^{\frac{9}{2}}\xi\right) = 0 \quad (14)$$

Equation (14) is dimensionally consistent. The units of $\ddot{\xi}$ are $1/s^2$ which are equivalent to the units of the parameter $QR$. Therefore, in this paper, we focus on the case involving a one-dimensional force field with magnitude of the following form:

$$F(\xi) = \sum_{i=0}^{j} c_i \xi^{i+\frac{3}{2}} \quad (15)$$

To visualize the force field described by Equation (15), the terms $\Xi_1$ and $\Xi_2$ were plotted comparatively in Figure 1, where

$$\Xi_1 = |\xi|^{\frac{1}{2}}\xi - \frac{1}{10}|\xi|^{\frac{3}{2}}\xi - \frac{1}{840}|\xi|^{\frac{5}{2}}\xi + \frac{11}{15,120}|\xi|^{\frac{7}{2}}\xi + \frac{1357}{6,652,800}|\xi|^{\frac{9}{2}}\xi \quad (16)$$

and,

$$\Xi_2 = |\xi|^{\frac{1}{2}}\xi \tag{17}$$

It is interesting to note that the force field described by Equation (15) presents an interesting behavior as also shown in Figure 1. For small deformations, the force is proportional to the term $|\xi|^{3/2}$, whereas for larger deformations the force–deformation curve tends to become linear. This behavior has been previously explained in terms of contact mechanics and arises by the behavior of the contact area between the interacting bodies [57]. For large deformations, the contact area tends to a limiting value. This is the reason for obtaining this force–deformation behavior [57].

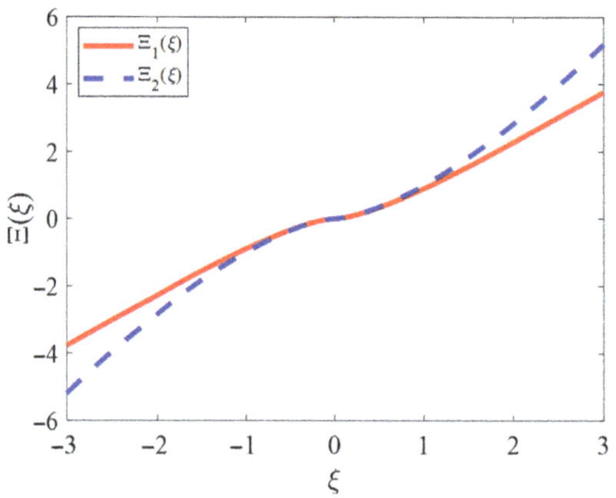


**Figure 1.** The $\Xi = f(\xi)$ curves were obtained using Equations (16) and (17). These curves were plotted to visualize both the classical Hertz force field and the force field described by Equation (15). For the latter case, the force is initially proportional to the term $|\xi|^{3/2}$ and tends to exhibit linear behavior for large deformations.

The differential equation to be solved, is described as below:

$$\ddot{\xi} + \sum_{i=0}^{j} c_i \xi |\xi|^{i+\frac{1}{2}} = 0 \tag{18}$$

Let $v(x) = \frac{d\xi}{dt}$, then using the chain rule we conclude in $\frac{d^2\xi}{dt^2} = \frac{dv}{dt} = \frac{dv}{d\xi}\frac{d\xi}{dt}$. Thus, Equation (18) is expressed as follows:

$$v\frac{dv}{d\xi} = -\sum_{i=0}^{j} c_i \xi |\xi|^{i+\frac{1}{2}} \Rightarrow vdv = -\sum_{i=0}^{j} c_i \xi |\xi|^{i+\frac{1}{2}} d\xi \Rightarrow \int vdv = -\int \sum_{i=0}^{j} c_i \xi |\xi|^{i+\frac{1}{2}} d\xi \Rightarrow$$

$$\frac{1}{2}\left(\frac{d\xi}{dt}\right)^2 + \sum_{i=0}^{j} \frac{2}{5+2i} c_i |\xi|^{i+\frac{5}{2}} = C \tag{19}$$

where C is the integration constant. Using the initial conditions, $\xi(0) = A$ and $\xi'(0) = 0$, we concluded in:

$$\frac{1}{2}\left(\frac{d\xi}{dt}\right)^2 + \sum_{i=0}^{j} \frac{2}{5+2i} c_i |\xi|^{i+\frac{5}{2}} = \sum_{i=0}^{j} \frac{2}{5+2i} c_i A^{i+\frac{5}{2}} \tag{20}$$

Equation (20) expresses the energy conservation of the oscillating system. The time interval required for the variable $\xi$ to change from $\xi$ = +$A$ (its value at $t$ = 0) to an arbitrary value $\xi$ is calculated as below:

$$\Delta t = t = -\int_{A}^{\xi} \frac{1}{\sqrt{\sum_{i=0}^{j} \frac{4}{5+2i} c_i A^{i+\frac{5}{2}} - \sum_{i=0}^{j} \frac{4}{5+2i} c_i |u|^{i+\frac{5}{2}}}} du \tag{21}$$

where $u$ is the dummy integration parameter. Therefore, the oscillation's period can be calculated as follows:

$$T = 4\int_{0}^{A} \frac{1}{\sqrt{\sum_{i=0}^{j} \frac{4}{5+2i} c_i A^{i+\frac{5}{2}} - \sum_{i=0}^{j} \frac{4}{5+2i} c_i |\xi|^{i+\frac{5}{2}}}} d\xi \tag{22}$$

For example, for $j = 4$ we conclude in:

$$T = 2\int_{0}^{A} \frac{1}{\sqrt{\frac{1}{5}c_0 A^{\frac{5}{2}} + \frac{1}{7}c_1 A^{\frac{7}{2}} + \frac{1}{9}c_2 A^{\frac{9}{2}} + \frac{1}{11}c_3 A^{\frac{11}{2}} + \frac{1}{13}c_4 A^{\frac{13}{2}} - \frac{1}{5}c_0|\xi|^{\frac{5}{2}} - \frac{1}{7}c_1|\xi|^{\frac{7}{2}} - \frac{1}{9}c_2|\xi|^{\frac{9}{2}} - \frac{1}{11}c_3|\xi|^{\frac{11}{2}} - \frac{1}{13}c_4|\xi|^{\frac{13}{2}}}} d\xi \tag{23}$$

where $c_0 = 4/3$, $c_1 = -4/30$, $c_2 = -4/2520$, $c_3 = 44/45{,}360$ and $c_4 = 14{,}927/19{,}958{,}400$ (we also considered $\frac{ER}{m(1-v^2)} = 1\ s^{-2}$ for simplicity and $t$, $T$ are measured in seconds. This parameter (i.e., $\frac{ER}{m(1-v^2)}$) serves only as a multiplicative factor and does not affect the results). Equation (23) can be solved numerically to provide the oscillation's period for any given amplitude.

An interesting question to be addressed concerns the $c_i$ coefficients in Equations (18)–(23), which originate from Equation (11). An analytical method for determining these coefficients is presented in [59]. The approach involves applying a Maclaurin series expansion to Equation (10) and subsequently solving the resulting expression with respect to $r_c/R$. By defining an equation of the form $r_c/R = f(x/R)$, we can substitute it into the general relation $\frac{dF}{dx} = 2E^* r_c$ for axisymmetric indenters and directly calculate the coefficient $c_1 = -4/30$. By including higher-order terms in the function $r_c/R = f(x/R)$, it is also possible to determine the remaining $c_i$ values so that our solution matches Sneddon's results. It is worth noting that several combinations of $c_i$ constants can lead to accurate solutions, as discussed in [59].

It is also important to note that as shown in Figure 1, $\xi$ takes both positive and negative values during oscillation. However, for analytical simplicity, the integration in Equation (19) was performed only over the positive range of $\xi$. Equation (19) represents the energy conservation principle, where the total energy, comprising kinetic and potential energy terms, remains constant throughout the motion. Since the oscillation is perfectly symmetric with respect to $\xi = 0$, it is sufficient to evaluate the motion over the positive half-cycle. The time interval $\Delta t$ corresponding to the change of $\xi$ from $+A$ to 0 represents one quarter of the total period ($T/4$). Thus, the complete oscillation period is obtained as $T = 4\Delta t$.

## 3. Analytical Approximate Approach

Let us re-examine the one-dimensional force field described by Equation (15). The force magnitude can be easily rewritten as follows:

$$F(\xi) = \sum_{i=0}^{j} c_i \xi^{i+\frac{3}{2}} = c_0 \xi^{\frac{3}{2}} \left(1 + \frac{c_1}{c_0}\xi + \frac{c_2}{c_0}\xi^2 + \ldots\right) = c_0 \xi^{\frac{3}{2}} \sum_{i=0}^{j} \frac{c_i}{c_0}\xi^i = c_0 \xi^{\frac{3}{2}} \Omega(\xi) \tag{24}$$

For $\Omega(\xi) = 1$, the oscillation's period as already mentioned in the introduction should be:

$$T = \frac{2\sqrt{2\pi}}{c_0^{\frac{1}{2}} A^{\frac{\frac{3}{2}-1}{2}} \sqrt{\frac{3}{2}+1}} \left[ \frac{\Gamma\left(\frac{1}{\frac{3}{2}+1}\right)}{\Gamma\left(\frac{3+\frac{3}{2}}{3+2}\right)} \right] = \frac{6.58}{\sqrt{c_0}} A^{-\frac{1}{4}} \tag{25}$$

An idea to account for cases where $\Omega(\xi) \neq 1$ is to consider an amplitude-dependent correction factor. The easiest approach is to consider the average value of $\Omega(\xi)$ over the domain $0 \leq \xi \leq A$. Therefore, we re-write Equation (24) in the following approximate form:

$$F(\xi) = c_0 \xi^{\frac{3}{2}} \Omega_{ave} \tag{26}$$

where

$$\Omega_{ave} = \frac{1}{A} \int_0^A \sum_{i=0}^{j} \frac{c_i}{c_0} \xi^i d\xi = \frac{1}{A} \int_0^A \left(1 + \frac{c_1}{c_0}\xi + \frac{c_2}{c_0}\xi^2 + \ldots\right) d\xi = 1 + \frac{c_1}{2c_0} A + \frac{c_2}{3c_0} A^2 + \ldots \tag{27}$$

Equation (27) can also be written in the generic form:

$$\Omega_{ave} = 1 + \sum_{i=1}^{j} \frac{c_i}{c_0(i+1)} A^i \tag{28}$$

Thus, by adding the correction factor $\Omega_{ave}$ to Equation (25) we finally conclude in:

$$T_{app.} = \frac{6.58}{\sqrt{c_0}\Omega_{ave}} A^{-\frac{1}{4}} = \frac{6.58}{\sqrt{c_0}\left[1 + \sum_{i=1}^{j} \frac{c_i}{c_0(i+1)} A^i\right]} A^{-\frac{1}{4}} \tag{29}$$

For $j = 4$:

$$T_{app.} = \frac{6.58}{\sqrt{c_0}\left(1 + \frac{c_1}{2c_0} A + \frac{c_2}{3c_0} A^2 + \frac{c_3}{4c_0} A^3 + \frac{c_4}{5c_0} A^4\right)} A^{-\frac{1}{4}} \tag{30}$$

In Figure 2a, we present the period as determined numerically using Equation (23) and as estimated using the approximate Equation (30) in the domain $0 \leq A \leq 1$. This is usually the domain of interest because an oscillation amplitude of $A = 1$ corresponds to $x = R$. For larger deformations, the linear elastic response of the contacting materials may no longer be valid (however, this case will also be discussed later). In this domain ($0 \leq A \leq 1$) the approximate Equation (30) leads to nearly identical results to Equation (23). The relative error is calculated using the following Equation:

$$\varepsilon(\%) = \frac{T_{app.} - T}{T} 100\% \tag{31}$$

The results are presented in Figure 2b. In Figure 2c the $\Omega$ parameter is presented in the domain $0 \leq \xi \leq 1$. It is interesting that it can be accurately approximated by a straight line as follows:

$$\Omega(\xi) \approx -0.1\xi + 1,\ 0 \leq \xi \leq 1 \tag{32}$$

Therefore, the average $\Omega$ value can also be expressed using the simple equation below:

$$\Omega_{ave} = \frac{1}{A} \int_0^A (-0.1\xi + 1) d\xi = -0.05A + 1 \tag{33}$$

Therefore, Equation (30) can be simplified as follows:

$$T_{app.} \approx \frac{6.58}{\sqrt{c_0}(1-0.05A)}A^{-\frac{1}{4}} = \frac{3.29\sqrt{3}}{1-0.05A}A^{-\frac{1}{4}} \tag{34}$$

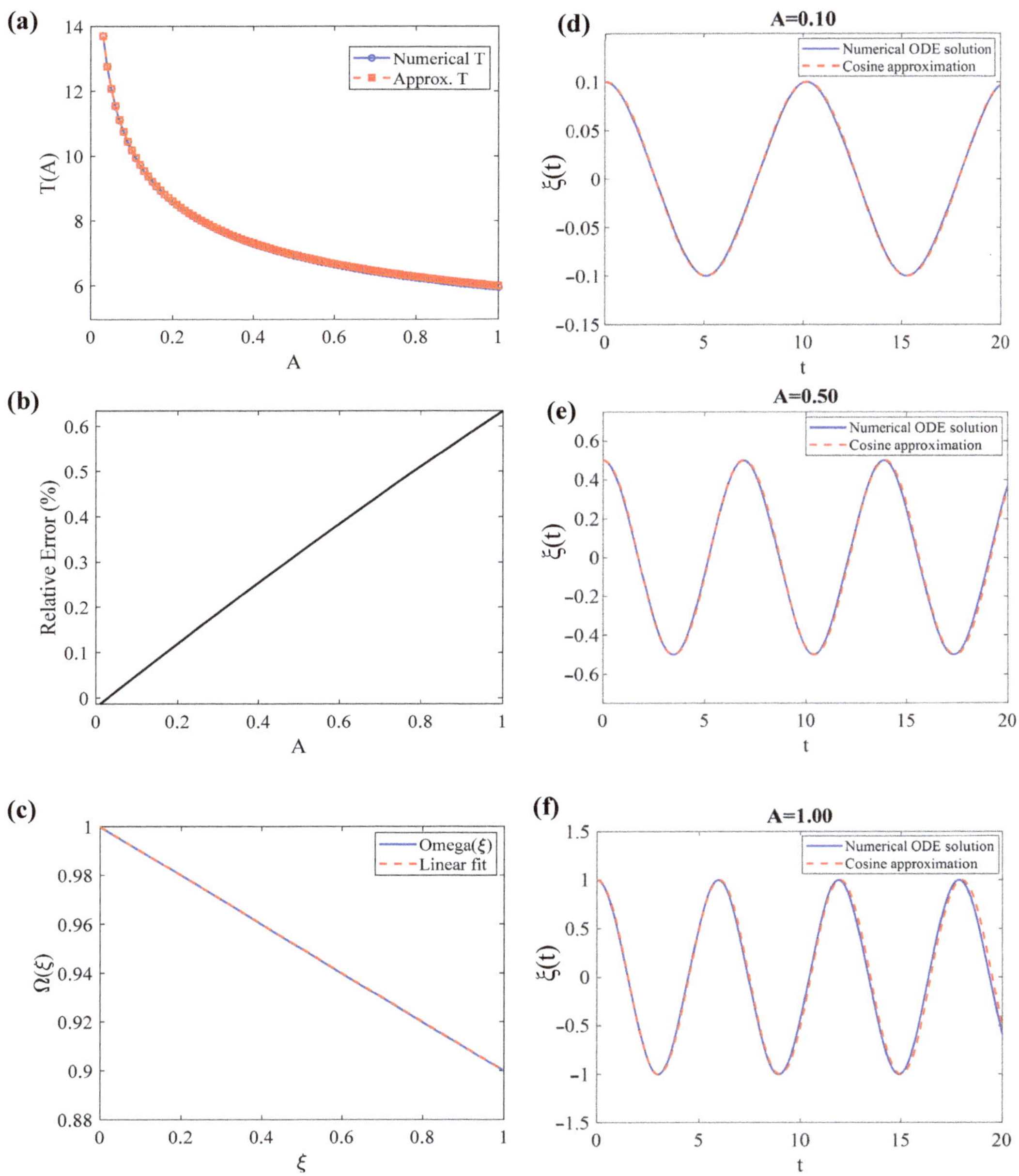


**Figure 2.** (**a**) The oscillation's period using Equations (23) and (30). (**b**) The differences between the numerical solution and the results provided by Equation (30) are small in the domain $0 \le A \le 1$. (**c**) The $\Omega(\xi)$ curve and a linear fit. The numerical solution $x = f(t)$ and the analytical solution based on Equation (35) are presented for comparison for (**d**) $A$ = 0.10, (**e**) $A$ = 0.50 and (**f**) $A$ = 1.00.

The linear averaging assumption used in Equation (32) is justified by the observation that, within the typical amplitude range of interest $0 \leq A \leq 1$, the function $\Omega(\xi)$ varies almost linearly with $\xi$. As shown in Figure 2c, $\Omega(\xi)$ can be closely approximated by a straight line, $\Omega(\xi) \approx -0.1\xi + 1$ (the R squared coefficient of the liner fit resulted in 1.0000), leading to a simple expression for the average, $\Omega_{ave} = 1 - 0.05A$. Higher-order terms in the expansion contribute increasingly smaller corrections because the coefficients $c_i/c_0$ decrease in magnitude for $i \geq 1$ and the powers of $\xi$ remain small in the domain of interest. Consequently, the integral of the full polynomial is very well approximated by its first-order (linear) term. This is confirmed numerically: the approximate period computed using the linear average in Equation (34) matches the period obtained from the exact numerical integration (Equation (23)) with small errors. The error associated with Equation (34) is approximately 0.63% for $A = 1$, which is similar to that obtained using Equation (30). Thus, the linear averaging is sufficient for practical purposes within the considered amplitude range, while retaining the simplicity and analytical tractability of the expression.

In addition, the numerical $\xi = f(t)$ functions (solutions of Equation (14)) are presented graphically in Figure 2d–f in the domain $0 \leq t \leq 20$ and compared with the trivial solution below:

$$\xi(t) = A\cos\left(\frac{2\pi}{T_{app.}}t\right) \tag{35}$$

In addition, we can further increase the accuracy of the analytical approach (Equation (30)). The curve described the relative error as shown in Figure 2b can be approximated to the following linear curve:

$$\varepsilon = \frac{0.63}{100}A \tag{36}$$

Therefore, we can add a correction factor to Equation (30) as follows:

$$T_{app.} = \left(1 - \frac{0.63}{100}A\right)\frac{6.58}{\sqrt{c_0}\left(1 + \frac{c_1}{2c_0}A + \frac{c_2}{3c_0}A^2 + \frac{c_3}{4c_0}A^3 + \frac{c_4}{5c_0}A^4\right)}A^{-\frac{1}{4}} \tag{37}$$

The idea on the semi-empirical factor is based in Figure 2b in the manuscript as already mentioned. Since the relative error when using the approximate approach is almost linear with respect to the oscillation's amplitude, we approximately determined the slope of the $\varepsilon(\%) = f(A)$ curve using the simple equation $\frac{\Delta\varepsilon(\%)}{\Delta A} \cong 0.63\%$. The period obtained using the numerical method was smaller than that obtained using the proposed method; therefore, we introduced the semi-empirical factor $1 - \frac{0.63}{100}A$ into our solution.

The numerical solution for the oscillation period (obtained using Equation (23) and the approximate solution with the correction factor introduced in Equation (36), i.e., Equation (37)), are presented comparatively in Figure 3a. The relative error of Equation (37) with respect to the exact numerical result is also shown in Figure 3b.

In conclusion, Equations (30) and (34) can be safely used for $\xi \leq 1$ (corresponding to a deformation $x \leq R$), while Equation (37) can be safely used for $\xi \leq 3$.

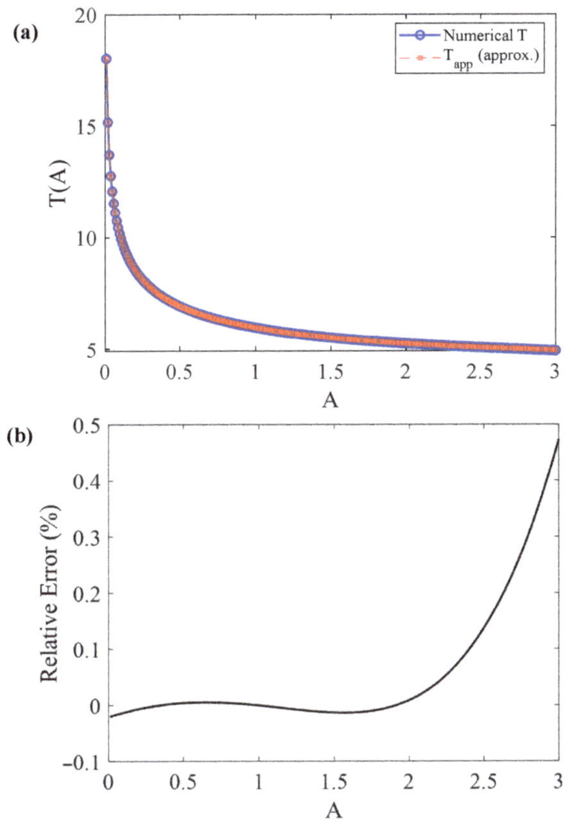


**Figure 3.** (**a**) Comparison between the oscillation period obtained from the exact numerical solution (Equation (23)) and the approximate solution including the correction factor (Equation (37)). (**b**) Relative error of the approximate solution with respect to the exact result.

## 4. Discussion

In this paper, the generic case of nonlinear oscillation under the influence of Hertzian-type restoring forces is examined (Equation (14)). The exact formulation of the oscillation period, expressed in Equation (23), requires the evaluation of a nonlinear integral that depends explicitly on the oscillation amplitude $A$. This formulation accounts for the full series expansion of the force law, which involves terms of the type $\xi^{i+\frac{3}{2}}$, where $i = 0, 1, 2, \ldots$. Although this approach is exact, the integral cannot be expressed in closed form for arbitrary $i$, and numerical evaluation is generally required. The analytical approximation introduced in Section 3 is based on two central ideas. First, the force law is separated into a dominant Hertzian-like contribution and a multiplicative correction factor $\Omega(\xi)$, as expressed in Equation (24). Second, the $\Omega(\xi)$ function is replaced by its average value over the oscillation range $0 \leq \xi \leq A$. This averaging procedure smooths out the variation of $\Omega(\xi)$ and effectively transforms the nonlinear problem into one with a closed-form solution similar to the classical Hertzian oscillator. It is important to note that this method works particularly well because the dominant scaling of the system is controlled by the Hertzian term ($c_0\xi^{3/2}$). To demonstrate this, we analyze the following formula:

$$R(\xi) = \frac{-\frac{1}{10}|\xi|^{\frac{3}{2}}\xi - \frac{1}{840}|\xi|^{\frac{5}{2}}\xi + \frac{11}{15{,}120}|\xi|^{\frac{7}{2}}\xi + \frac{1357}{6{,}652{,}800}|\xi|^{\frac{9}{2}}\xi}{|\xi|^{\frac{1}{2}}\xi} \tag{38}$$

The numerator of Equation (38) is the sum of the higher-order correction terms in the Hertzian equation, while the denominator is the Hertzian term. The absolute value of the $R(\xi)$ parameter with respect to $\xi$ in the domain $0.1 \leq \xi \leq 1$ is presented in Figure 4. For small $\xi$ values, the $\mid R(\xi) \mid$ parameter is small, leading to the result that in this regime the restoring force is dominated by the classic Hertzian term. As the $\xi$ variable increases, the higher-order terms significantly influence the type of restoring force. However, for $\xi \approx 1$, $\mid R(\xi) \mid \approx 0.1$. In this regime, employing the average value of $\Omega(\xi)$ parameter leads to small errors in period calculations.

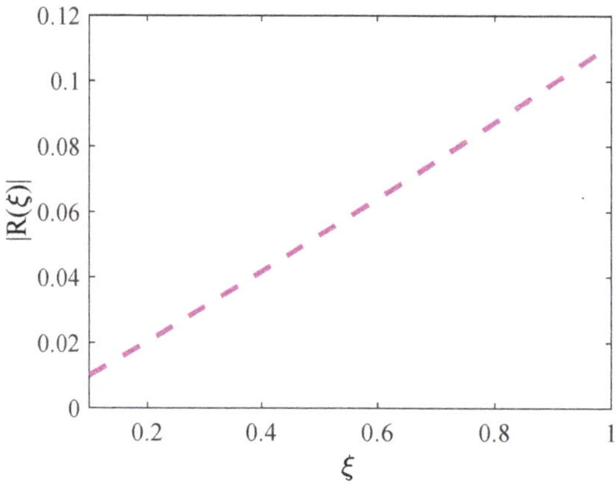


**Figure 4.** Absolute value of the $R(\xi)$ parameter as a function of $\xi$ in the domain $0.1 \leq \xi \leq 1$. For small $\xi$, $\mid R(\xi) \mid$ is negligible, and the restoring force is dominated by the classic Hertzian term. As $\xi$ increases, higher-order corrections become more significant. For $\xi \approx 1$, $\mid R(\xi) \mid$ is also small ($\mid R(\xi) \mid \approx 0.1$), indicating that using the average value of $\Omega(\xi)$ introduces only minor errors in period calculations.

Therefore, the higher-order corrections should enter as smooth polynomial contributions that do not significantly distort the overall dependence of the restoring force within the interval of interest. By replacing $\Omega(\xi)$ with its average, the main influence of these corrections is retained without introducing large errors in the calculation of the oscillation's period. The symmetry of the oscillation about $\xi = 0$ further supports the validity of the averaging approach, since no systematic phase bias is introduced. More specifically, the oscillation is symmetric with respect to $\xi = 0$, meaning that the system moves equally between positive and negative displacements of the same magnitude. This symmetry ensures that the restoring force acts in a balanced way during both halves of the cycle. As a result, when the correction factor $\Omega(\xi)$ is replaced by its average value, the overall dynamics are not distorted in one direction or the other. In other words, the averaging procedure does not introduce any systematic phase shift, and the resulting motion remains consistent with the natural cosine-like symmetry of the exact solution.

The comparison between the exact and approximate formulations confirms the accuracy of the proposed approach. For amplitudes in the range $0 \leq A \leq 1$, the discrepancy between Equations (23) and (30) is negligible, as shown in Figure 2b. In this domain, $\Omega(\xi)$ varies approximately linearly and remains close to unity, with a simple approximation $\Omega(\xi) \approx 1 - 0.1\xi$. Consequently, the use of the average value in Equation (27) accurately captures its net effect over the oscillation cycle. The resulting approximate period is nearly identical to the numerical value obtained from the exact integral. The domain of validity of the method is therefore well defined. The approximation remains accurate as long as the amplitude does not exceed $A \approx 1$. This corresponds physically to the deformation scale where $x = R$, beyond which the assumption of linear elastic contact may no longer hold. At larger deformations, the higher-order polynomial terms grow in importance, the contact area tends to a limiting value, and the force–deformation law approaches linearity. In this regime, representing the force as a Hertzian-like term corrected by a smooth factor becomes inadequate, leading to larger errors.

In addition, the accuracy of the analytical approach can be further improved. The curve representing the relative error, as shown in Figure 2b, can be well approximated by a simple linear relation (Equation (36)). This observation allows the introduction of a correction factor, resulting in an improved expression for the approximate period (Equation (37)). By applying this correction, the systematic errors associated with larger amplitudes are significantly reduced, providing a more accurate estimate of the period in the regime where the Hertzian-like approximation alone becomes insufficient. It should be noted that the elastic half-space approximation, which assumes linearly elastic materials, may no longer hold for large deformations, except in special cases involving very soft materials. Nevertheless, from a purely mathematical perspective, it is worthwhile to present this semi-empirical solution (including the correction factor) because it fully captures the dependence of the oscillation period on the amplitude, even for large deformations. This approach provides a complete analytical description of the system's behavior, which can be useful for theoretical studies and comparisons with numerical results.

It is also instructive to compare the results obtained using the method presented in this paper with those derived from the Harmonic Balance Method (HBM) [45], which is one of the most widely used and classical techniques for solving nonlinear second-order differential equations. Consider Equation (18) for $j = 4$:

$$\ddot{\xi} + \frac{4}{3}\left( |\xi|^{\frac{1}{2}}\xi - \frac{1}{10}|\xi|^{\frac{3}{2}}\xi - \frac{1}{840}|\xi|^{\frac{5}{2}}\xi + \frac{11}{15,120}|\xi|^{\frac{7}{2}}\xi + \frac{1357}{6,652,800}|\xi|^{\frac{9}{2}}\xi \right) = 0 \quad (39)$$

In addition, consider the trial solution $\xi(t) = Asin(\omega t)$, with $\theta = \omega t$. The first-harmonic contributions of the five nonlinear terms are given below:

$$|\xi|^{\frac{1}{2}}\xi = A^{\frac{3}{2}}|sin\theta|^{\frac{1}{2}}sin\theta \approx A^{\frac{3}{2}}f_1 sin\theta \quad (40)$$

$$|\xi|^{\frac{3}{2}}\xi = A^{\frac{5}{2}}|sin\theta|^{\frac{3}{2}}sin\theta \approx A^{\frac{5}{2}}f_2 sin\theta \quad (41)$$

$$|\xi|^{\frac{5}{2}}\xi = A^{\frac{7}{2}}|sin\theta|^{\frac{5}{2}}sin\theta \approx A^{\frac{7}{2}}f_3 sin\theta \quad (42)$$

$$|\xi|^{\frac{7}{2}}\xi = A^{\frac{9}{2}}|sin\theta|^{\frac{7}{2}}sin\theta \approx A^{\frac{9}{2}}f_4 sin\theta \quad (43)$$

$$|\xi|^{\frac{9}{2}}\xi = A^{\frac{11}{2}}|sin\theta|^{\frac{9}{2}}sin\theta \approx A^{\frac{11}{2}}f_5 sin\theta \quad (44)$$

where

$$f_1 = \frac{1}{\pi}\int_0^{2\pi} |sin\theta|^{\frac{1}{2}} sin^2\theta d\theta \quad (45)$$

$$f_2 = \frac{1}{\pi}\int_0^{2\pi} |sin\theta|^{3/2} sin^2\theta d\theta \quad (46)$$

$$f_3 = \frac{1}{\pi}\int_0^{2\pi} |sin\theta|^{5/2} sin^2\theta d\theta \quad (47)$$

$$f_4 = \frac{1}{\pi}\int_0^{2\pi} |sin\theta|^{7/2} sin^2\theta d\theta \quad (48)$$

$$f_5 = \frac{1}{\pi}\int_0^{2\pi} |sin\theta|^{9/2} sin^2\theta d\theta(49) \tag{49}$$

By symmetry the integral over [0, 2π] equals twice the integral over [0, π], where $sin\theta \geq 0$. Thus, Equations (45)–(49) are modified as below:

$$f_1 = \frac{2}{\pi}\int_0^{\pi} sin^{5/2}\theta d\theta = \frac{2\sqrt{\pi}}{\pi}\frac{\Gamma\left(\frac{7}{4}\right)}{\Gamma\left(\frac{9}{4}\right)} \cong 0.9153 \tag{50}$$

$$f_2 = \frac{2}{\pi}\int_0^{\pi} sin^{7/2}\theta d\theta = \frac{2\sqrt{\pi}}{\pi}\frac{\Gamma\left(\frac{9}{4}\right)}{\Gamma\left(\frac{11}{4}\right)} \cong 0.7949 \tag{51}$$

$$f_3 = \frac{2}{\pi}\int_0^{\pi} sin^{9/2}\theta d\theta = \frac{2\sqrt{\pi}}{\pi}\frac{\Gamma\left(\frac{11}{4}\right)}{\Gamma\left(\frac{13}{4}\right)} \cong 0.7119 \tag{52}$$

$$f_4 = \frac{2}{\pi}\int_0^{\pi} sin^{11/2}\theta d\theta = \frac{2\sqrt{\pi}}{\pi}\frac{\Gamma\left(\frac{13}{4}\right)}{\Gamma\left(\frac{15}{4}\right)} \cong 0.6504 \tag{53}$$

$$f_5 = \frac{2}{\pi}\int_0^{\pi} sin^{13/2}\theta d\theta \frac{2\sqrt{\pi}}{\pi}\frac{\Gamma\left(\frac{15}{4}\right)}{\Gamma\left(\frac{17}{4}\right)} \cong 0.6024 \tag{54}$$

By substituting to Equation (39):

$$-\omega^2 A sin\theta + \frac{4}{3}\left(A^{\frac{3}{2}} f_1 - \frac{1}{10}A^{\frac{5}{2}} f_2 - \frac{1}{840}A^{\frac{7}{2}} f_3 + \frac{11}{15,120}A^{\frac{9}{2}} f_4 + \frac{1357}{6,652,800}A^{\frac{11}{2}} f_5\right) sin\theta = 0 \tag{55}$$

Therefore:

$$\omega^2 = \frac{4}{3}\left(A^{\frac{1}{2}} f_1 - \frac{1}{10}A^{\frac{3}{2}} f_2 - \frac{1}{840}A^{\frac{5}{2}} f_3 + \frac{11}{15,120}A^{\frac{7}{2}} f_4 + \frac{1357}{6,652,800}A^{\frac{9}{2}} f_5\right) \tag{56}$$

Finally, the oscillation's period using the HBM results in:

$$T_{HBM} = \frac{2\pi}{\sqrt{\frac{4}{3}\left(A^{\frac{1}{2}} f_1 - \frac{1}{10}A^{\frac{3}{2}} f_2 - \frac{1}{840}A^{\frac{5}{2}} f_3 + \frac{11}{15,120}A^{\frac{7}{2}} f_4 + \frac{1357}{6,652,800}A^{\frac{9}{2}} f_5\right)}} \tag{57}$$

In Figure 5a, the numerical results for the oscillation period (Equation (23)) are presented for comparison with the results from the proposed Equation (37) and those from the harmonic balance method (Equation (57)) in the domain $0.1 \leq A \leq 3$.

In Figure 5b, the errors associated with Equation (37) (i.e., the proposed method) and the HBM (when terms higher than the first order are omitted, as explained in Equations (40)–(44)) are also presented for comparison.

The numerical results for the oscillation period obtained using the numerical solution (Equation (23)), the proposed approach (Equation (37)), and the HBM (Equation (57)) are presented for comparison in Table 1 for different amplitudes. The errors associated with Equations (37) and (57) are presented in Table 2.

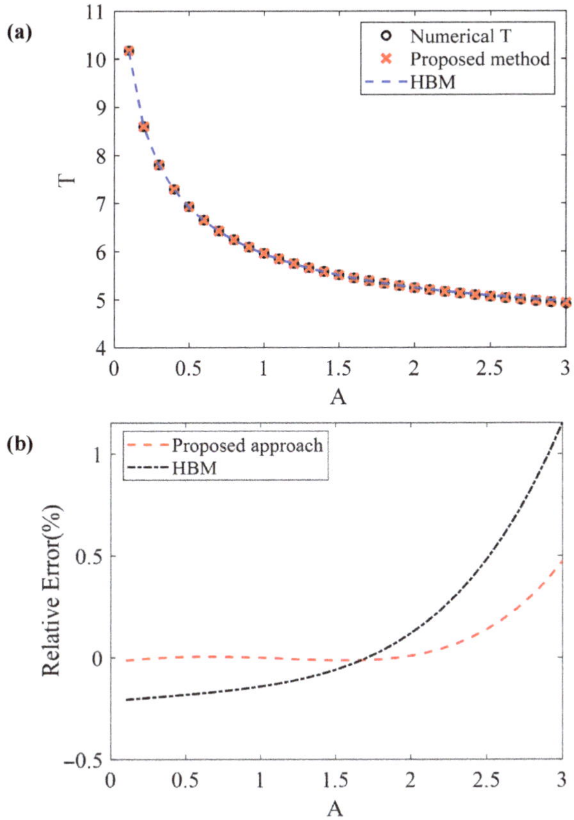


**Figure 5.** (**a**) Numerical results for the oscillation period (Equation (23)) compared with the proposed method (Equation (37)) and the harmonic balance method (Equation (57)) for $0.1 \leq A \leq 3$. (**b**) Corresponding errors for the proposed method and the HBM (with terms higher than first order omitted, as explained in Equations (40)–(44)).

**Table 1.** Oscillation Period Values.

| A | Numerical (Equation (23)) | Proposed Approach (Equation (37)) | HBM (Equation (57)) |
|---|---|---|---|
| 0.2 | 8.5971 | 8.5965 | 8.5799 |
| 0.4 | 7.2934 | 7.2936 | 7.2797 |
| 0.6 | 6.6500 | 6.6503 | 6.6383 |
| 0.8 | 6.2453 | 6.2455 | 6.2353 |
| 1.0 | 5.9612 | 5.9612 | 5.9528 |
| 1.2 | 5.7488 | 5.7484 | 5.7421 |
| 1.4 | 5.5831 | 5.5824 | 5.5785 |
| 1.6 | 5.4498 | 5.449 | 5.4479 |
| 1.8 | 5.3398 | 5.3394 | 5.3414 |
| 2.0 | 5.2470 | 5.2474 | 5.2532 |
| 2.2 | 5.1669 | 5.1691 | 5.1790 |

**Table 1.** *Cont.*

| A | Numerical (Equation (23)) | Proposed Approach (Equation (37)) | HBM (Equation (57)) |
|---|---|---|---|
| 2.4 | 5.0960 | 5.1010 | 5.1159 |
| 2.6 | 5.0317 | 5.0410 | 5.0613 |
| 2.8 | 4.9717 | 4.9869 | 5.0133 |
| 3.0 | 4.9139 | 4.9371 | 4.9705 |

**Table 2.** Percentage Errors (%).

| A | Equation (37) vs. Numerical | HBM (Equation (57)) vs. Numerical |
|---|---|---|
| 0.2 | −0.0066257 | −0.19966 |
| 0.4 | 0.0019226 | −0.18786 |
| 0.6 | 0.0052896 | −0.17505 |
| 0.8 | 0.0042107 | −0.16008 |
| 1.0 | −0.00014308 | −0.14122 |
| 1.2 | −0.0060995 | −0.11614 |
| 1.4 | −0.011415 | −0.081803 |
| 1.6 | −0.013203 | −0.034445 |
| 1.8 | −0.00787 | 0.030501 |
| 2.0 | 0.0089437 | 0.11844 |
| 2.2 | 0.042401 | 0.23565 |
| 2.4 | 0.098477 | 0.38927 |
| 2.6 | 0.18393 | 0.58728 |
| 2.8 | 0.30621 | 0.83836 |
| 3.0 | 0.47332 | 1.1518 |

It should be noted that the elastic half-space approximation, which assumes linearly elastic materials, may no longer be valid for large deformations, except in special cases involving very soft materials. Despite this limitation, the overall analysis in this work assumes a linear elastic response in all studied cases. If the material exhibits a weakly nonlinear elastic response, the method proposed in this paper can still be applied; however, the exponents in the differential Equation (18) will differ, since in the power law describing the applied force on the sphere, both the geometry and the material nonlinearities contribute. In such cases, the method can be applied again using the classical generic solution (Equation (7)) with a new average $\Omega$ factor tailored to the specific system of interest. A representative example is the work by Ding et al. [60], where the Hertzian equation was modified through the introduction of a correction factor. In conclusion, for cases of weak nonlinearities, such as those in which the force–deformation relationship can be expressed as $F = cx^a\Omega(x)$, where $\Omega(x)$ is a correction factor that does not significantly deviate from linearity, the approach presented in this paper can be applied. Nevertheless, it should be noted that for strongly nonlinear materials, the $\Omega$-function may deviate significantly from linear behavior. In these cases, a careful examination is required to assess the accuracy of the proposed method. Therefore, for cases of strong nonlinearities, involving special forms of the $\Omega(x)$ function, further research is required to establish the limits of validity of the proposed approach.

In addition, it is also important to note that the general behavior of the force–deformation relationship when a rigid sphere interacts with an elastic half-space follows Hertz's law, $F \sim x^{3/2}$, which gradually approaches a linear relation $F \sim x$ for large deformations. This linear behavior becomes evident for $x > 3R$, as presented in reference [57]. Therefore, the part of the oscillation corresponding to very large indentations tends to be harmonic. However, there is no universal equation that accurately captures the force–deformation relationship over the entire range from very small to extremely large deformations ($x \to \infty$). An attempt in this direction was made in [57], where an equation suitable for large indentations was derived. Nevertheless, it introduces a small error for shallow deformations, where the classical Hertz equation remains valid. In contrast, Equation (11) is accurate for $x \leq 3R$ and coincides with the Hertzian solution for very small indentations. In conclusion, there is no available equation for $j > 4$ in Equation (15) due to the mathematical complexity of Sneddon's nonlinear system and because, at very large deformations, the assumption of linear elasticity is no longer valid.

It is interesting to also note at this point that a practical example illustrating the relevance of the present solution is in AFM experiments with a spherical tip. While AFM typically involves forced oscillations to probe the mechanical properties of a sample, evaluating the free oscillation frequency of the tip provides valuable information for a rough estimation of the resonance frequency. It should be noted that the free oscillation frequency is not identical to the resonance frequency; however, in nonlinear systems, both exhibit similar amplitude-dependent shifts, allowing the free oscillation analysis to serve as a practical and efficient approximation. This approach provides insight into the dynamic behavior of the system and helps guide experimental interpretation without the need for full forced-vibration simulations. Future research will focus on cases involving nonlinear force oscillations of the AFM tip, aimed at developing advanced characterization methods for nanomechanical property mapping.

## 5. Conclusions

In summary, the proposed method accurately captures nonlinear oscillations under Hertzian-type restoring forces by preserving the dominant Hertzian scaling and incorporating higher-order corrections through an averaged factor. For small amplitudes, the approximation closely matches the exact period, while for larger amplitudes, averaging the $\Omega(\xi)$ function and introducing a semi-empirical linear adjustment (Equation (36)) reduces systematic errors. Although the elastic half-space assumption may break down for large deformations except in very soft materials, the corrected solution remains valuable mathematically, fully capturing the dependence of the oscillation period on amplitude and providing a useful analytical framework for theory and numerical comparisons.

The novelty of the present work lies primarily in extending the analysis of nonlinear oscillations to regimes that go beyond classical approximations, particularly for oscillatory systems involving large deformations in elastic contact problems, which are highly relevant in experimental contexts such as Atomic Force Microscopy (AFM). In particular, the classical problem of oscillations under elastic forces of the form $F = c_\lambda x|x|^{a-1}$ has been previously studied. However, when considering elastic forces acting on spheres, the restoring force involves a series of power-law terms. To the best of our knowledge, this is the first time such a case has been analyzed. In conclusion, we present a novel method for determining the oscillation frequency and validate the results through comparison with numerical solutions and the Harmonic Balance Method (HBM), which is widely regarded as one of the classical approaches for solving nonlinear oscillations.

**Author Contributions:** Conceptualization, S.V.K.; methodology, S.V.K., G.M.I., A.M., and A.S.; software, S.V.K., G.M.I., A.M.; validation, S.V.K., G.M.I., A.M., and A.S.; investigation, S.V.K., G.M.I., A.M., and A.S.; resources, S.V.K., G.M.I., A.M., and A.S.; writing—original draft, S.V.K.; writing—review and editing, S.V.K., G.M.I., A.M., A.S.; visualization, S.V.K., A.M.; supervision, S.V.K., A.S.; project administration, S.V.K., A.S.; funding acquisition, S.V.K., A.S. All authors have read and agreed to the published version of the manuscript.

**Funding:** Funded by the European Union. Views and opinions expressed are however those of the author(s) only and do not necessarily reflect those of the European Union or European Research Executive Agency. Neither the European Union nor the granting authority can be held responsible for them.

**Data Availability Statement:** All data are available within the paper.

**Conflicts of Interest:** Author Anna Malamou was employed by the company BioNanoTec Ltd. The remaining authors declare that the research was conducted in the absence of any commercial or financial relationships that could be construed as a potential conflict of interest.

## References

1. Abouelregal, A.E.; Mohammad-Sedighi, H.; Faghidian, S.A.; Shirazi, A.H. Temperature-dependent physical characteristics of the rotating nonlocal nanobeams subject to a varying heat source and a dynamic load. *Facta Univ. Ser. Mech.* **2021**, *19*, 633–656. [CrossRef]
2. Sedighi, H.M.; Shirazi, K.H. Dynamic pull-in instability of double-sided actuated nano-torsional switches. *Acta Mech. Solida Sin.* **2015**, *28*, 91–101. [CrossRef]
3. Mikhlin, Y.; Avramov, K. Nonlinear Normal Modes of Vibrating Mechanical Systems: 10 Years of Progress. *Appl. Mech. Rev.* **2024**, *76*, 050801. [CrossRef]
4. Abohamer, M.K.; Amer, T.S.; Galal, A.A.; Darweesh, M.A.; Arab, A.; Bahnasy, T.A. Nonlinear Oscillations of a Lumped System with Series Spring, Piezoelectric Device, and Feedback Controller. *Sci. Rep.* **2025**, *15*, 14642. [CrossRef]
5. Niu, J.-Y. A Universal Application of the General Integral Transform: New Technique for the Analysis of Nonlinear Oscillator Systems. *J. Low Freq. Noise Vib. Act. Control* **2024**, *44*, 428–436. [CrossRef]
6. Qie, N.; Houa, W.F.; He, J.H. The fastest insight into the large amplitude vibration of a string. *Rep. Mech. Eng.* **2021**, *2*, 1–5. [CrossRef]
7. Hosen, M.A.; Chowdhury, M.S.H. A new reliable analytical solution for strongly nonlinear oscillator with cubic and harmonic restoring force. *Results Phys.* **2015**, *5*, 111–114. [CrossRef]
8. Kontomaris, S.V.; Malamou, A. Exploring oscillations with a nonlinear restoring force. *Eur. J. Phys.* **2022**, *43*, 015006. [CrossRef]
9. Xu, L. Application of He's parameter-expansion method to an oscillation of a mass attached to a stretched elastic wire. *Phys. Lett. A* **2007**, *368*, 259–262. [CrossRef]
10. Li, S.; Niu, J.; Li, X. Primary resonance of fractional-order Duffing–van der Pol oscillator by harmonic balance method. *Chin. Phys. B* **2018**, *27*, 120502. [CrossRef]
11. Cveticanin, L.; Herisanu, N.; Ismail, G.M.; Zukovic, M. Vibration of the Liénard Oscillator with Quadratic Damping and Constant Excitation. *Mathematics* **2025**, *13*, 937. [CrossRef]
12. Ju, P.; Xue, X. Global residue harmonic balance method to periodic solutions of a class of strongly nonlinear oscillators. *Appl. Math. Model.* **2014**, *38*, 6144–6152. [CrossRef]
13. Wu, B.; Liu, W.; Chen, X.; Lim, C. Asymptotic analysis and accurate approximate solutions for strongly nonlinear conservative symmetric oscillators. *Appl. Math. Model.* **2017**, *49*, 243–254. [CrossRef]
14. He, J.H.; Yang, Q.; He, C.H.; Khan, Y. A simple frequency formulation for the tangent oscillator. *Axioms* **2021**, *10*, 320. [CrossRef]
15. Tian, Y. Frequency formula for a class of fractal vibration system. *Rep. Mech. Eng.* **2022**, *3*, 55–61. [CrossRef]
16. Mickens, R.E. A generalization of the method of harmonic balance. *J. Sound Vib.* **1986**, *111*, 515–518. [CrossRef]
17. Mickens, R.E. *Truly Nonlinear Oscillations*; World Scientific Publishing: Singapore, 2010.
18. He, J.H.; Wu, X.H. Variational iteration method: New development and applications. *Comput. Math. Appl.* **2007**, *54*, 881–894. [CrossRef]
19. Wu, H.G.; Hu, Y. On variational iteration method for fractional calculus. *Therm. Sci.* **2017**, *21*, 1707–1712. [CrossRef]
20. Liao, S.J.; Cheung, A.T. Application of homotopy analysis method in nonlinear oscillations. *ASME J. Appl. Mech.* **1998**, *65*, 914–922. [CrossRef]
21. Zhang, G.Q.; Wu, Z.Q. Homotopy analysis method for approximations of Duffing oscillator with dual frequency excitations. *Chaos Solitons Fractals* **2019**, *127*, 342–353. [CrossRef]

22. Wu, Y.; He, J.H. Homotopy perturbation method for nonlinear oscillators with coordinate-dependent mass. *Results Phys.* **2018**, *10*, 270–271. [CrossRef]
23. He, J.H.; El-Dib, Y.O.; Mady, A.A. Homotopy perturbation method for the fractal Toda oscillator. *Fractal Fract.* **2021**, *5*, 93. [CrossRef]
24. Anjum, N.; He, J.H.; Ain, Q.T.; Tian, D. Li–He's modified homotopy perturbation method for doubly-clamped electrically actuated microbeams-based microelectromechanical system. *Facta Univ. Ser. Mech.* **2021**, *19*, 601–612. [CrossRef]
25. He, J.H.; El-Dib, Y.O. The enhanced homotopy perturbation method for axial vibration of strings. *Facta Univ. Ser. Mech.* **2021**, *19*, 735–750. [CrossRef]
26. Belendez, A.; Hernandez, A.; Belendez, T. Asymptotic representation of the period for the nonlinear oscillator. *J. Sound. Vib.* **2007**, *299*, 403–408. [CrossRef]
27. Cveticanin, L.; Kovacic, I.; Rakaric, Z. Asymptotic methods for vibrations of the pure non-integer order oscillator. *Comput. Math. Appl.* **2010**, *60*, 2616–2628. [CrossRef]
28. Molla, M.H.U.; Alam, M.S. Higher accuracy analytical approximations to nonlinear oscillators with discontinuity by energy balance method. *Results Phys.* **2017**, *7*, 2104–2110. [CrossRef]
29. Ebaid, A.E. Approximate periodic solutions for the non-linear relativistic harmonic oscillator via differential transformation method. *Commun. Nonlinear Sci. Numer. Simul.* **2010**, *15*, 1921–1927. [CrossRef]
30. Wang, S.Q.; He, J.H. Nonlinear oscillator with discontinuity by parameter expansion method. *Chaos Soliton Fract.* **2008**, *35*, 688–691.
31. Sedighi, H.M.; Shirazi, K.H.; Noghrehabadi, A.R.; Yildirim, A.H. Asymptotic investigation of buckled beam nonlinear vibration. *Iran. J. Sci. Technol. Trans. Mech. Eng.* **2012**, *36*, 107–116.
32. He, J.H.; Anjum, N.; Skrzypacz, P. A variational principle for a nonlinear oscillator arising in the microelectromechanical system. *J. Appl. Comput. Mech.* **2021**, *7*, 78–83.
33. He, J.H.; Houa, W.F.; Qie, N.; Gepreel, K.A.; Shirazi, A.H.; Sedighi, H.M. Hamiltonian-based frequency-amplitude formulation for nonlinear oscillators. *Facta Univ. Ser. Mech.* **2021**, *19*, 199–208. [CrossRef]
34. Bonisoli, E.; Dimauro, L.; Venturini, S.; Cavallaro, S.P. Experimental detection of nonlinear dynamics using a laser profilometer. *Appl. Sci.* **2023**, *13*, 3295. [CrossRef]
35. Feldman, M. Non-linear system vibration analysis using Hilbert transform I. Free vibration analysis method 'Freevib'. *Mech. Syst. Signal Process.* **1994**, *8*, 119–127. [CrossRef]
36. Ruzzene, M.; Fasana, A.; Garibaldi, L.; Piombo, B. Natural frequencies and dampings identification using wavelet transform: Application to real data. *Mech. Syst. Signal Process.* **1997**, *11*, 207–218. [CrossRef]
37. Hosen, M.A.; Ismail, G.M.; Yildirim, A.; Kamal, M.A.S. A modified energy balance method to obtain higher-order approximations to the oscillators with cubic and harmonic restoring force. *J. Appl. Comput. Mech.* **2020**, *6*, 320–331.
38. El-Dib, Y.O.; Matoog, R.T. The rank upgrading technique for a harmonic restoring force of nonlinear oscillators. *J. Appl. Comput. Mech.* **2021**, *7*, 782–789.
39. Kontomaris, S.V.; Mazi, I.; Chliveros, G.; Malamou, A. Generic numerical and analytical methods for solving nonlinear oscillators. *Phys. Scr.* **2024**, *99*, 025231. [CrossRef]
40. Kontomaris, S.V.; Chliveros, G.; Malamou, A. Approximate solutions for undamped nonlinear oscillations using He's formulation. *J.* **2023**, *6*, 140–151. [CrossRef]
41. Big-Alabo, A. Approximate periodic solution for the large-amplitude oscillations of a simple pendulum. *Int. J. Mech. Eng. Educ.* **2020**, *48*, 335–350. [CrossRef]
42. He, J.H. The simplest approach to nonlinear oscillators. *Results Phys.* **2019**, *15*, 102546. [CrossRef]
43. Chen, B.; Lu, J.; Xia, Z. Numerical investigation of the fractal capillary oscillator. *J. Low Freq. Noise Vib. Act. Control* **2023**, *42*, 579–588. [CrossRef]
44. Jin, X.; Liu, M.; Pan, F.; Li, Y.; Fan, J. Low frequency of a deforming capillary vibration, part 1: Mathematical model. *J. Low Freq. Noise Vib. Act. Control* **2019**, *38*, 1676–1680.
45. Cveticanin, L.; Ismail, G.M. Higher-order approximate periodic solution for the oscillator with strong nonlinearity of polynomial type. *Eur. Phys. J. Plus* **2019**, *134*, 266. [CrossRef]
46. Kontomaris, S.V.; Mazi, I.; Malamou, A. A Note on a Simple Equation for Solving Nonlinear Undamped Oscillations. *J. Vib. Eng. Technol.* **2024**, *12*, 8235–8248. [CrossRef]
47. Cveticanin, L. Oscillator with Fractional Order Restoring Force. *J. Sound Vib.* **2009**, *320*, 1064–1077. [CrossRef]
48. Lekka, M. Discrimination Between Normal and Cancerous Cells Using AFM. *BioNanoScience* **2016**, *6*, 65–80. [CrossRef]
49. Kontomaris, S.V.; Malamou, A.; Stylianou, A. The Hertzian Theory in AFM Nanoindentation Experiments Regarding Biological Samples: Overcoming Limitations in Data Processing. *Micron* **2022**, *155*, 103228. [CrossRef]
50. Popov, V.L. *Contact Mechanics and Friction: Physical Principles and Applications*; Springer: Berlin/Heidelberg, Germany, 2017.

51. Kontomaris, S.V.; Malamou, A. Calculating the Duration of Impact When a Solid Sphere Collides on a Flat Rigid Wall. *Phys. Teach.* **2021**, *59*, 696–699. [CrossRef]
52. Kontomaris, S.V.; Stylianou, A. Atomic force microscopy for university students: Applications in biomaterials. *Eur. J. Phys.* **2017**, *38*, 033003. [CrossRef]
53. Kontomaris, S.V.; Psychogios, I.; Alimisis, V.; Malamou, A.; Dimas, C. Generic solutions to oscillations with fractional-order restoring forces: Positive and negative exponents. *Nonlinear Sci.* **2025**, *3*, 100030. [CrossRef]
54. Wu, C.-E.; Lin, K.-H.; Juang, J.-Y. Hertzian Load–Displacement Relation Holds for Spherical Indentation on Soft Elastic Solids Undergoing Large Deformations. *Tribol. Int.* **2016**, *97*, 71–76. [CrossRef]
55. Koruk, H.; Pouliopoulos, A.N. Elasticity and Viscoelasticity Imaging Based on Small Particles Exposed to External Forces. *Processes* **2023**, *11*, 3402. [CrossRef]
56. Puricelli, L.; Galluzzi, M.; Schulte, C.; Podestà, A.; Milani, P. Nanomechanical and topographical imaging of living cells by atomic force microscopy with colloidal probes. *Rev. Sci. Instrum.* **2015**, *86*, 033705. [CrossRef]
57. Kontomaris, S.V.; Malamou, A. A novel approximate method to calculate the force applied on an elastic half space by a rigid sphere. *Eur. J. Phys.* **2021**, *42*, 025010. [CrossRef]
58. Müller, P.; Abuhattum, S.; Möllmert, S.; Ulbricht, E.; Taubenberger, A.V.; Guck, J. Nanite: Using machine learning to assess the quality of atomic force microscopy-enabled nano-indentation data. *BMC Bioinform.* **2019**, *20*, 465. [CrossRef]
59. Kontomaris, S.-V.; Malamou, A.; Ismail, G.M.; Katsiki, A.; Stylianou, A. Beyond Hertz: Accurate Analytical Force–Indentation Equations for AFM Nanoindentation with Spherical Tips. *Metrology* **2025**, *5*, 63. [CrossRef]
60. Ding, Y.; Wang, J.; Xu, G.-K.; Wang, G.-F. Are Elastic Moduli of Biological Cells Depth Dependent or Not? Another Explanation Using a Contact Mechanics Model with Surface Tension. *Soft Matter* **2018**, *14*, 7534–7541. [CrossRef]

*vibration*

MDPI

*Article*

# Balancing Accuracy and Efficiency in Wire-Rope Isolator Modeling: A Simplified Beam-Element Framework

**Claudia Marin-Artieda**

Department of Civil and Environmental Engineering, Howard University, Washington, DC 20059, USA; cmarin@howard.edu

**Abstract:** Wire-rope isolators (WRIs) are widely used in vibration and seismic protection due to their multidirectional flexibility and amplitude-dependent hysteretic damping. However, their complex nonlinear behavior, especially under inclined and combined-mode loading, poses challenges for predictive modeling. This study presents a simplified finite-element modeling framework using constant-property Timoshenko beam elements with tuned Rayleigh damping to simulate WRI behavior across various configurations. Benchmark validation against analytical ring deformation confirmed the model's ability to capture geometric nonlinearities. The framework was extended to five WRI types, with effective cross-sectional properties calibrated against vendor-supplied quasi-static data. Dynamic simulations under sinusoidal excitation demonstrated strong agreement with experimental force-displacement loops in pure modes and showed moderate accuracy (within 29%) in inclined configurations. System-level validation using a rocking-control platform with four inclined WRIs showed that the model reliably predicts global stiffness and energy dissipation under base accelerations. While the model does not capture localized nonlinearities such as pinched hysteresis due to interstrand friction, it offers a computationally efficient tool for engineering design. The proposed method enables rapid evaluation of WRI performance in complex scenarios, supporting broader integration into performance-based seismic mitigation strategies.

**Keywords:** wire-rope isolators; geometric nonlinearity; rate independent; flexibility; energy dissipation; finite-element; Timoshenko beam elements; hysteresis; isolation

## 1. Introduction

Wire-rope isolators (WRIs) are versatile helical spring–damper devices made of steel cable bundles attached between rigid brackets. Their unique geometry provides them with multidirectional flexibility and significant hysteretic damping through interwire friction. These attributes have made WRIs popular for protecting sensitive equipment from shock and vibration in industries ranging from military and aerospace to infrastructure and seismic engineering. WRIs exhibit strongly nonlinear behavior. Under deflection, the ring-shaped geometry of the isolator deforms and uncoils, causing its effective stiffness to harden or soften, while friction between the strands dissipates energy. Early research demonstrated that the damping in the helical wire-rope isolator is governed by sliding friction among the cable wires [1].

Early experimental studies established that WRI responses are highly nonlinear and hysteretic due to their geometry and internal friction. Tinker and Cutchins [1] observed pronounced hysteresis loops in WRI force–displacement cycles, attributing the dissipation to Coulomb friction as wires slide against each other. Later works quantified these effects.

For example, Salvatore et al. [2] conducted dynamic tests on vertical WRIs and noted asymmetric restoring forces; the isolator softens in compression but stiffens in tension as the cables straighten. This tension–compression asymmetry grows with larger vibration amplitudes, whereas shear and roll modes tend to produce more symmetric loops. Notably, the energy dissipation in WRIs is largely amplitude-dependent and frequency-independent. The loop shape remains similar across loading rates, which is a desirable property for broadband vibration isolation. Comprehensive cyclic tests by Leblouba et al. [3] confirm rate-independent behavior of WRI hysteresis loops. This study highlights the influence of WRI geometry on its multi-directional behavior; the effective damping of an isolator is affected by the height-to-width ratio and the displacement amplitude. Overall, the experimental literature presents a consistent picture of WRIs as rate-independent, nonlinear spring–dampers whose properties evolve with amplitude and accumulated travel but remain stable across frequency. Any reliable WRI model must therefore incorporate geometric nonlinearity and amplitude-dependent hysteresis.

Purely analytical modeling of WRIs has generally been confined to idealized or single-strand cases, since closed-form mechanics struggle to capture the full helical geometry and interstrand contact. The most widely used model to capture the WRIs' behavior is the Bouc–Wen model [4], originally formulated by Wen for smooth hysteretic systems. Demetriades et al. [5] were among the first to adopt a Bouc–Wen formulation to simulate WRI loops in seismic analyses, treating the isolator as an equivalent nonlinear spring with parameters calibrated from experiments. Ni et al. [6] later performed cyclic loading tests to identify Bouc–Wen parameters for WRIs. Recent efforts have extended the classical Bouc–Wen model to capture the asymmetric, preload-dependent hysteresis loops observed in frictional isolators. For example, Maleki et al. [7] introduced a modified Bouc–Wen formulation in which the stiffness parameter varies exponentially with the instantaneous normal force. These Bouc–Wen-type models efficiently fit loops in individual directions, but usually require direction-specific calibration and careful tuning to reproduce pinching, mixed hardening/softening, or combined-mode effects.

Complementing phenomenological hysteresis, rate-independent, macro-models for WRIs along principal transverse directions (roll and shear) have been proposed by Vaiana et al. [8]. Their algebraic formulation employs five parameters of clear mechanical meaning and captures kinematic hardening efficiently, with validation in conventional modes; however, recalibration is needed when axial force (or inclination) changes, indicating sensitivity to loading conditions.

Mechanical beam-based models in the literature aim to enhance physical interpretability by representing ropes as hierarchical, helically wound structures with frictional interfaces. Foti et al. [9] proposed a mechanical modeling approach that treats the wire rope as a beam-like element composed of helically wound strands and wires, capturing its internal hierarchy and frictional interactions. Central to this approach is a nonlinear moment-curvature law that accounts for the transition between full-stick and full-slip states of interwire contact, which governs the hysteretic bending response. This constitutive behavior is embedded within a corotational beam finite-element (FE) framework. However, the formulation assumes that the curvature threshold for sliding is constant and unaffected by axial force. This simplification limits the model's accuracy, particularly under compressive loading. As a result, the model may underestimate energy dissipation and misrepresent tangent stiffness in compression scenarios. Despite this limitation, the mechanical model represents a promising foundation for the mechanical modeling of WRIs. Bonisoli et al. [10] investigated the nonlinear dynamic behavior of cantilever beams with magnetic interactions and non-holonomic contact, using laser profilometry and time–frequency analysis. Their approach interprets the nonlinear responses through multiple linearized Euler–Bernoulli configurations and model-order reduction, demonstrating that a sequence of linearized models can effectively capture the global dynamics while revealing nonlinear fingerprints such as superharmonics and time-varying frequencies. These findings support the use of linear elastic beam representations at the macroscopic scale, while acknowledging localized nonlinear mechanisms at interfaces, and reinforce the value of simplified modeling strategies for complex systems.

Although full 3D contact FE simulations of wire-rope isolators (WRIs), in which every strand and frictional interface is explicitly meshed, may reproduce both elastic stiffness and static hysteresis, their computational demands and contact-model calibration requirements make them impractical for routine design or for rapid exploration of inclined/combined configurations. FE models for simple straight wire-rope strands were developed by Jiang et al. [11] and Nawrocki and Labrosse [12]. Elata et al. [13] defined a modeling framework based on explicit parametrization of each wire's double-helix trajectory around an independent wire-rope core, where the helix angle and phase of every wire are defined by simple trigonometric functions of its axial and circumferential positions. More recently, Gai et al. [14] demonstrated that a carefully parameterized full-pitch 3D FE model using reduced-integration hexahedral elements can predict both intact and broken-wire behavior with high fidelity. Together, modern FE approaches, when coupled with precise geometric parameterization, can reproduce small-strain stiffness and large-deformation hysteresis of WRIs with high fidelity, substantially reducing reliance on exhaustive laboratory testing, albeit at the cost of increased computational effort.

The combined nonlinear stiffness and rate-independent damping allow WRIs to perform over broad frequency ranges and large displacements without performance degradation. However, these same nonlinearities pose challenges for design. Performance-based design of WRI systems demands nonlinear analyses that quantify isolator deformations under diverse excitations, orientations, and loading scenarios. WRI properties depend on the load direction, amplitude, and cable geometry, and manufacturers usually provide only standard quasi-static curves (e.g., vertical compression, horizontal shear) for a few configurations. Practitioners lack detailed force–displacement relationships under dynamic and non-standard loading; for instance, WRIs are inclined to resist the rocking of equipment. As a result, the integration of WRIs into performance-based designs for rocking control can be limited, since predictive models for arbitrary orientations or different dynamic excitations have not been readily available.

Applications underscore the need for improved WRI modeling. Marin-Artieda and Han [15], for example, investigated WRI "energy dissipation platforms" to control the overturning of equipment under earthquake excitations. Accurate prediction of WRI performance in energy-dissipation platforms demands characterization under combined-mode and inclined conditions, which add more challenges for predictive modeling and design integration. Unfortunately, solutions for the full multistrand helical geometry, together with interstrand contact, under combined working modes do not exist. Traditional modeling approaches, such as the Bouc–Wen formulation, have been employed to simulate WRI behavior, but they often require extensive parameter tuning and struggle to capture the full range of nonlinearities. While high-fidelity FE simulations that explicitly model each cable strand and its frictional interfaces may reproduce this complex dynamic behavior, their steep computational cost and the need for expert contact-model calibration severely limit their use in practical design workflows.

These observations motivate the development of compact models that preserve geometric realism, accurately predict global stiffness and energy across orientations, and require minimal calibration making them cost-effective tools for use in engineering practice. In this spirit, the present study proposes a simplified yet robust FE modeling framework experimentally validated. The FE modeling framework efficiently captures the nonlinear stiffness and damping of WRIs in various loading modes, including tension-compression, shear, roll, and inclined configurations. The model leverages Timoshenko beam elements to represent the cable helix with constant-effective bending, shear, and axial sectional properties, incorporates Rayleigh-type damping calibrated to mimic frictional energy dissipation, and implements large displacements and rotations. This approach dramatically reduces model complexity while preserving the key geometric nonlinearities of the isolator. The accuracy of the proposed model is demonstrated by comparison to dynamic tests, and its utility is illustrated in predicting the response of WRI-based isolation systems that would be difficult to test experimentally. This approach captures the essential geometric nonlinearities and energy dissipation mechanisms of WRIs while maintaining computational efficiency. The model is validated against a benchmark analytical solution and experimental data across multiple WRI types and loading configurations. This work contributes new knowledge by demonstrating that a reduced-complexity beam-based model can reliably simulate global stiffness and damping behavior of WRIs under both component-level and system-level conditions. It enables efficient evaluation of WRI performance in non-standard configurations, such as rocking-control platforms, and supports broader integration of WRIs into performance-based seismic and vibration mitigation designs.

## 2. Definition of the WRI FE Model

Figure 1 presents two photographs of a typical WRI used in this study, indicating the conventional working modes shear and roll in the horizontal direction and tension-compression in the vertical direction, together with the analogy of its cross-section to an elliptical ring.

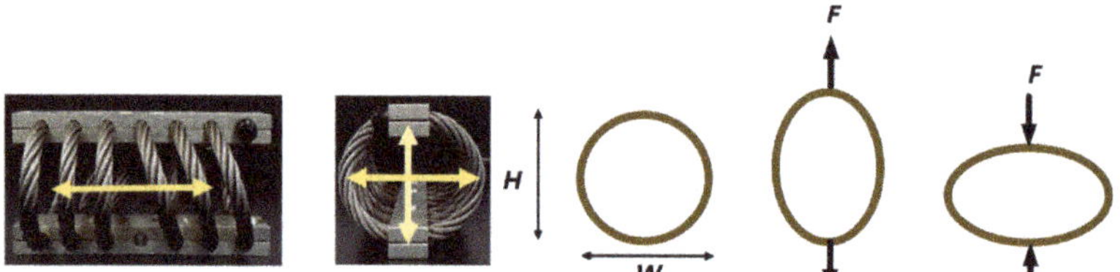


**Figure 1.** Photographs of a typical WRI showing shear, roll, and tension-compression modes (arrows indicating working mode), with the analogy of the WRI as an elliptical ring.

### 2.1. Identifying Basic Elements of the FE Models of WRIs

The variation of overall stiffness in WRIs arises primarily from their helical geometry. As the isolator deforms under tension or compression, its effective height $H$ changes, altering the global area moments of inertia. Modeling the WRI cross-section as an elliptical ring (Figure 1) with major radius $W$ and minor radius $H$, the principal moments of inertia are given by the following:

$$I_{xx} = \frac{\pi}{64} W H^3, \qquad I_{yy} = \frac{\pi}{64} H W^3. \tag{1}$$

Subsequently, the strain energy at any instant is modified for the corresponding height of the helical ring at a specific deformation level.

In this study, the FE model (FEM) for WRIs is developed using Timoshenko beam theory, which extends the classical Euler–Bernoulli beam theory by incorporating transverse shear deformation. Unlike Euler–Bernoulli theory, which assumes that plane sections remain perpendicular to the neutral axis, the Timoshenko theory allows these sections to rotate independently [16]. Timoshenko beam elements are particularly advantageous in modeling complex geometries and short-span structures, as they can capture higher-order cross-sectional deformation modes. Furthermore, the beam element selected for constructing the FE models should allow the analysis of large rotations and displacements.

To evaluate the capability of Timoshenko beam elements in capturing ring-type geometric nonlinearity, this study employs the Abaqus (version 2016) B31 element with nonlinear geometry enabled (*NLGEOM) and a linear elastic material model [17]. As a benchmark, the analytical work of Breslavsky et al. [18] is utilized, which provides solutions for both linearly elastic and large-strain Neo–Hookean circular rings, which serve as a reference for verifying FE implementations across both small- and large-deformation regimes. To replicate the benchmark scenario, a quarter-ring segment is modeled under uniform radial pressure, with symmetric boundary conditions applied to reflect the full ring behavior (Figure 2a). This configuration enables direct comparison between the FEM results and the analytical solutions, thereby validating the geometric nonlinearity captured by the B31 formulation. Using only ten B31 elements, a mesh-sensitivity study confirmed convergence. Figure 2b compares the FE-generated pressure-normalized deformation curve ($w/h$, $w$ radial displacement normalized by its thickness $h$) to the axisymmetric analytical solutions of [18].

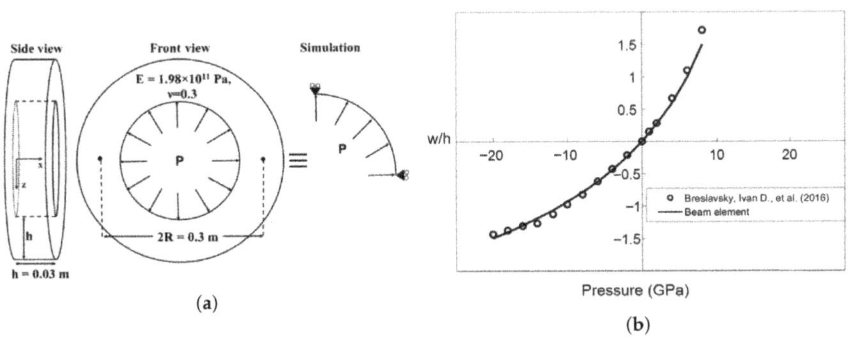


**Figure 2.** Timoshenko beam FE model against solutions for circular rings under pressure [18]. (**a**) Ring geometry. (**b**) FE beam-model (Timoshenko B31 elements) against the benchmark solution [18].

Figure 2b shows strong agreement throughout the pressure ($P$) spectrum between the normalized radial deflection $w/h$ from Breslavsky et al. [18] and the Timoshenko-beam FEM approximation. Under the considered pressures, the FEM results lie within only a few percent of the analytical benchmark, confirming that even a coarse mesh of Timoshenko beam elements reproduces the ring's nonlinear stiffness and large-strain response. In the small-strain regime ($|P| \lesssim 1$ GPa), the behavior is nearly linear. Beyond this threshold, the tensile branch ($P > 0$) exhibits rapid hardening, whereas the compressive branch ($P < 0$) softens more gradually, yielding an inherent asymmetry in $w/h$. Without any additional calibration, the beam-element model captures both features solely from the ring's geometry and material properties. At the highest pressures examined (around 8 GPa), the FEM solution underestimates the normalized deflection by about 12%. However, the FEM results remains close to the analytical curve, and the observed deviation is largely attributed to the steepness of the curve in that region, which amplifies small differences in deflection. This benchmark thus confirms that a Timoshenko-beam approximation with nonlinear geometry analysis can reliably capture large-strain ring deformations, giving confidence that the same methodology will extend effectively to the more complex helical geometry of WRIs.

### 2.2. Effective Cross-Section Definition and Damping Parameters

The cross-section of a rope consisting of twisted wires and strands is simplified to a generalized section with a constant effective area and an effective area moment of inertia, neglecting friction between individual wires. These effective properties are estimated and tuned using vendor-supplied force-deformation data for quasi-static tests under conventional working modes. Figure 3 depicts this process, and the geometry of a WRI and its typical cross section consisting of multiple twisted strands around a core are shown schematically. The rope is modeled as a beam element of effective diameter $d_{\text{eff}}$, whose cross-sectional properties derive from $n$ symmetrically intertwined strands (see Figure 3). Consequently, the effective area becomes the sum of individual strand areas, and the effective inertia about the centroid is the sum of each strand's inertia plus parallel-axis contributions. For a rope with $n$ strands, the generalized section carries the combined properties of all strands.

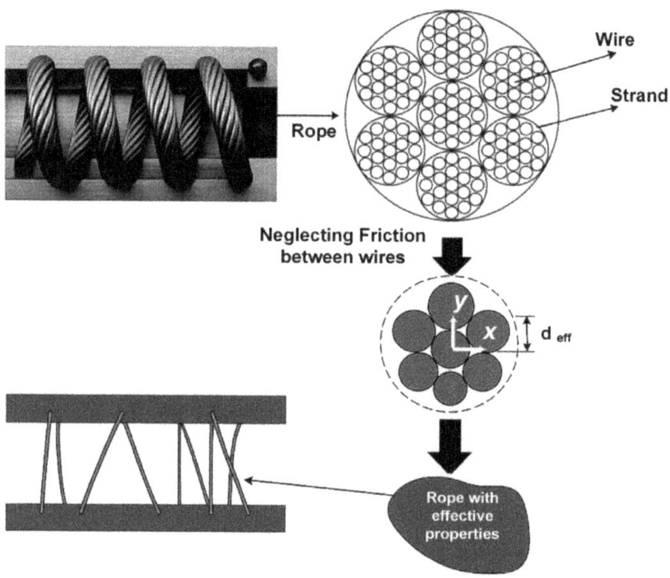


**Figure 3.** Effective cross-section properties of WRIs.

A rope's multi-wire cross-section, consisting of $n$ twisted strands, is represented by a single circular section of effective diameter $d_{\text{eff}}$. This equivalent section is calibrated to match vendor-supplied quasi-static load-displacement data. Neglecting interwire friction, each strand's contribution is defined by the following:

$$A_s = \frac{\pi\, d_{\text{eff}}^2}{4}, \quad I_s = \frac{\pi\, d_{\text{eff}}^4}{64}. \tag{2}$$

where $A_s$ and $I_s$ are the strand area and area moment of inertia. Summing over the $n$ strands located at coordinates $(x_i, y_i)$ relative to the section centroid yields the generalized section properties:

$$A_{\text{eff}} = \sum_{i=1}^{n} A_s = n\,A_s, \tag{3}$$

$$I_{xx,\text{eff}} = I_{yy,\text{eff}} = \sum_{i=1}^{n} (I_s + A_s\,y_i^2) = n\,I_s + \sum_{i=1}^{n} A_s\,y_i^2, \tag{4}$$

$$I_{zz,\text{eff}} = I_{xx,\text{eff}} + I_{yy,\text{eff}} \tag{5}$$

where the effective area and area moment of inertia values—$A_{\text{eff}}$, $I_{xx,\text{eff}}$, and $I_{yy,\text{eff}}$—are then assigned to the beam elements to represent the rope under static and dynamic loads.

The effective properties in Equations (3)–(5) assume a constant cross-section and linear elasticity at the wire-bundle level.

As is well documented in the literature and confirmed through experimental studies, WRIs exhibit damping that varies significantly with displacement amplitude, reflecting their nonlinear and hysteretic nature [19]. This behavior arises primarily from two mechanisms: Coulomb friction, when individual strands slide against one another, and viscous stick-slip effects, in which strands temporarily adhere before slipping [1]. Both mechanisms evolve with deformation, causing the effective stiffness to change with increasing displacement and making direct strand-level modeling computationally demanding and difficult to calibrate. To enable practical implementation while still capturing the essential energy dissipation, the combined damping behavior is approximated using Rayleigh damping. Rayleigh damping inherently neglects the amplitude-dependent and hysteretic characteristics of WRIs. Nonetheless, it remains a convenient classical formulation for simplified approaches, and is employed here to estimate the equivalent energy dissipation effects of WRIs.

$$C = \alpha M + \beta K, \tag{6}$$

where $C$ is the damping matrix, $M$ and $K$ are the mass and stiffness matrices, and $\alpha$ and $\beta$ are the mass- and stiffness-proportional damping coefficients, respectively.

In the search for the most appropriate Rayleigh damping parameters ($\alpha$ and $\beta$), it is important to recognize that, when applied in systems undergoing stiffness softening, such as WRIs, this formulation can introduce artificial effects. In particular, the stiffness-proportional term may generate damping forces that are unrealistically large relative to the restoring forces, resulting in excessive damping and, in some cases, unconservative predictions [20,21]. Likewise, in base-isolated systems or in cases dominated by rigid-body motions, the mass-proportional damping component can suppress low-frequency response and distort system dynamics, resulting in unrealistic damping forces [20].

To address these issues, preliminary calibration trials were conducted using experimental force–displacement data. Based on these evaluations, a Rayleigh damping model with $\alpha = 0$ and tuned $\beta$ was adopted to reproduce the observed energy dissipation in sinusoidal load–deflection tests. This approach is consistent with recommendations in the literature, which recommend eliminating the mass-proportional term in base-isolated systems to avoid unrealistic low-frequency damping [21,22]. While this assumption is appropriate for the conditions considered here, the selection of $\alpha$ and $\beta$ should be system-specific.

Table 1 lists the five helical isolator types used in this study, including their designations and key geometric dimensions. Figure 4 illustrates the typical WRI geometry, with dimension labels corresponding to those introduced in Table 1.

**Table 1.** Geometric dimensions of WRI types.

| Type | H (mm) | W (mm) | L (mm) | S (mm) | D (mm) |
|---|---|---|---|---|---|
| Type 1 (CB1280-35) | 54 | 80 | 146 | 13.8 | 13 |
| Type 2 (CB1300-20) | 60 | 75 | 168.8 | 22.3 | 14 |
| Type 3 (CB1300-30) | 64 | 91 | 169 | 22.3 | 16 |
| Type 4 (CB1400-15) | 64 | 102 | 216 | 30.2 | 19 |
| Type 5 (CB61400-17) | 70 | 93 | 178 | 22.3 | 20 |

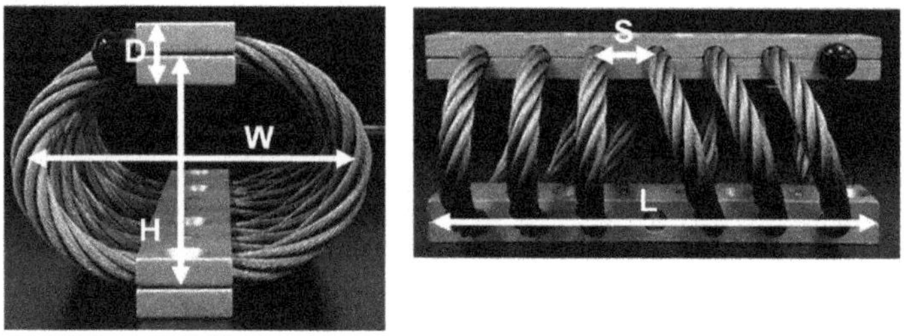


**Figure 4.** Dimensions of WRI used in the study.

Using the basic geometry illustrated in Figure 4, the WRI's helical centerlines as line sketches can be defined in any CAD software (2019). Exporting these sketches as standard geometry files allows direct import into an FE environment, where the centerlines can be used to generate the mesh of beam elements that represent the isolator's helical shape.

The next step in the modeling workflow is to determine the constant WRI's effective cross-sectional properties to assign to the beam elements of the FEM. The effective cross-sectional area and moments of inertia are calculated using Equations (3)–(5). The simplification of constant sectional properties is consistent with the global-response objectives of the study, namely capturing multimode stiffness and energy trends with minimal calibration effort.

Calibration of the effective properties is performed against static force–displacement curves obtained from vendor catalogs for each WRI type in its standard working modes, pure shear, pure roll, vertical tension-compression, and a 45° inclined orientation. The effective area and moments of inertia are tuned so that the FE predictions reproduce the cataloged curves, ensuring that the simplified beam representation accurately captures the quasi-static stiffness of the isolator. For the WRIs considered in this study, only minor tuning was required to obtain reasonable agreement across all working modes. Nevertheless, it is difficult to identify a single set of constant section properties that achieves equally high accuracy under all loading conditions. In line with the intended use of this FEM framework for engineering design practice, one representative set of effective properties is consistently maintained throughout the analyses.

Figure 5 illustrates the calibration procedure for WRI Type 3. A single set of effective cross-sectional properties is defined that closely reproduces the quasi-static stiffness in all conventional modes. A comparison of experimental (vendor data) and FEM force–displacement curves shows that:

- In pure tension–compression, the FEM reproduces the experimental stiffness within 5%.
- In pure shear and roll, the FEM and experimental curves are nearly identical across the entire displacement range.
- In the 45° inclined mode, which combines tension–compression and roll (or shear), the FEM matches the experimental curves up to about $\pm 10$ mm, but underestimates tensile stiffness by about 10% and compressive stiffness by about 5%. At larger displacements

(around ±25 mm), the experimental tangent slope exceeds the FEM prediction by roughly 15–20%.

These discrepancies between FEM and vendor curves for the 45° inclined mode reflect the difficulty of representing coupled geometric nonlinearity and interstrand friction with a single, constant-property cross-section. Nonetheless, even in combined-mode (45°) loading at large displacements, the FEM peak predictions remain within 20% of the experimental stiffness values, and the FEM result remains close to the experimental curves.

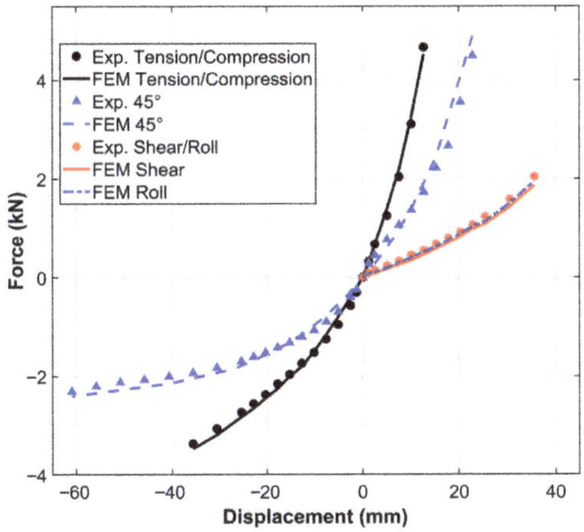


**Figure 5.** Static analysis of WRI Types 3 for tuning effective properties.

## 3. Responses of Individual WRIs Under Sinusoidal Excitations

### 3.1. Description of the WRI Settings and Assessment Parameters

The FEM in this study was developed based on the available experimental data, which provided force–displacement relationships necessary for evaluating model performance. However, the experimental setup was constrained by the load capacity of the equipment and the design of the fixtures used to secure WRIs. Experimental testing was conducted at the Applied Mechanics and Materials Research Laboratory at Howard University, using an axial-torsion system to measure the forces exerted on the specimens and generate force-displacement loops. To accommodate different loading configurations, including inclined angle loading, custom steel fixtures were designed to mount the WRIs securely on the test rig. Test protocols were defined to characterize the WRIs under each configuration, taking into account both the isolators' mechanical properties and the limitations of the testing apparatus and fixtures.

The FEM of the five types of WRIs in conventional working modes and with different angles of inclination (0°, 30°, 45°, and 60°) were subjected to sinusoidal displacement series. The displacement histories are applied to the top bar of the WRIs, while the bottom bar is fixed in all six degrees of freedom to mimic the experimental conditions. All applied displacements can be described by a sinusoidal function with a constant amplitude of 3, 6, 10, and 20 mm and with periods varying between 1, 10, and 100 s. The force–deformation relationships of the WRIs are analyzed to identify the stiffness and damping of the individual isolators. The effective stiffness and damping of the systems were calculated using the expressions below:

$$k_{\text{eff}} = \frac{|F^+| + |F^-|}{|D^+| + |D^-|}, \tag{7}$$

$$Tk_{\text{eff}} = \frac{|F^+|}{|D^+|}, \tag{8}$$

$$Ck_{\text{eff}} = \frac{|F^-|}{|D^-|}, \tag{9}$$

$$\beta_{\text{eff}} = \frac{2}{\pi} \frac{A_{\text{hyst}}}{k_{\text{eff}}\left(|D^+| + |D^-|\right)^2}. \tag{10}$$

where $k_{\text{eff}}$ is the overall effective stiffness, $F^+$ and $F^-$ are the maximum tensile and compressive forces, respectively, $D^+$ and $D^-$ are the maximum tensile and compressive displacements, respectively, $Tk_{\text{eff}}$ is the tensile effective stiffness, $Ck_{\text{eff}}$ is the compressive effective stiffness, $\beta_{\text{eff}}$ is the effective damping, and $A_{hyst}$ is the total area of the hysteresis loop.

The numerical predictions were compared with available experimental responses. Figure 6 presents photographs of the experimental setup at the top and the corresponding numerical models of the WRIs at the bottom.

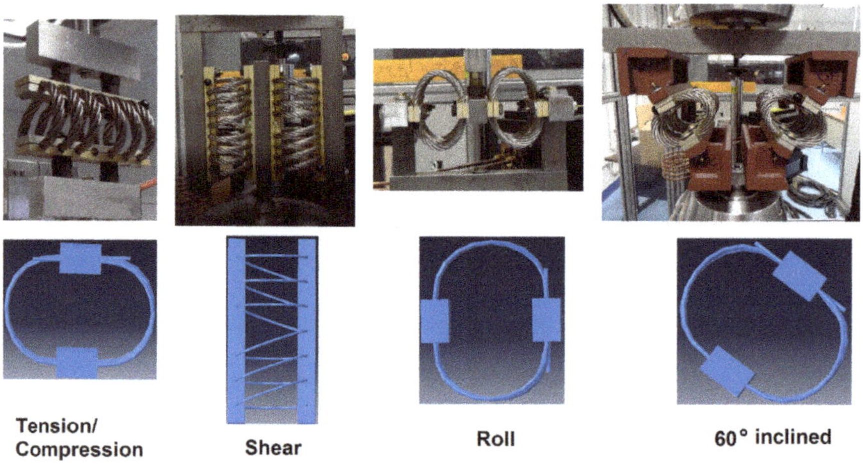


**Figure 6.** (**Top**): Photograph of the experimental setup. (**Bottom**): Corresponding FEM of the WRIs.

### 3.2. *Results of Individual WRIs Under Sinusoidal Excitations Under Tension-Compression*

This subsection illustrates the FE model's performance for WRI Type 3 under pure tension–compression sinusoidal loading (10 s period, 6 mm and 10 mm amplitudes), as plotted in Figure 7. At 6 mm amplitude, the FEM underestimates the experimental effective stiffness by 5% and underpredicts damping by approximately 6%. The compressive branch is captured almost exactly, while tensile stiffness is underestimated by about 10%. Increasing the amplitude to 10 mm yields identical effective stiffness in both experiment and FEM, with the damping ratio differing by 4%. In this case, the FEM slightly over-predicts tensile stiffness by 4% and compressive stiffness by 7%. Qualitatively, the FE hysteresis loops closely match the experimental curves in both slope and overall shape, though the model exhibits minor deviations at the tension peak for 6 mm and at the compression peak for 10 mm. These results confirm that, although the FEM reproduces overall stiffness at larger amplitudes, it still tends to smooth out the subtle asymmetries and energy dissipation mechanisms observed in the physical isolator.

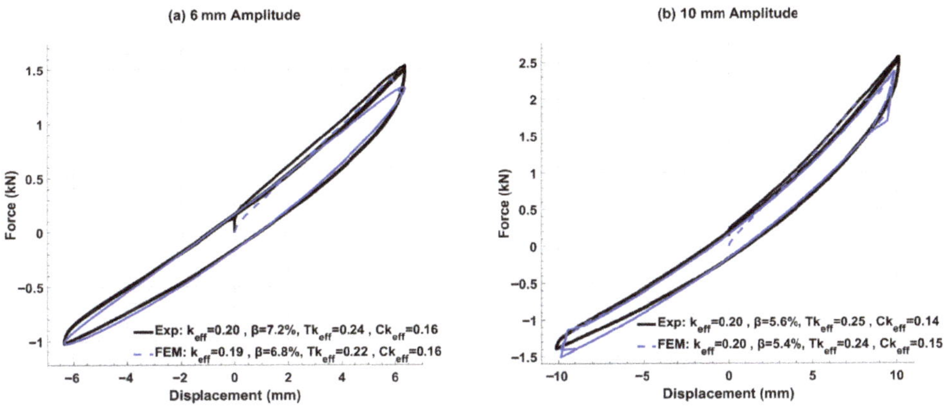


**Figure 7.** Force–displacement curves for WRI Type 3 under pure tension–compression sinusoidal loading (10 s period).

FEM of WRIs rotated by 30°, 45°, and 60° were developed to capture the combined tension–compression–roll response introduced by inclination (see the left side of Figure 6). Representative force–displacement loops for rotations of 30° and 45° on WRI Types 3 and 4 are shown in Figures 8 and 9, respectively.

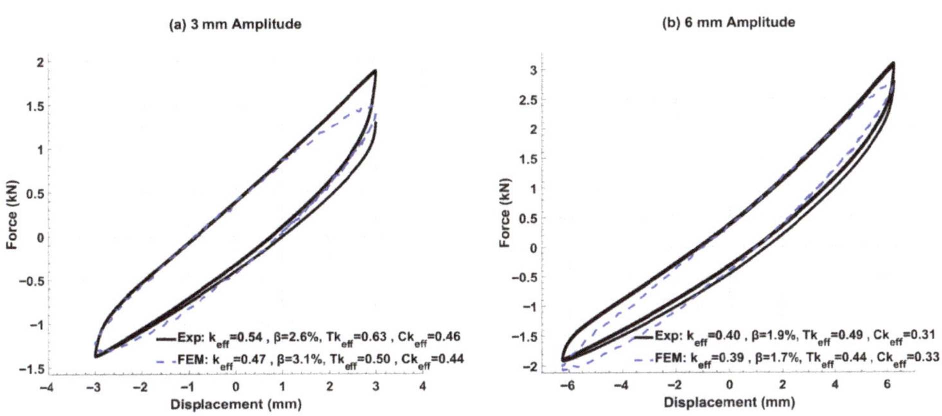


**Figure 8.** FEM vs. experimental force–displacement loops for rotated Type 3 WRI at 30° (tension, compression, and roll, 100 s period).

The assessment parameters for WRI Type 3 rotated by 30° are summarized in Figure 8. At a 3 mm amplitude (Figure 8a), the FEM predicts an overall effective stiffness $k_{\text{eff}}$ that is approximately 15% softer than the experimental value, and an effective damping that is 16% higher. The tensile stiffness is under-predicted by 26% and the compressive stiffness by 5%. Despite these offsets, the small-amplitude force–displacement curve closely matches the experimental shape and slope, with the largest deviation occurring in the peak tensile branch. At a 6 mm amplitude (Figure 8b), the overall hysteresis loop area from the FEM aligns well with the experiment. In this case, the FE stiffness is 12% softer in tension, while the predicted effective damping is 11% lower than measured. Similar results to the above were obtained for the WRI Type 2 rotated by 60°; therefore, those results are omitted here.

The assessment parameters for WRI Type 4 rotated by 45° are summarized in Figure 9. At a 3 mm amplitude (Figure 9a), the FEM predicts an overall effective stiffness $k_{\text{eff}}$ that is 13% softer and an effective damping that is 8% higher than the experimental values.

Examining the branches separately, the model underestimates the tensile stiffness $Tk_{eff}$ by 14% and the compressive stiffness $Ck_{eff}$ by 29%, differences that manifest clearly in the force–displacement curve shapes. At a 6 mm amplitude (Figure 9b), the FEM stiffness remains 14% softer, while the predicted effective damping is 7% higher than measured. Despite these differences, the overall FEM loop closely follows the experimental curve, capturing both the global slope and loop area with relatively good fidelity.

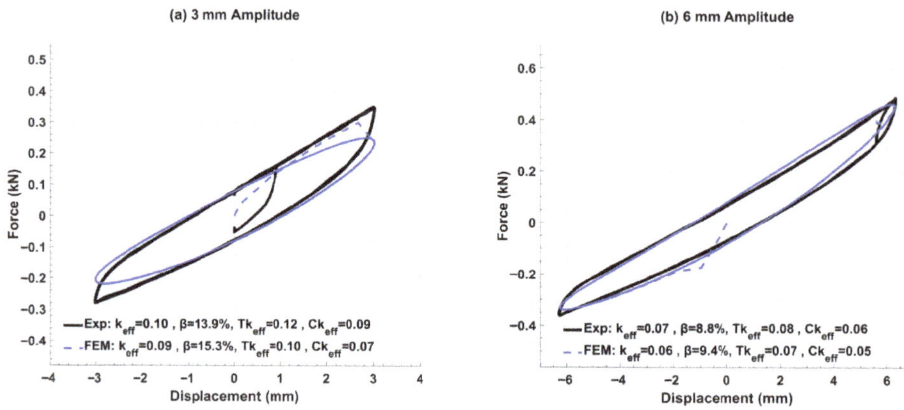


**Figure 9.** FEM vs. experimental force–displacement loops for rotated Type 4 WRI at 45° (tension, compression, and roll, 1 s period).

The largest deviations occur under combined-mode deformation when the isolator is inclined and the response explores the extremes of the tensile/compressive branches. Quantitatively, for Type 3 at 30°, the model is 15% softer in overall stiffness at 3 mm and 12% softer in tensile stiffness at 6 mm; for Type 4 at 45°, the model is 13–14% softer in overall stiffness, while the compressive branch reaches 29% underprediction at 3 mm, reducing at 6 mm. At the system level (four WRIs at 45°, 3 Hz), force–displacement and moment–rotation metrics are generally within 10–17%, with the largest differences near peak moments/rotations. These trends show the error is configuration-dependent (angle, amplitude) and branch-specific, rather than systematic across all inclined dynamics. The observed bias is consistent with the model's constant-section assumption and equivalent-viscous dissipation. Because the local interstrand stick-slip and amplitude-dependent hardening in the compressive branch are not explicitly modeled, the discrepancies grow when axial and bending effects couple most strongly (45°) and at branch extremes.

### *3.3. Results of Individual WRIs Under Sinusoidal Excitations Under Shear and Roll*

Figure 10 presents the pure-roll hysteresis loops for the Type 1 isolator at 12 mm and 20 mm amplitudes. In both cases, the beam-based FEM closely reproduces the effective roll stiffness. The damping ratio, however, is overpredicted by 31% at 12 mm amplitude and by 25% at 20 mm amplitude. Despite this overestimation of damping, the shape and area of the hysteresis loops remain almost identical between FEM and experiment for the 20 mm amplitude, with some minor discrepancies on the peaks of the loops.

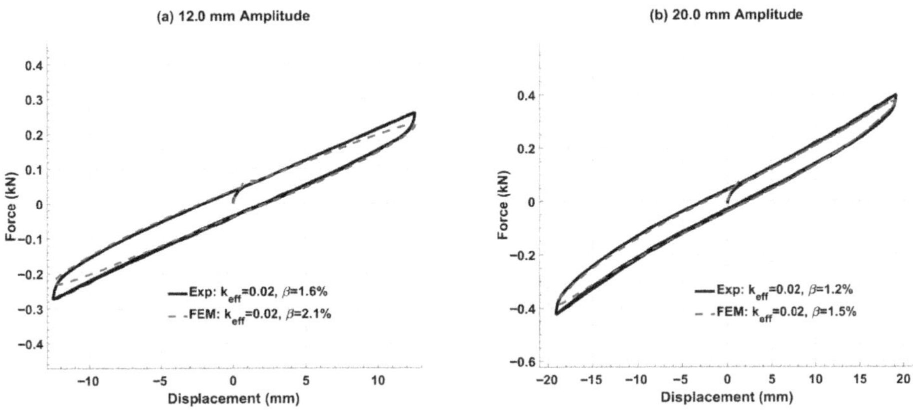


**Figure 10.** Pure-roll hysteresis loops for the Type 1 isolator (10 s period).

Figure 11 presents the pure-shear force–displacement curve for the Type 1 isolator at a 6 mm amplitude. The FEM exactly reproduces the experimental effective stiffness. The effective damping is overpredicted by 52%. Nonetheless, the loop's overall shape and slope align closely with the experiment, with minor differences in the tensile branch.

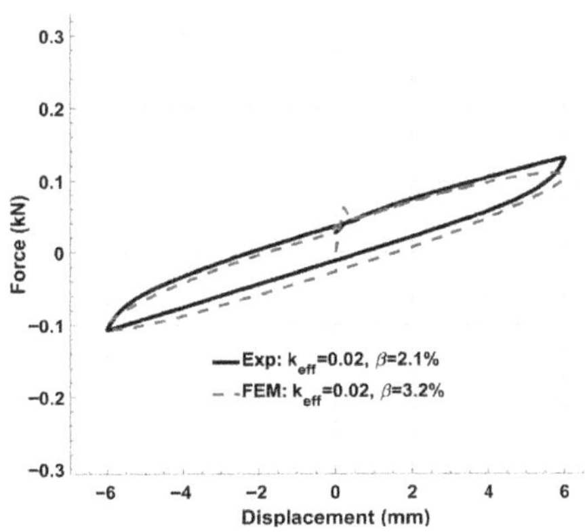


**Figure 11.** Pure-shear hysteresis loop for the Type 1 isolator (6 mm amplitude, 100 s period).

## 4. FEM Response of a WRI-Based Isolation System Under Sinusoidal Excitations

The results of the FEM simulations of a WRI base platform subjected to sinusoidal accelerations are compared against the experimental results of Marin-Artieda and Han [15]. The isolation platform studied here employs four Type 5 WRIs inclined at 45°, forming a testbed to investigate how WRIs can be arranged to counteract seismic rocking of slender equipment. In the experiments, steel plates were sandwiched between wire-rope assemblies in multiple geometric arrangements to explore how the relative location of the platform's center of mass (CM) and center of stiffness (CS) controls rocking behavior; these plates were modeled as rigid blocks in the FEM. By tilting the isolators "inward," the platform's CM is deliberately offset relative to its CS, producing a negative rotational stiffness that opposes overturning moments. Although the results presented in this section focus on the platform's global force–displacement and moment–rotation loops rather than direct rocking measurements, they nonetheless support the evaluation of the FEM's ability to capture complex, non-standard configurations where combined tension–compression–roll responses govern the system's behavior. Figure 12 shows one of the experimental prototype platforms fitted with four Type 5 inclined WRIs at 45° and its corresponding FE model in front and plan views.

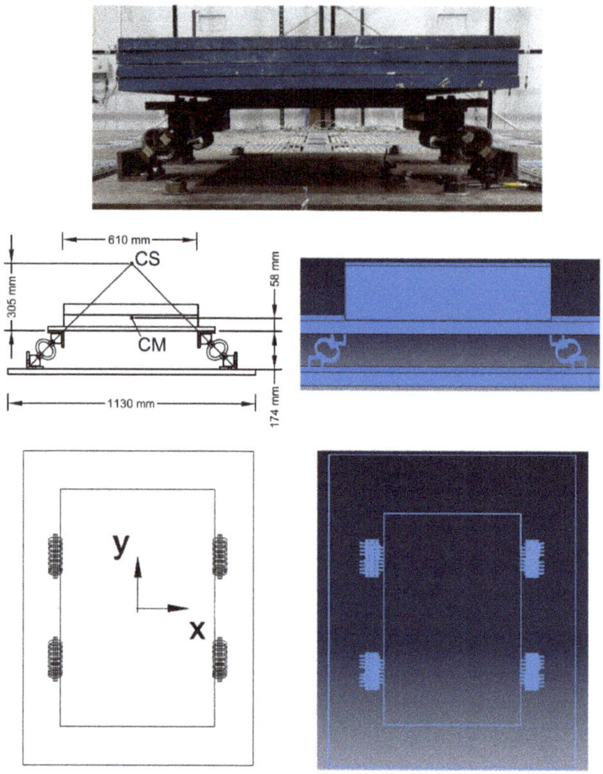


**Figure 12.** WRI-based isolation system with inclined isolators.

Figure 13 presents the average horizontal ($x$-direction) force-displacement response of the WRI platform under two unidirectional $x$-direction, 3 Hz sinusoidal excitations. At the lower amplitude (0.55 g; Figure 13a), the FEM reproduces the experimental slope and curve shape with good fidelity: effective stiffness is within 5%, and hysteresis area differs by 10%. The FEM loop appears slightly smoother with reduced "pinching" at reversals. Under the higher excitation (1.25 g; Figure 13b), discrepancies grow at the negative-load branch. The FEM underestimates peak forces and loop area by approximately 15%, reflecting diminished accuracy in stiffness hardening and energy dissipation at large deformations.

Figure 14 compares the FEM and experimental moment–rotation responses of the 45° tension-compression-roll platform (CM–CS offset = −247 mm, counter-rocking configuration) under 3 Hz sinusoidal excitation at amplitudes of 0.55 g (Figure 14a) and 1.25 g (Figure 14b).

At the lower amplitude (Figure 14a), the FEM predicts the negative rotational stiffness to within 10% of the experimental value. Although the model reproduces the overall hysteretic envelope, it does not capture the distinctive "waist" of the experimental loops where the branches are widest at the peak loads, but pinch sharply near zero rotation. Its largest discrepancy appears at the peak moments and rotations, which are slightly overestimated. The total hysteresis area differs by 15%, indicating that proportional damping provides a reasonable estimate of energy dissipation for small-to-moderate rotations but fails to reproduce the waisted loop shape. Moreover, the experimental loops transition smoothly at load reversals, whereas the FEM loops exhibit pronounced pinching at those points. Under the higher amplitude excitation (Figure 14b), the differences become more pronounced near the loop extremities. The model overestimates peak moments by 14% and peak rotations by approximately 18%, and the enclosed hysteresis area is approximately 17% larger. This discrepancy reflects reduced accuracy in both stiffness hardening and damping at large rotations. Additionally, the experimental loops exhibit stiffening in the positive-moment branch, a behavior that is not captured by the FEM.

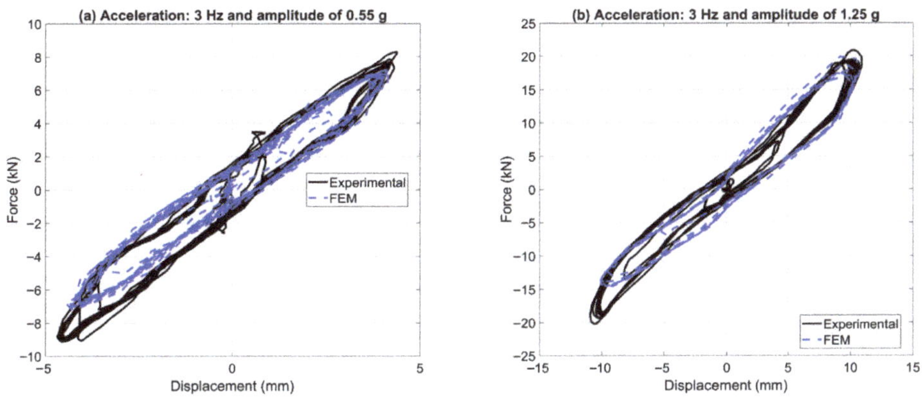


**Figure 13.** Platform force–displacement loops.

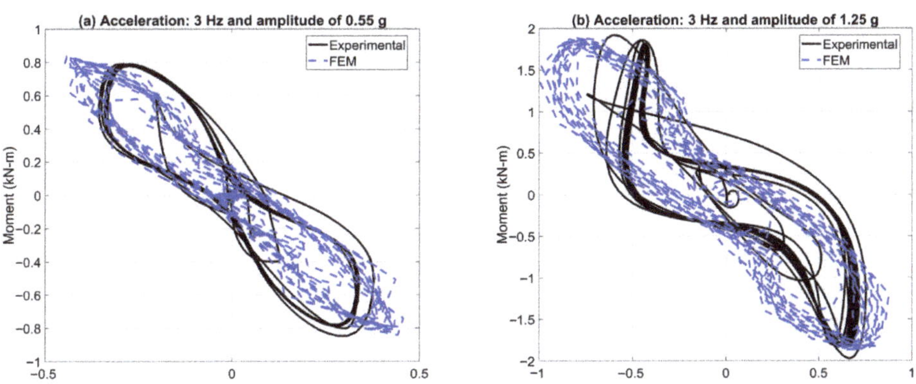


**Figure 14.** Moment-rotation loops with negative CM–CS offset.

## 5. Vertical Travel Distance of WRIs

One critical design parameter for WRI-based platforms is the vertical travel distance *d*, defined as the peak vertical deformation of the WRIs under the platform load (see Figure 15). This travel becomes especially important when WRIs are inclined, since excessive deformation can lead to contact between WRIs and attachments, potentially causing mechanical interference and impact.

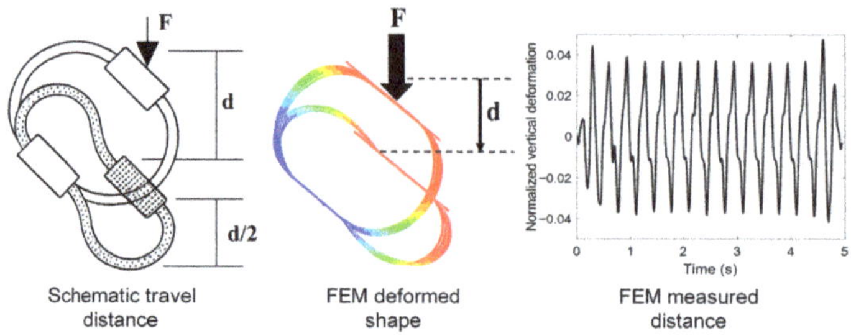


**Figure 15.** Vertical travel distance of WRIs.

By quantifying the travel distance in the FEM during design and prior to fabrication, designers can ensure adequate clearance between attachment bars and avoid collisions. FEM can facilitate the monitoring of the travel distances across different loading amplitudes and inclinations, enabling rapid evaluation of safety margins and serviceability limits without resorting to costly prototype testing.

## 6. Discussion

Building upon foundational validation efforts, this study demonstrates that a constant-property Timoshenko beam FEM, with tuned Rayleigh damping, offers a computationally efficient yet sufficiently accurate framework for simulating wire-rope isolator behavior under diverse loading conditions. Initial benchmarking against the work of Breslavsky et al. [18], using the Abaqus B31 beam element with nonlinear geometry (NLGEOM) and a linear-elastic material, confirmed the model's ability to capture asymmetric ring nonlinearities. The FEM results remain close to the analytical curve, and the observed 12% deviation at the maximum normalized deflection $w/h$ is largely due to the steepness of the curve in that region. Although this underestimation maintained acceptable accuracy across the large-deformation regime, it validates the model's suitability for further application.

Leveraging the calibrated framework, the modeling approach was extended to full helical WRI geometries encompassing five distinct types. A single set of effective cross-sectional properties, including area and moments of inertia, was calibrated against vendor-provided quasi-static force–displacement curves in standard working modes. This representative set was then applied uniformly across all beam elements. The simplified beam representation reproduced the nonlinear stiffness characteristics of the three principal deformation modes with good fidelity, requiring only minor tuning of the section properties. In the more complex 45° inclined configuration, which couples shear/roll with tension–compression, discrepancies emerged at larger displacements. Here, the FEM stiffness predictions deviated from the experimental curves by approximately 15–20%. These offsets reflect the difficulty of capturing combined geometric nonlinearity and interstrand friction with a single, constant-property cross-section. It is important to acknowledge that no single set of constant section properties can reproduce all loading configurations with equal accuracy; however, the model remains a practical and accessible tool for global stiffness estimation. Overall, the constant-property beam approximation proved to be a reliable and computationally efficient surrogate for representing WRI behavior, particularly in the standard working modes.

Dynamic analyses under sinusoidal excitation further underscored the FE model's balance between predictive fidelity and computational tractability. In pure deformation modes, the FEM force–displacement responses closely matched experimental data in both stiffness and hysteretic loop morphology. In inclined configurations, however, the model underestimated stiffness and did not capture well loop shapes at peak locations under low-amplitude (3 mm) excitation. At moderate amplitudes (6 mm), the predicted loops aligned more closely with experimental observations. It should be noted that dynamic validation in this study was restricted to sinusoidal excitation tests performed within the limits of the available equipment. To strengthen the generalization of the framework, future work may incorporate independent experimental datasets and publicly available dynamic test results for cross-validation. Such efforts will broaden the evaluation of the model's predictive capability across a wider range of loading rates, amplitudes, and orientations.

To evaluate the model's performance in a system-level context, a platform testbed supported by four 45°-inclined Type 5 WRIs was analyzed under a non-standard "rocking-control" configuration. When subjected to acceleration excitations of 0.55 g and 1.25 g, the model reproduced the global platform stiffness and energy dissipation within 10–20% of experimental measurements. Although it did not capture the detailed hysteretic loop shapes governed by interstrand friction, the model demonstrated robustness and predictive adequacy for global response metrics.

The results obtained with the platform testbed are particularly valuable for designing WRI isolation systems that include inclined configurations. Experimental evidence indicates that localized deformations and force–displacement relationships in inclined WRIs are complex and not fully captured by conventional WRI modeling approaches. Compounding this challenge, experimental data for inclined configurations are often unavailable, making it difficult to establish the parameters required for conventional WRI models. In such cases, the ability of the present framework to provide reliable global performance predictions becomes especially useful.

The proposed approach delivers robust estimates of system-level stiffness and energy dissipation, while also offering estimates of local deformation demands that are difficult to capture with traditional WRI modeling strategies. Although the simplified beam formulation does not reproduce detailed hysteretic morphology, it achieves a practical balance between computational efficiency and predictive fidelity. This makes the framework well-suited for engineering design applications when overall platform performance is the primary concern and detailed loop shapes are of secondary importance.

The inability of the FEM to replicate pinched hysteretic loops observed experimentally is primarily due to unmodeled physical mechanisms such as interstrand stick–slip, contact reconfiguration, geometric uncoiling, and axial realignment. These nonlinear effects intensify with increasing amplitude and contribute to hardening behavior not captured by the constant-parameter Rayleigh damping formulation. As a result, the FEM produces smoother hysteresis loops and underestimates stiffness at higher deformation levels.

The constant-property Timoshenko beam model with tuned damping offers a pragmatic compromise between modeling simplicity and predictive capability. While it does not capture all local nonlinearities, it effectively simulates global stiffness and energy dissipation across a range of configurations and loading regimes. Defining upper and lower performance bounds and validating them against experimental data can further enhance the model's robustness, ensuring its continued utility as a cost-effective tool for engineering analysis and design.

Overall, the present FEM framework, based on constant-property Timoshenko beam elements with tuned Rayleigh damping, provides a practical compromise between simplicity and predictive capability. It reliably captures global stiffness and energy dissipation across modes, with deviations from experiments typically within 10–20%. However, the model cannot reproduce pinched hysteresis loops or branch-specific hardening effects observed experimentally, particularly in inclined configurations. These discrepancies stem from unmodeled physical mechanisms such as interstrand stick–slip, contact reconfiguration, geometric uncoiling, and axial realignment, which intensify with amplitude and generate

rate-independent hysteresis features not captured by viscous damping. As a result, the FEM produces smoother loops and underestimates stiffness at large deformations. Future improvements could incorporate variable section properties that evolve with geometry and embedded frictional/contact laws (wire–wire and rope–seat interfaces, preload-dependent hysteresis) to better capture local morphology. Mechanical formulations that resolve the rope's hierarchical structure [9] or hybrid strategies using piecewise-linearized beam models [10] offer promising directions. In addition, to support long-term prediction, we outline a bounded damage index driven by cumulative dissipated energy and cycle count that scales the effective section properties and damping surrogate. Future possible work may calibrate these coefficients using repeated-cycle tests at representative amplitudes and frequencies, enabling the model to simulate progressive stiffness degradation and dissipation loss.

## 7. Conclusions

This study introduced and validated a simplified finite-element modeling framework for wire-rope isolators, employing constant-property Timoshenko beam elements with Rayleigh-type damping to capture the essential nonlinear stiffness and hysteretic behavior of WRIs under diverse loading conditions. The model was benchmarked against analytical solutions for ring-type geometries and demonstrated strong agreement, with less than 12% deviation in normalized deflection, thereby confirming its ability to capture large-strain geometric nonlinearities.

The model was extended to five full-scale helical WRI types, using a single set of effective cross-sectional properties calibrated against vendor-supplied quasi-static data. In conventional working modes, pure tension-compression, shear, and roll, the model reproduced experimental stiffness within 5%, and in inclined configurations, it maintained accuracy within 15–20% at moderate displacements. However, deviations increased at larger displacements, particularly under combined-mode loading, reflecting the limitations of the constant-property approximation in capturing coupled geometric and frictional nonlinearities.

Dynamic simulations under sinusoidal excitation further validated the model's predictive capability. In pure modes, the FEM closely matched experimental force–displacement loops in both shape and energy dissipation. In inclined configurations, the model underestimated stiffness and failed to capture loop pinching at low amplitudes (3 mm), though agreement improved at moderate amplitudes (6 mm). These results confirm the model's reliability for service-level deformations while highlighting its limitations in representing nonlinear hardening and asymmetric hysteresis at extreme or multimode deformation.

At the system level, the model was applied to a WRI-based isolation platform with four 45°-inclined WRIs subjected to base accelerations. The FEM reproduced platform stiffness and energy dissipation within 10–20% of experimental values across both 0.55 g and 1.25 g excitations. While the model did not capture detailed loop morphologies such as pinching and asymmetric branch stiffness, it successfully predicted global response trends, including negative rotational stiffness in counter-rocking configurations.

This study also demonstrated the model's utility in estimating vertical travel distances of inclined WRIs, a critical design parameter for avoiding mechanical interference. By enabling rapid evaluation of deformation limits across loading amplitudes and orientations, the model supports practical design workflows without the need for exhaustive physical testing.

The constant-property Timoshenko beam model with tuned damping offers a computationally efficient and sufficiently accurate tool for simulating WRI behavior in both component-level and system-level applications. While it does not resolve all local nonlinearities, particularly those arising from interstrand friction and contact reconfiguration, it provides a robust foundation for performance-based design.

Compared with phenomenological approaches such as Bouc-Wen and generalized hysteresis models, the present beam-based framework offers a computationally efficient and physically interpretable alternative that captures multimode global stiffness with minimal calibration effort. At the same time, the constant-property assumption limits its ability to reproduce local phenomena such as pinched loops, asymmetric branch stiffness, and nonlinear hardening in inclined configurations. These discrepancies are linked to unmodeled interstrand stick–slip, contact reconfiguration, and coupled axial–rotational effects.

It is worth noting that the accuracy of the framework can be further improved by incorporating direction-dependent stiffness corrections, rate-independent hysteresis laws, and geometric nonlinear strengthening rules to broaden validation. Additional refinements, such as degradation indices for long-term cyclic behavior, sensitivity and uncertainty analyses, and the inclusion of environmental effects such as temperature and corrosion, would also enhance the robustness and generalization of the model. While these refinements can increase accuracy, they also introduce modeling and computational effort. In engineering practice, the essential requirement is to balance accuracy with efficiency, depending on the application and design needs. For this reason, the design practice framework should be maintained as a simple and cost-effective tool that is straightforward to implement, ensuring accessibility for practical design use. In summary, this study establishes a simplified beam-based framework that balances accuracy with computational efficiency, providing a reliable baseline for performance-based design. At the same time, it offers the flexibility to incorporate higher-fidelity extensions when practitioners determine that additional accuracy, such as detailed hysteretic morphology or long-term degradation effects, is justified.

**Funding:** This material is based on work supported by the National Science Foundation (NSF) under Grant No. CMMI 1150462.

**Data Availability Statement:** FE models are available on GitHub: https://github.com/cmarin1/FEM_WRIs. accessed on 1 July 2025.

**Acknowledgments:** The author gratefully acknowledges Hamid Ghasemi, a former student, for his contributions to finite-element modeling during his academic training. His engagement with the modeling concepts presented in this work was instrumental in shaping the early stages of the research. Although Ghasemi initially expressed interest in co-authorship, the corresponding author was unable to locate or contact him during the manuscript preparation and revision process. In accordance with authorship guidelines and due to the absence of his review and approval of the final manuscript, he is recognized here with appreciation and respect for his contributions. The author also acknowledges the VMC Group for generously donating the wire-rope isolators used in this study. Any opinions, findings, and conclusions or recommendations expressed in this paper are those of the author and do not necessarily reflect the views of the supporting organizations.

**Conflicts of Interest:** The author declares no conflicts of interest. The sponsors had no role in the design of the study; in the collection, analysis, or interpretation of the data; in the writing of the manuscript; or in the decision to publish the results.

## Abbreviations

The following abbreviations are used in this manuscript:

| | |
|---|---|
| WRI | Wire-Rope Isolator |
| FE | finite-element |
| FEM | Finite-Element Model |
| $w/h$ | Normalized deformation |
| $P$ | Pressure |
| $A_s$ | Strand area |
| $I_s$ | Area moment of inertia |
| $I_{xx}$ | principal moments of inertia $xx$ |
| $I_{yy}$ | principal moments of inertia $yy$ |
| $k_{\mathrm{eff}}$ | Overall effective stiffness |
| $\mathrm{TK}_{\mathrm{eff}}$ | Tensile effective stiffness |
| $\mathrm{CK}_{\mathrm{eff}}$ | Compressive effective stiffness |
| $\beta_{\mathrm{eff}}$ | Effective damping |
| $F^+$ | Maximum tensile force |
| $F^-$ | Maximum compressive force |
| $D^+$ | Maximum tensile displacement |
| $D^-$ | Maximum compressive displacement |
| $\beta$ | stiffness-proportional damping coefficient |
| CM | Center of Mass |
| CS | Center of Stiffness |
| Hz | Hertz (unit of frequency) |
| g | Acceleration due to gravity (9.81 $\mathrm{m/s^2}$) |
| F | Force |
| mm | Millimeter |
| kN | Kilonewton |
| s | Second |
| $d$ | Vertical travel distance of WRI |
| $H$ | Isolator Height |
| $W$ | Isolator Width |
| $L$ | Isolator Length |
| $S$ | Spacing, center-to-center distance between ropes |

## References


1. Tinker, M.L.; Cutchins, M.A. Damping phenomena in a wire rope vibration isolation system. *J. Sound Vib.* **1992**, *157*, 7–18,. [CrossRef]
2. Salvatore, A.; Carboni, B.; Chen, L.Q.; Lacarbonara, W. Nonlinear dynamic response of a wire rope isolator: Experiment, identification and validation. *Eng. Struct.* **2021**, *238*. [CrossRef]
3. Leblouba, M.; Balaji, P.S.; Rahman, M.E. Quasi-static cyclic behavior of wire rope isolators: Comprehensive experimental study and improved mathematical modeling. *Heliyon* **2022**, *8*, e10944. [CrossRef] [PubMed]
4. Wen, Y.K. Method for random vibration of hysteretic systems. *J. Eng. Mech. Div.* **1976**, *102*, No. 102, 249–263. [CrossRef]
5. Demetriades, G.F.; Constantinou, M.C.; Reinhorn, A.M. Study of wire rope systems for seismic protection of equipment in buildings. *Eng. Struct.* **1993**, *15*, 321–334. [CrossRef]
6. Ni, Y.; Ko, J.M.; Wong, C.W.; Zhan, S. Modelling and identification of a wire-cable vibration isolator via a cyclic loading test Part 2: Identification and response prediction. *Proc. Inst. Mech. Eng. Part I J. Syst. Control Eng.* **1999**, *213*, 173–182. [CrossRef]
7. Maleki, M.; Ahmadian, H.; Rajabi, M. A modified Bouc–Wen model to simulate asymmetric hysteresis loop and stochastic model updating in frictional contacts. *Int. J. Solids Struct.* **2023**, *269*, 112212. [CrossRef]
8. Vaiana, N.; Marmo, F.; Sessa, S.; Rosati, L. Modeling of the Hysteretic Behavior of Wire Rope Isolators Using a Novel Rate-Independent Model. In *Nonlinear Dynamics of Structures, Systems and Devices*; Lacarbonara, W., Balachandran, B., Ma, J., Machado, J.A.T., Stepan, G., Eds.; Springer Nature: Cham, Switzerland, 2020; pp. 309–317. [CrossRef]
9. Foti, F.; Galeazzi, J.; Martinelli, L. On the modelling of the hysteretic behaviour of wire rope isolators. In Proceedings of the AIMETA 2019—XXIV Conference of the Italian Association of Theoretical and Applied Mechanics, Rome, Italy, 15–19 September 2019.
10. Bonisoli, E.; Dimauro, L.; Venturini, S.; Cavallaro, S.P. Experimental Detection of Nonlinear Dynamics Using a Laser Profilometer. *Appl. Sci.* **2023**, *13*, 3295. [CrossRef]

11. WJiang, G.; Yao, M.S.; Walton, J.M. A concise finite-element model for simple straight wire rope strand. *Int. J. Mech. Sci.* **1999**, *41*, 143–161. [CrossRef]
12. Nawrocki, A.; Labrosse, M. A finite-element model for simple straight wire rope strands. *Comput. Struct.* **2000**, *77*, 345–359. [CrossRef]
13. Elata, D.; Eshkenazy, R.; Weiss, M.P. The mechanical behavior of a wire rope with an independent wire rope core. *Int. J. Solids Struct.* **2004**, *41*, 1157–1172. [CrossRef]
14. Gai, J.; Yan, K.; Deng, Q.; Sun, M.; Ye, F. A finite-element model for a 6 × K31WS + FC wire rope and a study on its mechanical responses with or without wire breakage. *Appl. Sci.* **2023**, *13*, 8407. [CrossRef]
15. Marin-Artieda, C.; Han, X. Energy dissipation platforms based on wire rope isolators for the seismic protection of equipment. In Proceedings of the 16th World Conference on Earthquake, 16WCEE 2017, Santiago, Chile, 9–13 January 2017; pp. 1–12.
16. Sapountzakis, E.J.; Dourakopoulos, J.A. Nonlinear dynamic analysis of Timoshenko beams by BEM. Part I: Theory and numerical implementation. *Nonlinear Dyn.* **2009**, *58*, 295–306. [CrossRef]
17. Dassault Systèmes. *Abaqus 6.14 Documentation*; Dassault Systèmes Simulia Corp.: Providence, RI, USA, 2016.
18. Breslavsky, I.D.; Amabili, M.; Legrand, M.; Alijani, F. Axisymmetric deformations of circular rings made of linear and Neo-Hookean materials under internal and external pressure: A benchmark for finite-element codes. *Int. J. Nonlinear Mech.* **2016**, *84*, 39–45. [CrossRef]
19. Foss, G.C. Modal damping estimates from static load-deflection curves. *Shock tVib. Dig.* **2006**, *38*, 535–536.
20. Hall, J.F. Problems encountered from the use (or misuse) of Rayleigh damping in nonlinear analysis. *Earthq. Eng. Struct. Dyn.* **2006**, *35*, 525–545. [CrossRef]
21. Alipour, A.; Zareian, F. Study Rayleigh damping in structures: Uncertainties and treatments. In Proceedings of the 14th World Conference on Earthquake Engineering, Beijing, China, 12–17 October 2008; Paper No. 14-0243.
22. Jehel, P.; Léger, P.; Ibrahimbegovic, A. Initial versus tangent stiffness-based Rayleigh damping in inelastic time history seismic analyses. *Earthq. Eng. Struct. Dyn.* **2014**, *43*, 467–484. [CrossRef]

vibration

MDPI

Article

# Modeling Hysteretically Nonlinear Piezoelectric Composite Beams

**Abdulaziz H. Alazemi * and Andrew J. Kurdila**

Department of Mechanical Engineering, Virginia Tech, Blacksburg, VA 24061, USA; kurdila@vt.edu
* Correspondence: alduwaihees@vt.edu

**Abstract:** This paper presents a modeling framework for hysteretically nonlinear piezoelectric composite beams using functional differential equations (FDEs). While linear piezoelectric models are well established, they fail to capture the complex nonlinear behaviors that emerge at higher electric field strengths, particularly history-dependent hysteresis effects. This paper develops a cascade model that integrates a high-dimensional linear piezoelectric composite beam representation with a nonlinear Krasnosel'skii–Pokrovskii (KP) hysteresis operator. The resulting system is formulated using a state-space model where the input voltage undergoes a history-dependent transformation. Through modal expansion and discretization of the Preisach plane, we derive a tractable numerical implementation that preserves essential nonlinear phenomena. Numerical investigations demonstrate how system parameters, including the input voltage amplitude, and hysteresis parameters significantly influence the dynamic response, particularly the shape and amplitude of limit cycles. The results reveal that while the model accurately captures memory-dependent nonlinearities, it depends on numerous real and distributed parameters, highlighting the need for efficient reduced-order modeling approaches. This work provides a foundation for understanding and predicting the complex behavior of piezoelectric systems with hysteresis, with potential applications in vibration control, energy harvesting, and precision actuation.

**Keywords:** piezoelectric composites; hysteresis modeling; functional differential equations; Krasnosel'skii–Pokrovskii operator; nonlinear dynamics; limit cycles; smart materials

## 1. Introduction

Active materials respond to external stimuli by changing their properties or converting energy from one form to another. Among them, piezoelectric materials generate electrical energy under mechanical load and deform under applied electric fields, making them essential in sensing and actuation applications [1]. In contrast, shape-memory alloys (SMAs) respond to thermal changes and are widely used in medical devices and actuators, while magnetostrictive, electrorheological, and magnetorheological materials have diverse industrial applications [1,2].

This work focuses on active structural systems based on piezoelectric materials, which enable bidirectional energy conversion through the direct and converse piezoelectric effects [1–5]. Due to their high sensitivity to mechanical and electrical variations, piezoelectric composites are extensively used in vibration control, sensing, and energy harvesting. They play a central role in transducers, such as force, pressure, and acoustic sensors, and in actuators for microelectromechanical systems (MEMSs) and energy harvesters [1–6].

The theory of linear piezoelectric materials is well established and can be found in standard works [4,5,7]. However, modeling piezoelectric systems becomes significantly more complex under nonlinear effects arising from large electric fields, mechanical strain, or thermal conditions. These nonlinearities often require higher-order terms or nonlinear constitutive laws to accurately describe the system dynamics [3,4,8]. While some models treat the electromechanical coupling as linear, nonlinear modeling often yields more accurate predictions and can exploit beneficial system behaviors.

A broad class of nonlinear models incorporates polynomial terms into the constitutive relationships. For instance, a first-principles model that includes higher-order elastic effects, nonlinear damping, and circuit coupling was developed in [9], where experimental validation confirmed its accuracy for frequency response prediction. Similarly, quadratic nonlinearities were shown in [10] to effectively capture the dynamic behavior of piezoelectric bimorph cantilevers.

Other works explore nonlinear piezoelectricity in the context of smart damping and control. The use of nonlinear models to evaluate the frequency response of partially covered cantilevers connected to SSDI circuits is demonstrated in [11], while the authors of [12] compared two nonlinear piezoelectric models to improve vibration attenuation at low frequencies. In both cases, experimental measurements validate the enhanced performance predicted by nonlinear modeling. In addition, the authors of [13] developed and validated a nonlinear smart beam energy harvester with a shunt controller.

Another important class of nonlinearities in piezoelectric electromechanical systems arises from hysteresis, where the system output depends not only on the current input but also on its past history [14]. Hysteresis is observed in mechanical, electrical, and electromechanical coupling elements and can introduce significant challenges such as reduced control accuracy, increased power consumption, and instability [15]. Consequently, modeling hysteresis is essential for effective control and compensation strategies.

Several mathematical models have been developed to describe hysteresis behavior in electromechanical systems. These include the Preisach model, based on a superposition of elementary hysterons; the Bouc-Wen model, formulated using nonlinear differential equations; the Prandtl–Ishlinskii model, which uses series of play and stop operators; and the Krasnosel'skii–Pokrovskii (KP) model, which defines hysteresis via integral operators with specific kernels [14–16]. The choice among these models typically depends on the trade-off between accuracy and computational complexity.

From a broader mathematical perspective, history-dependent behavior can also be modeled using functional differential equations, differential inclusions, or history-dependent operators [14,16–19]. These frameworks are particularly suited for capturing the long-term memory effects inherent in piezoelectric systems but introduce further complexity in analysis and simulation.

Recent studies have incorporated such hysteresis models into experimental and control frameworks. For instance, the authors of [20] introduced a robust output-feedback controller that compensates for actuator hysteresis, while the authors of [21] proposes a modified Prandtl–Ishlinskii model integrated with sliding mode control to enhance the tracking performance of piezoelectric actuators.

Foundational texts on piezoelectricity and smart material systems provide the theoretical framework for modeling electromechanical structures, covering constitutive theory, parameter estimation, Galerkin approximation, and finite dimensional control synthesis [1,3–5,8]. In the context of energy harvesting, Erturk and Inman [6] presented analytical models of cantilever-based piezoelectric systems, validating them under various dynamic loading conditions.

Although the principles of linear piezoelectricity have long been used to derive models for active composites [1,5–7], it is well known that nonlinear effects, particularly under high electric fields, can significantly influence system behavior [6,17,22]. These nonlinearities, which are often poorly understood, complicate the modeling process. Two primary classes of nonlinear modeling approaches have emerged: those addressing polynomial-type nonlinearities and those focused on hysteretic, history-dependent behavior. This paper is concerned with the latter.

Numerous studies have explored the complex hysteretic behavior inherent in piezoelectric actuators, emphasizing the diversity of modeling strategies and the challenges posed by rate dependence, asymmetric loop formation, and multi-scale memory effects. Sabarianand et al. [15] provided a comprehensive review of classical and contemporary hysteresis models, including the Preisach, Bouc–Wen, Prandtl–Ishlinskii and Krasnosel'skii–Pokrovskii operators, highlighting their respective mathematical formulations, implementation trade-offs, and applicability to various actuation scenarios. Their survey underscores the need for tailored models that balance accuracy, generalizability, and computational efficiency, depending on the target application. More recently, Dai et al. [23] surveyed nonlinear modeling efforts specifically for piezoelectric ceramic actuators, emphasizing the limitations of traditional models in capturing dynamic and coupled electromechanical behaviors. They outline modern approaches incorporating neural networks, fractional calculus, and hybrid physics-based and data-driven techniques to better reflect real-world complexities such as temperature sensitivity and loading history. Together, these reviews reflect the growing understanding that hysteresis in piezoelectric systems is not only nonlinear and memory-dependent but also deeply context-specific, necessitating advances in modeling frameworks that can be generalized yet tractable for control and analysis.

Recent studies continue to advance the modeling and control of hysteretically nonlinear behavior in piezoelectric actuators, particularly for precision applications. Xu et al. [24] proposed a fractional-order model that captures both hysteresis and creep effects using non-integer derivatives, offering improved accuracy and reduced parameter complexity compared with classical models like the Bouc–Wen or Preisach models. Their coupled model, validated experimentally, shows significant improvements in tracking performance under sinusoidal and triangular excitations. Similarly, Li et al. [25] developed a hybrid hysteresis model by combining the Preisach–Ishlinskii formulation with a GA-optimized BP neural network. This approach, paired with a fuzzy PID controller, effectively compensates for nonlinearity while enhancing stability and precision across a wide range of frequencies. Baziyad et al. [26] addressed rate-dependent hysteresis by modifying the Preisach model through least squares support vector machines and particle swarm optimization, integrated within a two-degree-of-freedom $H_\infty$ robust control framework. Their experiments confirm superior tracking accuracy and robustness under dynamic loading. Although these works emphasize low-dimensional control performance, they rely on lumped-parameter or black box representations, which limits their ability to capture the distributed-parameter dynamics of real composite structures. This highlights the need for high-fidelity, spatially distributed models that integrate hysteresis effects directly into the governing equations of the structural system, a gap that the present study aims to address.

The primary objective of this study is to develop a modeling framework that captures the hysteretically nonlinear behavior of piezoelectric composite beams using a discretized Preisach-type operator integrated into distributed-parameter dynamics. While classical hysteresis models (e.g., Preisach, Bouc–Wen, and Prandtl–Ishlinskii) are widely used, they are rarely embedded directly into high-dimensional structural models, especially for composite configurations. Moreover, the influence of hysteresis parameters, such as the loop smoothness control (scaling factor) and weighting parameter, on system dynamics and limit cycle behavior remains poorly understood. Existing approaches often trade accuracy for simplicity or become computationally impractical. This work addresses these gaps by offering a tractable yet detailed framework suitable for analysis and future reduction or data-driven extensions.

The novelty of this work lies in three key aspects. First, we present a modeling framework that integrates rate-independent hysteresis, described by a discretized Preisach-type operator, directly into the dynamics of a distributed-parameter piezoelectric composite beam. This coupling results in a high-dimensional system that captures both the structural response and memory-dependent nonlinear effects. Second, we perform a systematic numerical study to investigate how variations in the hysteresis parameters, such as the loop smoothness control (scaling factor) and weighting parameter (kernel weight) of hysteresis and switching thresholds, influence the steady-state behavior of the system, including the formation and shape of limit cycles. Finally, the resulting model takes the form of a nonlinear functional differential equation (FDE). This formulation provides a general framework for modeling hysteretically nonlinear piezoelectric systems and highlights the computational challenges that motivate future reduced-order or data-driven modeling approaches. The mathematical formulation and coupling methodology are described in detail in Section 2, where the hysteresis operator is introduced into the governing PDE through the external forcing term, resulting in a nonlinear functional differential equation.

## 2. The Nonlinear System

In this section we discuss the structure of the nonlinear system studied in this paper, which has the overall form shown in Figure 1. It combines a high-dimensional linear model of a piezoelectric composite and a hysteretically nonlinear model.

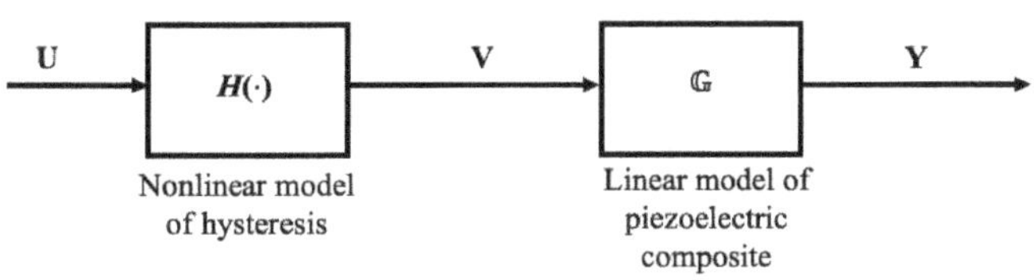


**Figure 1.** The nonlinear system that is a cascade of a nonlinear hysteresis model and a linear piezoelectric model.

The cascade model is $\mathbb{G} \circ H$, where $\mathbb{G}$ is a model of a linearly piezoelectric composite acting on the hysteresis operator $H$. In Section 2.1.3, we explicitly derive the state-space representation for the linear model $\mathbb{G}$:

$$\begin{aligned} \dot{\mathbf{X}}(t) &= \mathbb{A}\mathbf{X}(t) + \mathbb{B}\mathbf{V}(t) \\ \mathbf{Y}(t) &= \mathbb{C}\mathbf{X}(t), \end{aligned} \tag{1}$$

so that

$$\mathbb{G}(s) = \mathbb{C}(s\mathbb{I} - \mathbb{A})^{-1}\mathbb{B}. \tag{2}$$

Here, the state vector $X(t) \in \mathbb{R}^{N\times 1}$, the vector of measurements $Y(t) \in \mathbb{R}^{L\times 1}$, the system matrix $\mathbb{A} \in \mathbb{R}^{N\times N}$, the control influence matrix $\mathbb{B} \in \mathbb{R}^{N\times 1}$, the observation matrix $\mathbb{C} \in \mathbb{R}^{L\times N}$, and $N$ is the number of states.

We finally combine the state space and the hysteresis model to obtain the governing equations:

$$\begin{aligned} \dot{\mathbf{X}}(t) &= \mathbb{A}\mathbf{X}(t) + \mathbb{B}\mathbf{H}(\mathbf{U})(t) \\ \mathbf{Y}(t) &= \mathbb{C}\mathbf{X}(t), \end{aligned} \tag{3}$$

where the state $\mathbf{X}(t) \in \mathbb{R}^{N\times 1}$, the input voltage $\mathbf{U}(t) \in \mathbb{R}$, the measured outputs $\mathbf{Y}(t) \in \mathbb{R}^{L\times 1}$, the state matrix $\mathbb{A} \in \mathbb{R}^{N\times N}$, the control influence operator $\mathbb{B} \in \mathbb{R}^{N\times 1}$, and the observation matrix $\mathbb{C} \in \mathbb{R}^{L\times N}$. The history dependence is modeled in terms of the hysteresis operator $\mathbf{H}$, which maps the past history of the input $\{\mathbf{U}(\tau)\}_{\tau\in[0,t]}$ to its output $(\mathbf{HU})(t) \in \mathbb{R}$. This means that the operator representing the hysteresis is a mapping $\mathbf{H} : C([0,t),\mathbb{R}) \mapsto \mathbb{R}$, where $C([0,t),\mathbb{R})$ is the set of real-valued continuous functions on $[0,t)$. As depicted in Figure 1, this model can be understood as a linear or nonlinear cascade of models. Because of the type of hysteresis operator employed in this paper, these equations are examples of nonlinear FDEs and not ordinary differential equations (ODEs). The hysteresis operator H is described in Section 2.2.

### 2.1. Linear Piezoelectric System

In this paper, a composite cantilever beam, shown in Figure 2, is subjected to an input voltage V. An active composite beam having this structure is commonly known also as a bimorph. The bimorph cantilever beam in Figure 2 has a length $L$, width $b$, and thickness $h$. It consists of a central linearly elastic layer with a thickness $h_c$ and two piezoelectric layers with a thickness $h_p$ on either side. The structure is cantilevered at one end and free at the other, and it is subject to a voltage $2V(t)$ across the piezoelectric layers. The electric potential at the top surface of the piezoelectric layer is $\Phi^+$, while the potential for the bottom surface is $\Phi^-$.

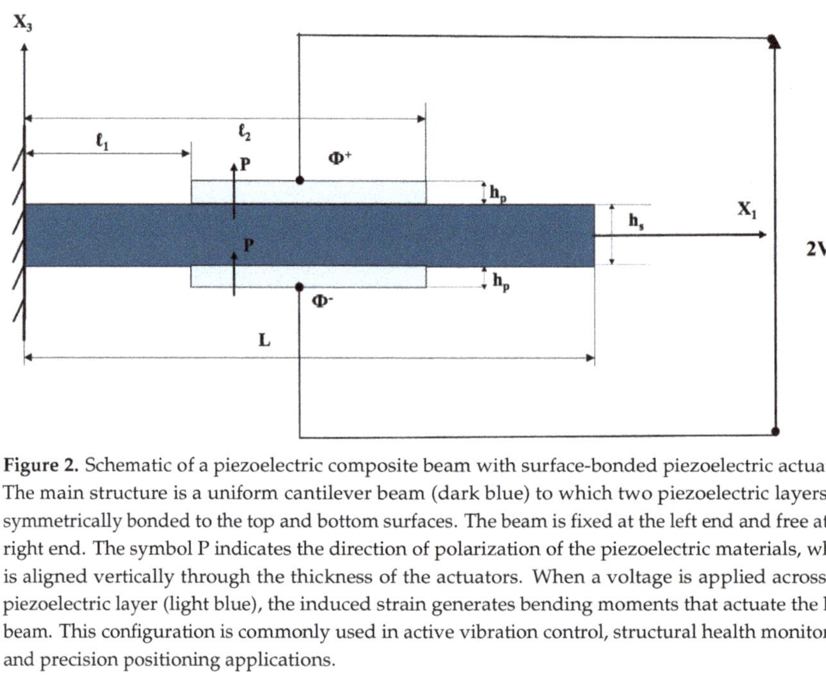


**Figure 2.** Schematic of a piezoelectric composite beam with surface-bonded piezoelectric actuators. The main structure is a uniform cantilever beam (dark blue) to which two piezoelectric layers are symmetrically bonded to the top and bottom surfaces. The beam is fixed at the left end and free at the right end. The symbol P indicates the direction of polarization of the piezoelectric materials, which is aligned vertically through the thickness of the actuators. When a voltage is applied across the piezoelectric layer (light blue), the induced strain generates bending moments that actuate the host beam. This configuration is commonly used in active vibration control, structural health monitoring, and precision positioning applications.

#### 2.1.1. Background Theory for Linear Piezoelectricity

In this section, we review a beam model for an active piezoelectric composite. We can write the approximate form of the constitutive law for the piezoelectric as follows:

$$\begin{pmatrix} T_{11} \\ D_3 \end{pmatrix} = \begin{pmatrix} C_{11}^E & -e_{31} \\ e_{31} & \varepsilon_{33}^S \end{pmatrix} \begin{pmatrix} S_{11} \\ E_3 \end{pmatrix}, \tag{4}$$

where $T_{11}$ is the bending stress in the $x_1$ direction, $D_3$ is the electric displacement in the $x_3$ direction, $E_3$ is the electric field in the $x_3$ direction, $S_{11}$ is the bending strain in the $x_1$

direction, $C_{11}^E$ is the piezoelectric stiffness, $e_{31}$ is the piezoelectric coupling constant, and $\varepsilon_{33}^S$ is the electric permittivity. We approximate the electric field to be

$$E_3(x_1, x_3) = \begin{cases} -\frac{V(t)}{h_p} & \text{if } (x_1, x_3) \in [\ell_1, \ell_2] \times [h_s, h_s + h_p] \\ \frac{V(t)}{h_p} & \text{if } (x_1, x_3) \in [\ell_1, \ell_2] \times [-h_s - h_p, -h_s] \end{cases}, \tag{5}$$

Hamilton's principle for an electromechanical system is used to derive the governing equation of the piezoelectric composite. The governing equations satisfy the variational condition

$$\delta \int_{t_0}^{t_1} (T - V_H) dt + \int_{t_0}^{t_1} \delta W dt = 0, \tag{6}$$

where $T$ represents the kinetic energy, $V_H$ denotes the electrical enthalpy, and $\delta W$ is the virtual work due to non-conservative electromechanical loads. The kinetic energy is expressed as follows:

$$T = \frac{1}{2}\rho A \int_0^L \left(\frac{\partial w(x,t)}{\partial t}\right)^2 dx, \tag{7}$$

where $\rho$ is the mass density of the beam, $A$ represents the cross-sectional area of the beam, and $w(t, x)$ denotes the deflection of the beam as a function of time $t$ and position $x$. The electric enthalpy density is approximated as follows:

$$H \approx \frac{1}{2} C_{11}^E S_{11}^2 - e_{31} S_{11} E_3 - \frac{1}{2} \varepsilon_{33}^S E_3^2, \tag{8}$$

and for a Bernoulli–Euler beam, we approximate the bending stress as follows:

$$S_{11} = -x_3 \frac{\partial^2 w}{\partial x_1^2}. \tag{9}$$

We integrate the electric enthalpy over the volume

$$V_H = \int\int\int_V \left(\frac{1}{2} C_{11}^E S_{11}^2 - e_{31} S_{11} E_3 - \frac{1}{2} \varepsilon_{33}^S E_3^2\right) dx_1\, dx_2\, dx_3, \tag{10}$$

and by further simplifying Equation (9), we get

$$V_H = \frac{1}{2} \int_0^L \left\{ C_{11}^E I \left(\frac{\partial^2 w}{\partial x_1^2}\right)^2 - (h_s + h_p) b \frac{\partial^2 w}{\partial x_1^2} \chi_{[\ell_1, \ell_2]} V - \frac{\varepsilon_{33}^S b \chi_{[\ell_1,\ell_2]}^2 V^2}{2h_p} \right\} dx. \tag{11}$$

In this expression, $I$ is the second area moment of inertia, and $\chi_{[\ell_1,\ell_2]}$ is the characteristic function of the domain. It equals one when $\chi$ is within the interval $[\ell_1, \ell_2]$ and is zero otherwise. By applying Hamilton's principle to the kinetic energy, we have

$$\int_{t_0}^{t_1} \delta T = \int_{t_0}^{t_1} \int_0^L -\rho A \ddot{w} \delta w \, dx\, dt. \tag{12}$$

For the electrical enthalpy, we consider $V$ to be prescribed, and therefore $\delta V = 0$. After setting the parameters $\ell_1 = 0$ and $\ell_2 = L$, we align our analysis with the methodology and results presented in [6]. We consequently get

$$\begin{aligned}
\delta \int_{t_0}^{t_1} V_H =& C_{11}^E I \int_{t_0}^{t_1} (w''\delta w')\Big|_0^L - C_{11}^E I \int_{t_0}^{t_1} (w'''\delta w)\Big|_0^L \\
&+ C_{11}^E I \int_{t_0}^{t_1} \int_0^L w''''\delta w \, dx \, dt \\
&- \frac{1}{2}(h_s + h_p) b e_{31} V \int_{t_0}^{t_1} \delta w'\Big|_0^L dt \\
&+ \frac{1}{2}(h_s + h_p) b e_{31} V \int_{t_0}^{t_1} \frac{\partial}{\partial x}\chi_{[0,L]}\delta w\Big|_0^L dt \\
&- \frac{1}{2}(h_s + h_p) b e_{31} V \int_{t_0}^{t_1} \int_0^L \frac{\partial^2}{\partial x^2}\chi_{[0,L]}\delta w \, dx \, dt.
\end{aligned} \tag{13}$$

Considering the damping coefficient C which captures dissipative forces proportional to the velocity, the Hamiltonian becomes

$$\begin{aligned}
\delta \int_{t_0}^{t_1} (T - V_H)\, dt =& \int_{t_0}^{t_1} \int_0^L \left\{ -\rho A \ddot{w} - c\dot{w} - C_{11}^E I w'''' \right\} \delta w \, dx \, dt \\
&+ \frac{1}{2} \int_{t_0}^{t_1} \int_0^L (h_s + h_p) b e_{31} V \frac{\partial^2}{\partial x^2}\chi_{[0,L]}\delta w \, dx \, dt \\
&- C_{11}^E I \left( w''\delta w'\big|_0^L - w'''\delta w\big|_0^L \right) \\
&+ \frac{1}{2}(h_s + h_p) b e_{31} V \chi_{[0,L]}\, \delta w'\big|_0^L \\
&- \frac{1}{2}(h_s + h_p) b e_{31} V \frac{\partial}{\partial x}\chi_{[0,L]}\delta w\Big|_0^L = 0.
\end{aligned} \tag{14}$$

Therefore, we have the equation of motion

$$\rho A \ddot{w} + c\dot{w} + C_{11}^E I w'''' - \frac{1}{2}(h_s + h_\rho) b e_{31} V \frac{\partial^2}{\partial x^2}\chi_{[0,L]} = 0, \tag{15}$$

subject to the boundary conditions

$$C_{11}^E I\, w''\delta w'\big|_0^L + \frac{1}{2}(h_s + h_p) b e_{31} V \chi_{[0,L]}\, \delta w'\big|_0^L = 0, \tag{16}$$

and

$$C_{11}^E I\, w'''\delta w\big|_0^L - \frac{1}{2}(h_s + h_p) b e_{31} V \frac{\partial}{\partial x}\chi_{[0,L]}\delta w\Big|_0^L = 0, \tag{17}$$

and the initial conditions $w(0,x) = w_0(x)$ and $\dot{w}(0,x) = v_0(x)$ for all $x \in \Omega$. This system has been studied in many works [5,6,9].

### 2.1.2. Approximation of the PDE

The partial differential equation in Equation (15) must be approximated in practice. The assumed mode methods model the transverse deflection function $w(x,t)$ in terms of a summation of $n$ generalized coordinate displacement functions $q_i(t)$ multiplied by the basis function $P_i(x)$. Common choices include finite element shape functions or eigenfunctions [5]. We then have

$$w(x,t) = \sum_{i=1}^{n} q_i(t) P_i(x), \tag{18}$$

where $n$ is the number of modes and the length of $q(t)$, where $q(t) \in \mathbb{R}^{n\times 1}$.

In this paper, we choose for the basis functions to be eigenfunctions $P_i(x) \triangleq \psi_i(x)$ ([27] Table 6.6, p. 539) which satisfy the orthogonality conditions

$$\int_0^L \psi_i\psi_j dx = \begin{cases} 0 & \text{for } i \neq j, \\ 1 & \text{for } i = j, \end{cases} \tag{19}$$

Using the eigenfunctions, the approximation of the governing PDEs generates the set of decoupled ODEs

$$M\ddot{q}(t) + C\dot{q}(t) + Kq(t) = BV(t). \tag{20}$$

where the mass matrix $M \in \mathbb{R}^{n\times n}$, the stiffness matrix $K \in \mathbb{R}^{n\times n}$, and the control influence vector $B \in \mathbb{R}^{n\times 1}$ are, respectively, as follows:

$$M_{ij} = \int_0^L \rho_e A_t \psi_i \psi_j \, dx = \rho_e A_t diag(m_i), \tag{21}$$

$$K_{ij} = \int_0^L C_e I_t \frac{d^2\psi_i}{dx^2}\frac{d^2\psi_j}{dx^2}\, dx = C_e I_t diag(k_i), \tag{22}$$

$$B_i = \int_0^L \frac{Ke_{31}\chi_{[0,L]}}{h_p}\frac{d^2\psi_i}{dx^2}\, dx. \tag{23}$$

In these equations, $\rho_e$ is the effective density, $A_t$ is the total area, $C_e$ is the effective modulus of elasticity, $I_t$ is the total inertia, $m_i$ is the $i^{\text{th}}$ model mass, and $k_i$ is the $i^{\text{th}}$ model stiffness. We assume that $C_{ij} = \text{diag}(c_i)$, where $c_i$ is the $i^{\text{th}}$ model damping.

### 2.1.3. State Space Representation

When considering a mechanical system bounded by piezoelectric material and influenced by a control voltage that exhibits hysteresis, we propose the following state-space representation. The approximate dynamic behavior of the system by the system of second-order ODEs is

$$\mathbf{M}\ddot{\mathbf{q}}(t) + \mathbf{C}\dot{\mathbf{q}}(t) + \mathbf{K}\mathbf{q}(t) = \mathbf{B}V(t)$$

where the degrees of freedom $q(t) \in \mathbb{R}^{n\times 1}$, $\mathbf{M} \in \mathbb{R}^{n\times n}$ denotes the mass matrix, $\mathbf{C} \in \mathbb{R}^{n\times n}$ is the damping matrix, $\mathbf{K} \in \mathbb{R}^{n\times n}$ is the stiffness matrix, and $B \in \mathbb{R}^{n\times 1}$ is the control influence vector. $V$ is the control voltage. The state vector $\mathbf{X}$ is defined in the usual way:

$$\mathbf{X}(t) = \begin{bmatrix} q(t) \\ \dot{q}(t) \end{bmatrix},$$

We set N = 2n, and we define the matrices

$$\mathbb{A} = \begin{bmatrix} 0 & \mathbb{I} \\ -\mathbf{M}^{-1}\mathbf{K} & -\mathbf{M}^{-1}\mathbf{C} \end{bmatrix} \in \mathbb{R}^{n\times n}, \quad \mathbb{B} = \begin{bmatrix} 0 \\ \mathbf{M}^{-1}\mathbf{B} \end{bmatrix} \in \mathbb{R}^{n\times 1},$$

These choices yield the state-space equations

$$\dot{\mathbf{X}}(t) = \mathbb{A}\mathbf{X}(\mathbf{t}) + \mathbb{B}V(t).$$

Since $n$ is the number of modes $\psi_i$, therefore, the number of states $N = 2n$ may be large. On the other hand, the number of measurements $L$ is ordinarily small. We define the $L$ outputs in $\mathbf{Y}(t)$ as follows:

$$\mathbf{Y}(t) = \begin{bmatrix} y_1(t) \\ y_2(t) \\ \vdots \\ y_L(t) \end{bmatrix} = \mathbb{C}\mathbf{X}(t), \quad \text{where } \mathbb{C} \in \mathbb{R}^{L \times N}.$$

### 2.2. Hysteretically Nonlinear Model

Hysteresis is a well-documented nonlinear phenomenon [14,16]. It presents intricacies in its modeling due to complex relationship between the input and output. Hysteresis is a rate-independent, history-dependent nonlinearity in the system, and there are many approaches to modeling it. Modeling can be based on analytical methods such as ordinary or partial differential equations [14,16].

In this paper, we define a specific history-dependent operator $H$. Hysteresis is defined in terms of integral operators constructed from certain "building block" history-dependent kernels acting on the input as described in [19,28].

In mathematical terms, the input–output relationship can be described by the operator equation

$$V(t) = (HU)(t),$$

where $U(t)$ is the input at time $t$, $V(t)$ is the output at time $t$, and the operator $H$ is a mapping $H : C([0,t], \mathbb{R}) \to \mathbb{R}$. In this study, hysteresis is modeled as a nonlinear transformation of the external input history $\{U(\tau) \mid \tau \in [0,t]\}$ into the output voltage $V(t) \in \mathbb{R}$, which acts as a forcing term in a linear system. This approach separates hysteretic effects from the state dynamics of the piezoelectric material, focusing on systems where hysteresis primarily affects the actuation signal. By treating $V(t)$ as an effective representation of the input after being shaped by the hysteresis operator H, we simplify the modeling process while retaining essential nonlinear characteristics. This cascade formulation assumes that hysteresis is dominated by the input $U(t)$, with minimal feedback from the system state $\mathbf{X}(\mathbf{t})$ to the operator. While this limits the ability to capture state-dependent memory effects in autonomous systems, it provides a practical framework for studying input–output interactions and facilitates reduced-order modeling. The novelty lies in the integration of hysteresis as a history-dependent transformation of the forcing term, enabling efficient state-space representation and the application of linear control techniques. In other words, the operator $H$ maps the history of the input $\{U(\tau) \mid \tau \in [0,t]\}$ up to time $t$ to generate the output $V(t) \in \mathbb{R}$.

The hysteresis operator $H$ is constructed from the superposition of hysteresis kernels $k(s_1, s_2, t, U)$, an example of which is depicted in Figure 3a. The function $k$ denotes the kernel and is a function of four variables: $t$, $s_1$,$s_2$, and $U$. The term $U$ is shorthand for the input history $\{U(\tau) \mid \tau \in [0,t]\}$, and the terms $s_1$ and $s_2$ are a pair of switching values in the Preisach plane, which is shown Figure 3b. We think of $s_1$ as describing when the hystereic output "jumps down" and $s_2$ as describing when the output "jumps up". Each kernel depends on two distinct switches $s_1$ and $s_2$ and the associated bounding functions $\gamma_{s_1}$ and $\gamma_{s_2}$. The bounding functions describe how smoothly the output jumps up or down at the switching values, as shown in Figure 3a. The value of $k$ is given by

$$k(s_1, s_2, t_n, U) = \begin{cases} \max\{k^{(n-1)}, \gamma_{s_1}(U(t))\} & \text{if } U \text{ is increasing over } [t_{n-1}, t_n], \\ \min\{k^{(n-1)}, \gamma_{s_2}(U(t))\} & \text{if } U \text{ is decreasing over } [t_{n-1}, t_n]. \end{cases}$$

The bounding functions in this paper, $\gamma_{s_1}$ and $\gamma_{s_2}$, are derived from a single envelope function $\gamma$ and are defined, respectively, as follows:

$$\gamma_{s_1}(\cdot) = \gamma(\cdot - s_1),$$
$$\gamma_{s_2}(\cdot) = \gamma(\cdot - s_2).$$

For the Krasnoselskii–Pokrovskii model [19], the envelope function $\gamma(u)$ is represented by

$$\gamma(u) = tanh(U/a).$$

Here, the parameter $a > 0$ defines the scaling factor applied to the envelope function.

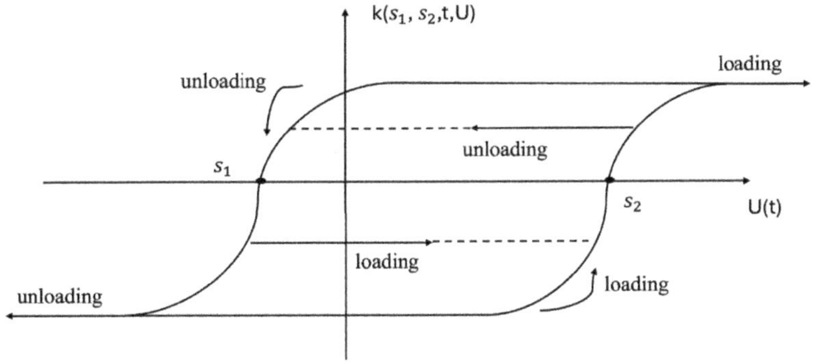


(**a**) A typical hysteresis kernel $k(s_1, s_2, t, U)$ for the part of switches $(s_1, s_2) \in S_\Omega$, where $S_\Omega$ is the Preisach plane.

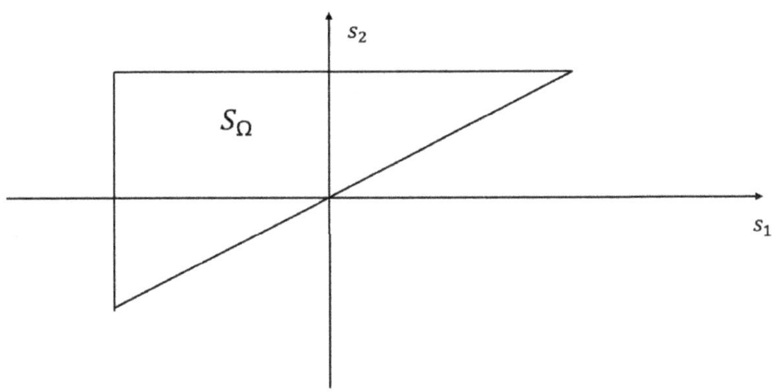


(**b**) The Preisach plane $S_\Omega$. Any point $(s_1, s_2) \in S_\Omega$ defines a hysteresis kernel $k(s_1, s_2, t, U)$.

**Figure 3.** Illustrations of the Preisach hysteresis modeling framework. (**a**) A typical elementary hysteresis kernel $k(s_1, s_2, t, U)$ associated with a single pair of switching thresholds $(s_1, s_2)$ on the Preisach plane. The kernel switches between two states (loading and unloading) based on the current value of the input $U(t)$ relative to the thresholds. The dashed arrows indicate memory-dependent switching behavior and the formation of hysteresis loops. (**b**) The Preisach plane $\Omega$, which is the set of all valid switching threshold pairs satisfying $s_1 \leq s_2$. Each point $(s_1, s_2) \in \Omega$ defines a hysteresis kernel. The full hysteresis response is obtained by integrating the weighted contributions of these kernels across the Preisach plane.

The operator $H$, when acting on the input $U$, is mathematically represented by

$$V(t) = (HU)(t) = \iint_{S_\Omega} k(s_1, s_2, t, U)\mu(\mathrm{d}s_1\,\mathrm{d}s_2). \tag{24}$$

where $k(s_1, s_2, t, U)$ is the fundamental hysteresis kernel introduced in [19,28] and $\mu$ is some given measure over the subset $S_\Omega$ of the Preisach plane. The subset $S_\Omega$ of the Preisach plane is depicted in Figure 3a. This model is sometimes referred to as the generalized play operator or the Krasnoselskii–Porovskii (KP) operator [16].

It should be noted that this hysteresis operator is quite general and depends on a number of real parameters as well as distributed (or functional) parameters. Distributed parameters are not real numbers; rather, they are functions. Distributed parameters reside in a generally infinite dimensional space of functions. The real parameters include the scaling factor $a$ in the definition above of the envelope function $\gamma(u)$, as well as the thresholds $\alpha, \beta$ that define the subset $S_\Omega$. The distributed parameters include the choice of the envelope function $\gamma(u)$ and the weight $\mu_{jk}(ds_1, ds_2) \triangleq \mu(s_1, s_2)ds_1 ds_2$.

Given the inherent complexity of the KP model, as outlined in Equation (24), approximations must be used for practical algorithms. This is accomplished by discretizing the Preisach plane as depicted in Figure 4 [19].

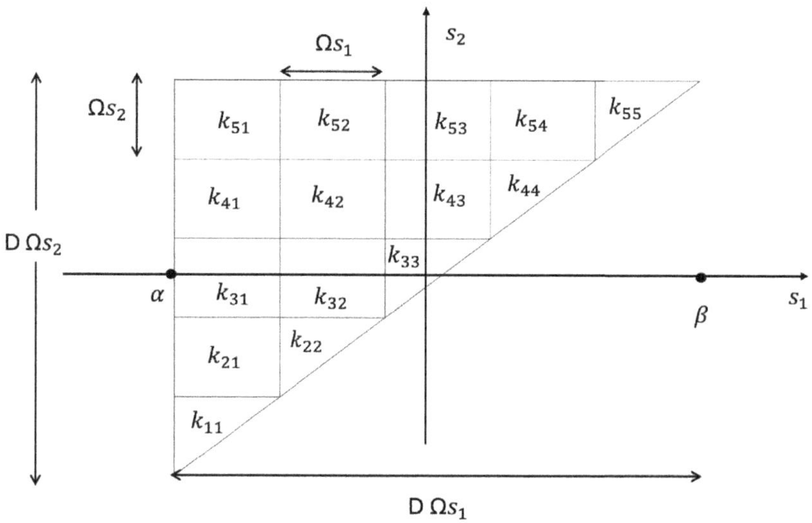


**Figure 4.** Discretization of the Preisach plane $\Omega$ illustrating the distribution of elementary hysteresis switches for a resolution parameter $D = 5$. The plane $\Omega \subset \mathbb{R}^2$ is defined by all pairs $(s_1, s_2)$ satisfying $s_1 \leq s_2$. It is partitioned into triangular subregions, each associated with a hysteresis kernel $k_{ij}$. The horizontal and vertical axes correspond to the switching thresholds $s_1$ and $s_2$, respectively, and the total number of kernel elements is determined by the discretization level $D$. The thresholds $\alpha$ and $\beta$ denote the lower and upper bounds of the input domain. The diagonal boundary $s_1 = s_2$ separates admissible kernel regions from non-physical configurations. This discretized structure enables efficient numerical implementation of the Preisach-type hysteresis operator.

We correspondingly introduce the discrete approximation

$$V_D(t) = \sum_{k=1}^{D} \sum_{j=k}^{D} k_{jk}(t)\mu_{jk}\Omega s_1 \Omega s_2. \tag{25}$$

The voltage $V_D$ is an approximation of the output voltage $V$ of the exact hysteresis operator above. The term $k_{jk}(t) := k((s_j, s_k), t, U)$ denotes the evaluation of the KP kernel at the pair $s = (s_j, s_k)$, while $\mu_{jk}\ \Omega s_1 \Omega s_2$ is a discrete weight used to approximate the integral over the subregion $\Omega_{jk}$ shown in Figure 4. The discrete parameters $\Omega s_1$ and $\Omega s_2$ are defined as follows:

$$\Omega s_1 = \Omega s_2 = \frac{\alpha - \beta}{D},$$

where $\alpha$ and $\beta$ are the upper and lower thresholds of the Preisach plane, respectively. This method of discretization is depicted in the Preisach plane in Figure 3b. This plane is divided into $\frac{D(D+1)}{2}$ distinct subsets $\Omega_{jk}$.

This means that the approximation of the hysteresis operator that generates $V_D$ depends on $D(D+1)/2 + 4$ real parameters $\mu_{jk}, \alpha, \beta, a$.

## 3. Numerical Examples

To illustrate the dynamic behavior of the piezoelectric system under various operating conditions, we present a series of numerical examples. These examples are simulated using the state-space representation and hysteresis model outlined in the previous sections. The numerical simulations were performed using Simulink, which was implemented with the fixed time-stepping algorithm to facilitate modeling nonlinear hysteresis effects; hysteresis modeling using adaptive time-step methods is considerably more complicated to implement. The fixed step size was chosen to be 0.001, and all simulations were carried out using the Matlab R2024b solver `ode14x`. The goal of these simulations was to understand how different input voltages and system parameters affected the tip displacement and velocity of the piezoelectric system and the effect on the hysteresis loop.

The fundamental parameters used for the simulations, including the dimensions and material properties of the substrate and piezoelectric components, are detailed in Table 1. The parameters defining the hysteresis operator $H$ are summarized in Table 2.

**Table 1.** Basic parameters for substrate and piezoelectric material. The parameters $I = bh^3/12$, $A = bh$.

| Parameter | Description | Substrate | Piezoelectric |
|---|---|---|---|
| $L_b$ | Length (mm) | 24.06 | 24.06 |
| b | Width (mm) | 6.4 | 6.4 |
| h | Thickness (mm) | 0.14 | 0.265 |
| $\rho$ | Density (kg/m$^3$) | 9000 | 7500 |
| $C^E$ | Modulus of Elasticity (GPa) | 105 | 60.6 |
| e | Piezoelectric Constant (C/m$^2$) | - | −16.6 |
| $\varepsilon$ | Permittivity Constant (nF/m) | - | 25.55 |

**Table 2.** Parameters that define $H$.

| Parameter | Description | Parameter Type | Value |
|---|---|---|---|
| $a$ | Scaling factor | Real | 0.4–20 |
| $\alpha$ | Upper bound of Preisach plane | Real | 50 |
| $\beta$ | Lower bound of Preisach plane | Real | −50 |
| $\gamma$ | Envelope function | Distributed | $\tanh(\cdot)$ |
| $\mu$ | Measure that defines hysteresis operator | Distributed | $\mu_{jk} * \Omega_{s_1} * \Omega_{s_2}$ |
| D | Dimension of discretization | Integer | 25 |
| $\Omega_s$ | Mesh width on Preisach plane | Real | $\frac{\alpha-\beta}{D}$ |
| $\mu_{jk}$ | $jk^{th}$ discrete measure over Preisach plane | Real | 0.001–0.005 |

In reviewing these tables, it is important to note that the FDE in its original form does indeed, as claimed in the introduction, depend on a large number of material, structural, and geometric real parameters. Even if the geometry and material parameters are well known or can be determined with relatively high accuracy, the measure $\mu$ and envelope function $\gamma$ are distributed parameters that are phenomenological and hence never known a priori. If we employ online identification or control methods, such as in [17,28], the governing evolution equations can be to the order of

$$O(N + 11 + D^2)$$

time-varying estimates. In the case study, the typical values of $N$ range between 10 and 100, while $D = 25$. We should also point out that it is usually problematic to determine the damping matrix $C$ by upscaling the (micro)mechanical material properties, and it is often considered an important source of uncertainty in the original equations.

To calculate the damping coefficient $C$, in this paper, we employed the simple proportional damping approach of Rayleigh damping, which is defined as follows:

$$C_{ij} = C_m \mathbb{M}_{ij} + C_k \mathbb{K}_{ij}.$$

In this equation, $C_m$ is a constant number that multiplies the entries of the mass matrix $\mathbb{M}$, and $C_k$ is a constant that multiplies the entries of the stiffness matrix $\mathbb{K}$. This method allows for a combination of mass-proportional and stiffness-proportional damping.

In our assessment via this case study, the system is driven by a range of input voltages to demonstrate the effects on the tip displacement and velocity. The input voltage to the system was chosen to be

$$U = V_{\text{amp}} \sin(\omega t), \tag{26}$$

where $V_{amp}$ is the input voltage and $\omega$ is the angular frequency. The limit cycles observed in the tip displacement versus tip velocity plots provide insights into the energy transfer within the system as well as the inherent damping due to the piezoelectric material and the structural dynamics. Table 3 provides an overview of the system parameters used in the simulation.

In this study, we chose the output variables of the dynamic system to represent the tip behavior of the piezoelectric composite beam. Specifically, the first output is defined as the tip displacement, and the second output is the corresponding tip velocity:

$$y_1(t) := \text{tip displacement}, \tag{27}$$

$$y_2(t) := \text{tip velocity}. \tag{28}$$

These outputs are used throughout the numerical examples to characterize the dynamic response of the system. In particular, the limit cycle plots are generated by projecting the system trajectory onto the $(y_1, y_2)$ phase plane, providing insight into the energy dissipation and nonlinear memory effects induced by the hysteresis operator.

**Table 3.** System parameters used in simulation.

| Symbol | Parameter | Value |
|---|---|---|
| $V_{\text{amp}}$ | Input voltage amplitude (volts) | 90–230 |
| $\omega$ | Angular velocity (rad/s) | 1 |
| $C_m$ | Constant number that multiplies the entries of the mass matrix (1/s) | 10 |
| $C_k$ | Constant number that multiplies the entries of the stiffness matrix (s) | $10^{-7}$ |

### 3.1. Effect of Maximum Input Voltage

To understand the effect of the input voltage, we changed the voltage amplitude in Table 3 from 90 volts to 230 volts in increments of 20 volts. As the voltage increased, the system exhibited a higher tip velocity, indicating a larger energy input into the system. This is evident from the expansion of the limit cycles in the displacement-velocity phase plane, depicted in Figure 5. The limit cycles became progressively larger with an increased input voltage, suggesting that the system moves further from its equilibrium position before the restoring forces bring it back, consistent with higher energy states. The hysteresis effect was more apparent at higher voltages because the input voltage $U(t)$ was more significantly shaped by the hysteresis operator $H(U)$, which led to a larger history-dependent response. This caused the limit cycles to become larger and more pronounced as the voltage increased, highlighting the stronger shaping effect of the hysteresis. Figure 5 illustrates the limit cycles for four selected input voltages, capturing the system's transition through different dynamic regimes.

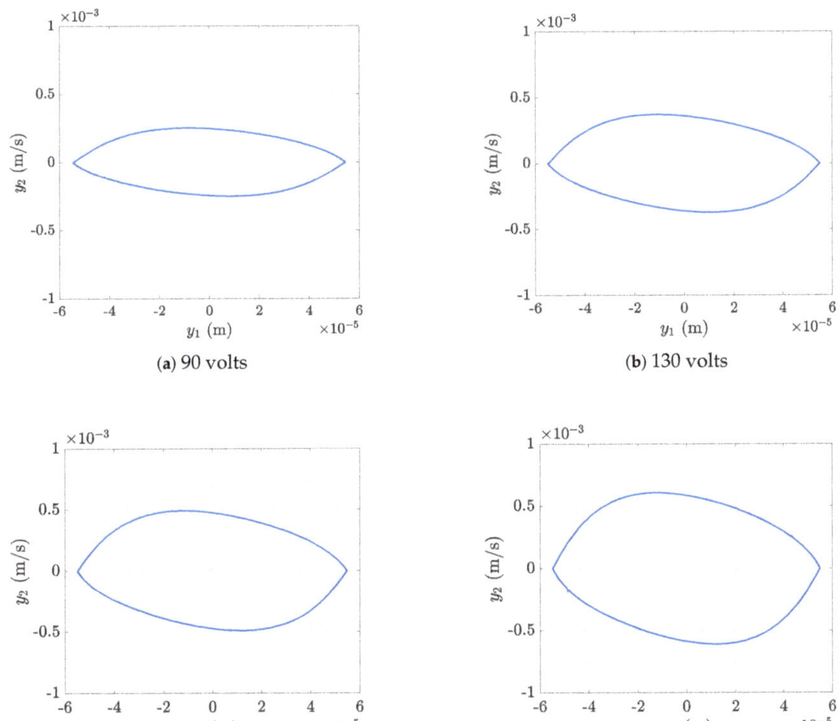


(a) 90 volts (b) 130 volts (c) 170 volts (d) 210 volts

**Figure 5.** Phase-plane representation of limit cycle behavior in the output response of the piezoelectric beam system under sinusoidal input. Each subfigure shows the trajectory in the $(y_1, y_2)$ plane, where $y_1$ is the tip displacement and $y_2$ is the tip velocity. The plots correspond to different input voltage amplitudes: (**a**) 90 volts, (**b**) 130 volts, (**c**) 170 volts, and (**d**) 210 volts. As the input voltage increased, the size of the limit cycle grew, reflecting stronger nonlinear effects induced by the hysteresis operator.

### 3.2. Effect of Scaling Factor

The scaling factor $a$ in Table 2 that defines the size of the envelope of the hysteresis kernel is crucial for modulating the hysteresis loop, particularly in terms of its effect on the smoothness of the loop and, consequently, the system's limit cycle properties. The variation in the configuration of the hysteresis loop with respect to different scaling factor values is depicted in Figure 6. It was observed that an increase in the scaling factor resulted in a more gradual transition within the hysteresis loop, giving it a smoother appearance. Conversely, smaller values of $a$ imparted a pronounced, ridge-like structure to the loop. This phenomenon is inherently linked to the formulation of the enveloping function $\gamma(u) = \tanh(U/a)$. As $a$ decreases, $\gamma$ asymptotically approaches either 1 or $-1$, signifying a binary state. On the other hand, an increase in $a$ allows $\gamma$ to adopt a continuum of values within the interval $[-1, 1]$, facilitating a more nuanced modulation of the hysteresis loop.

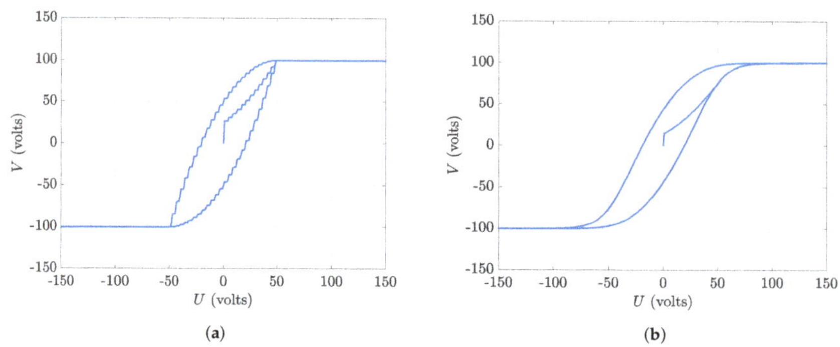


**Figure 6.** Hysteresis loops generated by the Preisach-type operator for different values of the scaling factor $a$. Here, $U(t)$ is the input voltage, and $V(t) = \mathbf{H}(U)(t)$ is the hysteresis operator output. Subfigure (**a**) corresponds to $a = 0.4$, and (**b**) corresponds to $a = 20$. As the scaling factor increases, the hysteresis loop becomes smoother and more gradual due to the influence of the envelope function $\gamma(U) = \tanh(U/a)$, which governs the range of admissible kernel weights. Smaller values of $a$ yield sharper transitions and a more ridge-like appearance, approximating binary switching behavior. This modulation of smoothness impacts the overall dynamics and limit cycle shape of the system.

In Figure 7, the effect of the scaling factor $a$ on the system's limit cycles is illustrated, with the voltage $V$ and kernel weight $\mu_{jk}$ remaining constant. At smaller values of $a$, the limit cycle exhibits significant nonlinearity, as reflected by the extensive velocity amplitude, indicative of a dynamic system response. As $a$ increases, the limit cycle transitions to a smoother profile, indicative of reduced nonlinearity. This transition reveals the smoothing effect of the scaling factor on the system's oscillations, leading to a stable and periodic limit cycle that signifies a more steady state of the dynamical system.

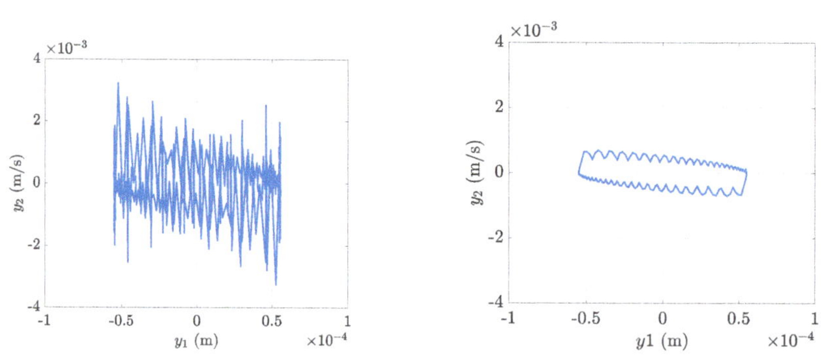


**Figure 7.** *Cont.*

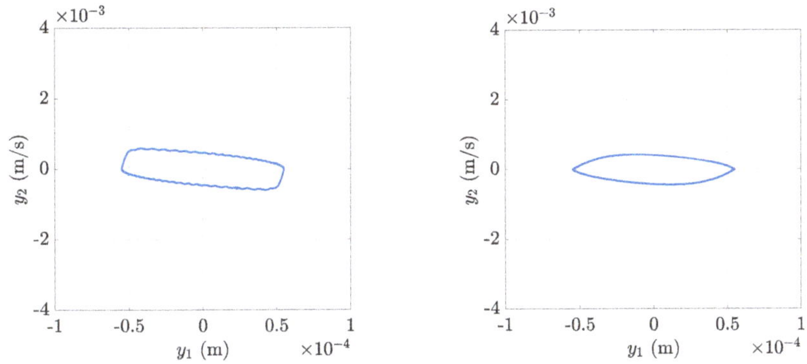


(**c**) $a = 3$ (**d**) $a = 20$

**Figure 7.** Limit cycle responses in the phase plane $(y_1, y_2)$ for a fixed input voltage $U = 150$ volts and constant kernel weight $\mu_{jk} = 0.02$ under varying values of the hysteresis scaling factor $a$. Subfigures (**a–d**) correspond to $a = 1, 2, 3$, and 20, respectively. As $a$ increases, the hysteresis effect smooths out, resulting in more stable limit cycles.

### 3.3. *Effect of Kernel Weight* $\mu_{jk}$

The modulation of hysteresis behavior and associated limit cycles by the kernel weight is depicted in Figure 8. Variations in the kernel weight $\mu_{jk}$ in Table 2 influence the amplitude of the hysteresis operator's output, directly influencing the dynamic system's response and thereby altering the observed limit cycles.

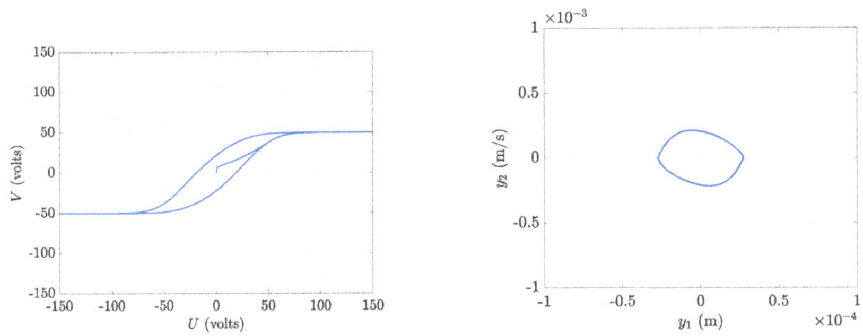


(**a**) $\mu = 0.01$ (**b**) Phase portrait for $\mu = 0.01$

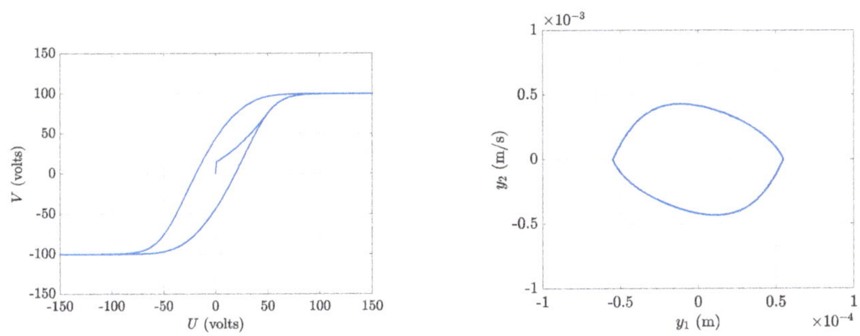


(**c**) $\mu = 0.02$ (**d**) Phase portrait for $\mu = 0.02$

**Figure 8.** *Cont.*

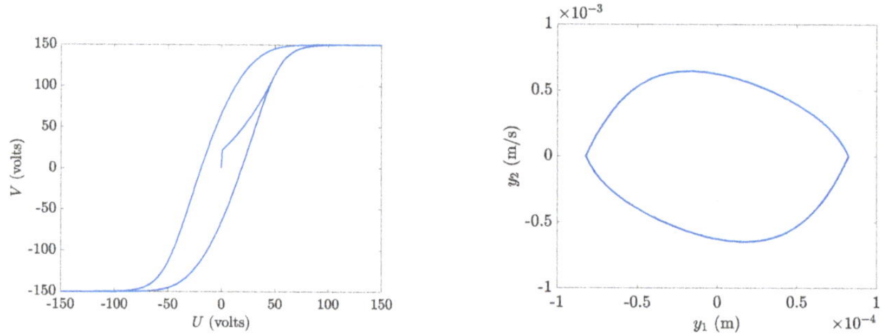


**(e)** $\mu = 0.03$ **(f)** Phase portrait for $\mu = 0.03$

**Figure 8.** Effect of the kernel weight $\mu_{jk}$ on the hysteresis loop (left column) and corresponding limit cycle response (right column) in the phase plane $(y_1, y_2)$. The input voltage is $U = 150$ volts, and the scaling factor is fixed.

### 3.4. Effect of Switching Parameters $\alpha$ and $\beta$

The switching parameters $\alpha$ and $\beta$ play a fundamental role in the behavior of the KP hysteresis model. These parameters define the thresholds at which individual relay operators within the Preisach plane switch states in response to changes in the input voltage.

The function $\mu$ represents a distributed measure over the Preisach plane, which incorporates the kernel weight, kernel size, and other scaling characteristics of the model. A component of this distributed measure, the kernel weight $\mu_{jk}$, determines the contribution of a specific Preisach cell located at $(s_1^j, s_2^k)$ to the overall hysteretic output.

While $\mu$ collectively governs the intensity and shape of the hysteresis loop, the switching parameters $\alpha$ and $\beta$ determine which regions of the Preisach plane are active during the input cycle. Consequently, fixing the kernel weight $\mu_{jk}$ while varying $\alpha$ and $\beta$ can significantly influence the hysteresis response. In particular, if the switching thresholds are too narrow relative to the input signal range, the hysteretic memory may be underutilized, leading to a response that does not fully represent the physical system.

To explore this effect, we fixed the input voltage amplitude at $U = 170$ volts and set the kernel weights uniformly as $\mu_{jk} = 0.02$. We then varied the switching thresholds $\alpha$ and $\beta$ to examine their impact on both the hysteresis loop and the associated limit cycle, as shown in Figure 9.

Figure 9 illustrates the influence of the switching parameters $\alpha$ and $\beta$ on both the hysteresis loop (left) and the resulting limit cycle (right) for a fixed input amplitude of $U = 170$ V. In case (a), with a wide switching range ($\alpha = -40$, $\beta = 80$), the hysteresis loop exhibited broad saturation regions. Specifically, the output voltage $V$ remained flat at approximately $-200$ V when the input $U$ was between $-170$ V and $-80$ V and saturated at approximately $+200$ V when $U$ was between $+100$ V and $+170$ V. The corresponding limit cycle exhibited a relatively large amplitude in terms of both displacement and velocity, with $y_1$ ranging between $-1 \times 10^{-4}$ m and $1 \times 10^{-4}$ m and $y_2$ ranging between $-0.7 \times 10^{-3}$ m/s and $0.7 \times 10^{-3}$ m/s.

In contrast, case (b) used narrower switching thresholds ($\alpha = -20$, $\beta = 30$), which confined the active Preisach region. The resulting hysteresis loop was less saturated, with the output voltage $V$ plateauing at approximately $-25$ V for inputs $U \in [-170, -40]$ and at $+25$ V for $U \in [50, 170]$. This limited hysteresis effect led to a significantly smaller limit cycle, with the displacement $y_1$ ranging from $-1.5 \times 10^{-5}$ m to $1.5 \times 10^{-5}$ m and the velocity $y_2$ ranging from $-1.5 \times 10^{-4}$ m/s to $1.5 \times 10^{-4}$ m/s.

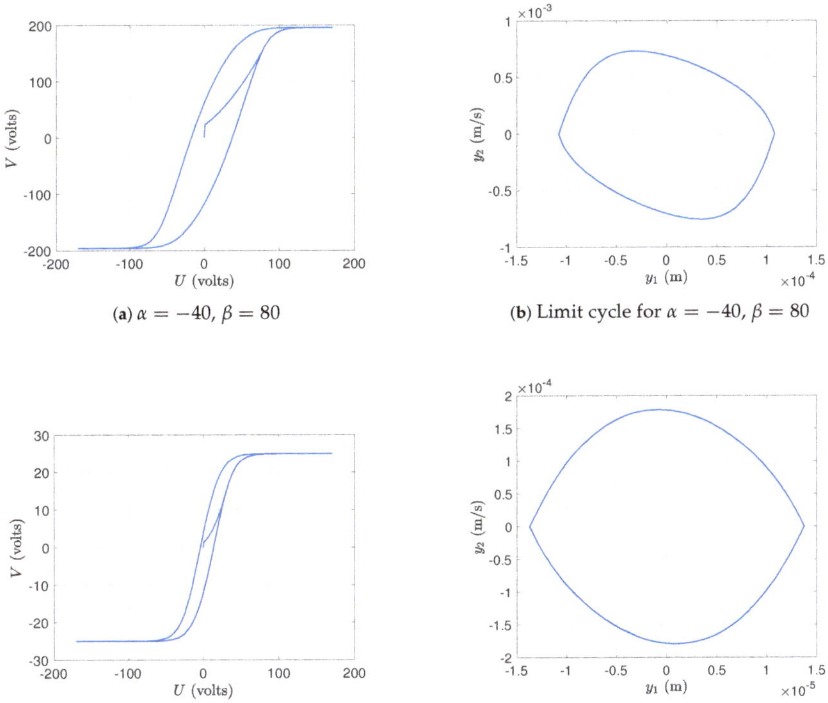


(a) $\alpha = -40$, $\beta = 80$

(b) Limit cycle for $\alpha = -40$, $\beta = 80$

(c) $\alpha = -20$, $\beta = 30$

(d) Limit cycle for $\alpha = -20$, $\beta = 30$

**Figure 9.** Comparison of hysteresis loops (left) and corresponding limit cycle responses (right) for a fixed input voltage amplitude $U = 170$ volts under two different sets of hysteresis switching thresholds $\alpha$ and $\beta$. Subfigures (**a**,**b**) correspond to $\alpha = -40$, $\beta = 80$, and subfigures (**c**,**d**) correspond to $\alpha = -20$, $\beta = 30$. Wider thresholds resulted in broader loops and larger oscillations, while narrower thresholds localized the response.

These results highlight how reducing the switching interval can dramatically compress the output range and attenuate the system's nonlinear dynamic response.

## 4. Discussion

The numerical examples presented in the previous section illustrate how the parameters of the discretized hysteresis operator influenced the dynamic behavior of a piezoelectric beam system. Several key insights emerged from these simulations:

- The system exhibited strongly nonlinear behavior, including memory-dependent limit cycles, under sinusoidal actuation. This arose not from geometric or material nonlinearities in the beam itself but from the hysteretic transformation of the input signal by the KP operator.
- The shape and amplitude of the limit cycles were highly sensitive to the parameters of the hysteresis model, including the scaling factor $a$, kernel weight $\mu_{jk}$, and switching thresholds $\alpha$, $\beta$. These parameters controlled not only the strength of the nonlinearity but also its frequency-dependent shaping effect on the system response.
- Because the hysteresis operator was discretized over the Preisach plane, the resulting system depended on a large number of real and distributed parameters. Combined with the high-dimensional state-space representation from modal reduction, the model

has a complexity that poses practical challenges for identification, control, and real-time simulation.

Experimental identification of such systems is particularly challenging due to several intrinsic complexities:

1. History-dependent and nonlinear input–output behavior;
2. High-dimensional state-space representations resulting from modal discretization;
3. Dependence on a large number of real-valued parameters in the hysteresis model;
4. Reliance on generally infinite-dimensional distributed parameters.

In particular, the parameters that define phenomenological hysteresis operators, such as Preisach or KP-type kernels, are rarely known a priori and are difficult to infer directly from measured data. These factors justify a modeling approach centered on numerical simulations and motivate the use of surrogate or data-driven modeling strategies.

Overall, while the proposed model captures nonlinear memory effects, it remains computationally intensive and highly sensitive to parameter selection. These challenges underscore the need for reduced-order modeling techniques that preserve essential hysteretic dynamics while lowering the computational burden. One promising direction is the use of operator-theoretic frameworks such as Koopman operator theory to approximate the system behavior with fewer degrees of freedom. These directions are pursued in follow-up work.

The modeling framework developed in this work is well suited for a range of practical applications where accurate representation of hysteretically nonlinear piezoelectric behavior is critical. In particular, the ability to capture memory-dependent effects in distributed-parameter systems enables more reliable prediction and control in engineering scenarios such as precision motion control, where sub-micrometer accuracy is required in nanopositioning systems and scanning probe microscopy; vibration suppression in smart structures used in aerospace and civil engineering, where piezoelectric layers are employed for active damping; and energy harvesting systems, such as MEMS devices and structural harvesters, where accurate modeling is necessary to optimize power output under nonlinear dynamic conditions. By embedding hysteresis effects directly into the governing equations, the model forms a foundation for reduced-order modeling, control synthesis, and data-driven extensions aimed at real-world deployment in smart material systems.

Several limitations of the present study suggest opportunities for future work. First, the hysteresis operator parameters were selected to illustrate qualitative behaviors rather than being fitted to experimental data; parameter identification and data-driven model calibration remain important open challenges. Second, the analysis was limited to sinusoidal actuation signals; evaluating the system's response under broadband or stochastic inputs would offer a more complete understanding of its dynamic behavior. Third, the current model is restricted to a one-dimensional beam; extending the framework to plate-like or three-dimensional piezoelectric structures would pose new challenges in both discretization and computational efficiency. Lastly, integrating this modeling approach with feedback control schemes and investigating model reduction or hybrid modeling techniques for real-time implementation remain promising directions for future investigation.

## 5. Conclusions

This work presents a comprehensive modeling framework for piezoelectric composite beams with hysteresis by coupling a distributed-parameter structural model with a nonlinear hysteresis operator. A Krasnosel'skii–Pokrovskii (KP) formulation is used to represent rate-independent hysteresis in a physically meaningful and flexible way. By discretizing the Preisach plane and applying modal expansion to the beam dynamics, the full system is expressed as a high-dimensional functional differential equation (FDE) that captures key hysteretic phenomena such as memory, path dependence, and saturation. The resulting model depends on a large number of structural modes and a distributed set of parameters defining the hysteresis operator, making it computationally intensive and challenging for control design, identification, and real-time implementation.

These observations underscore the need for effective reduced-order modeling strategies that can preserve the essential dynamics while significantly lowering computational cost and model complexity. In particular, data-driven techniques such as Koopman operator theory offer promising avenues for constructing low-dimensional approximations of complex nonlinear systems like the one studied here. Future work will investigate the use of operator-theoretic methods to learn predictive models directly from simulated or experimental data, with the aim of developing efficient surrogate models that retain the memory effects and input–output behavior of the original FDE system.

**Author Contributions:** Conceptualization, A.H.A. and A.J.K.; methodology, A.H.A. and A.J.K.; writing—original draft preparation, A.H.A.; writing—review and editing, A.H.A. and A.J.K.; supervision, A.J.K. All authors have read and agreed to the published version of the manuscript.

**Funding:** This research received no external funding.

**Data Availability Statement:** The data supporting the findings of this study were generated via numerical simulation and are available from the corresponding author upon reasonable request.

**Conflicts of Interest:** The authors declare no conflicts of interest.

## References

1. Banks, H.; Smith, R.; Wang, Y. *Smart Material Structures Modeling, Estimation, and Control*; Wiley: Mason, Paris, 1996.
2. Tichy, J.; Erhart, J.; Kittinger, E.; Privratska, J. *Fundamentals of Piezoelectric Sensorics*; Springer, Inc.: Berlin, Germany, 2010.
3. Yang, J. *Introduction to the Theory of Piezoelectricity*; Springer: New York, NY, USA, 2005.
4. Yang, J. *Mechanics of Piezoelectric Structures*; Springer: New York, NY, USA, 2020.
5. Kurdila, A.J.; Tarazaga, P.A. *Vibrations of Linear Piezostructures*; Wiley: Blacksburg, VA, USA, 2021.
6. Erturk, A.; Inman, D.J. *Piezoelectric Energy Harvesting*; Wiley: Hoboken, NJ, USA, 2011.
7. Tiersten, H.F. *Linear Piezoelectric Plate Vibrations: Elements of the Linear Theory of Piezoelectricity and the Vibrations of Piezoelectric Plates*; Plenum Press: New York, NY, USA, 1969.
8. Maugin, G.A. *Nonlinear Electromechanical Effects and Applications*; Series in Theoretical and Applied Mechanics; World Scientific Pub: Singapore, 1985; Volume 1.
9. Stanton, S.C.; Erturk, A.; Mann, B.P.; Inman, D.J. Nonlinear piezoelectricity in electroelastic energy harvesters: Modeling and experimental identification. *J. Appl. Phys.* **2010**, *108*, 074903. [CrossRef]
10. Leadenham, S.; Erturk, A. Unified Nonlinear Electroelastic Dynamics of a Bimorph Piezoelectric Cantilever for Energy Harvesting, Sensing, and Actuation. *Nonlinear Dyn.* **2015**, *79*, 1727–1743. [CrossRef]
11. Asanuma, H.; Komatsuzaki, T. Nonlinear Piezoelectricity and Damping in Partially-Covered Piezoelectric Cantilever with Self-Sensing Synchronized Switch Damping on Inductor Circuit. *Mech. Syst. Signal Process.* **2020**, *144*, 106867. [CrossRef]
12. Asanuma, H. Selecting Nonlinear Piezoelectricity for Fully Autonomous Self-Sensing Synchronized Switch Damping on Inductor Technique. *Mech. Syst. Signal Process.* **2021**, *159*, 107846. [CrossRef]
13. Lumentut, M.F.; Shu, Y.C. Shunted Optimal Vibration Energy Harvesting Control of Discontinuous Smart Beams. *Compos. Struct.* **2020**, *242*, 112126. [CrossRef]
14. Krasnosel'skii, M.A.; Pokrovskii, A.V. *Systems with Hysteresis*; Springer: Berlin, Germany, 1989.
15. Sabarianand, D.V.; Karthikeyan, P.; Muthuramalingam, T. A Review on Control Strategies for Compensation of Hysteresis and Creep on Piezoelectric Actuators Based Micro Systems. *Mech. Syst. Signal Process.* **2020**, *140*, 106634. [CrossRef]
16. Visintin, A. *Differential Models of Hysteresis*; Springer: Berlin, Germany, 1994.
17. Paruchuri, S.; Guo, J.; Kurdila, A. Reproducing kernel Hilbert space embedding for adaptive estimation of nonlinearities in piezoelectric systems. *Nonlinear Dyn.* **2020**, *101*, 1397–1415. [CrossRef]
18. Gan, J.; Zhang, X. A review of nonlinear hysteresis modeling and control of piezoelectric actuators. *AIP Adv.* **2019**, *9*, 040702. [CrossRef]
19. Li, Z.; Shan, J.; Gabbert, U. Inverse Compensation of Hysteresis Using Krasnoselskii-Pokrovskii Model. *IEEE/ASME Trans. Mechatronics* **2018**, *23*, 966–971.

20. Salah, M.; Saleem, A. Hysteresis Compensation-Based Robust Output Feedback Control for Long-Stroke Piezoelectric Actuators at High Frequency. *Sens. Actuators A Phys.* **2021**, *319*, 112542. [CrossRef]
21. Yu, Z.; Wu, Y.; Fang, Z.; Sun, H. Modeling and Compensation of Hysteresis in Piezoelectric Actuators. *Heliyon* **2020**, *6*, e03999. [CrossRef] [PubMed]
22. von Wagner, U.; Hagedorn, P. Nonlinear effects of piezoceramics excited by weak electric fields. *Nonlinear Dyn.* **2003**, *31*, 133–149. [CrossRef]
23. Dai, Y.; Li, D.; Wang, D. Review on the Nonlinear Modeling of Hysteresis in Piezoelectric Ceramic Actuators. *Actuators* **2023**, *12*, 442. [CrossRef]
24. Xu, Y.; Luo, Y.; Luo, X.; Chen, Y.; Liu, W. Fractional-Order Modeling of Piezoelectric Actuators with Coupled Hysteresis and Creep Effects. *Fractal Fract.* **2024**, *8*, 3. [CrossRef]
25. Li, H.; Tong, Y.; Li, C. Modeling and Control of a Linear Piezoelectric Actuator. *Actuators* **2024**, *13*, 55. [CrossRef]
26. Baziyad, A.G.; Ahmad, I.; Salamah, Y.B. Precision Motion Control of a Piezoelectric Actuator via a Modified Preisach Hysteresis Model and Two-Degree-of-Freedom H-Infinity Robust Control. *Micromachines* **2023**, *14*, 1208. [CrossRef] [PubMed]
27. Inman, D.J. *Engineering Vibration*, 4th ed.; Pearson: Boston, MA, USA, 2014.
28. Dadashi, S.; Bobade, P.; Kurdila, A.J. Online Estimation and Adaptive Control for a Class of History Dependent Functional Differential Equations. *Nonlinear Dyn.* **2018**, *92*, 1431–1451. [CrossRef]

*vibration*

MDPI

*Article*

# Machine Learning Approach to Nonlinear Fluid-Induced Vibration of Pronged Nanotubes in a Thermal–Magnetic Environment

**Ahmed Yinusa [1,*], Ridwan Amokun [1], John Eke [1], Gbeminiyi Sobamowo [1], George Oguntala [2], Adegboyega Ehinmowo [3], Faruq Salami [1], Oluwatosin Osigwe [1], Adekunle Adelaja [1], Sunday Ojolo [1] and Mohammed Usman [4]**

[1] Department of Mechanical Engineering, University of Lagos, Akoka 101017, Lagos State, Nigeria
[2] Department of Biomedical Engineering, Birmingham City University, Birmingham B4 7BD, UK
[3] Department of Petroleum and Gas Engineering, University of Lagos, Akoka 101017, Lagos State, Nigeria
[4] Department of Chemical Engineering, University of Lagos, Akoka 101017, Lagos State, Nigeria
* Correspondence: aayinusa@unilag.edu.ng

**Abstract:** Exploring the dynamics of nonlinear nanofluidic flow-induced vibrations, this work focuses on single-walled branched carbon nanotubes (SWCNTs) operating in a thermal–magnetic environment. Carbon nanotubes (CNTs), renowned for their exceptional strength, conductivity, and flexibility, are modeled using Euler–Bernoulli beam theory alongside Eringen's nonlocal elasticity to capture nanoscale effects for varying downstream angles. The intricate interactions between nanofluids and SWCNTs are analyzed using the Differential Transform Method (DTM) and validated through ANSYS simulations, where modal analysis reveals the vibrational characteristics of various geometries. To enhance predictive accuracy and system stability, machine learning algorithms, including XGBoost, CATBoost, Random Forest, and Artificial Neural Networks, are employed, offering a robust comparison for optimizing vibrational and thermo-magnetic performance. Key parameters such as nanotube geometry, magnetic flux density, and fluid flow dynamics are identified as critical to minimizing vibrational noise and improving structural stability. These insights advance applications in energy harvesting, biomedical devices like artificial muscles and nanosensors, and nanoscale fluid control systems. Overall, the study demonstrates the significant advantages of integrating machine learning with physics-based simulations for next-generation nanotechnology solutions.

**Keywords:** branched SWCNT; data wrangling; extreme gradient boosting; random forest; category boosting; artificial neural network; differential transform method

## 1. Introduction

CNTs are carbon allotropes possessing a tube-like or cylindrical lattice structure. The diameter of CNTs could be estimated to be around 0.5–100 nm. The cylindrical structure has an aspect ratio around one thousand (1000) [1]. However, the actual length of CNTs can vary considerably, with some exceeding 10,000 times their diameter [2]. These lattices are subdivided into layers, and as such, they are sometimes classified as Single-Walled Carbon Nanotubes (SWCNTs) when there is only one layer [3]. The diameters for SWCNTs could typically vary from 0.78 nm to 1.4 nm [4]. Double-Walled Carbon Nanotubes (DWCNTs) are titled as such when there are two cylindrical layers and Multi-Walled Carbon Nanotubes (MWCNTs) when there are multiple layers. Aside from being lightweight, they are also known for their strength and toughness and thermal and electrical conductivity. They are less dense than copper and have shown that they are highly conductive [5]. Since the discovery of these allotropes of carbon by Sumio Iijima in 1991, the uniqueness and explorable capabilities in their properties have made researchers conduct different studies on how they could be applied to improve systems, processes, and mechanisms [6]. Cui et al. [7] demonstrated that they can be used to improve the strength of cement-based properties, like temperature resistance, can greatly be increased using physical dispersion technology. However, some detriments accompany this process. When CNTs are subjected to a great amount of compressive force, vibrational frequency in gigahertz and terahertz is experienced [8]. Consequently, it would be interesting to understand the effect of the presence of a magnetic field on CNTs conveying nanofluids. Carbon nanotubes have been used in the field of biotechnology as artificial neurons. However, this application has come with the shortcoming of the vibration response of this material to impacted force. The use of this material can be optimized by the optimization of vibration parameters to control vibrational frequency. Machine learning algorithms like Extreme Gradient Boosting (XGBoost), Category Boosting (CATBoost), Random Forest (RF), and Neural Networks, such as Artificial Neural Networks (ANNs), can be used to build a responsive model that can help optimize parameters to be studied. Also, this can be used to understand relationships between the physical parameters in this context and then predict the optimum parameters to be used. Hence, this research focuses on optimizing vibrational parameters with consideration for the magneto-thermal stability of a branched-chain carbon nanotube conveying nanofluid. Also, this work focuses on applying machine learning algorithms. such as Neural Networks and tree-based algorithms, to optimize thermo-magnetic and vibrational parameters. The discovery of CNTs has been accompanied by various applications and research. Although the unique properties of carbon nanotubes cannot be overemphasized, there have been shortcomings, and researchers in the field of nanotechnology have conducted several studies to understand and improve on some of these properties.

While prior studies have analyzed thermal and magnetic effects on nanotube dynamics independently, few have explored the interdependent influence of branching geometry, magnetic damping, and fluid flow velocity in a unified framework. Moreover, traditional physics-based models have limited predictive capacity in identifying optimal parameter combinations across a high-dimensional input space. The existing literature often lacks tools to quantify the sensitivity of each input or predict vibrational outcomes under varying configurations efficiently. This study addresses this gap by integrating physics-informed modeling with supervised machine learning algorithms, enabling both accurate prediction and feature-level interpretation of key stability parameters.

A few highlights of recent work and research conducted in the study of CNTs and applications are as follows:

Mahmure et al. [9] examined the nonlinear free vibration behavior of thin-walled composite shells reinforced with carbon nanotubes. The shells sit on the Winkler-Pasternak Foundation. This model considers a variety of uniform and nonuniform CNT distributions, as well as Von-Karman nonlinearities. Equations accounted for the composite shell, CNTs, foundation, and nonlinearities. The equations were solved using the Galerkin and Grigolyuk methods. This approach provides valuable insights for designing composite structures with CNTs. Huang and Yao [10] presented the analysis of SWCNTs modeled with beam theory under thermal and electrostatic conditions. They showed that, unlike the Euler–Bernoulli theory, this approach presented independent stiffness. Mechanical phenomena, such as static bending, buckling, and nonlinear vibration, were also examined. Their finding suggested that there might be a need to re-evaluate classical beam theories.

The vibration of SWCNTs resting on Winkler foundation with magnetic effect was studied by modeling the system using Euler–Bernoulli beam theory, and for the bending of a pipe, fluid velocity has the most significant impact, followed by the nonlocal parameter and magnetic field intensity, using a one-parameter finite element and the Newmark temporal integration method [11].

Yinusa and Sobamowo [12] studied the mechanics of SWCNT using Galerkin decomposition, considering the impact of a nonlinear Winkler foundation. The acquired solutions are then processed with after-treatment procedures to capture a vast temporal domain during dynamic reaction. Parametric studies confirmed the veracity of nonlocal elastic theory and the frequency-increasing influence of the foundation parameter. The dynamics of nanosensors have also been considered via Hamilton's principle with the weighted residual approach used for the analysis of both classical and higher-order boundary conditions for a simply supported beam. The study examined the impact of various factors, including the vibration modes, on the dynamic response of the system [13]. Chen et al. [14] studied vibration analysis of functionally graded CNT in a thermal environment. The potential applications of functionally graded carbon nanotube-reinforced composites (FG-CNTRC) in the transportation of natural gas and oil in various industries were investigated. In a heated setting, the study examined the nonlinear free vibration properties called FG-CNTRC. The fundamental frequencies of the deforming modes in SWCNTs exhibit a dependence on their aspect ratio. While it can be adequately described by Bernoulli-Euler beam theory for aspect ratios exceeding 30, accurate determination for shorter nanotubes necessitates employing modal analysis simulations [15]. An analytical approach was presented by Dat et al. [16] on the vibration of smart, rolled plates reinforced with CNTs. The relationships between close parameters were shown using the Galerkin and harmonic equilibrium method. The effect of CNTs on overall volume was studied from the numerical results obtained. De Rosa et al. [17] employed DQM to analyze free vibration of nonuniform SWCNTs with nonlocal effects, considering a clamped-elastic support and a concentrated mass at the free end. They investigated the influences of different control parameters on the system's vibration and validated the method employed [17]. This confirms that DQM is a powerful tool for CNT's vibration analyses even when analytical solutions are unavailable. Precise control of sound and movement in nano-scale devices via tailored vibration frequency manipulation can be optimally achieved.

Oveissi et al. [18] evaluated the effects of nanofluid flow on vibrations of SWCNT. This study considered water nanoflow and air nanoflow. Equations are resolved by the Galerkin technique, with descriptions made with the help of the nanorod model. The first mode divergence has the lowest value, and a drop in critical flow velocity with increasing wave number is one of the key results. Particularly in airflow over SWCNTs, the Knudsen effect greatly affects critical velocities. Additionally, although an increase in characteristic length associated with the strain-inertia gradient theory leads to a decrease in critical flow velocity, an increase in nonlocal parameter and Knudsen number. Chang et al. [19] analyzed stability and the nonlinear vibration in CNS-reinforced pipes conveying fluids. The governing equations were derived using Euler beam theory, the Von-Karman nonlinear stress-strain principle, the Hamilton variation principle, and the Galerkin method, and it was discovered that the influence of the maximum displacement is more vivid at the second-order frequency. A similar analysis has been performed for micropipes to enhance their biomedical applications, where an analytical solution of micropipes conveying fluids in an elastic medium was performed and the nonlinear differential equation was discretized using the Galerkin method, where the result shown was that the optimal convergence parameter greatly reduced the error of the homotopy solution [20]. Moborukoje et al. [21] investigated nanotubes that transport fluid. The nanotube was on different foundations.

Parametric experiments utilizing approximate analytical solutions yielded numerous notable findings. The results showed that surface stress influences vibration frequencies, with a drop associated with positive values and a rise with negative values. Furthermore, nonlocal parameters influenced the surface effect and the nanotube's natural frequency. The findings provide important insights on controlling and creating carbon nanotubes in thermomagnetic settings. The nonlinear vibration of CNT composites was investigated via Hamilton's principle and Timoshenko beam theory, Karman's geometric nonlinear theory, and the resulting partial differential equation solved by discretization using FEM [22]. This unveils the characteristics of function between various parameters considered in this analysis. Similarly, the use of functionally graded carbon nanotubes in drug delivery systems has also been analyzed with the mathematical model for analysis generated using the Euler–Bernoulli beam theory, Hamilton principle, and Erigen elasticity principle, and the result shows that increasing the magnetic field dramatically improves stability (critical flow velocity increases by 227%), while loading the CNTs with drugs weakens it (velocity drops by 17–21%), but a "hybrid load distribution" can optimize the material, balancing stability and drug delivery [23]. The impacts of electrical voltage, Lorentz force, and modulus on the nanosystem's stability were examined. The simultaneous application of these force fields is discussed for the first time, revealing potential applications in smart structure design [24]. It has also been shown that energy can be harvested from CNTs in a magnetic field due to the vibration produced while conveying nanofluids. This analysis was achieved using the Euler–Bernoulli theory, Navier–Stokes equation, piezo-elastic theory, and Vander Pol equation to obtain the governing equation, and it was observed that the output voltage is increased when there is a reduction in the density of the fluids in the modeled system [25].

Most recently, Yinusa et al. [26] performed a vibration study on SWCNTs and MWCNTs resting on nonlinear foundations. The research focused on vibrational analysis of nanotubes conveying nanofluids. The study, which focused on CNTs' unique properties, developed fully coupled equations of motion for nanotube vibrations. The study employed multi-dimensional numerical PDE solvers in MATLAB (R2024a). The study provides insights into nanotube design and serves as a reference for future research in the field.

Through a relatively new approach, the use of machine learning in predicting the mechanical properties, fault detection in structures, and analysis of mechanical parameters can be made possible with the aid of mathematical modeling and simulation of different occurrences or physical states. Vivanco-Benavides et al. [27] discussed how machine learning can be used to analyze the properties of CNTs. The research reviewed and explored the powerful synergy between machine learning and the investigation of CNT's intricate physical properties. It also highlighted the versatility of machine learning in tackling the complexities of CNTs and emphasized the importance of algorithm selection based on data availability and the number of parameters involved. Typical use of deep learning algorithms may not be efficient with a limited amount of data. Algorithms such as Artificial Neural Networks (ANNs) for deep learning approaches, Support Vector Machines (SVM), and Random Forest (Decision Trees) have proven particularly effective in analyzing uncontrolled CNT properties. Random Forest and Support Vector Machines could be employed, as they have proven to be able to help detect noise in data. While machine learning has been proven to successfully evaluate and predict mechanical properties, further research is needed to refine vibrational frequency predictions based on chiral parameters. For thermal, electrical, and electronic properties, complementary approaches combining machine learning with molecular dynamics and density functional theory hold significant promise. Intriguingly, a strong link has been found between defect detection in CNTs and the number of machine learning iterations required to analyze their properties, suggesting that understanding vibrational behavior in defective CNTs could be essential for nanosensor development. Overall, this review concludes that machine learning offers a promising avenue for predicting and comprehending CNT properties, but further studies are needed to tailor algorithms and deepen our understanding of specific areas like vibrational behavior and defect detection. Crucially, the integration of machine learning with simulations holds the potential to reduce research costs and accelerate the exploration of these remarkable nanomaterials, which is made possible with the existence of software that can be used to accurately represent real materials. The modeling of Stealth Carbon Nanotubes (S-CNTs) for radio wave shielding and attenuation in next-generation aircraft was the main emphasis of Pollayi and Rao [28]. A Python-based artificial intelligence/machine learning (AI/ML) framework was used, and CNTs were simulated using CNT bands (V2.7.3). The K-Means Clustering Algorithm is used to estimate the optimal diameter of CNTs. A deep learning framework for carbon nanotubes was presented by CanaCija [29], with an emphasis on modeling techniques and mechanical characteristics. Tensile experiments were performed on SWCNTs considering all possible conformations. The analysis revealed that nanotube chirality significantly affected the physical and structural properties. Using the molecular dynamics dataset, an Artificial Neural Network was created that can accurately predict mechanical properties while reducing the effects of temperature fluctuations. The study also thoroughly examined the impact of dataset size on prediction quality, providing future researchers with useful modeling techniques. Similarly, Yang et al. [30] conducted a machine learning-aided uncertainty analysis of nanotubes, while Akbarzadeh et al. [31] suggested a computational approach. One hidden-layer perceptron was used in the ANN model, which optimized it using the Levenberg–Marquardt algorithm. Input variables included MWCNT's mass concentrations. Khanam et al. [32] aimed to predict physical and chemical properties in graphene composites. Different concentrations of graphene were extruded with varying speeds. Thermal conductivity, crystallization, degradation, and tensile strength were optimized. Tensile strength peaked at 4% graphene weight and 150 rpm. Artificial Neural Networks effectively predicted these properties, demonstrating potential cost and time savings in development and manufacturing. Hajilounezhad et al. [33] anticipated the mechanical attributes of CNT forestry by using image-based machine learning and physics-based simulations. Because there are large experimental parameter spaces that make it difficult to comprehend and manage the self-assembly of CNT forests, simulations were performed. They present CNTNet, a deep learning classifier module that is based on images and was trained using synthetic imagery. To forecast the strength of carbon CNT-reinforced concrete, Kekez and Kubica [34] investigated the application of ANNs. The paper discusses the importance of concrete, which has strong mechanical qualities, is long-lasting, multipurpose, and aesthetically pleasing. The research underscores the importance of ANNs as an effective tool for improving concrete mixtures based on their intended purposes, with an emphasis on environmentally friendly and versatile materials. To close the gap between the ANN approach and other approaches, experimental findings are used for the learning process.

The investigation of the simultaneous impacts of surface energy, initial stress, and nonlocality on the nonlinear vibration of carbon nanotubes conveying fluid has been investigated by Sunday et al. [35]. The nanotubes in this study rested on both linear and nonlinear elastic foundations and operated in a thermo-magnetic environment. The derived equations governing the behavior were solved using Galerkin's decomposition-Adomian decomposition method. The study explored the concurrent impacts of surface elasticity, initial stress, residual surface tension, and nonlocality on the nonlinear vibration. The effects of various parameters, such as surface energy and nonlocality, on the dynamic behavior of the nanostructure, are investigated, presented, and discussed [36].

Abubakar et al. [37] conducted a theoretical exploration into the dynamic response of nanostructures, specifically focusing on SWCNTs. The study emphasizes the vibration behavior of SWCNTs on an elastic foundation, considering the impacts of magnetic and thermal effects in the presence of electrostatic force. Employing Eringen's nonlocal elastic theory and Euler–Bernoulli beam theory, Hamilton's principle was used to derive the nonlinear governing equation of motion. Utilizing the Galerkin separation technique, the researchers decomposed the resulting nonlinear mathematical model into spatial and temporal equations. The temporal aspect was then addressed using the Homotopy Perturbation Method (HPM) to obtain an approximate analytical solution. The investigation systematically explores the influences of magnetic term, thermal term, nonlocal parameter, linear elastic foundation, nonlinear elastic foundation, Pasternak foundation, and electrostatic force on the amplitude-frequency response and time-deflection curves of SWCNTs. Through parametric studies, the research contributed valuable insights, enhancing the comprehension and better understanding of the dynamic behavior of SWCNTs under specified conditions [37].

Selim and Musa [38] studied the nonlinear vibration of CNTs filled with water and simply supported at both ends. A mathematical model was obtained using Donnel's shear theory. The mathematical model produced was solved numerically, and the results depicted that there is an increase in the nonlinear frequency with an increase in the parameters of CNTs, like radius and thickness. Also, the nonlocal parameter influence is higher at the lower modes compared to the higher modes.

Strozzi et al. [39] conducted a comparative analysis of three classical shell theories—Donnell, Sanders, and Flügge—in modeling the linear vibrations of single-walled carbon nanotubes (SWCNTs). The study utilized a continuous homogeneous cylindrical shell model with equivalent thickness and surface density to represent the discrete SWCNT. An anisotropic elastic shell model, accounting for the intrinsic chirality of carbon nanotubes (CNTs), was employed. The investigation applied simply supported boundary conditions and utilized a complex method to solve the equations of motion and determine natural frequencies. Comparisons with molecular dynamics simulations confirmed the superior accuracy of the Flügge shell theory. The study further conducted a parametric analysis, evaluating the influence of diameter, aspect ratio, and the number of waves along longitudinal and circumferential directions on SWCNTs' natural frequencies under the three shell theories. Results indicated inaccuracies in the Donnell shell theory for specific conditions, whereas the Sanders shell theory demonstrated consistent accuracy across all geometries and wave numbers. The study recommended the adoption of the Sanders shell theory as a simpler yet accurate alternative to the Flügge shell theory for modeling the vibrations of SWCNTs.

The study by Olaleye et al. [40] employed the variation iteration method to explore the simultaneous impacts of surface elasticity, initial stress, residual surface tension, and nonlocality on the nonlinear vibration of single-walled carbon nanotubes conveying fluid. The nanotubes rest on both linear and nonlinear elastic foundations and operate in a thermo-magnetic environment. The equation of motion governing nanotube vibration is derived using Erigen's theory, Euler–Bernoulli's theory, and Hamilton's principle. Through parametric studies, the research reveals significant insights, including the influence of surface stress on frequency ratios and the gradual approach of nanotube natural frequency to the nonlinear Euler–Bernoulli beam limit at high values of nonlocal parameters and nanotube length. The study contributes valuable information for controlling and designing carbon nanotubes in thermo-magnetic environments resting on elastic foundations [40].

Data for model building in the present study is to be obtained from simulation using Fluid Fluent in ANSYS R1 2023. These would be obtained by varying parameters and imitating the condition of flow. A grid-dependence test would be conducted by varying necessary parameters to understand the sensitivity of various parameter changes. This would serve as a data evaluation measure, as no fluent solution can be trusted without an appropriate grid-independence test for confirmation. Ansys software provides an environment that can be used to imitate actual occurrences. Actual results are computed using FEM. This points out the need to specify an accurate boundary condition. The machine learning algorithms to be used for this research are as follows: XGBoost also referred to as eXtreme Gradient Boosting, is a powerful tool used for regression and classification tasks. It works using the principle of gradient boosting. XGBoost operates by combining multiple weak learners (simple models) to create a strong ensemble model as an aggregate of all other models, thereby capturing and combining their strengths to make correct predictions [35]. XGBoost has consistently been ranked among the most accurate algorithms, often outperforming other algorithms like Random Forest and Support Vector Machines. It has inbuilt regularization techniques that prevent overfitting, ensuring the model can generalize well and the model build doesn't just memorize the training data. Figure 1a illustrates the ensembling method utilized by XGBoost and its capability in handling large datasets efficiently, making it suitable for real-world applications.

CATBoost is another state-of-the-art gradient-boosting algorithm focusing on efficiency and accuracy for both classification and regression tasks. It was developed by Yandex (the Russian tech giant). It has some unique features that differentiate it from popular counterparts like XGBoost. It excels in data where there are features with categorical values, and as shown in Figure 1b, it has a strategic mechanism for handling categorical data. It automatically detects and interprets category levels, leading to better performance on datasets with a mix of numerical and categorical features. It also leverages GPUs for faster training, making it a great choice for large datasets and complex problems. Its API provides built-in visualization and interpretation tools to understand how the model makes predictions, valuable for debugging and gaining insights into your data. Similarly to XGBoost, CATBoost utilizes various regularization methods to prevent overfitting and improve generalizability. CATBoost often competes with and even surpasses XGBoost in terms of accuracy on many tasks, especially those with categorical data. GPU support makes CATBoost significantly faster for training large models.

The Random Forest algorithm is a machine learning problem widely used for regression and classification problems. The algorithm can be viewed as an ensemble of several decision trees to produce a single decision tree that represents the average of all decision trees, as shown in Figure 1c. The averaging technique helps to control overfitting, thereby presenting a model that would be able to generalize. However, the danger of overfitting lingers when the number of trees in the ensemble becomes too much.

Neural Networks, especially Artificial Neural Networks (ANNs), have emerged as powerful tools in various scientific disciplines, offering a versatile approach to model complex relationships and patterns within data. In the context of this study, Neural Networks find application in understanding the intricate dynamics and behaviors of these nanostructures of SWCNTs. Artificial Neural Networks consist of interconnected nodes, which mimic the structure of biological neurons, and they excel at capturing nonlinear relationships and patterns within datasets. Figure 1d shows three different layers (input, hidden, and output layers) of a multilayer perception. In the realm of CNTs, where the vibrational properties are influenced by numerous factors such as chirality, length, diameter, and environmental conditions, Neural Networks provide a valuable means to model and predict these intricate relationships.

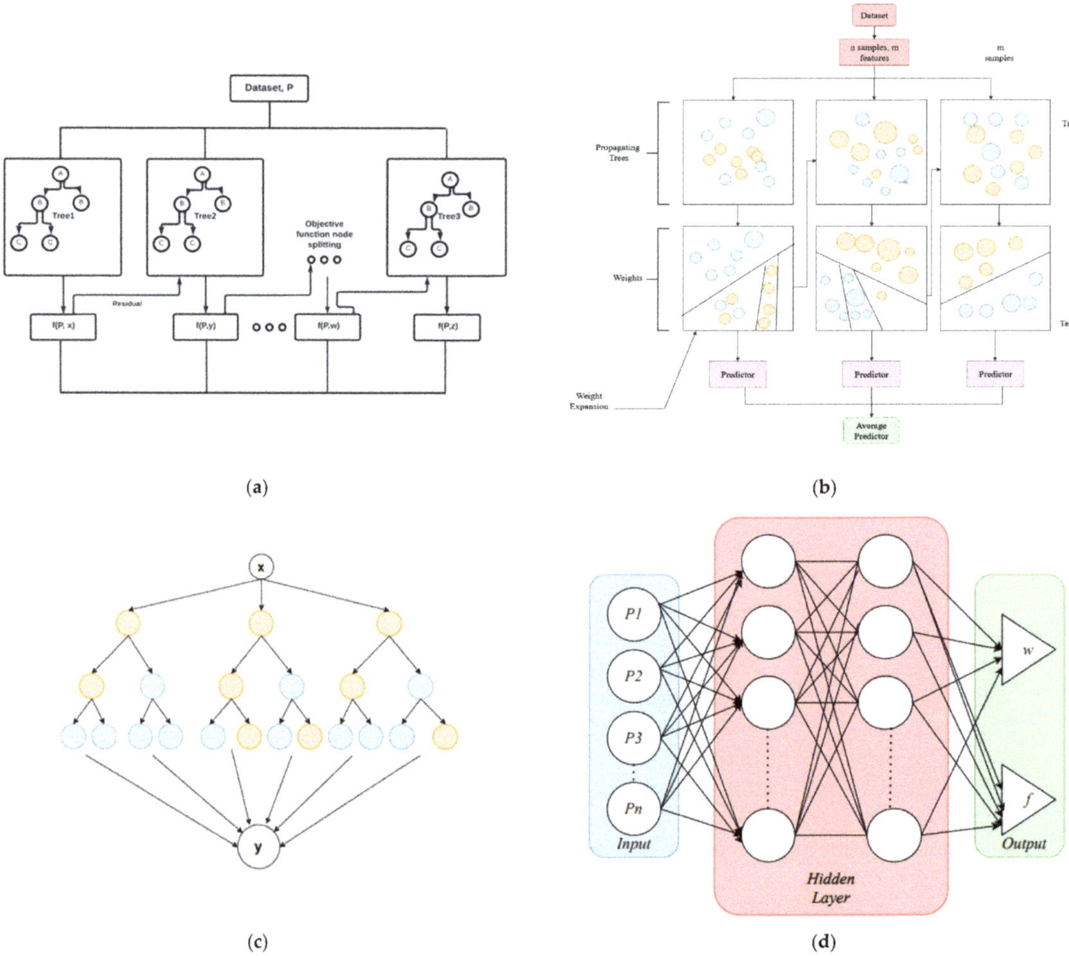


**Figure 1.** Architecture of the machine learning models used in this study: (**a**) Random Forest; (**b**) XGBoost; (**c**) CATBoost; and (**d**) Artificial Neural Network. These models were selected to explore a range of regression methods for predicting vibrational and frequency responses under varying physical parameters.

## 2. Model Formulation

The following assumptions are made in the mathematical modeling of the vibration of a branched SWCNT: the nanotube is resting on a Winkler's foundation; the CNT's lateral displacement and velocity have zero beginning conditions; and the material being considered is assumed to be isotropic and homogeneous.

### 2.1. Model Formulation for Velocity of Flow of Nanotube

To formulate the equation for the nanomagnetic flow (NMF), the following modified Navier–Stokes equation is employed:

We have the continuity equation described as follows:

$$\nabla U = 0 \tag{1}$$

$$\frac{\partial\psi}{\partial t} + U \cdot \nabla\psi + \frac{e_{np}}{\rho_{np}} = 0 \tag{2}$$

The momentum equation can also be expressed as follows:

$$\rho_f \frac{\partial U}{\partial t} = -\nabla P + \mu_{eff}\nabla^2 U + (J \times B) \tag{3}$$

The energy equation is expressed as follows:

$$\rho_c \frac{\partial T}{\partial t} + \rho_c(v \cdot \nabla T) = h \cdot e_n n \tag{4}$$

The following assumptions underlie the development of this thermo-fluidic model: a low diluted mixture, a time-invariant gradient of nanoparticle volumetric fraction, and a Newtonian, incompressible fluid with a uniform magnetic field specified. Additionally, we see that the diffusion of the mass flux term and transient and inertia terms are neglected here. Based on these assumptions, Equation (1) remains unchanged, and Equations (2)–(4) reduce to the following:

$$\nabla U = 0$$

$$U \cdot \nabla\psi = 0 \tag{5}$$

$$-\nabla P + \mu_{eff}\nabla^2 U + (J \times B) = 0 \tag{6}$$

$$\rho_c(v \cdot \nabla T) = h \cdot e_{np} \tag{7}$$

The velocity distribution is to be obtained by solving the momentum equation (assuming slip conditions) which can be expressed as follows:

$$\mu_{eff}\nabla^2 U + (J \times B) = \nabla P \tag{8}$$

Since the following is valid:

$$J = \sigma(U \times B) \tag{9}$$

Then we obtain the following:

$$\nabla^2 U + \frac{\sigma B^2}{\mu_{eff}} U = \frac{1}{\mu_{eff}} \nabla P \tag{10}$$

We know that the velocity is a function of the radius; we know that the fluid flow in the cylindrical coordinates can be described as follows:

$$\nabla^2 U = \frac{d^2 U}{dr^2} + \frac{1}{r}\frac{dU}{dr} \tag{11}$$

Using this cylindrical coordinate and inserting it into equation, we obtain the following:

$$\nabla^2 U = \frac{d^2 U}{dr^2} + \frac{1}{r}\frac{dU}{dr} \tag{12}$$

$$\frac{1}{\mu_{eff}} \nabla P = \frac{d^2 U}{dr^2} + \frac{1}{r}\frac{dU}{dr} - \frac{\sigma B^2}{\mu_{eff}} U \tag{13}$$

For slip conditions, we obtain the following:

$$\begin{array}{ll} r = 0, & \frac{dU(r)}{dr} = 0 \\ r = R, & \frac{dU(r)}{dr} = -\lambda U(r) \end{array} \tag{14}$$

Also, we obtain the following:

$$\lambda = f(R, b, Kn, \sigma)$$

The solution for the differential equation described can be viewed as follows:

$$U(r) = U_{CF} + U_{PI} \tag{15}$$

To solve the complimentary part, we obtain the following:

$$\frac{d^2U(r)}{dr^2} + \frac{1}{r}\frac{dU(r)}{dr} - \frac{\sigma B^2}{\mu_{eff}}U(r) = 0 \tag{16}$$

The solution can then be described to be in the following form:

$$U_{CF} = \left(A \cdot I_0\left(\sqrt{\frac{\sigma B^2}{\mu_{eff}}}r\right) + C \cdot K_0\left(\sqrt{\frac{\sigma B^2}{\mu_{eff}}}r\right)\right) \tag{17}$$

To solve the PI (particular integral), we obtain the following:

$$\frac{1}{\mu_{eff}}\nabla P = \frac{d^2U_{PI}(r)}{dr^2} + \frac{1}{r}\frac{dU_{PI}(r)}{dr} - \frac{\sigma B^2}{\mu_{eff}}U_{PI}(r) \tag{18}$$

Then, we obtain the following:

$$\frac{1}{\mu_{eff}}\nabla P = \frac{d^2(\beta_r)}{dr^2} + \frac{1}{r}\frac{d(\beta_r)}{dr} - \frac{\sigma B^2}{\mu_{eff}}(\beta_r) \tag{19}$$

$$\beta_r = U_{PI}\text{and } U_{PI} = \frac{1}{\sigma B_l^2}\nabla P$$

Which gives the generalized solution for the velocity profile, written as follows:

$$U_{PI} = \frac{1}{\sigma B_l^2}U(r) = \left(A \cdot I_0\left(\sqrt{\frac{\sigma B^2}{\mu_{eff}}}r\right) + C \cdot K_0\left(\sqrt{\frac{\sigma B^2}{\mu_{eff}}}r\right)\right) \tag{20}$$

where $C = 0$, then $A = \frac{\nabla P}{\sigma B_l^2}\left(\frac{1}{I_0\left(\sqrt{\frac{\sigma B_l^2}{\mu_{eff}}}R_i\right) + \frac{I_1}{\lambda}\sqrt{\frac{\sigma B_l^2}{\mu_{eff}}}\cdot\left(\sqrt{\frac{\sigma B_l^2}{\mu_{eff}}}R_i\right)}\right)$

By substituting for $A$, and then $\mu_{eff} = \frac{\mu_b}{(1+aKn)}$ and $\xi = \sigma B_l^2$, an expression for the observed velocity profile can be obtained as follows:

$$U(r) = \left(\frac{1}{I_0\left(\sqrt{\frac{(1+aK_n)\xi}{\mu_b}}R_i\right) + \frac{1}{\lambda}\sqrt{\frac{(1+aK_n)\xi}{\mu_b}}I_1\left(\sqrt{\frac{(1+aK_n)\xi}{\mu_b}}R_1\right)}\right)\frac{\nabla P}{\xi}.I_0\left(\sqrt{\frac{(1+aK_n)\xi}{\mu_b}}r\right) - \frac{1}{\xi}\nabla P \tag{21}$$

By factoring out the common term $\nabla P/\xi$, we can develop a versatile model that can be applied to various boundary conditions to determine the velocity profile, written as follows:

$$U(r) = \frac{\nabla P}{\xi}\left(\frac{I_0\left(\sqrt{\frac{(1+aK_n)\xi}{\mu_b}}r\right) - \frac{1}{\xi}\nabla P}{I_0\left(\sqrt{\frac{(1+aK_n)\xi}{\mu_b}}R_i\right) + \frac{1}{\lambda}\sqrt{\frac{(1+aK_n)\xi}{\mu_b}}I_1\left(\sqrt{\frac{(1+aK_n)\xi}{\mu_b}}R_1\right)}\right) \tag{22}$$

$$U(r)=\frac{\nabla P}{\xi}\left(\frac{I_0\left(\sqrt{\frac{(1+aK_n)\xi}{\mu_b}}r\right)}{I_0\left(\sqrt{\frac{(1+aK_n)\xi}{\mu_b}}R_i\right)+\frac{1}{\lambda}\sqrt{\frac{(1+aK_n)\xi}{\mu_b}}I_1\left(\sqrt{\frac{(1+aK_n)\xi}{\mu_b}}R_1\right)}-1\right) \tag{23}$$

We must remember the fundamental boundary conditions for slip and no-slip to calculate the value for $1/\lambda$, specifically the following:

$$U\big|_{r=R_i}=0$$

$$\text{For slip conditions, we have } U\bigg|_{r=R_i}=-R_i\left(\frac{2-\sigma_v}{\sigma_v}\right)\left(\frac{Kn}{1-bKn}\right)\frac{\partial U}{\partial r}\bigg|_{r=R_i} \tag{24}$$

Then we obtain the following:

$$\frac{\partial U}{\partial r}\bigg|_{r=R_i}=-\frac{U}{R_i\left(\frac{2-\sigma_v}{\sigma_v}\right)\left(\frac{Kn}{1-bKn}\right)}=-\lambda U \tag{25}$$

$$\text{Where } \frac{1}{\lambda}=R_i\left(\frac{2-\sigma_v}{\sigma_v}\right)\left(\frac{Kn}{1-bKn}\right) \tag{26}$$

The NMF velocity distribution in the SWCNT, expressed in terms of the parameters described and the Bessel functions and can be fully expressed as follows:

$$U(r)=\frac{\nabla P}{\xi}\left(\frac{I_0\left(\sqrt{\frac{(1+aKn)\xi}{\mu_b}}r\right)}{I_0\left(\sqrt{\frac{(1+\alpha Kn)\xi}{\mu_b}}R_i\right)+R_i\left(\frac{2-\sigma_v}{\sigma_v}\right)\left(\frac{Kn}{1-bKn}\right)\sqrt{\frac{(1+aKn)\xi}{\mu_b}}.I_1\left(\sqrt{\frac{(1+aKn)\xi}{\mu_b}}R_i\right)}-1\right) \tag{27}$$

## 2.2. Model Formulation for Influence of Nanomagnetic Fluid at Nanotube Junction

To analyze the forces acting on the fluid in the transverse and axial directions, we can apply the principle of momentum balance to the control volume depicted in Figure 2.

$$q_x=\rho_f Q_1 U_1-\rho_f Q_2 U_2-\rho_f Q_3 U_3 \tag{28}$$

Assuming the discharges from the exits are the same, then we obtain the following:

$$q_x=\rho_f Q_1 U_1-2\rho_f Q_2 U_2(\because Q_2=Q_3) \tag{29}$$

$$q_x=\rho_f Q_1(U_1-U_2) \tag{30}$$

$$q_x=\rho_f A_f U_1(U_1-U_2) \tag{31}$$

$$q_x=m_f U_1^2\left(1-\frac{U_2}{U_1}\right) \tag{32}$$

Also, by making substitutions based on the relations, we obtain the following:

$$q_x=m_f(\Gamma U)^2(1-\cos\phi_n)\left(\because 2Q_2=Q_1,\frac{U_2}{U_1}=\cos\phi_n,\frac{U_i}{U}=\Gamma\right) \tag{33}$$

$$q_y=\rho_f Q_2 U_2+\rho_f Q_3 U_3 \tag{34}$$

$$q_y=\rho_f Q_2 U_1\sin\phi_n-\rho_f Q_3 U_1\sin\phi_n=0 \tag{35}$$

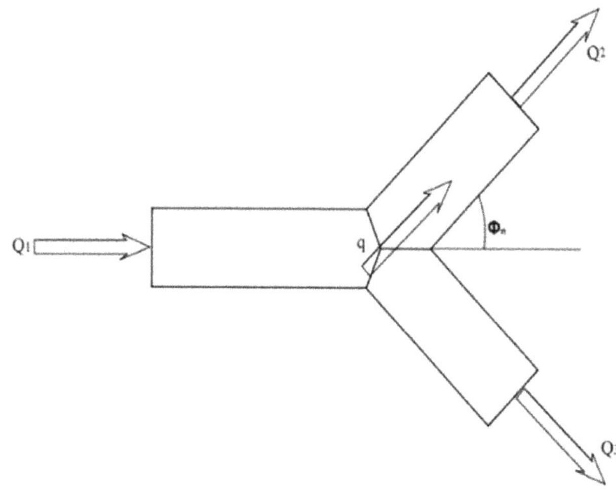


**Figure 2.** Nanofluid flows in branched nanotube.

### 2.3. Model Formulation for Vibration of Carbon-Nanotube

Considering the assumptions in Section 2 above and given a nanotube resting on an elastic foundation and a simple support, as shown in Figure 3.

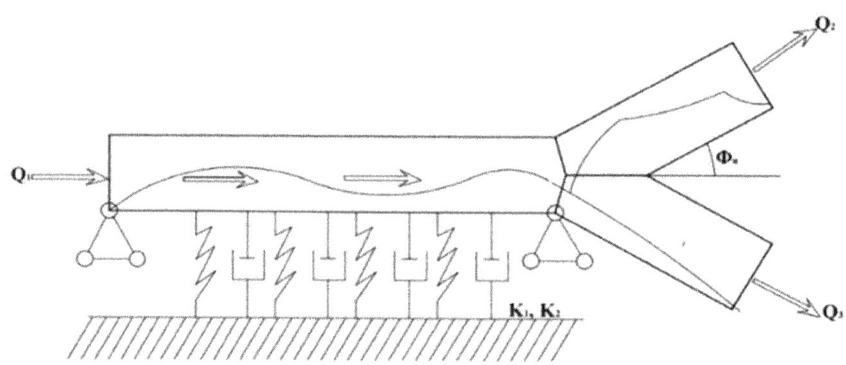


**Figure 3.** A simple supported branched SWCNT resting on an elastic foundation.

The model is therefore formulated as follows:

The deflection of the nanotube after considering surface effect can be described by Equation (36).

$$EI\frac{\partial^4 w}{\partial x^4} + E_s I_s \frac{\partial^4 w}{\partial x^4} + \left[-\frac{\partial Q}{\partial x}\right] = (e_0 a)^2 \frac{\partial^2}{\partial x^2}\left[-\frac{\partial Q}{\partial x}\right] \tag{36}$$

The acting forces include weight, magnetic force, foundation force, elastic force, surface force, pressure force, support force, and fluid flow forces.

The net force can then be calculated as follows:

$$-\frac{\partial Q}{\partial x} = q_{weight} + q_{magnetic} + q_{elastic} + q_{fluidflow} + q_{\text{support}} + q_{foundation} + q_{surface} + q_{pressure} + q_x \tag{37}$$

The terms for each of these forces are expressed as follows:

$$\text{Weight term}: q_{weight} = ma = m\frac{\partial^2 w}{\partial x^2} \quad (38a)$$

$$\text{Magnetic term}: q_{magnetic} = -\mu A H_x \frac{\partial^2 w}{\partial x^2} \quad (38b)$$

$$\text{Elastic force term}: q_{elastic} = -\eta A \xi_x \frac{\partial^2 w}{\partial x^2} \quad (38c)$$

$$\text{Surface force term}; q_{surface} = \delta\sigma_0 A \frac{\partial^2 w}{\partial x^2} \quad (38d)$$

$$\text{Pressure force term}; q_{pressure} = PA \frac{\partial^2 w}{\partial x^2} \quad (38e)$$

$$\text{Support force term}; q_{\text{support}} = \left[\frac{E}{2L}\int_0^L \left(\frac{\partial^2 w}{\partial x^2}\right) dx\right] \frac{\partial^2 w}{\partial x^2} \quad (38f)$$

$$\text{Foundation force term}; q_{winkler-foundation} = k_1 w + k_2 w^3 \quad (38g)$$

The fluid flow force comprises three constituent forces which are the centripetal force, centrifugal force, and the force due to the Coriolis effect, written as follows:

$$q_{fluidflow} = q_{centripetal} + q_{centrifugal} + q_{corolis} \quad (38h)$$

The fluid flows can thus be expressed as follows:

$$q_{fluidflow} = m_f \frac{\partial^2 w}{\partial t^2} + m_f U^2 \frac{\partial^2 w}{\partial x^2} + 2m_f U \frac{\partial^2 w}{\partial x \partial t} \quad (38i)$$

By adding the force term, we obtain the following:

$$-\frac{\partial Q}{\partial x} = q_{weight} + q_{magnetic} + q_{elastic} + q_{fluidflow} + q_{\text{support}} + q_{foundation} + q_{surface-energy} + q_x \quad (39)$$

The equation can be fully expressed as follows:

$$\begin{aligned} -\frac{\partial Q}{\partial x} &= m\frac{\partial^2 w}{\partial x^2} + \left[-\mu A H_x \frac{\partial^2 w}{\partial x^2}\right] + \left[-\eta A \xi_x \frac{\partial^2 w}{\partial x^2}\right] + m_f \frac{\partial^2 w}{\partial t^2} + m_f U^2 \frac{\partial^2 w}{\partial x^2} \\ &+ 2m_f U \frac{\partial^2 w}{\partial x \partial t} + \left[\frac{E}{2L}\int_0^L \left(\frac{\partial^2 w}{\partial x^2}\right) dx\right] \frac{\partial^2 w}{\partial x^2} + k_1 w + k_2 w^3 + \delta\sigma_0 A \frac{\partial^2 w}{\partial x^2} + PA \frac{\partial^2 w}{\partial x^2} \\ &+ m_f (\Gamma U)^2 (1 - \cos\phi_n) \end{aligned} \quad (40)$$

By factorizing Equation (40) the term above the equation reduces to the following:

$$\begin{aligned} -\frac{\partial Q}{\partial x} &= \begin{bmatrix} m - \mu A H_x - \eta A \xi_x + m_f + m_f U^2 + 2m_f U \\ + \left[\frac{E}{2L}\int_0^L \left(\frac{\partial^2 w}{\partial x^2}\right) dx\right] + \delta\sigma_0 A + PA \end{bmatrix} \frac{\partial^2 w}{\partial x^2} + k_1 w + k_2 w^3 \\ &+ m_f (\Gamma U)^2 (1 - \cos\phi_n) \end{aligned} \quad (41)$$

Inserting Equation (41) in Equation (36), we obtain the following:

$$
\begin{aligned}
& EI\frac{\partial^4 w}{\partial x^4} + E_s I_s \frac{\partial^4 w}{\partial x^4} \\
& + \left[ \begin{array}{l} \left[ \begin{array}{l} m - \mu A H_x - \eta A \xi_x + m_f + m_f U^2 + 2m_f U + \left[ \frac{E}{2L} \int_0^L \left( \frac{\partial^2 w}{\partial x^2} \right) dx \right] + \delta\sigma_0 A + PA \\ + m_f (\Gamma U)^2 (1 - \cos\phi_n) \end{array} \right. \end{array} \right] \frac{\partial^2 w}{\partial x^2} + k_1 w + k_2 w^3 \Bigg] = \\
& (e_0 a)^2 \frac{\partial^2}{\partial x^2} \left[ \begin{array}{l} \left[ m - \mu A H_x - \eta A \xi_x + m_f + m_f U^2 + 2m_f U + \left[ \frac{E}{2L} \int_0^L \left( \frac{\partial^2 w}{\partial x^2} \right) dx \right] + \delta\sigma_0 A + PA \right] \frac{\partial^2 w}{\partial x^2} \\ + k_1 w + k_2 w^3 + m_f (\Gamma U)^2 (1 - \cos\phi_n) \end{array} \right]
\end{aligned}
\tag{42}
$$

The governing equation for a branched, single-wall carbon nanotube under thermo-magnetic influence conveying nanofluid with the specified assumptions can be represented as the following:

$$
\begin{aligned}
& EI\frac{\partial^4 w}{\partial x^4} + E_s I_s \frac{\partial^4 w}{\partial x^4} \\
& + \left[ \begin{array}{l} \left[ m - \mu A H_x - \eta A \xi_x + m_f + m_f U^2 + 2m_f U + \left[ \frac{E}{2L} \int_0^L \left( \frac{\partial^2 w}{\partial x^2} \right) dx \right] + \delta\sigma_0 A \right] \frac{\partial^2 w}{\partial x^2} + \\ k_1 w + k_2 w^3 + m_f (\Gamma U)^2 (1 - \cos\phi_n) \end{array} \right] = \\
& (e_0 a)^2 \frac{\partial^2}{\partial x^2} \left[ \begin{array}{l} \left[ m - \mu A H_x - \eta A \xi_x + m_f + m_f U^2 + 2m_f U + \left[ \frac{E}{2L} \int_0^L \left( \frac{\partial^2 w}{\partial x^2} \right) dx \right] + \delta\sigma_0 A \right] \frac{\partial^2 w}{\partial x^2} + k_1 w + k_2 w^3 \\ + m_f (\Gamma U)^2 (1 - \cos\phi_n) \end{array} \right]
\end{aligned}
\tag{43}
$$

### 2.4. Galerkin Decomposition Method

Galerkin's approach is essential for converting the governing equation into an ordinary differential equation (ODE). This conversion simplifies the governing equation, making it easier to solve numerically.

As a result, the system's vibrational response may be described using the generalized coordinate $q(t)$ and the first Eigen function of a simply supported beam Φ(t) as follows:

$$w(x,t) = \Phi(x) \cdot q(t) \tag{44}$$

Inserting Equation (44) into Equation (43), we obtain the following:

$$
\left[ \begin{array}{l}
EI\frac{\partial^4(\Phi(x)\cdot q(t))}{\partial x^4} - (e_0 a)^2 \left[ \begin{array}{l} \left(m_n + m_f\right)\frac{\partial^4(\Phi(x)\cdot q(t))}{\partial x^2 \partial t^2} + \left(\mathrm{PA} - \mathrm{T} + m_f v^2 - \eta H_x^2 \mathrm{A}\right)\frac{\partial^4(\Phi(x)\cdot q(t))}{\partial x^4} \\ +2m_f v \frac{\partial^4(\Phi(x)\cdot q(t))}{\partial x^3 \partial t} + \left(k_1 + 3(\Phi(x)\cdot q(t))^2 k_2\right)\frac{\partial^2(\Phi(x)\cdot q(t))}{\partial x^2} \\ +6(\Phi(x)\cdot q(t))k_2\left[\frac{\partial(\Phi(x)\cdot q(t))}{\partial x}\right]^2 \end{array} \right] \\
+ \left[ \begin{array}{l} \left(m_n + m_f\right)\frac{\partial^2(\Phi(x)\cdot q(t))}{\partial t^2} + \left(\mathrm{PA} - \mathrm{T} + m_f v^2 - \eta H_x^2 A\right)\frac{\partial^2(\Phi(x)\cdot q(t))}{\partial x^2} + 2m_f v \frac{\partial^2(\Phi(x)\cdot q(t))}{\partial x \partial t} \\ +k_1(\Phi(x)\cdot q(t)) + k_2(\Phi(x)\cdot q(t))^3 \end{array} \right]
\end{array} \right] = 0 \tag{45}
$$

Arranging the terms in Equation (45) we obtain the following:

$$\left[\begin{array}{l} \mathrm{EI}\frac{\partial^4\Phi(x)}{\partial x^4}\cdot q(t)-(e_0a)^2\left[\begin{array}{l}\left(m_n+m_f\right)\frac{\partial^2\Phi(x)}{\partial x^2}\cdot\frac{\partial^2 q(t)}{\partial t^2}\\ +\left(\mathrm{PA}-\mathrm{T}+m_fv^2-\eta H_x^2A\right)\frac{\partial^4\Phi(x)}{\partial x^4}\cdot q(t)\\ +2m_fv\frac{\partial^3\Phi(x)}{\partial x^3}\cdot\frac{\partial q(t)}{\partial t}+k_1\frac{\partial^2\Phi(x)}{\partial x^2}\cdot q(t)\\ +3\Phi^2(x)k_2\frac{\partial^2\Phi(x)}{\partial x^2}\cdot q^3(t)+6\Phi(x)\cdot q^3(t)k_2\left[\frac{\partial\Phi(x)}{\partial x}\right]^2\end{array}\right]\\ +\left[\begin{array}{l}\left(m_n+m_f\right)\frac{\partial^2 q(t)}{\partial t^2}\cdot\Phi(x)+\left(\mathrm{PA}-\mathrm{T}+m_fv^2-\eta H_x^2A\right)\frac{\partial^2\Phi(x)}{\partial x^2}\cdot q(t)\\ +2m_fv\frac{\partial\Phi(x)}{\partial x}\cdot\frac{\partial q(t)}{\partial t}+k_1\Phi(x)\cdot q(t)+k_2\Phi^3(x)\cdot q^3(t)\end{array}\right]\end{array}\right]=0 \quad (46)$$

Rearranging Equation (46) and collecting like terms gives us the following:

$$\begin{array}{l}\left[\left(m_n+m_f\right)-(e_0a)^2\left(\left(m_n+m_f\right)\frac{\partial^2\Phi(x)}{\partial x^2}\right)\right]\frac{\partial^2 q(t)}{\partial t^2}\\ +\left[2m_fv\frac{\partial\Phi(x)}{\partial x}-(e_0a)^2\left(2m_fv\frac{\partial^3\Phi(x)}{\partial x^3}\right)\right]\frac{\partial q(t)}{\partial t}\\ +\left[\begin{array}{l}\mathrm{EI}\frac{\partial^4\Phi(x)}{\partial x^4}-(e_0a)^2\left(\begin{array}{l}\left(\mathrm{PA}-\mathrm{T}+m_fv^2-\eta H_x^2A\right)\frac{\partial^4\Phi(x)}{\partial x^4}\\ +k_1\frac{\partial^2\Phi(x)}{\partial x^2}\end{array}\right)\\ +\left(\mathrm{PA}-\mathrm{T}+m_fv^2-\eta H_x^2A\right)\frac{\partial^2\Phi(x)}{\partial x^2}+k_1\Phi(x)\end{array}\right]q(t)\\ +\left[k_2\Phi^3(x)-(e_0a)^2\left(3\Phi^2(x)k_2\frac{\partial^2\Phi(x)}{\partial x^2}+6\Phi(x)k_2\left[\frac{\partial\Phi(x)}{\partial x}\right]^2\right)\right]q^3(t)=0\end{array} \quad (47)$$

This then can be re-written as follows:

$$\begin{array}{l}\left[\left(m_n+m_f\right)-(e_0a)^2\left(\left(m_n+m_f\right)\frac{\partial^2\Phi(x)}{\partial x^2}\right)\right]\ddot{q}(t)\\ +\left[2m_fv\frac{\partial\Phi(x)}{\partial x}-(e_0a)^2\left(2m_fv\frac{\partial^3\Phi(x)}{\partial x^3}\right)\right]\dot{q}(t)\\ +\left[\begin{array}{l}\mathrm{EI}\frac{\partial^4\Phi(x)}{\partial x^4}-(e_0a)^2\left(\begin{array}{l}\left(\mathrm{PA}-\mathrm{T}+m_fv^2-\eta H_x^2A\right)\frac{\partial^4\Phi(x)}{\partial x^4}\\ +k_1\frac{\partial^2\Phi(x)}{\partial x^2}\end{array}\right)\\ +\left(\mathrm{PA}-\mathrm{T}+m_fv^2-\eta H_x^2A\right)\frac{\partial^2\Phi(x)}{\partial x^2}+k_1\Phi(x)\end{array}\right]q(t)\\ +\left[k_2\Phi^3(x)-(e_0a)^2\left(3\Phi^2(x)k_2\frac{\partial^2\Phi(x)}{\partial x^2}+6\Phi(x)k_2\left[\frac{\partial\Phi(x)}{\partial x}\right]^2\right)\right]q^3(t)=0\end{array} \quad (48)$$

Each of the terms are multiplied by the Eigen function Φ(t), where $R_i(x,t)$ can be set to be equal to the governing Equation (48).

Then, we integrate the obtained equation over the interval (0, L) and apply the stepwise approach of Galerkin's method, which yields the following:

$$\int_0^L R_i(x,t)\Phi(x)dx = 0 \tag{49}$$

Substituting Equation (48) and expanding it gives us the following:

$$\int_0^L \left[ \begin{array}{l} \left[\left(m_n + m_f\right) - (e_0 a)^2\left(\left(m_n + m_f\right)\frac{\partial^2\Phi(x)}{\partial x^2}\right)\right]\ddot{q}(t) \\ + \left[2m_f v\frac{\partial\Phi(x)}{\partial x} - (e_0 a)^2\left(2m_f v\frac{\partial^3\Phi(x)}{\partial x^3}\right)\right]\dot{q}(t) \\ + \left[ \begin{array}{l} \mathrm{EI}\frac{\partial^4\Phi(x)}{\partial x^4} - (e_0 a)^2\left(\left(\mathrm{PA} - \mathrm{T} + m_f v^2 - \eta H_x^2 \mathrm{A}\right)\frac{\partial^4\Phi(x)}{\partial x^4} + k_1\frac{\partial^2\Phi(x)}{\partial x^2}\right) \\ + \left(\mathrm{PA} - \mathrm{T} + m_f v^2 - \eta H_x^2 \mathrm{A}\right)\frac{\partial^2\Phi(x)}{\partial x^2} + k_1\Phi(x) \end{array} \right] q(t) \\ + \left[-(e_0 a)^2\left(3\Phi^2(x)k_2\frac{\partial^2\Phi(x)}{\partial x^2} + 6\Phi(x)k_2\left[\frac{\partial\Phi(x)}{\partial x}\right]^2\right) + k_2\Phi^3(x)\right]q^3(t) \end{array} \right]\Phi(x)dx = 0 \tag{50}$$

The equation above is then converted into an ODE.

$$\begin{aligned} &\int_0^L \left[-(e_0 a)^2\left(\left(m_n + m_f\right)\frac{\partial^2\Phi(x)}{\partial x^2}\right) + \left(m_n + m_f\right)\right]\ddot{q}(t)\Phi(x)dx \\ &= \left[\int_0^L \left[-(e_0 a)^2\Phi(x)\left(\left(m_n + m_f\right)\frac{\partial^2\Phi(x)}{\partial x^2}\right) + \left(m_n + m_f\right)\Phi(x)\right]dx\right]\ddot{q}(t) \end{aligned} \tag{51}$$

$$\begin{aligned} &\int_0^L \left[-(e_0 a)^2\left(2m_f v\frac{\partial^3\Phi(x)}{\partial x^3}\right) + 2m_f v\frac{\partial\Phi(x)}{\partial x}\right]\dot{q}(t)\Phi(x)dx \\ &= \left[\int_0^L \left[-(e_0 a)^2\Phi(x)\left(2m_f v\frac{\partial^3\Phi(x)}{\partial x^3}\right) + 2m_f v\Phi(x)\frac{\partial\Phi(x)}{\partial x}\right]dx\right]\dot{q}(t) \end{aligned} \tag{52}$$

$$\begin{aligned} &\int_0^L \left[ \begin{array}{l} \mathrm{EI}\frac{\partial^4\Phi(x)}{\partial x^4} - (e_0 a)^2\left(\left(\mathrm{PA} - \mathrm{T} + m_f v^2 - \eta H_x^2 \mathrm{A}\right)\frac{\partial^4\Phi(x)}{\partial x^4} + k_1\frac{\partial^2\Phi(x)}{\partial x^2}\right) \\ + \left(\mathrm{PA} - \mathrm{T} + m_f v^2 - \eta H_x^2 \mathrm{A}\right)\frac{\partial^2\Phi(x)}{\partial x^2} + k_1\Phi(x) \end{array} \right] q(t)\Phi(x)dx \\ &= \left[\int_0^L \left[ \begin{array}{l} \mathrm{EI}\Phi(x)\frac{\partial^4\Phi(x)}{\partial x^4} - (e_0 a)^2\Phi(x)\left(\left(\mathrm{PA} - \mathrm{T} + m_f v^2 - \eta H_x^2 \mathrm{A}\right)\frac{\partial^4\Phi(x)}{\partial x^4} + k_1\frac{\partial^2\Phi(x)}{\partial x^2}\right) \\ + \left(\mathrm{PA} - \mathrm{T} + m_f v^2 - \eta H_x^2 \mathrm{A}\right)\Phi(x)\frac{\partial^2\Phi(x)}{\partial x^2} + k_1\Phi^2(x) \end{array} \right] dx\right] q(x) \end{aligned} \tag{53}$$

$$\begin{aligned} &\int_0^L \left[-(e_0 a)^2\left(3\Phi^2(x)k_2\frac{\partial^2\Phi(x)}{\partial x^2} + 6\Phi(x)k_2\left[\frac{\partial\Phi(x)}{\partial x}\right]^2\right) + k_2\Phi^3(x)\right]q^3(t)\Phi(x)dx \\ &= \left[\int_0^L \left[-(e_0 a)^2\left(3\Phi^3(x)k_2\frac{\partial^2\Phi(x)}{\partial x^2} + 6\Phi^2(x)k_2\left[\frac{\partial\Phi(x)}{\partial x}\right]^2\right) + k_2\Phi^4(x)\right]dx\right]q^3(t) \end{aligned} \tag{54}$$

By rearranging and grouping like terms, the Galerkin's method helps us yield the following Duffing equation:

$$M\ddot{q}(t) + C\dot{q}(t) + K_1 q(t) + K_2 q^3(t) = 0 \tag{55}$$

where we obtain the following:

$$\begin{aligned} M &= \int_0^L \left[ -(e_0 a)^2 \Phi(x) \left( \left( m_n + m_f \right) \frac{\partial^2 \Phi(x)}{\partial x^2} \right) + \left( m_n + m_f \right) \Phi(x) \right] dx \\ &= \int_0^L \Phi(x) \left[ -(e_0 a)^2 \left( \left( m_n + m_f \right) \frac{\partial^2 \Phi(x)}{\partial x^2} \right) + \left( m_n + m_f \right) \right] dx \end{aligned} \tag{56}$$

where $C$ is given as follows:

$$\begin{aligned} C &= \int_0^L \left[ -(e_0 a)^2 \Phi(x) \left( 2 m_f v \frac{\partial^3 \Phi(x)}{\partial x^3} \right) + 2 m_f v \Phi(x) \frac{\partial \Phi(x)}{\partial x} \right] dx \\ &= \int_0^L \Phi(x) \left[ -(e_0 a)^2 \left( 2 m_f v \frac{\partial^3 \Phi(x)}{\partial x^3} \right) + 2 m_f v \frac{\partial \Phi(x)}{\partial x} \right] dx \end{aligned} \tag{57}$$

Also, the value of $K_1$ is given as follows:

$$\begin{aligned} K_1 &= \int_0^L \left[ \begin{array}{l} \mathrm{EI}\Phi(x) \frac{\partial^4 \Phi(x)}{\partial x^4} - (e_0 a)^2 \Phi(x) \left( \left( \mathrm{PA} - \mathrm{T} + m_f v^2 - \eta H_x^2 \mathrm{A} \right) \frac{\partial^4 \Phi(x)}{\partial x^4} + k_1 \frac{\partial^2 \Phi(x)}{\partial x^2} \right) \\ + \left( \mathrm{PA} - \mathrm{T} + m_f v^2 - \eta H_x^2 \mathrm{A} \right) \Phi(x) \frac{\partial^2 \Phi(x)}{\partial x^2} + k_1 \Phi^2(x) \end{array} \right] dx \\ &= \int_0^L \Phi(x) \left( \begin{array}{l} \mathrm{EI} \frac{\partial^4 \Phi(x)}{\partial x^4} - (e_0 a)^2 \left( \left( \mathrm{PA} - \mathrm{T} + m_f v^2 - \eta H_x^2 \mathrm{A} \right) \frac{\partial^4 \Phi(x)}{\partial x^4} + k_1 \frac{\partial^2 \Phi(x)}{\partial x^2} \right) \\ + \left( \mathrm{PA} - \mathrm{T} + m_f v^2 - \eta H_x^2 \mathrm{A} \right) \frac{\partial^2 \Phi(x)}{\partial x^2} + k_1 \Phi(x) \end{array} \right) dx \end{aligned} \tag{58}$$

The value of constant $K_2$ is given as follows:

$$\begin{aligned} K_2 &= \int_0^L \left[ -(e_0 a)^2 \left( 3\Phi^3(x) k_2 \frac{\partial^2 \Phi(x)}{\partial x^2} + 6\Phi^2(x) k_2 \left[ \frac{\partial \Phi(x)}{\partial x} \right]^2 \right) + k_2 \Phi^4(x) \right] dx \\ &= \int_0^L \Phi(x) \left( -(e_0 a)^2 \left( 3\Phi^2(x) k_2 \frac{\partial^2 \Phi(x)}{\partial x^2} + 6\Phi(x) k_2 \left( \frac{\partial \Phi(x)}{\partial x} \right)^2 \right) + k_2 \Phi^3(x) \right) dx \end{aligned} \tag{59}$$

To determine the natural frequency of the model the following is obtained:

$$\omega_n = \sqrt{\frac{K_1}{M}} \tag{60}$$

Substituting Equations (58) and (56) into Equation (60) gives us the following:

$$\omega_n = \sqrt{\frac{\int_0^L \Phi(x) \left( \begin{array}{l} \mathrm{EI} \frac{\partial^4 \Phi(x)}{\partial x^4} - (e_0 a)^2 \left( \left( \mathrm{PA} - \mathrm{T} + m_f v^2 - \eta H_x^2 \mathrm{A} \right) \frac{\partial^4 \Phi(x)}{\partial x^4} + k_1 \frac{\partial^2 \Phi(x)}{\partial x^2} \right) \\ + \left( \mathrm{PA} - \mathrm{T} + m_f v^2 - \eta H_x^2 \mathrm{A} \right) \frac{\partial^2 \Phi(x)}{\partial x^2} + k_1 \Phi(x) \end{array} \right) dx}{\int_0^L \Phi(x) \left[ -(e_0 a)^2 \left( \left( m_n + m_f \right) \frac{\partial^2 \Phi(x)}{\partial x^2} \right) + \left( m_n + m_f \right) \right] dx}} \tag{61}$$

## 3. Analytical Solutions to the Developed Models

Finding an exact solution to the developed equations is extremely difficult due to the inclusion of nonlinear variables. Therefore, the use of numerical techniques is being considered. These nonlinear equations can now be handled and solved using a variety of approximation or semi-analytical methods. This work uses DTM to analyze the linked nonlinear equations analytically [41].

By applying DTM, Equation (55) becomes the following:

$$\begin{aligned} & M[(k+1)(k+2)q[k+2]] + C[(k+1)q[k+1]] \\ & + K_1[q[k]] - K_2\left[\sum_{p=0}^{k}\sum_{l=0}^{p} q[l]q[p-l]q[k-p]\right] = 0 \end{aligned} \tag{62}$$

This can be re-written as follows:

$$M[(k+1)(k+2)q[k+2]] = \begin{pmatrix} K_2\left[\sum_{p=0}^{k}\sum_{l=0}^{p} q[l]q[p-l]q[k-p]\right] \\ -C[(k+1)q[k+1]] - K_1[q[k]] \end{pmatrix}$$

We make the highest coefficient of q the subject of the following formula:

$$\therefore q[k+2] = \frac{K_2\left[\sum_{p=0}^{k}\sum_{l=0}^{p} q[l]q[p-l]q[k-p]\right] - C[(k+1)q[k+1]] - K_1[q[k]]}{M(k+1)(k+2)} \tag{63}$$

The initial conditions are as follows:

We let the initial displacement $q$ be $q = \beta$ and let the initial velocity $q$ be $q = 0$.

Given that the following is valid:

$$q = q[k]$$

At $k = 0$, we obtain the following:

$$\therefore q[0] = \beta$$

Also, we obtain the following:

$$\dot{q} = (k+1)q[k+1]$$

At $k = 0$, we obtain the following:

$$\therefore \dot{q} = q[1] = 0$$

Also, we obtain the following:

$$\delta(k) = \begin{cases} 1 & k = 0 \\ 0 & k \neq 0 \end{cases}$$

Iterating the governing Equation (63) for $k = 0, 1, 2, 3,$ and 4 gives us the following value for $q[k]$:

At $k = 1$, we have the following:

$$q_2 = \frac{K_2\beta^3 - K_1\beta}{2M} \tag{64}$$

At $k = 2$, we have the following:

$$q_3 = -\frac{1}{6M}\left[\frac{C\left(K_2\beta^3 - K_1\beta\right)}{M}\right] \tag{65}$$

At $k = 3$, we have the following:

$$q_4 = \frac{1}{12M}\left[\frac{3K_2\beta^2\left(K_2\beta^3 - K_1\beta\right)}{2M} - \frac{K_1\left(K_2\beta^3 - K_1\beta\right)}{2M} + \frac{C^2\left(K_2\beta^3 - K_1\beta\right)}{2M^2}\right] \tag{66}$$

At $k = 4$, we have the following:

$$q_5 = \frac{1}{20M}\left[-\frac{K_2\beta^2C\left(K_2\beta^3-K_1\beta\right)}{2M^2} + \frac{K_1C\left(K_2\beta^3-K_1\beta\right)}{6M^2} - \frac{C\left(\frac{3K_2\beta^2\left(K_2\beta^3-K_1\beta\right)}{2M} - \frac{K_1\left(K_2\beta^3-K_1\beta\right)}{2M} + \frac{C^2\left(K_2\beta^3-K_1\beta\right)}{2M^2}\right)}{3M}\right] \tag{67}$$

At $k = 5$, we have the following:

$$q_6 = \frac{1}{30M}\left[\begin{array}{l} K_2\left(\frac{\beta^2\left(\frac{3K_2\beta^2(K_2\beta^3-K_1\beta)}{2M} - \frac{K_1(K_2\beta^3-K_1\beta)}{2M} + \frac{C^2(K_2\beta^3-K_1\beta)}{2M^2}\right)}{4M} + \frac{3\beta(K_2\beta^3-K_1\beta)^2}{4M^2}\right) \\ -\frac{K_1\left(\frac{3K_2\beta^2\left(K_2\beta^3-K_1\beta\right)}{2M} - \frac{K_1\left(K_2\beta^3-K_1\beta\right)}{2M} + \frac{C^2\left(K_2\beta^3-K_1\beta\right)}{2M^2}\right)}{12M} \\ -\frac{C\left(-\frac{K_2\beta^2C\left(K_2\beta^3-K_1\beta\right)}{2M^2} + \frac{K_1C\left(K_2\beta^3-K_1\beta\right)}{6M^2} - \frac{C\left(\frac{3K_2\beta^2(K_2\beta^3-K_1\beta)}{2M} - \frac{K_1(K_2\beta^3-K_1\beta)}{2M} + \frac{C^2(K_2\beta^3-K_1\beta)}{2M^2}\right)}{3M}\right)}{4M} \end{array}\right] \tag{68}$$

At $k = 6$, we have the following:

$$q_7 = -\frac{1}{42M}\left[\begin{array}{l} K_2\left(-\frac{1}{20M}\left(3\beta^2\left(-\frac{K_2\beta^2C\left(K_2\beta^3-K_1\beta\right)}{2M^2} + \frac{K_1C\left(K_2\beta^3-K_1\beta\right)}{6M^2} - \frac{C\left(\frac{3K_2\beta^2\left(K_2\beta^3-K_1\beta\right)}{2M} - \frac{K_1\left(K_2\beta^3-K_1\beta\right)}{2M} + \frac{C^2\left(K_2\beta^3-K_1\beta\right)}{2M^2}\right)}{3M}\right)\right) - \frac{\beta C\left(K_2\beta^3-K_1\beta\right)^2}{2M^3}\right) \\ -\frac{1}{20M}\left(K_1\left(-\frac{K_2\beta^2C\left(K_2\beta^3-K_1\beta\right)}{2M^2} + \frac{K_1C\left(K_2\beta^3-K_1\beta\right)}{6M^2} - \frac{C\left(\frac{3K_2\beta^2\left(K_2\beta^3-K_1\beta\right)}{2M} - \frac{K_1\left(K_2\beta^3-K_1\beta\right)}{2M} + \frac{C^2\left(K_2\beta^3-K_1\beta\right)}{2M^2}\right)}{3M}\right)\right) \\ -\frac{1}{5M}\left(C\left(\begin{array}{l} K_2\left(\frac{\beta^2\left(\frac{3K_2\beta^2\left(K_2\beta^3-K_1\beta\right)}{2M} - \frac{K_1\left(K_2\beta^3-K_1\beta\right)}{2M} + \frac{C^2\left(K_2\beta^3-K_1\beta\right)}{2M^2}\right)}{4M} + \frac{3\beta\left(K_2\beta^3+K_1\beta\right)^2}{4M^2}\right) \\ -\frac{K_1\left(\frac{3K_2\beta^2\left(K_2\beta^3-K_1\beta\right)}{2M} - \frac{K_1\left(K_2\beta^3-K_1\beta\right)}{2M} + \frac{C^2\left(K_2\beta^3-K_1\beta\right)}{2M^2}\right)}{12M} \\ -\frac{1}{4M}\left(C\left(-\frac{K_2\beta^2C\left(K_2\beta^3-K_1\beta\right)}{2M^2} + \frac{K_1C\left(K_2\beta^3-K_1\beta\right)}{6M^2} - \frac{C\left(\frac{3K_2\beta^2\left(K_2\beta^3-K_1\beta\right)}{2M} - \frac{K_1\left(K_2\beta^3-K_1\beta\right)}{2M} + \frac{C^2\left(K_2\beta^3-K_1\beta\right)}{2M^2}\right)}{3M}\right)\right) \end{array}\right)\right) \end{array}\right] \tag{69}$$

The lateral displacement of the CNT is given as follows:

$$q[t] = \sum_{i=0}^{k} q_i t^i = q_0 + q_1 t + q_2 t^2 + q_3 t^3 + q_4 t^4 + \ldots + q_k t^k \tag{70}$$

For $k = 5$, we have the following:

$$q[t] = \sum_{i=0}^{5} q_i t^i = q_0 + q_1 t + q_2 t^2 + q_3 t^3 + q_4 t^4 + q_5 t^5 \tag{71}$$

Therefore, we have the following:

$$\begin{gathered} q[t] = \beta - \frac{(K_2\beta^3 - K_1\beta)t^2}{2M} - \frac{C(K_2\beta^3 - K_1\beta)t^3}{6M^2} \\ + \frac{\left(\frac{3K_2\beta^2(K_2\beta^3-K_1\beta)}{2M} - \frac{K_1(K_2\beta^3-K_1\beta)}{2M} + \frac{C^2(K_2\beta^3-K_1\beta)}{2M^2}\right)t^4}{12M} \\ + \frac{\left(-\frac{K_2\beta^2C(K_2\beta^3-K_1\beta)}{2M^2} + \frac{K_1C(K_2\beta^3-K_1\beta)}{6M^2} - \frac{C\left(\frac{3K_2\beta^2(K_2\beta^3-K_1\beta)}{2M} - \frac{K_1(K_2\beta^3-K_1\beta)}{2M} + \frac{C^2(K_2\beta^3-K_1\beta)}{2M^2}\right)}{3M}\right)t^5}{20M} \end{gathered} \tag{72}$$

Equation (72) is the DTM solution of the Duffing equation. The displacement equation $w(x,t)$ is then calculated as follows:

Recalling that the following is valid:

$$w(x,t) = \Phi(x) \cdot q(t) \tag{73}$$

Substituting the value of $q(t)$ in Equation (73) gives us the following:

$$w(x,t) = \left[\begin{gathered} \beta - \frac{(K_2\beta^3 - K_1\beta)t^2}{2M} - \frac{C(K_2\beta^3 - K_1\beta)t^3}{6M^2} \\ + \frac{\left(\frac{3K_2\beta^2(K_2\beta^3-K_1\beta)}{2M} - \frac{K_1(K_2\beta^3-K_1\beta)}{2M} + \frac{C^2(K_2\beta^3-K_1\beta)}{2M^2}\right)t^4}{12M} \\ + \frac{\left(-\frac{K_2\beta^2C(K_2\beta^3-K_1\beta)}{2M^2} + \frac{K_1C(K_2\beta^3-K_1\beta)}{6M^2} - \frac{C\left(\frac{3K_2\beta^2(K_2\beta^3-K_1\beta)}{2M} - \frac{K_1(K_2\beta^3-K_1\beta)}{2M} + \frac{C^2(K_2\beta^3-K_1\beta)}{2M^2}\right)}{3M}\right)t^5}{20M} \end{gathered}\right]\Phi(x) \tag{74}$$

The displacement Equation (74) is solved from the boundary and initial conditions of the CNT.

In this project, the CNT is subjected to a simple supported beam, written as follows:

$$\Phi(x) = \sin\left(\frac{n\pi}{L}x\right) \tag{75}$$

where initial conditions are given as follows:

$$\begin{gathered} w(x,0) = \beta \\ \dot{w}(x,0) = 0 \end{gathered}$$

where the boundary conditions are given as follows:

$$w(0,t) = w''(0,t) = 0$$
$$w(L,t) = w''(L,t) = 0$$

Substituting Equation (75) into Equation (73) we have the following:

$$K_2 = \frac{\begin{array}{l} 2\left(6(e_0a)^2n^2\left(k_2+\frac{1}{2}\right)\pi^2 + k_2L^2\right)\sin(n\pi)\cos(n\pi)^3 - \\ 5\left(\frac{6(e_0a)^2\left(k_2+\frac{5}{2}\right)n^2\pi^2}{5} + k_2L^2\right)\sin(n\pi)\cos(n\pi) \\ +3\pi n\left(-2(e_0a)^2\left(k_2-\frac{3}{2}\right)n^2\pi^2 + k_2L^2\right)\end{array}}{8Ln\pi}$$

$$w(x,t) = \left[\begin{array}{l} \beta - \frac{(K_2\beta^3-K_1\beta)t^2}{2M} - \frac{C(K_2\beta^3-K_1\beta)t^3}{6M^2} \\ + \frac{\left(\frac{3K_2\beta^2(K_2\beta^3-K_1\beta)}{2M} - \frac{K_1(K_2\beta^3-K_1\beta)}{2M} + \frac{C^2(K_2\beta^3-K_1\beta)}{2M^2}\right)t^4}{12M} \\ + \frac{\left(\begin{array}{l} -\frac{K_2\beta^2C(K_2\beta^3-K_1\beta)}{2M^2} + \frac{K_1C(K_2\beta^3-K_1\beta)}{6M^2} \\ -\frac{C\left(\frac{3K_2\beta^2(K_2\beta^3-K_1\beta)}{2M} - \frac{K_1(K_2\beta^3-K_1\beta)}{2M} + \frac{C^2(K_2\beta^3-K_1\beta)}{2M^2}\right)}{3M}\end{array}\right)t^5}{20M} \end{array}\right]\sin\left(\frac{n\pi}{L}x\right) \tag{76}$$

Therefore, the constants $M$, $C$, $K_1$, and $K_2$ are obtained as follows:

$$M = -\frac{\left(\left(\frac{\pi^2\sin(\pi n)a^2n^2}{2} + L^2\right)\cos(\pi n) - \frac{\pi^3a^2n^3}{2} - L^2\right)\left(m_n + m_f\right)}{\pi nL} \tag{77}$$

$$C = \frac{\sin(\pi n)^2 v m_f\left(\pi^2a^2n^2 + L^2\right)}{L^2} \tag{78}$$

$$K_1 = \frac{(-\cos(n\pi)\sin(n\pi) + n\pi)\left(\begin{array}{l} n^4\left(\left((H^2\eta - P)A - m_fv^2 + T\right)e_0a^2 + EI\right)\pi^4 \\ +\left(e_0a^2k_1 + (H^2\eta - P)A - m_fv^2 + T\right)L^2n^2\pi^2 + k_1L^4\end{array}\right)}{2L^3n\pi} \tag{79}$$

$$K_2 = \frac{\begin{array}{l} 2\left(6(e_0a)^2n^2\left(k_2+\frac{1}{2}\right)\pi^2 + k_2L^2\right)\sin(n\pi)\cos(n\pi)^3 - \\ 5\left(\frac{6(e_0a)^2\left(k_2+\frac{5}{2}\right)n^2\pi^2}{5} + k_2L^2\right)\sin(n\pi)\cos(n\pi) \\ +3\pi n\left(-2(e_0a)^2\left(k_2-\frac{3}{2}\right)n^2\pi^2 + k_2L^2\right)\end{array}}{8Ln\pi} \tag{80}$$

Then the natural frequency is solved and is obtained as follows:

$$\omega_n = \sqrt{\frac{\frac{(-\cos(n\pi)\sin(n\pi)+n\pi)\left(\begin{array}{c} n^4\left(\left((H^2\eta - P)A - m_f v^2 + T\right)e_0 a^2 + EI\right)\pi^4 \\ +\left(e_0 a^2 k_1 + (H^2\eta - P)A - m_f v^2 + T\right)L^2 n^2 \pi^2 + k_1 L^4 \end{array}\right)}{2L^3 n\pi}}{-\frac{(m_n+m_f)\left(\left(\frac{\sin(n\pi)(e_0 a)^2 n^2\pi^2}{2}+L^2\right)\cos(n\pi)-\frac{(e_0 a)^2 n^3\pi^3}{2}-L^2\right)}{Ln\pi}}} \tag{81}$$

### 3.1. *After Treatment Techniques*

In this work, the flow-induced damped vibration has a truncated series that is periodic only in a restricted domain. After treatment techniques are applied to treat the solution to enable it to capture a wider range of the independent variable during simulation.

#### 3.1.1. Cosine-After Treatment Technique (CAT Technique)

This is used for DTM transformation of a dynamic equation with even power. We consider the following series:

$$\varphi_N(t) = \sum_{k=0}^{N} q_k t^k \tag{82}$$

Expressing in even power terms gives us the following:

$$\varphi_N(t) = \sum_{k=0}^{N} q_{2k} t^{2k} \tag{83}$$

Which is subjected to the following:

$$\forall k = 0, 1, 2, \ldots, \frac{N}{2} - 1 \tag{84}$$

The equivalent Cosine's equation is then given as follows:

$$\varphi_N(t) = \sum_{k=0}^{n} \lambda_k \cos(\theta_k t) \tag{85}$$

Then, using the Fourier series expansion, it becomes the following:

$$\varphi_N(t) = \sum_{k=1}^{n} \lambda_k - \left(\sum_{k=1}^{n} \lambda_k \theta_k^2\right)\frac{t^2}{2!} + \left(\sum_{k=1}^{n} \lambda_k \theta_k^4\right)\frac{t^4}{4!} - \left(\sum_{k=1}^{n} \lambda_k \theta_k^6\right)\frac{t^6}{6!} + \left(\sum_{k=1}^{n} \lambda_k \theta_k^8\right)\frac{t^8}{8!} - \cdots \tag{86}$$

The series solution of Equation (85) is given as follows:

$$\varphi_N(t) = q_0 + q_2 t^2 + q_4 t^4 + q_6 t^6 + \cdots + q_N t^N \tag{87}$$

By comparing both solutions, we have the following:

$$q_0 + q_2 t^2 + q_4 t^4 + q_6 t^6 + \cdots + q_N t^N = \sum_{k=1}^{n} \lambda_k - \left(\sum_{k=1}^{n} \lambda_k \theta_k^2\right)\frac{t^2}{2!} + \left(\sum_{k=1}^{n} \lambda_k \theta_k^4\right)\frac{t^4}{4!} - \left(\sum_{k=1}^{n} \lambda_k \theta_k^6\right)\frac{t^6}{6!} + \cdots \tag{88}$$

By comparing coefficients, we have the following:

$$\begin{aligned} t^0 &: \sum_{k=1}^{n} \lambda_k = q_0 \\ t^2 &: \sum_{k=1}^{n} \lambda_k \theta^2 = -2! q_2 \\ t^4 &: \sum_{k=1}^{n} \lambda_k \theta^4 = 4! q_4 \\ t^6 &: \sum_{k=1}^{n} \lambda_k \theta^6 = -6! q_6 \end{aligned} \quad (89)$$

An approximation of the expression in the form of a Cos function is given in place of these terms.

$$\varphi_N(t) = \sum_{k=1}^{2} \lambda_k \cos(\theta_k t) = \lambda_1 \cos(\theta_1 t) + \lambda_2 \cos(\theta_2 t) \quad (90)$$

Applying the technique/approach to Equation (85) gives us the following:

$$\begin{aligned} t^0 &: \lambda_1 + \lambda_2 = q_0 \\ t^2 &: \lambda_1 {\theta_1}^2 + \lambda_2 {\theta_2}^2 = -2! q_2 \\ t^4 &: \lambda_1 {\theta_1}^4 + \lambda_2 {\theta_2}^4 = 4! q_4 \\ t^6 &: \lambda_1 {\theta_1}^6 + \lambda_2 {\theta_2}^6 = -6! q_6 \end{aligned} \quad (91)$$

The four unknowns $\lambda_1$, $\theta_1$, $\lambda_2$, and $\theta_2$ were found numerically.

#### 3.1.2. Sine-After Treatment Technique (SAT Technique)

Similarly, we can consider an Nth series given as follows:

$$\varphi_N(t) = \sum_{k=0}^{N} q_k t^k \quad (92)$$

Writing in odd power gives us the following:

$$\varphi_N(t) = \sum_{k=1}^{N} q_{1k} t^{1k} \quad (93)$$

Which is subject to the following:

$$\forall k = 0, 1, 2, \ldots, \frac{N}{2} - 1 \quad (94)$$

The equivalent Sine equation is then given as follows:

$$\varphi_N(t) = \sum_{k=1}^{n} \mu_k \sin(\psi_k t) \quad (95)$$

Then using Fourier series expansion, it becomes the following:

$$\varphi_N(t) = \left( \sum_{k=1}^{n} \mu_k {\psi_k}^1 \right) \frac{t^1}{1} - \left( \sum_{k=1}^{n} \mu_k {\psi_k}^3 \right) \frac{t^3}{3!} + \left( \sum_{k=1}^{n} \mu_k {\psi_k}^5 \right) \frac{t^5}{5!} - \left( \sum_{k=1}^{n} \mu_k {\psi_k}^7 \right) \frac{t^7}{7!} + \left( \sum_{k=1}^{n} \mu_k {\psi_k}^9 \right) \frac{t^9}{9!} \cdots \quad (96)$$

The series solution of Equation (95) is given as follows:

$$\varphi_N(t) = q_1 t^1 + q_3 t^3 + q_5 t^5 + q_7 t^7 + \cdots + q_N t^N \quad (97)$$

By comparing both solutions, we have the following:

$$q_1t^1 + q_3t^3 + q_5t^5 + q_7t^7 + \cdots + q_Nt^N = \left(\sum_{k=1}^{n} \mu_k\psi_k{}^1\right)\frac{t^1}{1} - \left(\sum_{k=1}^{n} \mu_k\psi_k{}^3\right)\frac{t^3}{3!} + \left(\sum_{k=1}^{n} \mu_k\psi_k{}^5\right)\frac{t^5}{5!} - \cdots \tag{98}$$

By comparing coefficients, we have the following:

$$\begin{aligned} t^1 &: \sum_{k=1}^{n} \mu_k\psi_k^1 = q_1 \\ t^3 &: \sum_{k=1}^{n} \mu_k\psi_k^3 = -3!q_3 \\ t^5 &: \sum_{k=1}^{n} \mu_k\psi_k^5 = 5!q_5 \\ t^7 &: \sum_{k=1}^{n} \mu_k\psi_k^7 = -7!q_7 \end{aligned} \tag{99}$$

An approximation of the expression in the form of a Sine function is given in place of these terms.

$$\varphi_N(t) = \sum_{k=1}^{2} \mu_k \sin(\psi_k t) = \mu_1 \sin(\psi_1 t) + \mu_2 \sin(\psi_2 t) \tag{100}$$

Applying the technique/approach to Equation (95) gives us the following:

$$\begin{aligned} t^1 &: \mu_1\psi_1^1 + \mu_2\psi_2^1 = q_1 \\ t^3 &: \mu_1\psi_1^3 + \mu_2\psi_2^3 = -3!q_3 \\ t^5 &: \mu_1\psi_1^5 + \mu_2\psi_2^5 = 5!q_5 \\ t^7 &: \mu_1\psi_1^7 + \mu_2\psi_2^7 = -7!q_7 \end{aligned} \tag{101}$$

Here the unknowns are solved numerically.

### *3.2. SAT and CAT Combinations*

The variables are present in the truncated series solution because of the damping that is available in the duffing equation.

$$\varphi(t) = \lambda_1 \cos(\theta_1 t) + \lambda_2 \cos(\theta_2 t) + \mu_1 \sin(\psi_1 t) + \mu_2 \sin(\psi_2 t) \tag{102}$$

This is then solved by applying the solutions obtained from CAT and SAT techniques using MATLAB software.

Therefore, we then obtain the displacement $w(x,t)$ by combining the special term obtained from the boundary condition and the temporal term obtained from after treatment.

$$w(x,t) = \Phi(x) \cdot \varphi(t) \tag{103}$$

Which is re-written as follows and is solved with the MATLAB software to obtain the displacement over a given period.

$$w(x,t) = \sin\left(\tfrac{n\pi}{L}x\right) \cdot \lambda_1 \cos(\theta_1 t) + \lambda_2 \cos(\theta_2 t) + \mu_1 \sin(\psi_1 t) + \mu_2 \sin(\psi_2 t) \tag{104}$$

### 3.3. Simulation and Data Collection

In simulating the vibrational response of a carbon nanotube in a thermal–magnetic environment, it was assumed that the surface of the tube was continuous and there was no discontinuity at any point across the entire surface of the tube. This simulation approach includes the following:

- Modeling;
- Engineering data;
- Meshing;
- Magnetostatic analysis;
- Static structural analysis;
- Modal analysis;
- Transient structural analysis;
- Data collection.

#### 3.3.1. Modeling

The geometry of the carbon nanotube (CNT) was modeled using Ansys SpaceClaim (ANSYS 2023 R1)). It provides an avenue to model structures on the nano-level. All nanotubes are modeled with a diameter of 30 nm and with a thickness of 1 nm. The geometric properties are presented in Table 1.

**Table 1.** Geometrical properties of nanotubes [8].

| S/N | Downstream Angle (ϕ deg) | Horizontal Length [nm] | Slant Length [nm] | Area [nm$^2$] |
|---|---|---|---|---|
| 1 | 0 | 200 | - | 18,849.5559 |
| 2 | 15 | 100 | 83.43 | 21,467.0328 |
| 3 | 30 | 100 | 80.00 | 22,463.1906 |
| 4 | 45 | 100 | 70.71 | 21,107.2913 |
| 5 | 60 | 100 | 80.00 | 22,945.5640 |
| 6 | 75 | 100 | 80.00 | 22,882.0359 |
| 7 | 90 | 100 | 80.00 | 24,023.8767 |

The geometry of the carbon nanotubes for different orientations were modeled on Ansys Space Claim and the following dimensions were assigned:

#### 3.3.2. Engineering Data

The material property of carbon nanotubes was represented by the creation of a new material. The material's properties were obtained by experiments and theoretical work. The electrical properties were obtained. These properties are presented in Table 2.

**Table 2.** Properties of nanotubes [29].

| S/N | Property | Values |
|---|---|---|
| 1 | Density | 1330 kg/m$^3$ |
| 2 | Isotropic Secant Coefficient of Expansion | $-1.5$ C$^{-1}$ |
| 3 | Melting Temperature | 3550 C |
| 4 | Young's Moduli | 3.6 TPa |
| 5 | Poisson's Ratio | 0.273 |
| 6 | Bulk's moduli | 2.6432 TPa |
| 7 | Shear's Moduli | 1.414 TPa |
| 8 | Strength Coefficient | 63 GPa |

**Table 2.** *Cont.*

| S/N | Property | Values |
|---|---|---|
| 9 | Strength Exponent | −0.18 |
| 10 | Ductility Coefficient | 0.05 |
| 11 | Ductility Exponent | −0.5 |
| 12 | Cyclic Strength Coefficient | 63 GPa |
| 13 | Cyclic Strength Hardening Exponent | 0.15 |
| 14 | Tensile Yield Strength | 630 GPa |
| 15 | Compressive Yield Strength | 150 GPa |
| 16 | Tensile Ultimate Strength | 93 GPa |
| 17 | Compressive Ultimate Strength | 112 GPa |
| 18 | Isotropic Relative Permeability | 1.05 |
| 19 | Isotropic Resistivity | $3 \times 10^{-7}$ Ωm |

#### 3.3.3. Meshing

The meshing process in ANSYS typically employs the FEM to discretize the CNT model into smaller solvable elements. A fine mesh size of 0.04 nm is used to capture the detailed behavior of the CNT accurately. The choice of mesh size is very critical as it affects the accuracy of the simulation results, and thus grid independence tests were performed to ascertain the validity of the simulation. A finer mesh provides more precise results but at the cost of increased computational resources. The properties of the mesh are further described in Table 3 below:

**Table 3.** Mesh properties.

| S/N | Property | Value |
|---|---|---|
| 1 | Mesh Type | Adaptive Sizing |
| 2 | Element Quality | 0.95 |
| 3 | Element Order | Quadratic |
| 4 | Transition Ratio | 0.272 |
| 5 | Growth Rate | 1.2 |
| 6 | Number of Element | 8905 |
| 7 | Number of Nodes | 60,595 |
| 8 | Bounding Box Diagonal | $2.0445 \times 10^{-4}$ mm |
| 9 | Average Surface Area | $9.1458 \times 10^{-9}$ mm$^2$ |
| 10 | Minimum Edge Length | $8.7965 \times 10^{-5}$ mm |

#### 3.3.4. Mode and Mode Shape Analysis

Mode analysis is conducted with remote displacement to represent the pin support of the nanotube. This analysis determines the fundamental frequency per mode shape of CNT. The mode shapes and fundamental frequencies are essential for understanding the vibrational characteristics of the CNT. The vibrational frequency for 10 modes was obtained with the corresponding deformation for each of the modes shaped. In the analysis of random vibration, these parameters are key to determining the PSD values, which help in describing the variation of vibrational energy. The setup for the modal analysis is explained as follows:

#### 3.3.5. Transient Structural Analysis

The transient structural analysis is used to simulate the overall effect of varying forces on the nanotube under elastic conditions. An elastic support with a stiffness constant of 0.0000003 N/mm$^2$.

Cylindrical support is imposed on the tube. The magnetostatic force obtained from the magnetostatic analysis is imported using Ansys MAPDL scripts to account for the effect of magnetism. The transient structural analysis was run for 2 steps within 1 s. The initial time step was set at 0.01 s, with the maximum and minimum time steps at 0.01 s and 0.001 s, respectively.

#### 3.3.6. Grid Independence Test

To make sure that the mesh size has no discernible impact on the simulation results, a grid independence test is conducted. This test involves changing the parameters and comparing the outcomes to ensure consistency. To ensure that the mesh size selected yields dependable results without needless computational overhead, the grid independence test data validates the viability and accuracy of the results for the straight CNTs.

#### 3.3.7. Data Collection

The results obtained from the simulations were stored in formats compatible with the subsequent machine learning stages. Specifically, image outputs, such as mode shapes, stress contour maps, and bifurcation diagrams, were saved in JPEG and PNG formats. Numerical data, including displacement magnitudes, natural frequencies, and other feature values, were exported and stored in CSV (comma-separated values) format to facilitate preprocessing, feature extraction, and training of the machine learning models.

The collected data include mode shapes, natural frequencies, stress distributions, and displacement patterns, which are crucial for a comprehensive understanding of the vibrational behavior of CNTs.

The dataset shape of the dataset is 121 rows with 36 columns representing various physical measurements and conditions.

#### 3.3.8. Data Preparation

Before model building, the data was carefully preprocessed. Missing values were detected and processed using linear interpolation in a bid to maintain dataset integrity without bias. Statistical tests were used to detect outliers, which were corrected or removed based on physical reasonableness. Feature variables were scaled using the StandardScaler method in a way that each variable's mean was zero and its standard deviation was one, thus boosting model convergence and performance.

The data preparation process for building the machine learning models is as follows:

Data Cleaning: The dataset was inspected for null values, outliers, and inconsistencies. An interpolation technique was used to fill in any missing data, ensuring that the dataset was complete and ready for modeling. The columns were renamed to simpler, more descriptive names, facilitating easier analysis and interpretation.

Exploratory Data Analysis (EDA): Exploratory data analysis was performed to understand the data. Histograms are plotted for each of the features to understand the distribution of the features. About 68% of the features depict a Gaussian distribution (normal distribution). Also, the correlation plot shows that most of the features are correlated. However, the stability frequency and the mass coefficient are highly correlated.

Scaling and Normalization: The feature distributions were analyzed and found to be approximately Gaussian for about 68% of the features, which justified the application of StandardScaler to normalize the dataset effectively before model training. Features were standardized using StandardScaler to ensure that all input variables had similar scales, improving the performance of machine learning algorithms.

Data Splitting: After preprocessing, the dataset was randomly shuffled and split into training and testing sets in a 4:1 ratio (80% training, 20% testing). This split ensured that the models could be trained on most of the data at hand while being validated on an unseen subset to evaluate generalizability. No data leakage occurred between training and test sets. These preprocessing steps were implemented using Python libraries, including Pandas (version 2.1.4) and Scikit-learn (version 1.4.2). Being a supervised learning process, the key input parameters are shown in Table 4.

**Table 4.** Summary of key input parameters used in simulation.

| Parameter | Symbol | Value Range | Units |
|---|---|---|---|
| Magnetic Field Strength | B | 50–150 | mT |
| Axial Flow Velocity | U | 0–300 | m/s |
| Nanotube Diameter | D | 30 | nm |
| Nanotube Wall Thickness | t | 1 | nm |
| Young's Modulus | E | 3.6 | TPa |
| Density of Nanotube Material | P | 1330 | kg/m$^3$ |
| Magnetic Permeability | M | 1.05 | - |
| Winkler Foundation Stiffness | $k_w$ | $3 \times 10^{-7}$ | N/mm$^2$ |
| Branching Angle | Φ | 0–90° | Degrees |
| Thermal Environment Temperature | T | 300–400 | K |

### 3.4. Model Building

The model-building process is presented in Figure 4. Four different machine learning models were chosen to predict the displacement and frequency of each mode. These are XGBoost, CATBoost, Random Forest, and Neural Network.

XGBoost: The model is configured with hyperparameters such as learning rate, max depth, and number of estimators. These parameters control the complexity and learning process of the model. The model is trained using the training dataset. Common to the XGBoost algorithm is the utilization of gradient boosting techniques to minimize prediction error, which enhance the possibility of generating more accurate results.

Mathematically, we can observe the objective of the XGBoost algorithm as that which would always try to reduce the loss function and a regularization term.

$$\mathcal{L}(y,\hat{y}) = \sum_{i=1}^{n} L_a(y_i,\hat{y}_i) + \Omega_{XGB}(\mathbf{F}) \tag{105}$$

The regularization term which is a composition of L1 and L2 is given by the following:

$$\Omega_{XGB}(\mathbf{F}) = \gamma_L T_n + \frac{1}{2}\lambda_j \sum_{i=1}^{T}\sum_{j=1}^{d} (w_{ij})^2 \tag{106}$$

Due to the iterative nature of the XGBoost algorithm, the i-th instance of the predicted value can be given as follows:

$$\hat{y}_i^{(t)} = \hat{y}_i^{(t-1)} + f_t(x_i) \tag{107}$$

CATBoost: CATBoost is set up with parameters like XGBoost, including learning rate, depth, and number of iterations. CATBoost is particularly effective with categorical features, but here we focus on numerical features. The model is trained with the training data, leveraging its boosting capabilities to handle overfitting and improve prediction accuracy.

$$\mathcal{L}(y,\hat{y}) = \sum_{i=1}^{n} L(y_i,\hat{y}_i) + \Omega_{CAT}(\mathbf{F}) \tag{108}$$

The regularization term is like that of XGBoost, and as such, can be represented by Equation (106). While the loss function is similar, it is worth noting that the way categorical features are handled by each algorithm differs.

Random Forest: Random Forest is configured with parameters such as the number of trees and maximum depth. This model creates an ensemble of decision trees to make predictions. This model is trained by aggregating the predictions from multiple decision trees to reduce variance and improve robustness.

Where a Random Forest can be denoted as $F$, the predictions for a new instance may be defined as follows:

$$F(x) = \frac{1}{n_i} \sum_{n_i=1}^{n_i} T_b(x) \quad (109)$$

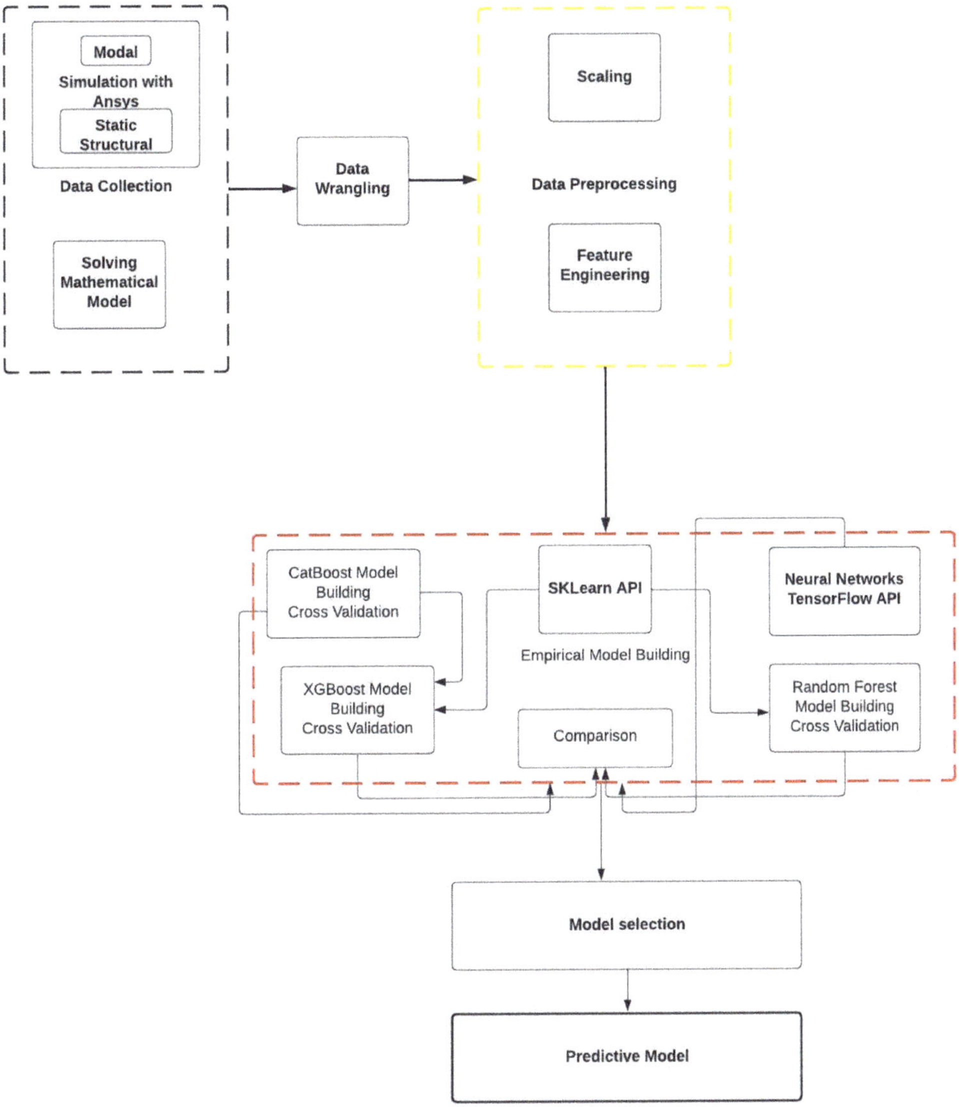


**Figure 4.** Model building flow chart.

Neural Network: The neural network architecture includes several dense layers with activation functions such as ReLU. Dropout layers are used to prevent overfitting. The model is trained with a specified number of epochs and a validation split to monitor performance. Early stopping is employed to halt training when no further improvement is observed, preventing overfitting.

Evaluation: Performance of each model is evaluated using metrics such as Mean Squared Error (MSE) and $R^2$. These metrics provide insight into how well the model predictions match the actual values. For models like XGBoost, CATBoost, and Random Forest, feature importance is computed to understand which features contribute the most to the predictions. This helps in identifying key factors affecting vibration modes. Therefore, we then obtain the displacement w(x,t) by combining the special term obtained from the boundary condition and the temporal term obtained from after treatment.

## 4. Results and Discussion

### 4.1. Results from Mathematical Modeling

The following outcomes are attained by focusing on the problem's physics and carrying out sufficient parametric research.

#### 4.1.1. Modal Numbers' Impact on Nanotube's Mode Shapes

Figure 5 shows how the boundary conditions and mode number affect the nanotube's nonlinear deformation patterns. For the first five mode configurations, this graphic displays the deflection of the nanotube along its dimensionless length. We can see from this image that for a given beam length, the stability of the nanotube decreases as the mode number upsurges. This behavior is explained by the trial function's reliance on the mode number.

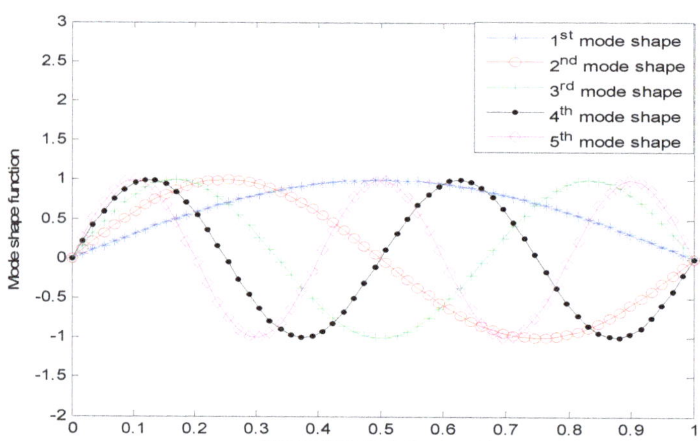


**Figure 5.** Modal number's impacts on the nanotube's mode shapes for pinned-pinned supports.

#### 4.1.2. Branch Angles' Impacts on Stability

The relationship between dimensionless frequency and dimensionless flow velocity is shown to be influenced by the downstream angle, which establishes the end shape of the nanotube (Figures 6 and 7). With increasing velocity, the frequency first falls parabolically for both linear and nonlinear foundations. This suggests that within this range, the nanotube stays stable. Nevertheless, the nanotube eventually reaches a critical dimensionless velocity, after which bifurcation makes it unstable. This suggests that the nanotube is more stable under nonlinear conditions. Furthermore, the region of bifurcation is shorter for the nonlinear analysis, meaning that the nanotube regains stability more quickly after becoming unstable. As the flow velocity continues to increase beyond the critical velocity, the system remains unstable with zero frequency. Eventually, another critical velocity is reached where the system temporarily stabilizes. However, further increases in velocity can lead to divergence and potentially flutter.

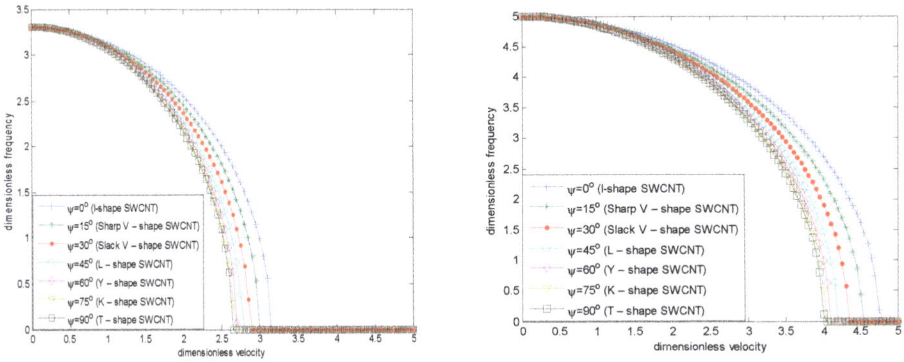


(a) (b)

**Figure 6.** Branch angle impact on linear stability and nonlinear stability: (**a**) linear; (**b**) nonlinear.

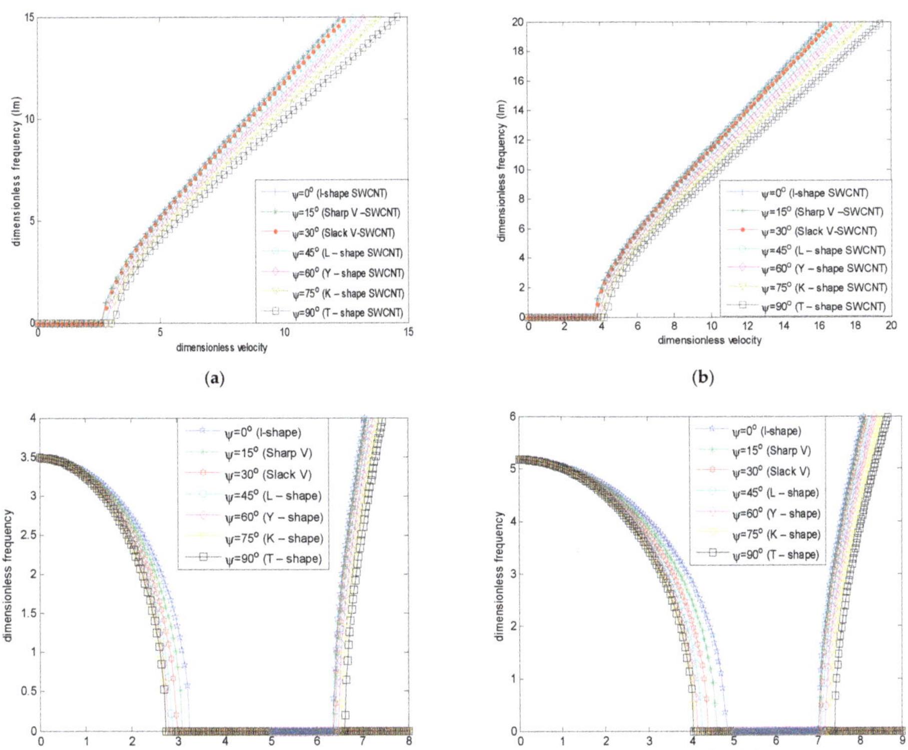


(a) (b)

(c) (d)

**Figure 7.** Branch angle impact on linear stability and nonlinear stability: (**a**) linear for post-bifurcation; (**b**) nonlinear for post-bifurcation; (**c**) linear pre- and post-bifurcation; (**d**) nonlinear pre- and post-bifurcation.

#### 4.1.3. Winkler's Foundations Effects on Stability of SWCNT

Figure 8 shows how the foundation affects the stability of the nanotube. The graphs show that raising the Winkler foundation coefficients causes the stiffness of the nanotube to grow in tandem, raising the system frequencies.

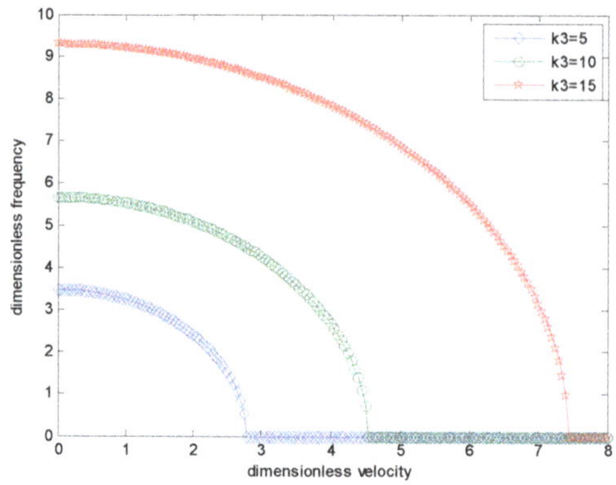


**Figure 8.** Winkler's foundations effects on stability of SWCNT.

#### 4.1.4. Vibrational Dynamic Responses of the SWCNT

The carbon nanotube's nonlinear and linear dynamics are contrasted in Figure 9. The divergence is caused by the nonlinear variables present in the Duffing equations, which are used to compute nonlinear frequency and frequency ratio. As the maximal dynamic vibration rises, this difference becomes more noticeable.

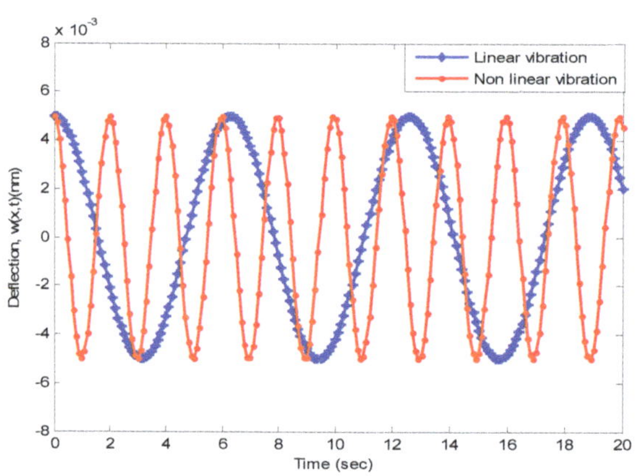


**Figure 9.** Vibrational dynamic responses of the SWCNT.

#### 4.1.5. Magnetic Term's Influence on SWCNT Response

The influences of the magnetic term on dynamic responses of the nanotube for modes 1 and 2 are shown in Figure 10a–d. It was discovered that mode 2 responses were completely out of sync with the mode 1 responses. Furthermore, it was found that as the magnetic influence grew, the system's dynamic responses decreased. This suggests that the dynamic activity of the nanotube is dampened by the magnetic characteristic.

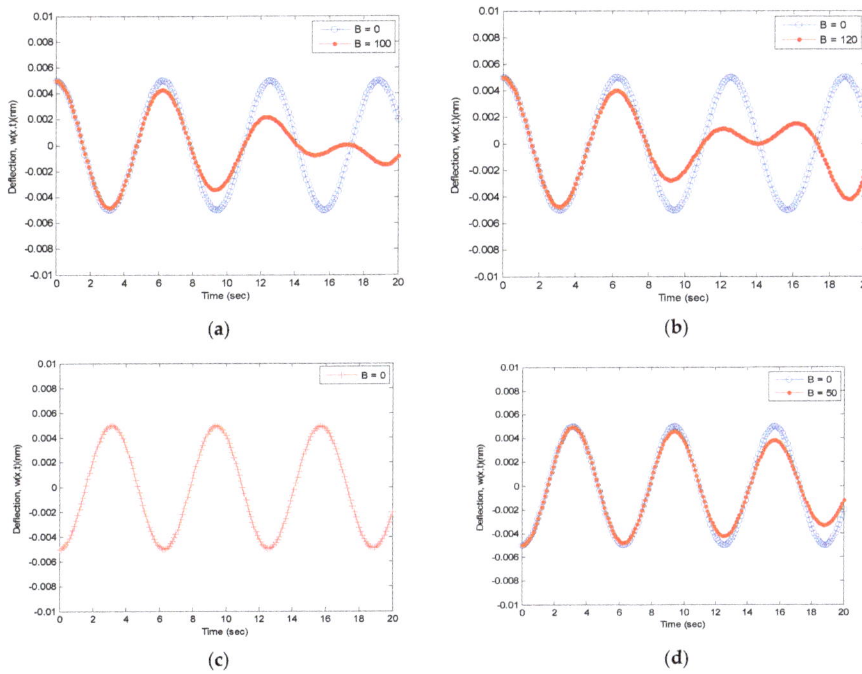


**Figure 10.** The linear response at different modes (**a**) Linear response at mode 1 for B = 100; (**b**) Linear response at mode 1 for B = 120; (**c**) Linear response at mode 2 for B = 50; (**d**) Nonlinear response at mode 1 for B = 50.

## 4.2. Simulation Results

### 4.2.1. Magnetostatic Analysis

The magnetic flux density (MFD) and magnetic flux intensity (MFI) vary along the periphery of the straight nanotube and are even more evident in the surface area, as shown in Figure 11 and Table 5 presents the actual values.

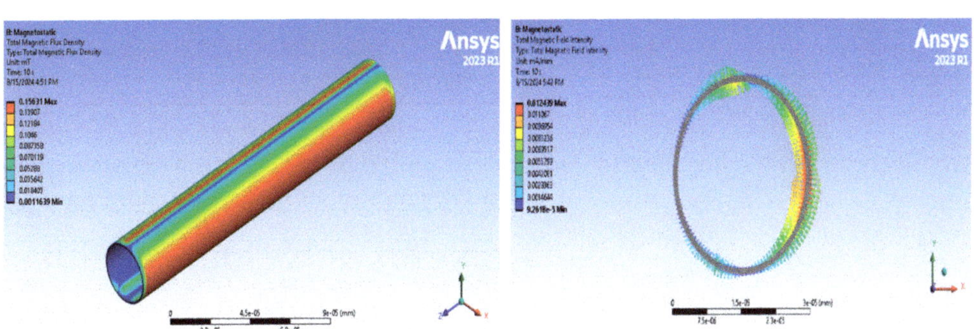


**Figure 11.** Values distribution for MFD and MFI for I-shaped nanotube: (**a**) magnetic flux density I-shaped; (**b**) vector representation of magnetic field intensity I-shaped.

**Table 5.** MFD and MFI for I-shaped nanotube.

| Time Steps | MiMFD (mT) | MaMFD (mT) | AvMFD (mT) | MiMFI (mA/mm) | MaMFI (mA/mm) | AvMFI (mA/mm) |
|---|---|---|---|---|---|---|
| 1. | 0 | 0 | 0 | 0 | 0 | 0 |
| 2. | $5.1727 \times 10^{-4}$ | $6.9473 \times 10^{-2}$ | $2.9487 \times 10^{-2}$ | $4.1163 \times 10^{-5}$ | $5.5285 \times 10^{-3}$ | $2.3465 \times 10^{-3}$ |
| 3. | $5.8193 \times 10^{-4}$ | $7.8157 \times 10^{-2}$ | $3.3173 \times 10^{-2}$ | $4.6309 \times 10^{-5}$ | $6.2195 \times 10^{-3}$ | $2.6398 \times 10^{-3}$ |
| 4. | $6.4659 \times 10^{-4}$ | $8.6841 \times 10^{-2}$ | $3.6858 \times 10^{-2}$ | $5.1454 \times 10^{-5}$ | $6.9106 \times 10^{-3}$ | $2.9331 \times 10^{-3}$ |
| 5. | $7.1125 \times 10^{-4}$ | $9.5525 \times 10^{-2}$ | $4.0544 \times 10^{-2}$ | $5.66 \times 10^{-5}$ | $7.6016 \times 10^{-3}$ | $3.2264 \times 10^{-3}$ |
| 6. | $7.7591 \times 10^{-4}$ | 0.10421 | $4.423 \times 10^{-2}$ | $6.1745 \times 10^{-5}$ | $8.2927 \times 10^{-3}$ | $3.5197 \times 10^{-3}$ |
| 7. | $9.0523 \times 10^{-4}$ | 0.12158 | $5.1602 \times 10^{-2}$ | $7.2036 \times 10^{-5}$ | $9.6748 \times 10^{-3}$ | $4.1063 \times 10^{-3}$ |
| 8. | $9.9144 \times 10^{-4}$ | 0.13316 | $5.6516 \times 10^{-2}$ | $7.8896 \times 10^{-5}$ | $1.0596 \times 10^{-2}$ | $4.4974 \times 10^{-3}$ |
| 9. | $1.0777 \times 10^{-3}$ | 0.14473 | $6.1431 \times 10^{-2}$ | $8.5757 \times 10^{-5}$ | $1.1518 \times 10^{-2}$ | $4.8885 \times 10^{-3}$ |
| 10. | $1.1639 \times 10^{-3}$ | 0.15631 | $6.6345 \times 10^{-2}$ | $9.2618 \times 10^{-5}$ | $1.2439 \times 10^{-2}$ | $5.2796 \times 10^{-3}$ |

The magnetic flux density and magnetic flux intensity for Y-shaped and T-shaped nanotubes exhibit a contrasting behavior to those seen in the straight nanotube, as they are almost uniform across the surface with slight areas having relatively higher values, as shown in Figures 12–17 and Tables 6–11 present the actual values.

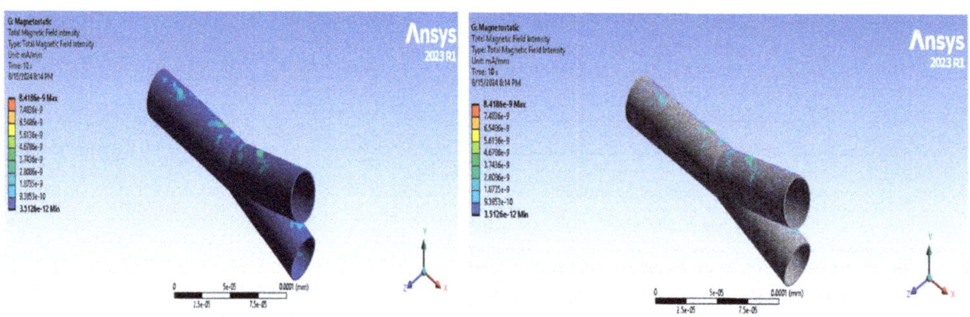


(**a**) (**b**)

**Figure 12.** Values distribution MFI for Y-shaped nanotube with branch angle of 15°: (**a**) MFI of nanotube; (**b**) MFI vector representation of nanotube.

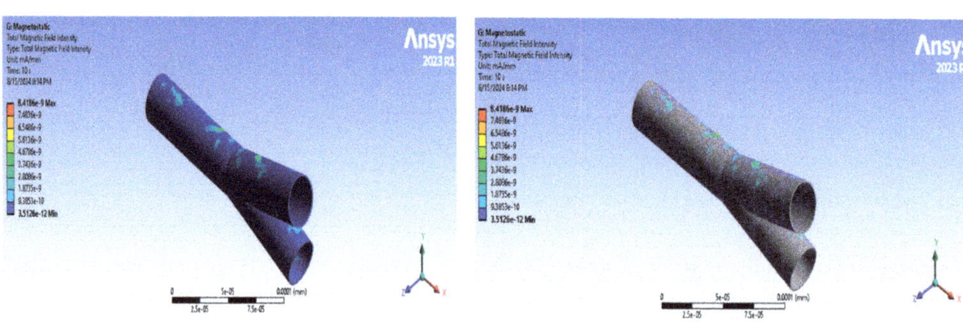


(**a**) (**b**)

**Figure 13.** Values distribution MFI for Y-shaped nanotube with branch angle of 30°: (**a**) MFI of nanotube; (**b**) MFI vector representation of nanotube.

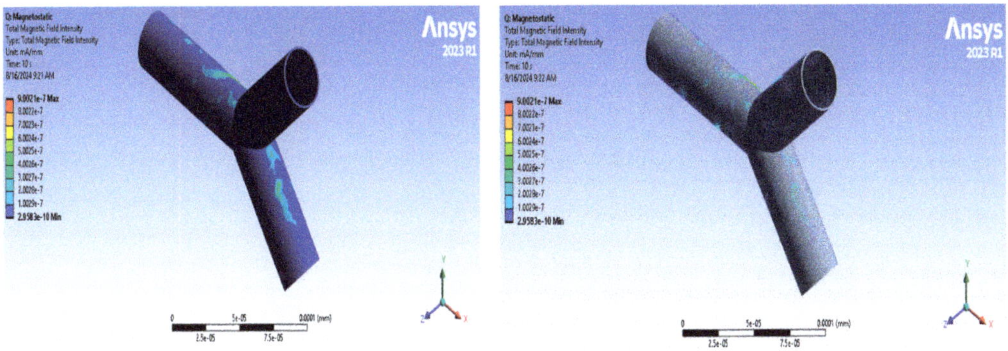


(a) (b)

**Figure 14.** Values distribution MFI for Y-shaped nanotube with branch angle of 45°: (**a**) MFI of nanotube; (**b**) MFI vector representation of nanotube.

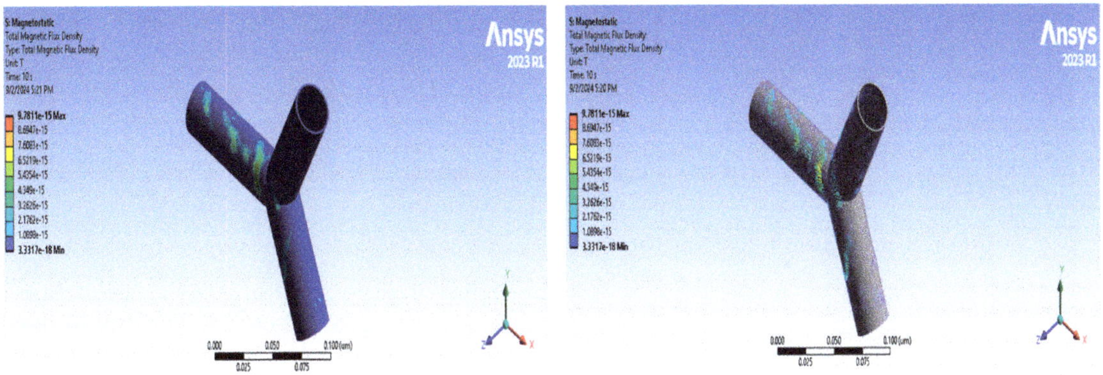


(a) (b)

**Figure 15.** Values distribution MFI for Y-shaped nanotube with branch angle of 60°: (**a**) MFI of nanotube; (**b**) MFI vector representation of nanotube.

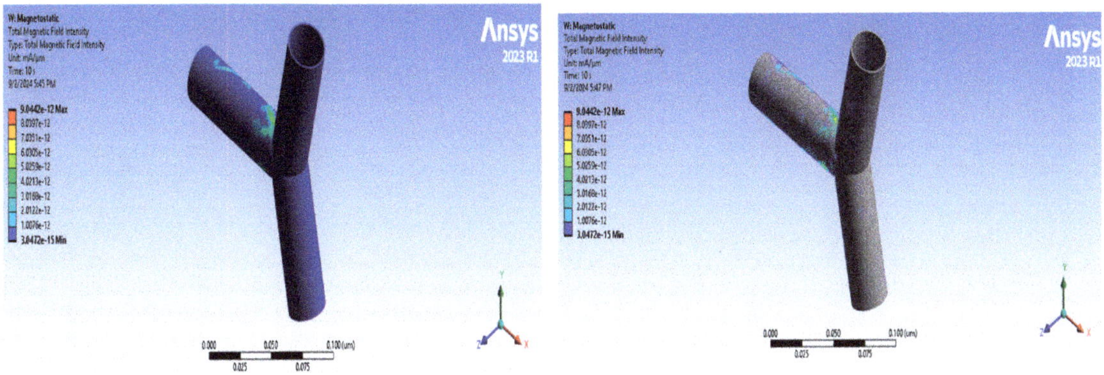


(a) (b)

**Figure 16.** Values distribution MFI for Y-shaped nanotube with branch angle of 75°: (**a**) MFI of nanotube; (**b**) MFI vector representation of nanotube.

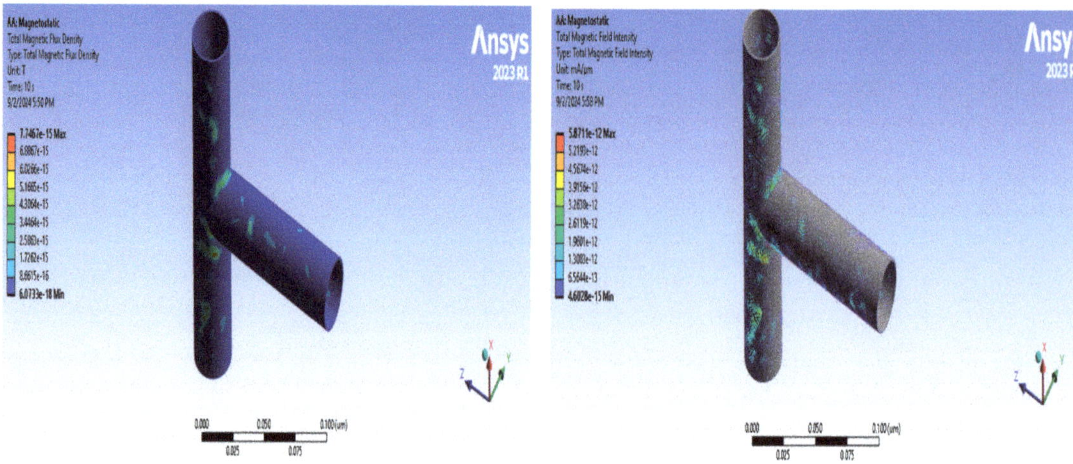


(a) (b)

**Figure 17.** Values distribution MFI for T-shaped nanotube with branch angle of 90°: (**a**) MFI of nanotube; (**b**) MFI vector representation of nanotube.

**Table 6.** MFD and MFI of nanotube with branch angle of 15°.

| Time Steps | MiMFD (mT) | MaMFD (mT) | AvMFD (mT) | MiMFI (mA/mm) | MaMFI (mA/mm) | AvMFI (mA/mm) |
|---|---|---|---|---|---|---|
| 1. | $4.414 \times 10^{-12}$ | $1.0579 \times 10^{-8}$ | $4.922 \times 10^{-10}$ | $3.5126 \times 10^{-13}$ | $8.4186 \times 10^{-10}$ | $3.9168 \times 10^{-11}$ |
| 2. | $1.1035 \times 10^{-11}$ | $2.6448 \times 10^{-8}$ | $1.2305 \times 10^{-9}$ | $8.7814 \times 10^{-13}$ | $2.1047 \times 10^{-9}$ | $9.792 \times 10^{-11}$ |
| 3. | $1.7215 \times 10^{-11}$ | $4.1259 \times 10^{-8}$ | $1.9196 \times 10^{-9}$ | $1.3699 \times 10^{-12}$ | $3.2833 \times 10^{-9}$ | $1.5275 \times 10^{-10}$ |
| 4. | $1.7656 \times 10^{-11}$ | $4.2317 \times 10^{-8}$ | $1.9688 \times 10^{-9}$ | $1.405 \times 10^{-12}$ | $3.3675 \times 10^{-9}$ | $1.5667 \times 10^{-10}$ |
| 5. | $2.4277 \times 10^{-11}$ | $5.8185 \times 10^{-8}$ | $2.7071 \times 10^{-9}$ | $1.9319 \times 10^{-12}$ | $4.6303 \times 10^{-9}$ | $2.1542 \times 10^{-10}$ |
| 6. | $2.8691 \times 10^{-11}$ | $6.8765 \times 10^{-8}$ | $3.1993 \times 10^{-9}$ | $2.2832 \times 10^{-12}$ | $5.4721 \times 10^{-9}$ | $2.5459 \times 10^{-10}$ |
| 7. | $3.0898 \times 10^{-11}$ | $7.4054 \times 10^{-8}$ | $3.4454 \times 10^{-9}$ | $2.4588 \times 10^{-12}$ | $5.893 \times 10^{-9}$ | $2.7417 \times 10^{-10}$ |
| 8. | $3.5312 \times 10^{-11}$ | $8.4633 \times 10^{-8}$ | $3.9376 \times 10^{-9}$ | $2.81 \times 10^{-12}$ | $6.7349 \times 10^{-9}$ | $3.1334 \times 10^{-10}$ |
| 9. | $3.9726 \times 10^{-11}$ | $9.5213 \times 10^{-8}$ | $4.4298 \times 10^{-9}$ | $3.1613 \times 10^{-12}$ | $7.5768 \times 10^{-9}$ | $3.5251 \times 10^{-10}$ |
| 10. | $4.414 \times 10^{-11}$ | $1.0579 \times 10^{-7}$ | $4.922 \times 10^{-9}$ | $3.5126 \times 10^{-12}$ | $8.4186 \times 10^{-9}$ | $3.9168 \times 10^{-10}$ |

**Table 7.** MFD and MFI of nanotube with branch angle of 30°.

| Time Steps | MiMFD (mT) | MaMFD (mT) | AvMFD (mT) | MiMFI (mA/mm) | MaMFI (mA/mm) | AvMFI (mA/mm) |
|---|---|---|---|---|---|---|
| 1. | $9.4821 \times 10^{-13}$ | $3.5836 \times 10^{-9}$ | $2.3093 \times 10^{-10}$ | $7.5456 \times 10^{-14}$ | $2.8517 \times 10^{-10}$ | $1.8377 \times 10^{-11}$ |
| 2. | $2.95 \times 10^{-12}$ | $1.1149 \times 10^{-8}$ | $7.1844 \times 10^{-10}$ | $2.3475 \times 10^{-13}$ | $8.8721 \times 10^{-10}$ | $5.7172 \times 10^{-11}$ |
| 3. | $4.9518 \times 10^{-12}$ | $1.8714 \times 10^{-8}$ | $1.206 \times 10^{-9}$ | $3.9405 \times 10^{-13}$ | $1.4892 \times 10^{-9}$ | $9.5967 \times 10^{-11}$ |
| 4. | $6.9535 \times 10^{-12}$ | $2.628 \times 10^{-8}$ | $1.6935 \times 10^{-9}$ | $5.5334 \times 10^{-13}$ | $2.0913 \times 10^{-9}$ | $1.3476 \times 10^{-10}$ |
| 5. | $8.9553 \times 10^{-12}$ | $3.3845 \times 10^{-8}$ | $2.181 \times 10^{-9}$ | $7.1264 \times 10^{-13}$ | $2.6933 \times 10^{-9}$ | $1.7356 \times 10^{-10}$ |
| 6. | $1.0957 \times 10^{-11}$ | $4.1411 \times 10^{-8}$ | $2.6685 \times 10^{-9}$ | $8.7194 \times 10^{-13}$ | $3.2954 \times 10^{-9}$ | $2.1235 \times 10^{-10}$ |
| 7. | $1.2959 \times 10^{-11}$ | $4.8976 \times 10^{-8}$ | $3.156 \times 10^{-9}$ | $1.0312 \times 10^{-12}$ | $3.8974 \times 10^{-9}$ | $2.5115 \times 10^{-10}$ |
| 8. | $1.4961 \times 10^{-11}$ | $5.6541 \times 10^{-8}$ | $3.6435 \times 10^{-9}$ | $1.1905 \times 10^{-12}$ | $4.4994 \times 10^{-9}$ | $2.8994 \times 10^{-10}$ |
| 9. | $1.6962 \times 10^{-11}$ | $6.4107 \times 10^{-8}$ | $4.131 \times 10^{-9}$ | $1.3498 \times 10^{-12}$ | $5.1015 \times 10^{-9}$ | $3.2874 \times 10^{-10}$ |
| 10. | $1.8964 \times 10^{-11}$ | $7.1672 \times 10^{-8}$ | $4.6185 \times 10^{-9}$ | $1.5091 \times 10^{-12}$ | $5.7035 \times 10^{-9}$ | $3.6753 \times 10^{-10}$ |

**Table 8.** MFD and MFI of nanotube with branch angle of 45°.

| Time Steps | MiMFD (mT) | MaMFD (mT) | AvMFD (mT) | MiMFI (mA/mm) | MaMFI (mA/mm) | AvMFI (mA/mm) |
|---|---|---|---|---|---|---|
| 1. | $3.3458 \times 10^{-10}$ | $1.0181 \times 10^{-6}$ | $4.7945 \times 10^{-8}$ | $2.6625 \times 10^{-11}$ | $8.1019 \times 10^{-8}$ | $3.8153 \times 10^{-9}$ |
| 2. | $7.1046 \times 10^{-10}$ | $2.1619 \times 10^{-6}$ | $1.0181 \times 10^{-7}$ | $5.6537 \times 10^{-11}$ | $1.7204 \times 10^{-7}$ | $8.1017 \times 10^{-9}$ |
| 3. | $1.0863 \times 10^{-9}$ | $3.3057 \times 10^{-6}$ | $1.5567 \times 10^{-7}$ | $8.6449 \times 10^{-11}$ | $2.6306 \times 10^{-7}$ | $1.2388 \times 10^{-8}$ |
| 4. | $1.4622 \times 10^{-9}$ | $4.4496 \times 10^{-6}$ | $2.0954 \times 10^{-7}$ | $1.1636 \times 10^{-10}$ | $3.5408 \times 10^{-7}$ | $1.6674 \times 10^{-8}$ |
| 5. | $1.8381 \times 10^{-9}$ | $5.5934 \times 10^{-6}$ | $2.634 \times 10^{-7}$ | $1.4627 \times 10^{-10}$ | $4.4511 \times 10^{-7}$ | $2.0961 \times 10^{-8}$ |
| 6. | $2.214 \times 10^{-9}$ | $6.7372 \times 10^{-6}$ | $3.1727 \times 10^{-7}$ | $1.7618 \times 10^{-10}$ | $5.3613 \times 10^{-7}$ | $2.5247 \times 10^{-8}$ |
| 7. | $2.5899 \times 10^{-9}$ | $7.881 \times 10^{-6}$ | $3.7113 \times 10^{-7}$ | $2.061 \times 10^{-10}$ | $6.2715 \times 10^{-7}$ | $2.9534 \times 10^{-8}$ |
| 8. | $2.9658 \times 10^{-9}$ | $9.0248 \times 10^{-6}$ | $4.2499 \times 10^{-7}$ | $2.3601 \times 10^{-10}$ | $7.1817 \times 10^{-7}$ | $3.382 \times 10^{-8}$ |
| 9. | $3.3417 \times 10^{-9}$ | $1.0169 \times 10^{-5}$ | $4.7886 \times 10^{-7}$ | $2.6592 \times 10^{-10}$ | $8.0919 \times 10^{-7}$ | $3.8106 \times 10^{-8}$ |
| 10. | $3.7175 \times 10^{-9}$ | $1.1312 \times 10^{-5}$ | $5.3272 \times 10^{-7}$ | $2.9583 \times 10^{-10}$ | $9.0021 \times 10^{-7}$ | $4.2393 \times 10^{-8}$ |

**Table 9.** MFD and MFI of nanotube with branch angle of 60°.

| Time Steps | MiMFD (mT) | MaMFD (mT) | AvMFD (mT) | MiMFI (mA/mm) | MaMFI (mA/mm) | AvMFI (mA/mm) |
|---|---|---|---|---|---|---|
| 1. | $3.3317 \times 10^{-19}$ | $9.7811 \times 10^{-16}$ | $4.5868 \times 10^{-17}$ | $2.525 \times 10^{-16}$ | $7.4129 \times 10^{-13}$ | $3.4763 \times 10^{-14}$ |
| 2. | $6.6633 \times 10^{-19}$ | $1.9562 \times 10^{-15}$ | $9.1737 \times 10^{-17}$ | $5.05 \times 10^{-16}$ | $1.4826 \times 10^{-12}$ | $6.9525 \times 10^{-14}$ |
| 3. | $9.995 \times 10^{-19}$ | $2.9343 \times 10^{-15}$ | $1.376 \times 10^{-16}$ | $7.575 \times 10^{-16}$ | $2.2239 \times 10^{-12}$ | $1.0429 \times 10^{-13}$ |
| 4. | $1.3327 \times 10^{-18}$ | $3.9125 \times 10^{-15}$ | $1.8347 \times 10^{-16}$ | $1.01 \times 10^{-15}$ | $2.9652 \times 10^{-12}$ | $1.3905 \times 10^{-13}$ |
| 5. | $1.6658 \times 10^{-18}$ | $4.8906 \times 10^{-15}$ | $2.2934 \times 10^{-16}$ | $1.2625 \times 10^{-15}$ | $3.7065 \times 10^{-12}$ | $1.7381 \times 10^{-13}$ |
| 6. | $1.999 \times 10^{-18}$ | $5.8687 \times 10^{-15}$ | $2.7521 \times 10^{-16}$ | $1.515 \times 10^{-15}$ | $4.4478 \times 10^{-12}$ | $2.0858 \times 10^{-13}$ |
| 7. | $2.3322 \times 10^{-18}$ | $6.8468 \times 10^{-15}$ | $3.2108 \times 10^{-16}$ | $1.7675 \times 10^{-15}$ | $5.1891 \times 10^{-12}$ | $2.4334 \times 10^{-13}$ |
| 8. | $2.6653 \times 10^{-18}$ | $7.8249 \times 10^{-15}$ | $3.6695 \times 10^{-16}$ | $2.02 \times 10^{-15}$ | $5.9304 \times 10^{-12}$ | $2.781 \times 10^{-13}$ |
| 9. | $2.9985 \times 10^{-18}$ | $8.803 \times 10^{-15}$ | $4.1281 \times 10^{-16}$ | $2.2725 \times 10^{-15}$ | $6.6716 \times 10^{-12}$ | $3.1286 \times 10^{-13}$ |
| 10. | $3.3317 \times 10^{-18}$ | $9.7811 \times 10^{-15}$ | $4.5868 \times 10^{-16}$ | $2.525 \times 10^{-15}$ | $7.4129 \times 10^{-12}$ | $3.4763 \times 10^{-13}$ |

**Table 10.** MFD and MFI of nanotube with branch angle of 75°.

| Time Steps | MiMFD (mT) | MaMFD (mT) | AvMFD (mT) | MiMFI (mA/mm) | MaMFI (mA/mm) | AvMFI (mA/mm) |
|---|---|---|---|---|---|---|
| 1. | $4.0206 \times 10^{-19}$ | $1.1934 \times 10^{-15}$ | $3.3235 \times 10^{-17}$ | $3.0472 \times 10^{-16}$ | $9.0442 \times 10^{-13}$ | $2.5188 \times 10^{-14}$ |
| 2. | $8.0412 \times 10^{-19}$ | $2.3867 \times 10^{-15}$ | $6.647 \times 10^{-17}$ | $6.0943 \times 10^{-16}$ | $1.8088 \times 10^{-12}$ | $5.0376 \times 10^{-14}$ |
| 3. | $1.2062 \times 10^{-18}$ | $3.5801 \times 10^{-15}$ | $9.9705 \times 10^{-17}$ | $9.1415 \times 10^{-16}$ | $2.7133 \times 10^{-12}$ | $7.5565 \times 10^{-14}$ |
| 4. | $1.6082 \times 10^{-18}$ | $4.7734 \times 10^{-15}$ | $1.3294 \times 10^{-16}$ | $1.2189 \times 10^{-15}$ | $3.6177 \times 10^{-12}$ | $1.0075 \times 10^{-13}$ |
| 5. | $2.0103 \times 10^{-18}$ | $5.9668 \times 10^{-15}$ | $1.6618 \times 10^{-16}$ | $1.5236 \times 10^{-15}$ | $4.5221 \times 10^{-12}$ | $1.2594 \times 10^{-13}$ |
| 6. | $2.4124 \times 10^{-18}$ | $7.1601 \times 10^{-15}$ | $1.9941 \times 10^{-16}$ | $1.8283 \times 10^{-15}$ | $5.4265 \times 10^{-12}$ | $1.5113 \times 10^{-13}$ |
| 7. | $2.8144 \times 10^{-18}$ | $8.3535 \times 10^{-15}$ | $2.3265 \times 10^{-16}$ | $2.133 \times 10^{-15}$ | $6.331 \times 10^{-12}$ | $1.7632 \times 10^{-13}$ |
| 8. | $3.2165 \times 10^{-18}$ | $9.5469 \times 10^{-15}$ | $2.6588 \times 10^{-16}$ | $2.4377 \times 10^{-15}$ | $7.2354 \times 10^{-12}$ | $2.0151 \times 10^{-13}$ |
| 9. | $3.6186 \times 10^{-18}$ | $1.074 \times 10^{-14}$ | $2.9912 \times 10^{-16}$ | $2.7424 \times 10^{-15}$ | $8.1398 \times 10^{-12}$ | $2.2669 \times 10^{-13}$ |
| 10. | $4.0206 \times 10^{-18}$ | $1.1934 \times 10^{-14}$ | $3.3235 \times 10^{-16}$ | $3.0472 \times 10^{-15}$ | $9.0442 \times 10^{-12}$ | $2.5188 \times 10^{-13}$ |

**Table 11.** MFD and MFI of nanotube with branch angle of 90°.

| Time Steps | MiMFD (mT) | MaMFD (mT) | AvMFD (mT) | MiMFI (mA/mm) | MaMFI (mA/mm) | AvMFI (mA/mm) |
|---|---|---|---|---|---|---|
| 1. | $6.0733 \times 10^{-19}$ | $7.7467 \times 10^{-16}$ | $4.8008 \times 10^{-17}$ | $4.6028 \times 10^{-16}$ | $5.8711 \times 10^{-13}$ | $3.6384 \times 10^{-14}$ |
| 2. | $1.2147 \times 10^{-18}$ | $1.5493 \times 10^{-15}$ | $9.6016 \times 10^{-17}$ | $9.2056 \times 10^{-16}$ | $1.1742 \times 10^{-12}$ | $7.2768 \times 10^{-14}$ |
| 3. | $1.822 \times 10^{-18}$ | $2.324 \times 10^{-15}$ | $1.4402 \times 10^{-16}$ | $1.3808 \times 10^{-15}$ | $1.7613 \times 10^{-12}$ | $1.0915 \times 10^{-13}$ |
| 4. | $2.4293 \times 10^{-18}$ | $3.0987 \times 10^{-15}$ | $1.9203 \times 10^{-16}$ | $1.8411 \times 10^{-15}$ | $2.3484 \times 10^{-12}$ | $1.4554 \times 10^{-13}$ |

**Table 11.** *Cont.*

| Time Steps | MiMFD (mT) | MaMFD (mT) | AvMFD (mT) | MiMFI (mA/mm) | MaMFI (mA/mm) | AvMFI (mA/mm) |
|---|---|---|---|---|---|---|
| 5. | $3.0366 \times 10^{-18}$ | $3.8734 \times 10^{-15}$ | $2.4004 \times 10^{-16}$ | $2.3014 \times 10^{-15}$ | $2.9355 \times 10^{-12}$ | $1.8192 \times 10^{-13}$ |
| 6. | $3.644 \times 10^{-18}$ | $4.648 \times 10^{-15}$ | $2.8805 \times 10^{-16}$ | $2.7617 \times 10^{-15}$ | $3.5227 \times 10^{-12}$ | $2.1831 \times 10^{-13}$ |
| 7. | $4.2513 \times 10^{-18}$ | $5.4227 \times 10^{-15}$ | $3.3606 \times 10^{-16}$ | $3.222 \times 10^{-15}$ | $4.1098 \times 10^{-12}$ | $2.5469 \times 10^{-13}$ |
| 8. | $4.8586 \times 10^{-18}$ | $6.1974 \times 10^{-15}$ | $3.8406 \times 10^{-16}$ | $3.6823 \times 10^{-15}$ | $4.6969 \times 10^{-12}$ | $2.9107 \times 10^{-13}$ |
| 9. | $5.466 \times 10^{-18}$ | $6.9721 \times 10^{-15}$ | $4.3207 \times 10^{-16}$ | $4.1425 \times 10^{-15}$ | $5.284 \times 10^{-12}$ | $3.2746 \times 10^{-13}$ |
| 10. | $6.0733 \times 10^{-18}$ | $7.7467 \times 10^{-15}$ | $4.8008 \times 10^{-16}$ | $4.6028 \times 10^{-15}$ | $5.8711 \times 10^{-12}$ | $3.6384 \times 10^{-13}$ |

#### 4.2.2. Modal Analysis

The modal deformation for different modes is presented below for the different orientations of nanotubes considered in this study. The straight carbon nanotubes consist of two cases, which were when there were simple supports at both ends and with and without support on the external surface area of the cylinder.

The modal shapes with the deformation at different modes for I-shaped CNT are presented as follows:

Figures 18 and 19 both represent the deformation at different modes. The deformation is seen to increase steadily for the first five modes.

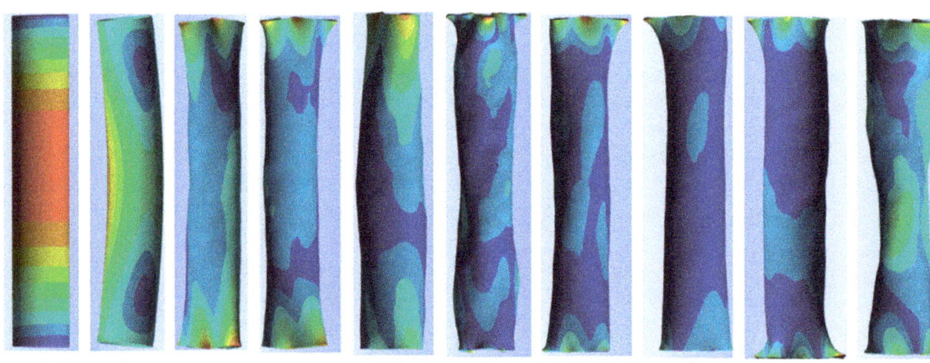

1 2 3 4 5 6 7 8 9 10

**Figure 18.** Deformation by modes for I-shaped CNT with no periphery support.

1 2 3 4 5 6 7 8 9 10

**Figure 19.** Deformation by modes for I-shaped CNT with elastic support.

The modal shapes with the deformation at different modes for Y-shaped CNT are presented in Tables 12 and 13.

**Table 12.** Modal frequency for CNT at varying downstream angle.

| | Frequency (MHz) | | | | | | |
|---|---|---|---|---|---|---|---|
| **Modes** | **0** | **15** | **30** | **45** | **60** | **75** | **90** |
| 1 | $2.353 \times 10^{-11}$ | $2.08 \times 10^{-12}$ | $9.08 \times 10^{-6}$ | $2.56 \times 10^{-13}$ | $-2.0479 \times 10^{-13}$ | $2.0312 \times 10^{-13}$ | $4.0616 \times 10^{-13}$ |
| 2 | $1.5417 \times 10^{-11}$ | $3.32 \times 10^{-7}$ | $3.33 \times 10^{-6}$ | $3.90 \times 10^{-7}$ | $-1.061 \times 10^{-9}$ | $1.061 \times 10^{-9}$ | $1.061 \times 10^{-9}$ |
| 3 | $1.2398 \times 10^{-10}$ | $4.02 \times 10^{-7}$ | $3.13 \times 10^{-5}$ | $1.47 \times 10^{-7}$ | $-1.061 \times 10^{-9}$ | $1.061 \times 10^{-9}$ | $1.061 \times 10^{-9}$ |
| 4 | $2.058 \times 10^{-10}$ | $3.43 \times 10^{-7}$ | $3.19 \times 10^{-5}$ | $5.67 \times 10^{-7}$ | $-1.061 \times 10^{-9}$ | $1.061 \times 10^{-9}$ | $1.061 \times 10^{-9}$ |
| 5 | $2.7809 \times 10^{-10}$ | $9.83 \times 10^{-7}$ | $4.75 \times 10^{-5}$ | $4.52 \times 10^{-7}$ | $-1.061 \times 10^{-9}$ | $1.061 \times 10^{-9}$ | $1.061 \times 10^{-9}$ |
| 6 | $3.9117 \times 10^{-10}$ | $3.70 \times 10^{-7}$ | $6.89 \times 10^{-5}$ | $7.40 \times 10^{-7}$ | $-1.061 \times 10^{-9}$ | $1.061 \times 10^{-9}$ | $1.061 \times 10^{-9}$ |
| 7 | $3.9954 \times 10^{-10}$ | $1.38 \times 10^{-6}$ | $5.18 \times 10^{-5}$ | $5.21 \times 10^{-7}$ | $-1.061 \times 10^{-9}$ | $1.061 \times 10^{-9}$ | $1.061 \times 10^{-9}$ |
| 8 | $4.0446 \times 10^{-10}$ | $5.30 \times 10^{-7}$ | $6.97 \times 10^{-5}$ | $9.04 \times 10^{-7}$ | $-1.061 \times 10^{-9}$ | $1.061 \times 10^{-9}$ | $1.061 \times 10^{-9}$ |
| 9 | $4.0489 \times 10^{-10}$ | $1.82 \times 10^{-6}$ | $7.64 \times 10^{-5}$ | $1.13 \times 10^{-6}$ | $-1.061 \times 10^{-9}$ | $1.061 \times 10^{-9}$ | $1.061 \times 10^{-9}$ |
| 10 | $4.0499 \times 10^{-10}$ | $6.90 \times 10^{-7}$ | $1.52 \times 10^{-4}$ | $1.26 \times 10^{-6}$ | $-1.061 \times 10^{-9}$ | $1.061 \times 10^{-9}$ | $1.061 \times 10^{-9}$ |

**Table 13.** Deformation for CNT at varying downstream angles.

| | Deformation (Чm) | | | | | | |
|---|---|---|---|---|---|---|---|
| **Modes** | **0** | **15** | **30** | **45** | **60** | **75** | **90** |
| 1 | 31.998 | 0.00018869 | $5.72 \times 10^{-5}$ | $3.39 \times 10^{-5}$ | $9.604 \times 10^{-3}$ | $9.2143 \times 10^{-3}$ | $8.2562 \times 10^{-3}$ |
| 2 | 347.69 | 1.732 | 2.512 | 1.6242 | 1.7744 | 1.8811 | 2.2076 |
| 3 | 430.27 | 1.7666 | 2.3358 | 2.1649 | 1.9492 | 2.4072 | 1.8218 |
| 4 | 567.02 | 1.6772 | 2.1682 | 1.9493 | 2.6251 | 2.2065 | 1.8914 |
| 5 | 614.73 | 1.7843 | 2.3358 | 2.3673 | 2.4339 | 2.2916 | 2.2211 |
| 6 | 561.82 | 1.684 | 2.1682 | 2.6993 | 2.6043 | 2.1477 | 1.7175 |
| 7 | 571.76 | 2.3413 | 2.3115 | 1.9122 | 2.5140 | 2.4863 | 1.7255 |
| 8 | 502.72 | 1.9917 | 2.2874 | 2.15218 | 1.6629 | 1.8826 | 1.9705 |
| 9 | 518.4 | 1.7508 | 2.1793 | 2.2192 | 2.9362 | 2.1940 | 1.9857 |
| 10 | 489.92 | 1.4758 | 2.2088 | 1.6817 | 2.0070 | 1.1628 | 1.9993 |

Figures 20–25 show large deformations for the branched region of the nanotubes. However, for the T-shaped nanotube, the deformation is prominent along the vertical section, which coincides with the unconstrained section. Figure 25 shows that for all five modes represented, there are large deflections from the branched part of the nanotubes.

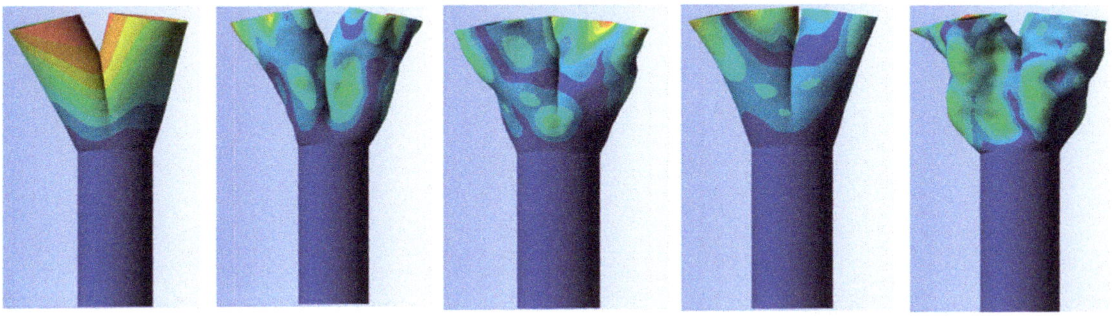

**1** **2** **3** **4** **5**

**Figure 20.** Deformation by modes for Y-shaped CNT (15°).

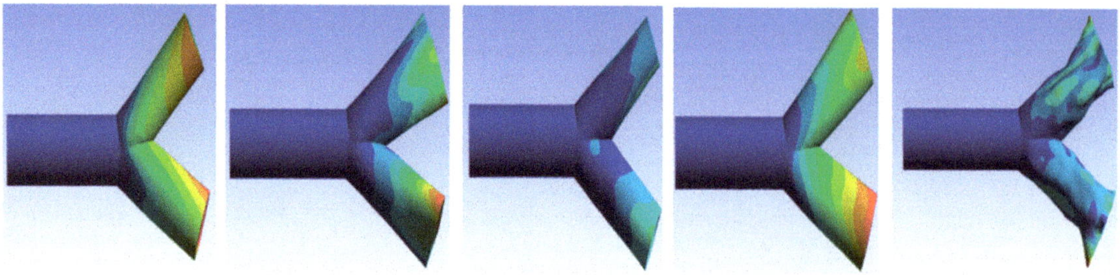

1 2 3 4 5

**Figure 21.** Deformation by modes for Y-shaped CNT (30°).

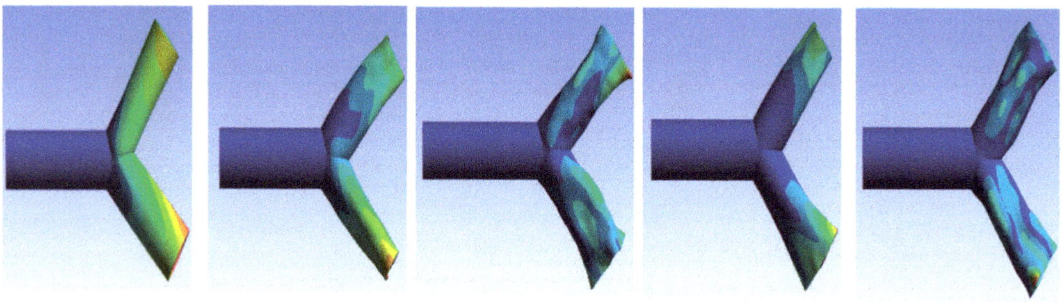

1 2 3 4 5

**Figure 22.** Deformation by modes for Y-shaped CNT (45°).

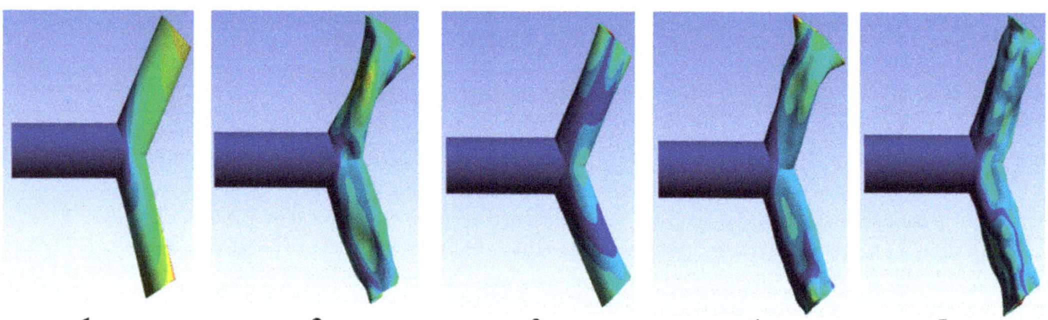

1 2 3 4 5

**Figure 23.** Deformation by modes for Y-shaped CNT (60°).

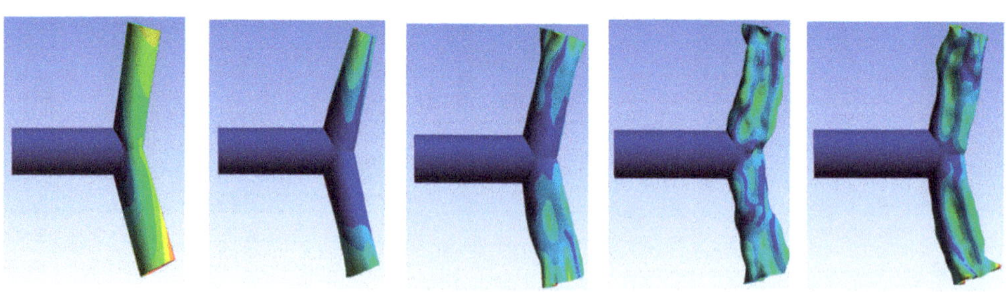

1 2 3 4 5

**Figure 24.** Deformation by modes for Y-shaped CNT (75°).

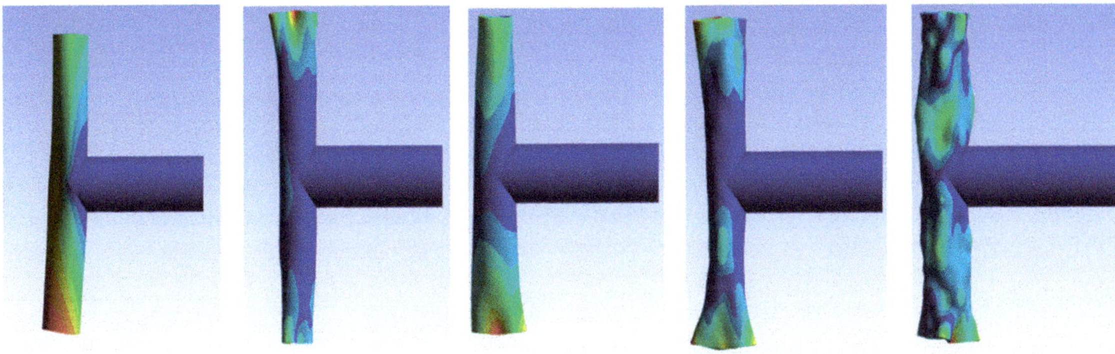


**Figure 25.** Deformation by modes for T-shaped CNT.

#### 4.2.3. Transient Structural Analysis

Figure 26a–d represents the result obtained from the transient structural analysis. For the simply supported I-shaped nanotube, it is observed that there is large radial deformation under the modeled thermo-magnetic environment. Furthermore, for the branched nanotubes, large stress occurs at unsupported edges, with the T-shaped nanotube an exception to this. Maximum stress for the T-shaped nanotube occurs at the inner walls of the tube.

### *4.3. Machine Learning Results*

Figures 27–31 showcase the performance of the ML models in the prediction of carbon nanotube vibrations for displacement and frequency for five modes. Each scatter plot represents the bond between the predicted values and true values for a specific feature. The first five features represent the amplitude for five modes, and the next five represent the frequency in the same order.

In Figure 27 it can be observed that Features 3, 4, 6, and 8 show $R^2$ values of 1.00, indicating an almost perfect fit between the predicted and true values. With similar performance quotas, features 1, 7, and 10 show $R^2$ values between 0.42 and 0.89, suggesting a reasonable fit. Feature 2 has an $R^2$ of 0.28, indicating a moderate fit. Feature 5 has an $R^2$ of 0.00, suggesting a very poor fit. The model is unable to capture the relationship between the predicted and true values for this feature. Feature 9 has an extremely low $R^2$ value of $-397.67$, indicating a very poor fit and potential issues with the data or model for this feature.

The XGBoost model shows promising results for several features but struggles with others.

Features 3, 4, 6, and 8 show $R^2$ values of 0.97, indicating a very good fit between the predicted and true values, as shown in Figure 28. However, feature 10 has an $R^2$ of 0.92, suggesting a good fit. On the contrary, Features 1 and 5 have $R^2$ values between 0.24 and 0.49, indicating a moderate fit. Feature 2 has an $R^2$ of $-0.26$, suggesting a poor fit. The model is not capturing the relationship between predicted and true values well for this feature.

Features 7 and 9 have extremely low $R^2$ values, indicating a very poor fitness and potential issues with the data or model for these features. Overall, the CATBoost model shows promising results for several features but struggles with others.

Features 1, 4, 6, and 9: These features show an $R^2$ of 1.00, indicating a perfect fit as shown in Figure 29. This suggests the Random Forest models are performing exceptionally well for these features.

Other features: The $R^2$ values for the remaining features are negative or close to zero, suggesting a poor fit. The model is not performing well in predicting these features.

Overall, the Random Forest model appears to be effective in predicting certain vibrational modes of the carbon nanotube (features 1, 4, 6, and 9) but struggles with others.

Based on the $R^2$ values, the ANN model is performing extremely poorly for all features, as depicted by Figure 30. The $R^2$ values are extremely large and negative, indicating a very poor fit between the predicted and true values. This suggests that the ANN model is not able to capture the underlying patterns in the data and is making very poor predictions.

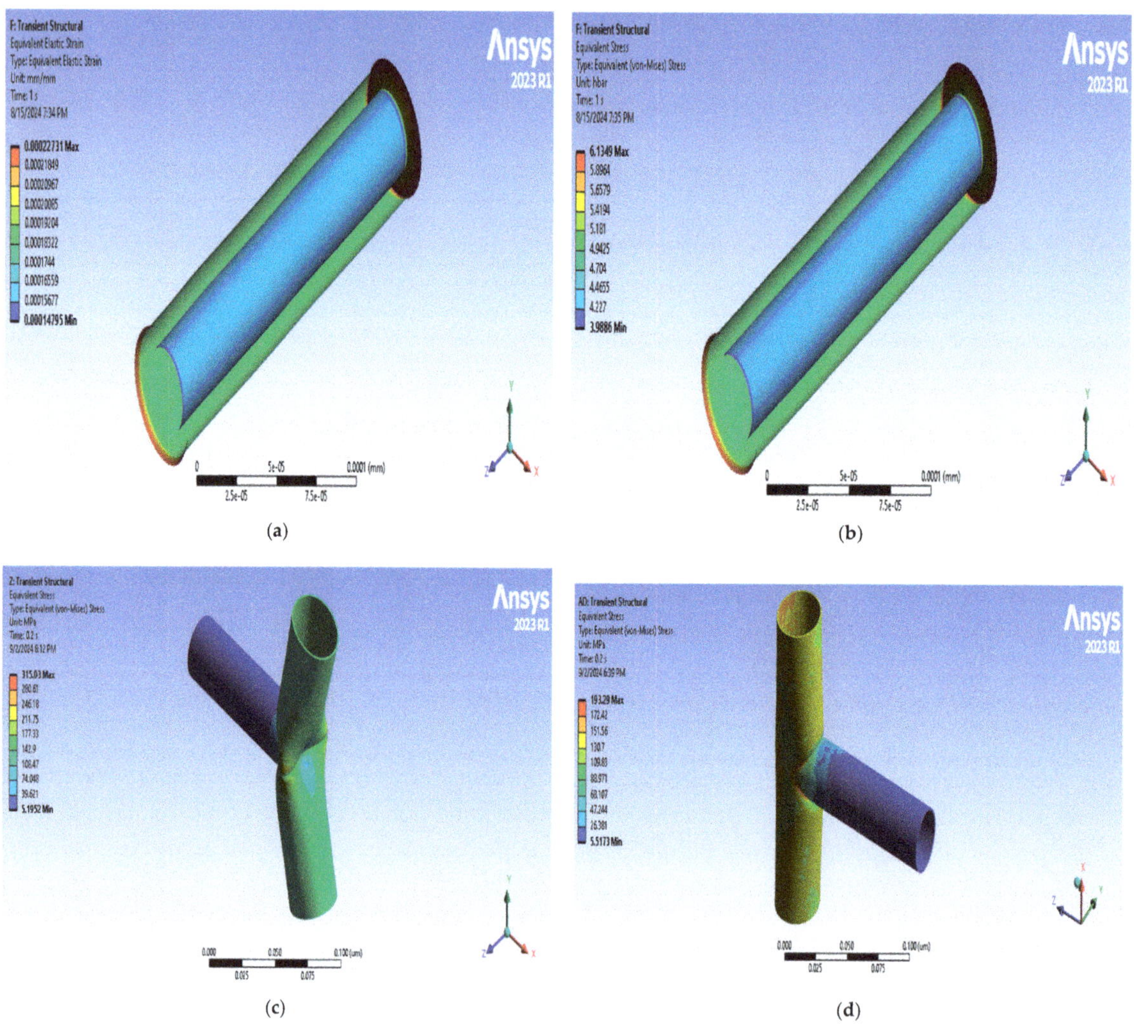


**Figure 26.** Values distribution of nanotubes for transient analysis: (**a**) transient structural I-shaped; (**b**) equivalent stress I-shaped; (**c**) equivalent stress Y-shaped; (**d**) equivalent stress T-shaped.

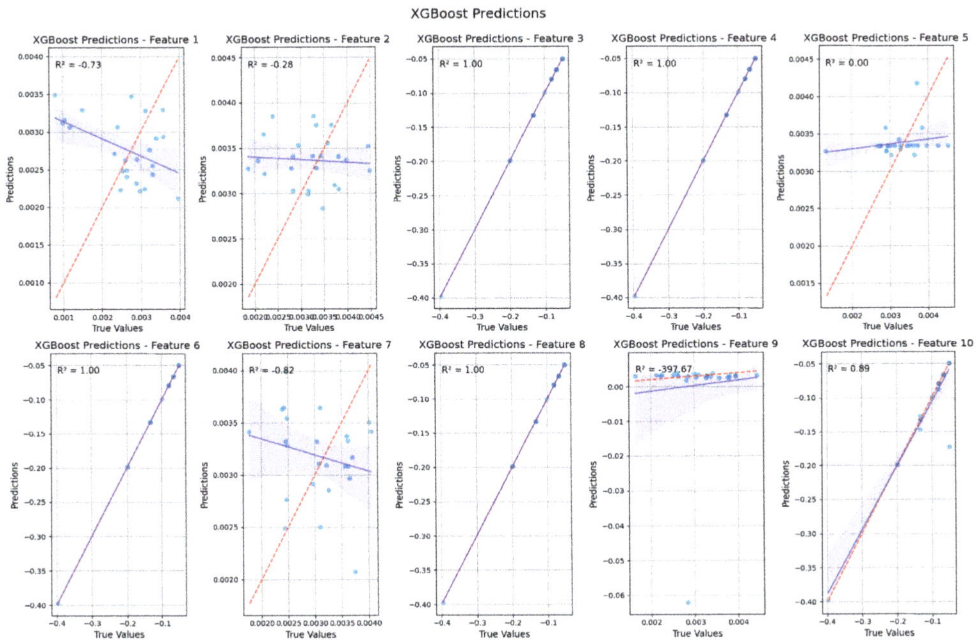


**Figure 27.** XGBoost performance evaluation.

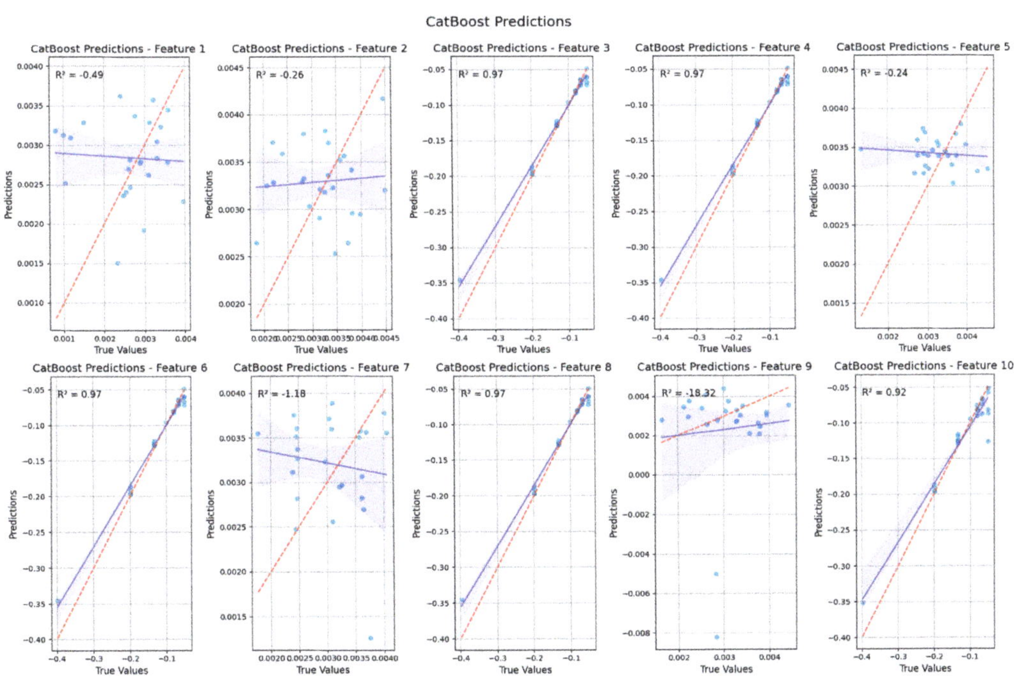


**Figure 28.** CATBoost performance evaluation.

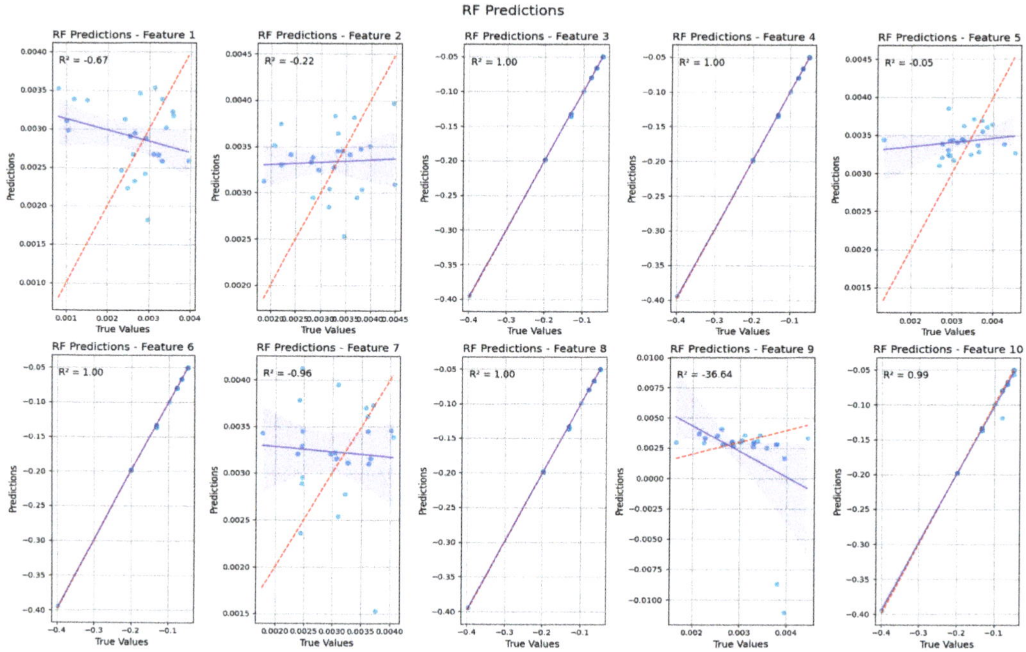


**Figure 29.** Random Forest performance evaluation.

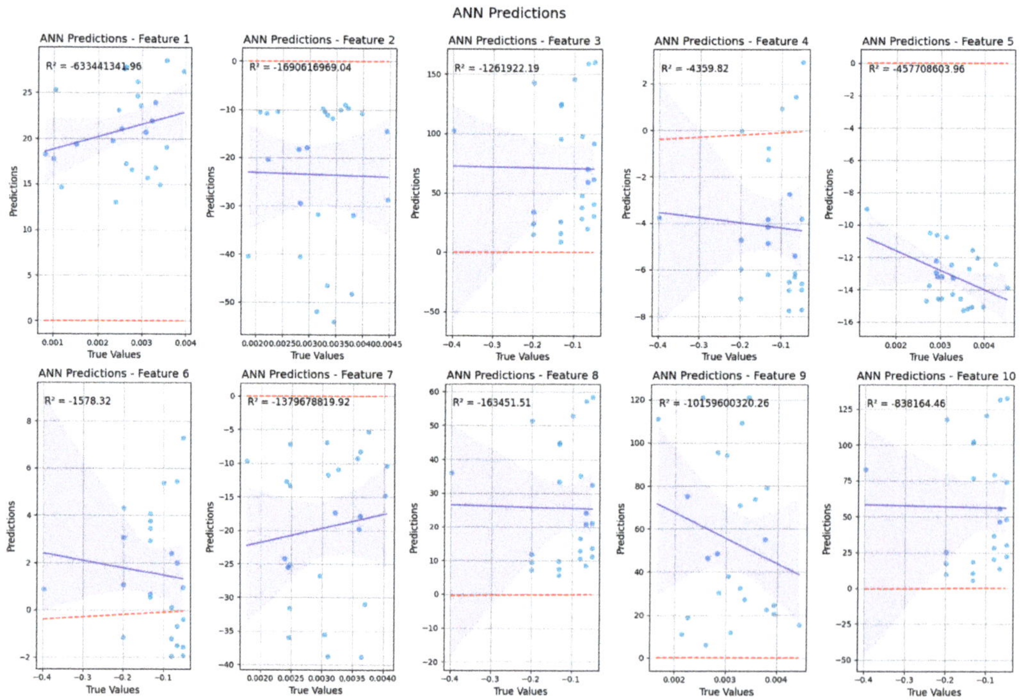


**Figure 30.** ANN performance evaluation.

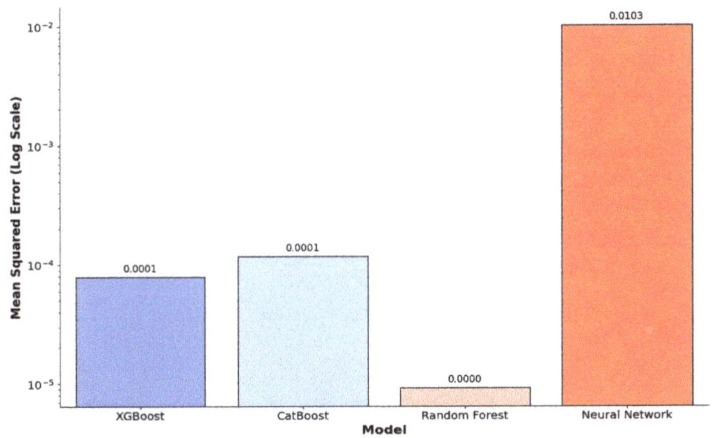


**Figure 31.** Comparing models performance evaluation.

## *4.4. Comparison of Machine Learning Models for Carbon Nanotube Vibration Prediction*

Having analyzed the performance of four models—Random Forest, XGBoost, CATBoost, and ANN—in predicting carbon nanotube vibrations based on provided visualizations provided in Figures 27–30. The following performance summary is presented.

From Figure 31, XGBoost and CATBoost exhibit significantly lower MSE values than Random Forest and Neural Network, indicating superior performance in predicting the target variable as depicted by Table 14. Random Forest demonstrates a moderate level of error, falling between the top-performing models and the Neural Network.

**Table 14.** Machine learning models performance summary.

| Model | Overall Performance | Strengths | Weaknesses |
|---|---|---|---|
| Random Forest | Moderate | Good performance on features 1, 4, 6, and 9 | Poor performance on other features |
| XGBoost | Good | Excellent performance on features 3, 4, 6, and 8; good performance on others 1, 7, and 10 | Fair performance on feature 2, poor performance on feature 5, very poor performance on feature 9 |
| CATBoost | Good | Excellent performance on several features (3, 4, 6, and 8); good performance on feature 10 | Fair performance on features 1 and 5, poor performance on feature 2, very poor performance on features 7 and 9 |
| ANN | Very Poor | None | Poor performance on all features |

While Neural Network shows the highest MSE, suggesting that the model struggles to capture the underlying patterns in the data.

In addition to $R^2$ and MSE, we also evaluated model performance using the mean absolute error (MAE) and root mean square error (RMSE). MAE provides an average magnitude of error without penalizing large deviations disproportionately, while RMSE emphasizes larger errors due to its quadratic nature. These metrics allowed a more complete assessment of the regression models. XGBoost consistently achieved the lowest MAE and RMSE values across prediction tasks, confirming its robustness in modeling the nonlinear vibrational behavior of the system.

Although the machine learning models demonstrated good predictive power, some limitations were apparent, particularly in the representation of higher-mode complicated nonlinear interactions. Several reasons may be attributable to this. To begin with, the relatively small dataset size, although it covered a broad matrix of physical parameters, might have constrained the generalizability of the models in all situations. Second, feature correlations and variation among inputs, especially between mass coefficient, stability frequency, and magnetic flux, might have introduced latent biases to model learning. In addition, simulation noise and boundary condition assumptions can lead to small differences between training data and physical reality. Feature importance analysis showed that certain features had a dominating impact on the model predictions, which meant that under-represented features (e.g., branch angle interactions at high velocities) would need to be further refined for future datasets. Refinement of these aspects, e.g., data volume expansion, feature engineering, and noise reduction, will continue to enhance the usability and robustness of ML models for difficult nanoscale vibration problems.

## 5. Conclusions

This study presented a comprehensive approach to modeling and analyzing the nonlinear fluid-induced vibration of pronged single-walled carbon nanotubes (SWCNTs) subjected to thermal and magnetic fields. Through a combination of analytical modeling using the Euler–Bernoulli and Eringen nonlocal theories, numerical simulation via ANSYS, and predictive modeling with machine learning algorithms (XGBoost, CATBoost, Random Forest, and Neural Networks), we identified the critical parameters influencing system stability, including branching angle, magnetic flux density, flow velocity, and foundation stiffness.

The parametric analysis led to the following key findings:

- The angle at which the nanotube branches significantly affects its stability. A larger angle has been demonstrated to reduce stability.
- The dynamic behavior of the SWCNT is dampened by the existing magnetic field introduced to the environment.
- The nonlocal term has a major impact on flow velocity and frequency. Furthermore, raising the foundation parameters and shear moduli increases the fundamental frequency of the nanotube.
- The fluid–structure mass ratio becomes particularly important in post-bifurcation regions, where frequencies and velocities increase with rising temperatures.
- The difference between linear and nonlinear vibration frequencies becomes more pronounced as flow velocity and amplitude rise.
- Increasing axial pre-tension on the nanotube reduces its stability, especially as it deviates further from linear behavior.

In addition to magnetic effects, temperature also plays a critical role in influencing the vibrational characteristics of branched SWCNTs. As temperature increases, thermal expansion leads to a reduction in the effective stiffness of the nanotube material, resulting in lower natural frequencies and greater susceptibility to instability under fluid flow. This softening effect reduces the critical flow velocities at which bifurcation or flutter phenomena are observed.

The novel integration of machine learning algorithms with physics-based models provides rapid, accurate predictions and optimization strategies for enhancing system stability at the nanoscale. This integration offers a new pathway for designing nanostructures in energy harvesting, biomedical devices, and nanofluidic systems where real-time vibrational control is essential.

Future work may involve expanding the dataset through further simulations and experimental validation, exploring additional environmental conditions such as multiphysics couplings (piezoelectric effects, electro-thermal fields), and testing alternative machine learning architectures, including ensemble deep learning methods, to enhance predictive accuracy.

**Author Contributions:** Conceptualization, A.Y., R.A., G.S., A.E., A.A., G.S., S.O. and M.U.; methodology, A.Y., R.A., G.S. and A.A.; software, J.E., A.E., F.S. and O.O.; formal analysis, F.S. and O.O.; resources, G.O.; data curation, A.E.; writing—original draft preparation, A.Y., R.A. and J.E.; writing—review and editing, A.Y., R.A., J.E., G.S. and G.O.; visualization, F.S. and O.O.; project administration, A.A., S.O. and M.U. All authors have read and agreed to the published version of the manuscript.

**Funding:** This research received no external funding.

**Data Availability Statement:** Data are available at request.

**Conflicts of Interest:** The authors declare no conflict of interest.

## Abbreviations

The following abbreviations are used in this manuscript:

| | |
|---|---|
| L | Length |
| $\rho$ | Mass Density of Beam |
| I | Moment of Inertia |
| $(e_0a)$ | Nonlocal Parameter |
| K | Spring Stiffness |
| A | Area of Cross-Section |
| $I_0, I_1$ | Bessel Terms |
| t | Time |
| t* | Dimensionless Time |
| P | Pressure |
| w | Lateral Displacement of Beam |
| $k_1, k_2$ | Winkler-Like Elastic Foundation Coefficient |
| T | Number of Trees in the Ensemble |
| B | 2D Magnetic Field |
| G | Shear Modulus |
| $R_i$ | Radius of CNT |
| t | Elastic Stiffness Matrix of Classical Isotropic Elasticity |
| MFI | Magnetic Field Intensity |
| MFD | Magnetic Field Density |
| $\rho_f$ | Fluid Density |
| $T_b(x)$ | Prediction of the b-th Tree at an Instance x |
| $w_{ij}$ | The Weight of the j-th Leaf in the i-th Tree |
| $\lambda_j$ | The L2 Regularization Parameter |
| $\Omega_{CAT}$ | The regularization Term for CATBoost |
| E | Youngs Modulus |
| U(r) | Fluid Velocity Distribution |
| $m_f$ | Mass of Nanofluid |
| $t_b$ | Thickness |
| $\sigma_{x\,x}$ | Nonlocal Stress Tensor |
| $I_s$ | Surface Moment of Inertia |
| $y_i$ | True Label of the i-th Instance |
| $\hat{y}_i$ | Predicted Label for the i-th Instance |
| Kn | Knusden Number |
| $\varnothing$ | Downstream Angle |
| d | The Number of Leaves in Each Tree |
| $\gamma_L$ | The L1-Regularisation Parameter |
| $L_a$ | The Loss Function |
| $n_i$ | The Number of Trees in the Forest |

## References


1. Chen, X.; Zhao, J.; She, G.-L.; Luo, J. Nonlinear free vibration analysis of functionally graded carbon nanotube reinforced fluid-conveying pipe in thermal environment. *Steel Compos. Struct.* **2022**, *35*, 641–652. [CrossRef]
2. Saifuddin, N.; Raziah, A.Z.; Junizah, A.R. Carbon Nanotubes: A Review on Structure and Their Interaction with Proteins. *J. Chem.* **2013**, *2013*, 676815. [CrossRef]
3. Rust, C.; Shapturenka, P.; Spari, M.; Jin, Q.; Li, H.; Bacher, A.; Guttmann, M.; Zheng, M.; Adel, T.; Hight Walker, A.R.; et al. The Impact of Carbon Nanotube Length and Diameter on their Global Alignment by Dead-End Filtration. *Small* **2023**, *19*, 2206774. [CrossRef]
4. Sakharova, N.A.; Pereira, G.A.F.; Antunes, J.M.; Fernandes, J.V. Mechanical Characterization of Multiwalled Carbon Nanotubes: Numerical Simulation Study. *Materials* **2020**, *13*, 4283. [CrossRef]
5. Xu, L.; Jiao, X.; Shi, C.; Wu, A.-P.; Hou, P.-X.; Liu, C.; Cheng, H.-M. Single-Walled Carbon Nanotube/Copper Core–Shell Fibers with a High Specific Electrical Conductivity. *ACS Nano* **2023**, *17*, 9245–9254. [CrossRef]
6. Iijima, S. Helical microtubules of graphitic carbon. *Nature* **1991**, *354*, 56–58. [CrossRef]
7. Cui, K.; Chuang, J.; Feo, L.; Chow, C.; Lau, D. Developments and Applications of Carbon Nanotube Reinforced Cement-Based Composites as Functional Building Materials. *Front. Mater.* **2022**, *9*, 861646. [CrossRef]
8. Yinusa, A.; Sobamowo, M.; Adelaja, A. Nonlinear vibration analysis of an embedded branched nanofluid-conveying carbon nanotube: Influence of downstream angle, temperature change and two-dimensional external magnetic fields. *Nano Mater. Sci.* **2020**, *2*, 323–332. [CrossRef]
9. Mahmure, A.; Tornabene, F.; Dimitri, R.; Kuruoglu, N. Free Vibration of Thin-Walled Composite Shell Structures Reinforced with Uniform and Linear Carbon Nanotubes: Effect of the Elastic Foundation and Nonlinearity. *Nanomaterials* **2021**, *11*, 2090. [CrossRef]
10. Huang, K.; Yao, J. Beam Theory of Thermal–Electro-Mechanical Coupling for Single-Wall Carbon Nanotubes. *Nanomaterials* **2021**, *11*, 923. [CrossRef]
11. Shaba, A.I.; Jiya, M.; Aiyesimi, Y.M.; Mohammed, A.A. Vibration of Single–Walled Carbon Nanotube (SWCNT) on a Winkler Foundation with Magnetic Field Effect. *Sci. World J.* **2021**, *16*, 402–406.
12. Yinusa, A.; Sobamowo, M. Mechanics of nonlinear internal flow-induced vibration and stability analysis of a pre-tensioned single-walled carbon nanotube using classical differential transform method with CAT and SAT after-treatment techniques. *Forces Mech.* **2022**, *7*, 100083. [CrossRef]
13. Yan, Y.; Li, J.; Ma, X.; Wang, W. Application and dynamical behaviour of CNTs as fluidic nanosensors based on the nonlocal strain gradient theory. *Sens. Actuators A Phys.* **2021**, *330*, 112836. [CrossRef]
14. Chen, Y.; Lyu, M.; Zhang, Z.; Yang, F.; Li, Y. Controlled Preparation of Single-Walled Carbon Nanotubes as Materials for Electronics. *ACS Cent. Sci.* **2022**, *8*, 1490–1505. [CrossRef] [PubMed]
15. Miyashiro, D.; Taira, H.; Hamano, R.; Reserva, R.L.; Umemura, K. Mechanical vibration of single-walled carbon nanotubes at different lengths and carbon nanobelts by modal analysis method. *Compos. Part C Open Access* **2020**, *2*, 100028. [CrossRef]
16. Dat, N.D.; Quan, T.Q.; Mahesh, V.; Duc, N.D. Analytical solutions for nonlinear magneto-electro-elastic vibration of the smart sandwich plate with carbon nanotube reinforced nanocomposite core in hygrothermal environment. *Int. J. Mech. Sci.* **2020**, *186*, 105906. [CrossRef]
17. De Rosa, M.A.; Lippiello, M.; Babilio, E.; Ceraldi, C. Nonlocal Vibration Analysis of a Nonuniform Carbon Nanotube with Elastic Constraints and an Attached Mass. *Materials* **2021**, *14*, 3445. [CrossRef]
18. Oveissi, S.; Eftekhari, S.A.; Toghraie, D. Longitudinal vibration and instabilities of carbon nanotubes conveying fluid considering size effects of nanoflow and nanostructure. *Phys. E Low-Dimens. Syst. Nanostruct.* **2016**, *83*, 164–173. [CrossRef]
19. Chang, X.; Hong, X.; Qu, C.; Li, Y. Stability and nonlinear vibration of carbon nanotubes-reinforced composite pipes conveying fluid. *Ocean. Eng.* **2023**, *281*, 114960. [CrossRef]
20. Sabahi, M.A.; Saidi, A.R.; Khodabakhsh, R. An analytical solution for nonlinear vibration of functionally graded porous micropipes conveying fluid in damping medium. *Ocean. Eng.* **2022**, *245*, 110482. [CrossRef]
21. Moborukoje, K.I.; Oladosu, S.; Kuku, R.; Sobamowo, G. Exploring the Significances of Surface Energy, Initial Stress and Nonlocality Effects on Vibration of Carbon Nanotubes Conveying Fluid Resting on Elastic Foundations in a Magneto-Thermal Environment. *World Sci. News* **2022**, *179*, 112–134.
22. Ghadirian, H.; Mohebpour, S.; Malekzadeh, P.; Daneshmand, F. Nonlinear free vibrations and stability analysis of FG-CNTRC pipes conveying fluid based on Timoshenko model. *Compos. Struct.* **2022**, *292*, 115637. [CrossRef]
23. Heidary, Z.; Ramezani, S.R.; Mojra, A. Exploring the benefits of functionally graded carbon nanotubes (FG-CNTs) as a platform for targeted drug delivery systems. *Comput. Methods Programs Biomed.* **2023**, *238*, 107603. [CrossRef] [PubMed]
24. Oveissi, S.; Ghassemi, A.; Salehi, M.; Eftekhari, S.A.; Ziaei-Rad, S. Hydro–Hygro–Thermo–Magneto–Electro elastic wave propagation of axially moving nano-cylindrical shells conveying various magnetic-nano-fluids resting on the electromagnetic-visco-Pasternak medium. *Thin-Walled Struct.* **2022**, *173*, 108926. [CrossRef]

25. Oveissi, S.; Salehi, M.; Ghassemi, A.; Ali Eftekhari, S.; Ziaei-Rad, S. Energy harvesting of nanofluid-conveying axially moving cylindrical composite nanoshells of integrated CNT and piezoelectric layers with magnetorheological elastomer core under external fluid vortex-induced vibration. *J. Magn. Magn. Mater.* **2023**, *572*, 170551. [CrossRef]
26. Yinusa, A.; Sobamowo, M.; Adelaja, A.; Ojolo, S.; Waheed, M.; Usman, M.; Siqueira, A.M.D.O.; Campos, J.C.C.; Ola-Gbadamosi, R. Numerical investigation on the thermal-nanofluidic flow induced transverse and longitudinal vibrations of single and multi-walled branched nanotubes resting on nonlinear elastic foundations in a magnetic environment. *Partial Differ. Equ. Appl. Math.* **2024**, *9*, 100602. [CrossRef]
27. Vivanco-Benavides, L.E.; Martínez-González, C.L.; Mercado-Zúñiga, C.; Torres-Torres, C. Machine learning and materials informatics approaches in the analysis of physical properties of carbon nanotubes: A review. *Comput. Mater. Sci.* **2021**, *201*, 110939. [CrossRef]
28. Pollayi, H.; Rao, P. Stealth Carbon Nano-tubes (S-CNTs): AI/ML Based Modelling of Nano-structured Composites for Attenuation and Shielding in Aircraft. In Proceedings of the International Symposium on Lightweight and Sustainable Polymeric Materials (LSPM23), Bangkok, Thailand, 17 February 2023; pp. 1–10. [CrossRef]
29. Čanađija, M. Deep learning framework for carbon nanotubes: Mechanical properties and modelling strategies. *Carbon* **2021**, *184*, 891–901. [CrossRef]
30. Yang, J.; Ni, Z.; Fan, Y.; Hang, Z.; Liu, H.; Feng, C. Machine learning aided uncertainty analysis on nonlinear vibration of cracked FG-GNPRC dielectric beam. *Structures* **2023**, *58*, 105456. [CrossRef]
31. Akbarzadeh, S.; Akbarzadeh, K.; Ramezanzadeh, M.; Naderi, R.; Mahdavian, M.; Olivier, M. Corrosion resistance enhancement of a sol-gel coating by incorporation of modified carbon nanotubes: Artificial neural network (ANN) modelling and experimental explorations. *Prog. Org. Coat.* **2022**, *174*, 107296. [CrossRef]
32. Khanam, P.N.; AlMaadeed, M.A.; AlMaadeed, S.; Kunhoth, S.; Ouederni, M.; Sun, D.; Hamilton, A.; Harkin Jones, E.; Mayoral, B. Optimization and Prediction of Mechanical and Thermal Properties of Graphene/LLDPE Nanocomposites by Using Artificial Neural Networks. *Int. J. Polym. Sci.* **2016**, *2016*, 5340252. [CrossRef]
33. Hajilounezhad, T.; Bao, R.; Palaniappan, K.; Bunyak, F.; Calyam, P.; Maschmann, M.R. Predicting carbon nanotube forest attributes and mechanical properties using simulated images and deep learning. *NPJ Comput. Mater.* **2021**, *7*, 134. [CrossRef]
34. Kekez, S.; Kubica, J. Application of Artificial Neural Networks for Prediction of Mechanical Properties of CNT/CNF Reinforced Concrete. *Materials* **2021**, *14*, 5637. [CrossRef] [PubMed]
35. Chen, T.; Guestrin, C. XGBoost: A Scalable Tree Boosting System. In Proceedings of the 22nd Acm Sigkdd International Conference on Knowledge Discovery and Data Mining, San Francisco, CA, USA, 13–17 August 2016. [CrossRef]
36. Sunday, G.J.; Oladosu, S.A.; Kuku, R.O.; Sobamowo, G.M.; Yinusa, A.A. Explorations of the combined effects of surface energy, initial stress and nonlocality on the dynamic behaviour of carbon nanotubes conveying fluid resting on elastic foundations in a thermo-magnetic environment. *J. Nanosci. Technol.* **2023**, *1*, 102.
37. Abubakar, E.H.; Sobamowo, M.G.; Sadiq, M.O.; Yinusa, A.A. Vibration of Single-walled Carbon Nanotubes Resting on Elastic Medium with Magnetic and Thermal Effects under the Influence of Electrostatic Force. *World Sci. News* **2024**, *188*, 93–118.
38. Selim, M.M.; Musa, A. Nonlinear Vibration of a Pre-Stressed Water-Filled Single-Walled Carbon Nanotube Using Shell Model. *Nanomaterials* **2020**, *10*, 974. [CrossRef]
39. Strozzi, M.; Elishakoff, I.E.; Bochicchio, M.; Cocconcelli, M.; Rubini, R.; Radi, E. A Comparison of Shell Theories for Vibration Analysis of Single-Walled Carbon Nanotubes Based on an Anisotropic Elastic Shell Model. *Nanomaterials* **2023**, *13*, 1390. [CrossRef]
40. Olaleye, T.I.; Oladosu, S.A.; Kuku, R.O.; Sobamowo, G.M. Variation iteration method to the investigations of surface energy, initial stress and nonlocality on vibration of carbon nanotubes conveying fluid resting on elastic foundations in a thermo-magnetic environment. *J. Mater. Environ. Sci.* **2023**, *14*, 520–538.
41. Brociek, R.; Pleszczyński, M. Differential Transform Method and Neural Network for Solving Variational Calculus Problems. *Mathematics* **2023**, *12*, 2182. [CrossRef]

*vibration*

MDPI

*Article*

# Nonlinear Dynamics of a Coupled Electromechanical Transmission

**Antonio Zippo [1,*], Moslem Molaie [2] and Francesco Pellicano [1]**

[1] Department of Engineering "Enzo Ferrari", Centre InterMech MoRe, University of Modena and Reggio Emilia, 41125 Modena, Italy; francesco.pellicano@unimore.it

[2] Department of Engineering "Enzo Ferrari", University of Modena and Reggio Emilia, 41125 Modena, Italy; moslem_molaie@unimore.it

* Correspondence: antonio.zippo@unimore.it

**Abstract:** The mechanical connection between a transmission system and an electric motor gives rise to a strong interaction between their respective dynamics. In particular, the coupling between an electric motor and a nonlinear spur gear transmission significantly influences the overall dynamic behavior of the integrated system. This study presents a detailed investigation into the electromechanical coupling effects between a permanent magnet synchronous machine (PMSM) and a nonlinear spur gear transmission. To focus on these effects, three configurations are analyzed: (i) a standalone gear pair model without motor interaction, (ii) a combined gear–motor system without dynamic coupling, and (iii) a fully coupled electromechanical system where the mechanical feedback influences motor control. The dynamic interaction between the motor's torsional vibrations and the gear transmission is captured using the derivative of the transmission error as a feedback signal, enabling a closed-loop electromechanical model. Numerical simulations highlight the critical role of this coupling in shaping system dynamics, offering insights into the stability and performance of electric drive–gear transmission systems under different operating conditions. It also underscores the limitations of traditional modeling approaches that neglect feedback effects from the mechanical subsystem. The findings contribute to a more accurate and comprehensive understanding of coupled motor–gear dynamics, which is essential for the design and control of advanced electromechanical transmission systems in high-performance applications.

**Keywords:** nonlinear vibration; electromechanical coupling; permanent magnet synchronous machine; nonlinear spur gear; dynamic modeling

## 1. Introduction

The automotive industry is undergoing a significant transformation with the rapid adoption of electrified powertrains, where permanent magnet synchronous motors play a pivotal role due to their high efficiency, power density, and controllability.

PMSMs are increasingly used in various applications, particularly in electric vehicles and industrial settings, due to their favorable characteristics, including high power density, high efficiency, and superior dynamic performance. PMSMs do not suffer from the drawbacks associated with DC motors, such as maintenance needs and lower robustness, which makes them a preferable choice for a multitude of applications [1,2]. The intrinsic advantages of PMSMs manifest in their high torque-to-weight ratio and the capability for precise speed control [3]. For instance, they demonstrate exceptional performance in servo applications where fast response times and high acceleration are critical [1,4].

Moreover, the application range of PMSMs is expanding, including their integration into renewable energy systems, aerospace technology, and medical devices, showcasing their versatility [5]. This increasing implementation necessitates a focus on improving efficiency and reducing operational noise, aspects that are vital in demanding applications like electric vehicles, where energy efficiency and user comfort are paramount [6,7]. However, integrating PMSMs with mechanical gear systems introduces complex electromechanical coupling phenomena that can adversely affect vehicle performance, particularly in terms of torque ripple, vibrations, and acoustic noise. Gear transmissions themselves are a source of noise and vibration, as with planetary gear transmission, where excessive noise and vibration can be indicators of poor load-sharing and phasing. These issues can lead to reduced system performance, increased wear, and potentially catastrophic failure if left unaddressed [8]. Electromechanical coupling refers to the bidirectional interaction between the electrical dynamics of the motor and the mechanical characteristics of the gear system. In automotive applications, this coupling becomes critical as factors such as gear backlash, meshing stiffness, and torsional vibrations can influence the motor's electromagnetic behavior, leading to challenges in control and noise mitigation. Therefore, a comprehensive understanding of this coupling is essential for accurate modeling, robust control design, and fault diagnosis in advanced electromechanical systems. One of the main disturbances is due to torque ripple in PMSMs, which is a well-known issue that leads to vibrations and acoustic noise, affecting the vehicle's noise, vibration, and harshness (NVH) performance. Various methods have been proposed to minimize torque ripple, including current modulation techniques and rotor design optimizations. Jedryczka et al. [9] presented a method for torque ripple minimization by modulating phase currents, demonstrating significant reductions in torque ripple levels.

In the context of automotive applications, torque ripple and radial electromagnetic (EM) vibrations are critical concerns. Dai et al. [10] focused on an 8-pole, 36-slot Integrated Starter Generator (ISG) motor optimized using a combination of Q-axis and magnetic bridge notching (QMC notch), as well as segmented rotor skewing, to reduce torque ripple and radial EM vibration; their results showed that the QMC notch significantly reduces torque ripple, while skewing designs greatly diminish radial EM vibrations. Another technique that is useful to suppress the dynamic load of the integrated electric drive system is an active damping control strategy [11]; in the considered work, the disturbances were caused by the sudden step electromagnetic torque during rapid acceleration. Overall, the interaction between PMSMs and gear systems introduces complex electromechanical coupling dynamics that cannot be neglected. Jiang et al. studied the electromechanical coupling torsional resonance characteristics of a multistage gear transmission system driven by a low-speed and high-power PMSM, considering the PMSM electromagnetic effect and bending–torsional vibration characteristics [12]. Their findings emphasized the importance of considering the effect of the electromechanical coupling on the design and analysis of such systems. In hybrid electric vehicles (HEVs), the dual-motor electric drive system (DEDS) presents additional challenges as the electromechanical coupling dynamics of the DEDS, including the PMSM and the gear transmission system, considering time-varying mesh stiffness of gears and the nonlinear characteristics of inverters [13]. The dynamic characteristics of the coupled electromechanical system were simulated and analyzed in single-motor and dual-motor drive modes under steady-state and impact conditions.

Furthermore, the influence of torque ripple on the motor–gear system, considering rotor-step skewing, was investigated. A coupled electromechanical model was proposed to investigate the vibration characteristics, highlighting the significance of rotor design in mitigating electromechanical coupling effects [14]. The key issue is the accurate modeling of electromechanical coupling that has become essential for developing effective control strategies. A new method was proposed to analyze electromechanical coupling vibration, emphasizing that speed and dynamic meshing force are the main factors affecting forced vibration [15]. Additionally, another study introduced a dynamic model of a gear transmission system driven by a motor, considering the lubrication and friction characteristics, to define the electromechanical dynamic behavior [16]. In [17], a passive control strategy is proposed to suppress torsional vibrations resulting from electromechanical coupling, focusing on a multi-stable nonlinear energy sink that mitigates these fluctuations; while in [18], a PMSM torque ripple suppression method based on a self-adaptive mutation algorithm (SMA)-optimized iterative learning control (ILC) was presented. The advantage of this method is that it is not necessary to have prior knowledge of the system and motor parameters. This approach effectively reduces torque ripples, leading to improved motor performance. Another relevant aspect is represented by experimental validation and simulation studies, which are crucial for understanding electromechanical coupling effects. An evaluation of torque ripple and tooth forces of a skewed PMSM was conducted using 2D and 3D finite-element (FE) simulations. The study compared various skewing configurations, analyzing torque ripple amplitudes and tooth forces, which significantly impact a vehicle's NVH behavior [19].

Recent advancements in modeling electromechanical coupling phenomena have significantly improved the understanding of integrated motor–gear systems. For instance, studies have demonstrated that coupling the electromagnetic characteristics of motors with mechanical vibrations of gears and shafts can reveal intricate dynamic behaviors such as resonance risks, nonlinear modal interactions, and time-varying excitation effects. A dynamic model combining a nonlinear permeance network of a squirrel-cage induction motor with a planetary gear system, for example, was used to explore how machine slotting, magnetic saturation, and mesh stiffness (MS) influence system behavior under internal excitations [20]. Similarly, other investigations have shown that the magnetic field of induction machines introduces additional vibration modes and alters the torsional response of gear components, emphasizing the importance of integrated analysis to predict resonance regions [21]. Particularly in electric vehicles and hybrid electric vehicles, electromechanical coupling is further complicated by nonlinearities arising from permanent magnet synchronous motors, control systems, and meshing interfaces. Research in this area has established that electromagnetic harmonics and mechanical excitations jointly shape the frequency content of both current signals and dynamic responses, often manifesting in complex phenomena such as mode transitions, resonance amplifications, and jump instabilities under transient operating conditions [13,22]. Furthermore, the presence of unbalanced magnetic pull (UMP) in PMSMs, due to field asymmetries and rotor misalignment, has been shown to induce softening stiffness behavior and unstable dynamic regions, thereby motivating the need for robust modeling frameworks to accurately predict and mitigate such effects [23]. Despite these advances, challenges remain in achieving a unified and computationally efficient framework that can simultaneously capture the electromagnetic nonlinearity, dynamic gear meshing behavior, and coupling effects under realistic working conditions. Many of the existing studies either rely on reduced-order representations of the electrical subsystem or overlook the back-coupling effects that mechanical deformations impose on motor performance. There is a growing demand for multibody simulation approaches that integrate high-fidelity motor models with flexible gear transmission dynamics, enabling better design, diagnosis, and control of advanced electromechanical systems.

This paper aims to investigate the electromechanical coupling between PMSMs and gear pairs, highlighting the differences through the comparison of three configurations in the modeling that accounts for both electrical and mechanical nonlinearities. The focus will be on capturing the dynamic interactions under varying load and speed conditions, exploring their implications for control performance. Particular attention will be paid to torque ripple and resonant phenomena.

## 2. Dynamical Model

The electromechanical coupling in spur gear systems plays a pivotal role in shaping the dynamic behavior and overall efficiency of these systems. This coupling primarily arises from the interaction between the torsional vibrations of the electric motor and those of the gear transmission. Gaining a clear understanding of this interaction is vital for evaluating the operational stability and performance of gear transmissions. In this work, we investigate a spur gear pair, as depicted in Figure 1, to explore the influence of electromechanical coupling on transmission dynamics. In particular, the mechanical output speed is employed as a feedback signal to regulate the motor's power output, thus establishing a closed-loop interaction between the electrical and mechanical subsystems. Although many earlier studies model the motor and gear train as a unified system, they often overlook the feedback effect exerted by the mechanical subsystem on the motor. In contrast, this study formulates the nonlinear governing equations of gear transmission, develops an independent model for the electrical drive, and subsequently integrates them to reflect the coupled dynamics. To emphasize the role of this interaction, we construct and compare three alternative models: (i) a purely mechanical gear model that excludes motor influence; (ii) a motor–gear system without dynamic coupling where the feedback from the gear transmission is not considered, called in this work combined model; and (iii) a fully coupled motor-gear system that considers the effect of the gear transmission. The dynamic responses of all three configurations are then analyzed to thoroughly assess the impact of electromechanical coupling.

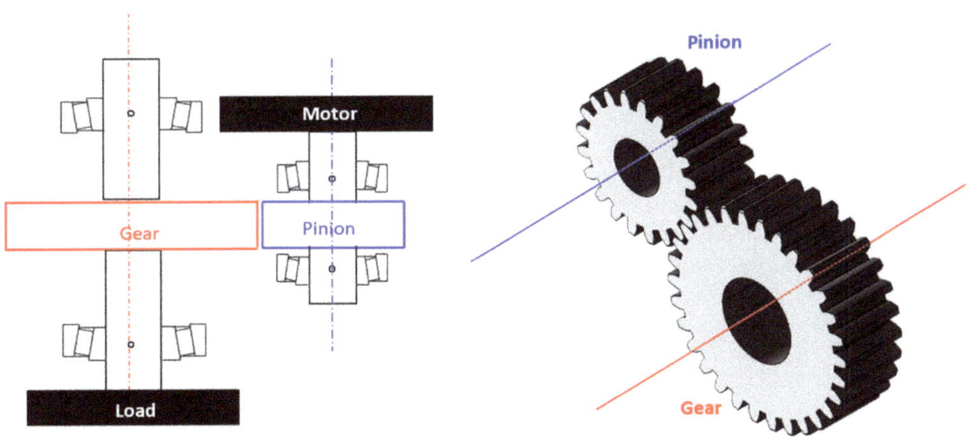


**Figure 1.** A model for an electromechanical gear transmission system.

### 2.1. Gear Pair Dynamical Model

Here, a spur gear pair is considered, see Figure 2. The translational degrees of freedom of both the driver and driven gears are fully constrained, as well as their tilt rotations; only the rotation about the gear axes is allowed. The pinion and gear are rigidly connected to the motor and the load, respectively.

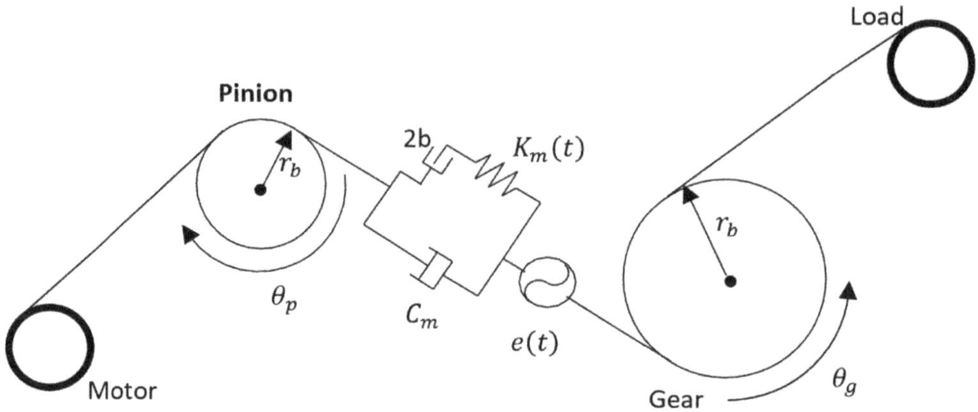


**Figure 2.** A dynamical model of a spur gear pair with a rigid connection to the motor and load.

The equations of motion in the "$\theta$" direction for the gear and pinion are:

$$I_g\ddot{\theta}_g = -r_g F_z - T_l \tag{1}$$

$$I_p\ddot{\theta}_p = r_p F_z + T_m \tag{2}$$

The dynamic transmission error can be defined as:

$$\lambda = \left(r_p\theta_p - r_g\theta_g\right) + e(t)\text{where } e(t) = \sum_{j=1}^{s} e_j\cos(j\omega_m(t)\,t) \tag{3}$$

$e(t)$ represents the geometric transmission error (GTE), it is also known as a kinematic transmission error; in spur gears, it refers to the local gaps or deviations between mating teeth, see Figure 3. These errors can result from various factors, including mounting and manufacturing errors, as well as modifications to the teeth profiles. The presence of GTE is important due to the smooth transmission of motion along the line of action between the teeth. The measurement and analysis of GTE are typically carried out through unloaded tooth contact analysis (UTCA): a very low torque is applied to the gear pair, allowing the measurement of the error without significant elastic deformation [24–26]. This analysis reflects the rigid body transmission error, which is primarily influenced by geometric factors rather than elastic deformation. The GTE is a critical factor in determining the mesh stiffness of the gear pair, which directly affects the dynamic behavior of the system. Indeed, when one performs the loaded tooth contact analysis and needs to calculate the elastic deformation, the GTE must be removed from the total deformation. When addressing the changes in GTE vs. time, a periodic function expresses its variation as time progresses. A common way to represent the GTE of the system is the Fourier series, where the fundamental frequency is the gear mesh frequency $\omega_m(t) = \frac{2\pi}{60}N_p n_s(t)$. Note that $\omega_m(t)$ is time-dependent for both coupled and combined models, $n_s$ is the input shaft speed (rpm), $N_p$ is the number of pinion teeth, $s = (N_1 - 1)/2$ is the number of harmonics of the Fourier series.

The dynamic load of the pinion and its component along the line of action $F_n$ is calculated as:

$$F_n = K_m(t)f(\lambda - e) + C_m\left(\dot{\lambda} - \dot{e}\right) \tag{4}$$

$$f(\lambda - e) = \begin{cases} \lambda + b - e, & \lambda - e > b \\ 0, & -b \le \lambda - e \le b \\ \lambda - b - e, & \lambda - e < -b \end{cases} \tag{5}$$

$f(\lambda - e)$ is the backlash function dependent on the linear displacement $\lambda(t)$, the gap $2 \cdot b$ and the GTE $e(t)$; whenever $\lambda - e$ is between $-b$ and $+b$, the contact loss happens (single-sided impact). In the case of forward motion, for $\lambda - e > b$, the mesh is expected to be in the forward contact (desired situation), while if $\lambda - e < -b$, undesired backside contact happens (double-sided impact) [27,28].

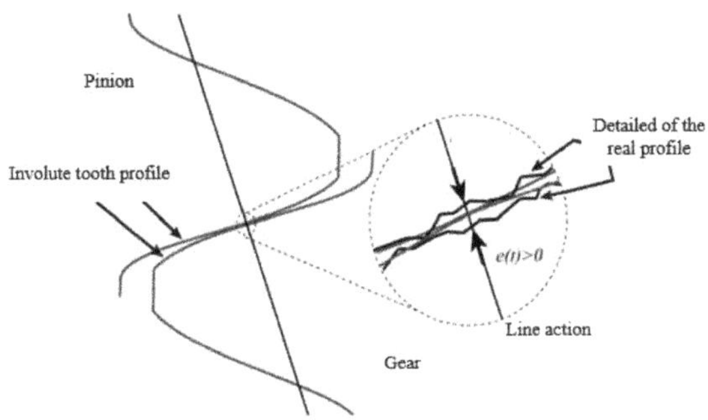


**Figure 3.** Geometric transmission error due to tooth profile imperfections.

The mesh stiffness of gear pairs varies over time as different teeth come into and out of contact during rotation. This variation is influenced by factors such as gear geometry, material properties, and load distribution. Mesh stiffness variations cause fluctuations in the dynamic mesh force, which, in turn, excite vibrations in the gear system. These vibrations might lead to parametric resonance in the system and may contribute to the premature failure of gear teeth due to fatigue. To calculate the MS of the system (see Figure 4), we need to obtain the elastic deflection of the gear pair. Conducting loaded and unloaded tooth contact analyses, namely, LTCA and UTCA, provides us with loaded static transmission errors, which contain not only elastic deflection but also GTE. Thus, the loaded static transmission errors must be subtracted from GTE obtained from unloaded tooth contact analysis. A periodic function is required to express its variation as time progresses. One commonly employed function to represent the MS of the system is the Fourier series with the fundamental mesh frequency; the Fourier series of the meshing stiffness is given by:

$$\begin{cases} \omega_m(t) = \frac{2\pi}{60} N_p n_s(t) \\ s = (N_1 - 1)/2 \end{cases} \Rightarrow K_m(t) = k_0 + \sum_{j=1}^{S} a_j \cos(j\omega_m(t)\,t) + \sum_{j=1}^{S} b_j \sin(j\omega_m(t)\,t) \tag{6}$$

From the pure elastic deformation obtained from the LTCA and UTCA, the meshing stiffness and GTE can be calculated. Therefore, by combining Equations (1)–(4), the final version of the governing equation for a single degree of freedom is written in terms of DTE:

$$m_{eq}\ddot{\lambda} + C_m\left(\dot{\lambda} - \dot{e}\right) + K_m(t) f(\lambda - e) = T_{eq} \tag{7}$$

$$m_{eq} = \left(\frac{(r_p)^2}{I_p} + \frac{(r_g)^2}{I_g}\right)^{-1}, \quad T_{eq} = \left(\frac{T_m}{r_p}\right) \tag{8}$$

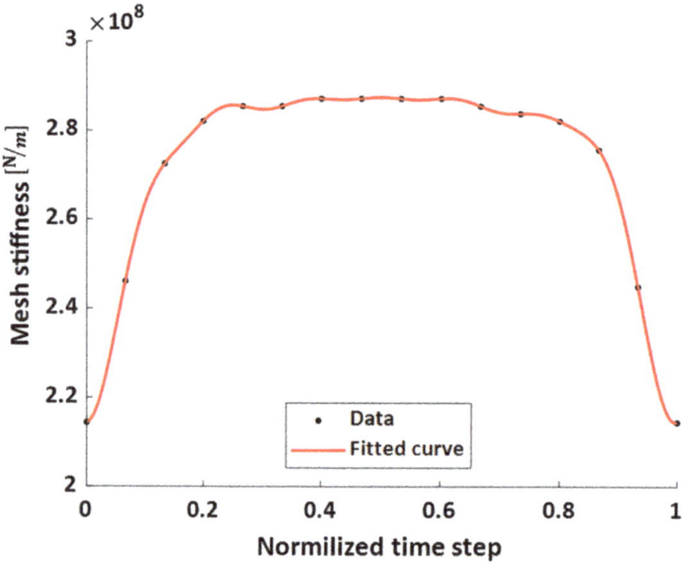


**Figure 4.** Mesh stiffness of the spur gear pair.

The governing Equation (7) is numerically solved by Simulink ODE23t solver.

Simulink's graphical environment allows for an intuitive representation of the dynamic interactions among mechanical elements, such as inertia, damping, backlash, and mesh stiffness. Figure 5 illustrates the Simulink model, which corresponds to the mathematical formulation given by Equation (7).

### 2.2. Model of the Electric Motor

In this study, the electrical drive system is modeled using the PMSM block available in Simulink R2024b [29–31]. This built-in block facilitates the simulation of both motor and generator modes, with customizable electrical and mechanical parameters such as stator resistance, inductance, rotor inertia, and torque constant. It supports machine configurations with either sinusoidal or trapezoidal back electromotive force (EMF) profiles and allows for the selection of round or salient-pole rotor types. By accurately reflecting the dynamic behavior of the PMSM under different load conditions and control strategies, this model enables a detailed investigation of electromechanical coupling effects in the system. Figure 6 illustrates the Simulink model for the motor under investigation.

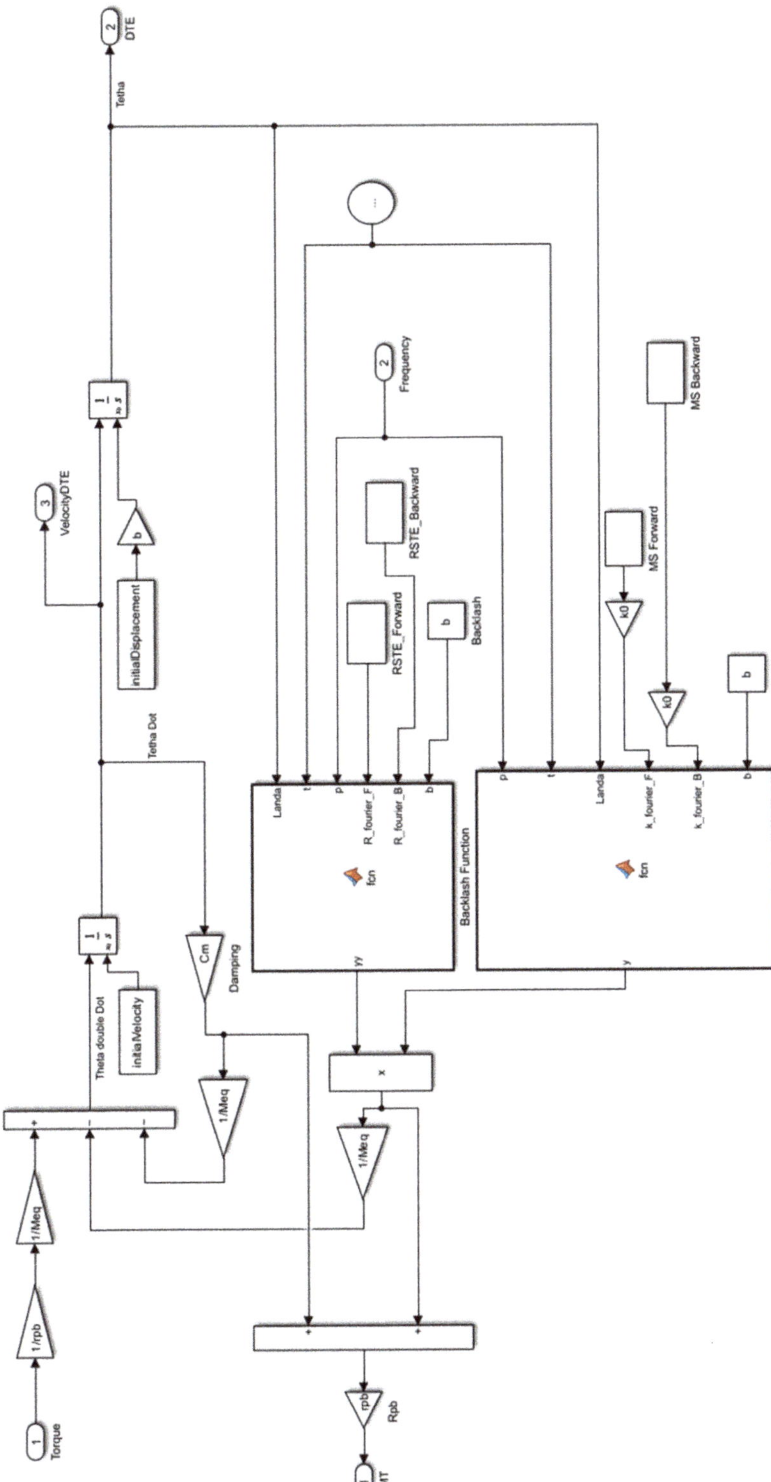


Figure 5. The Simulink model of the gear pair transmission system.

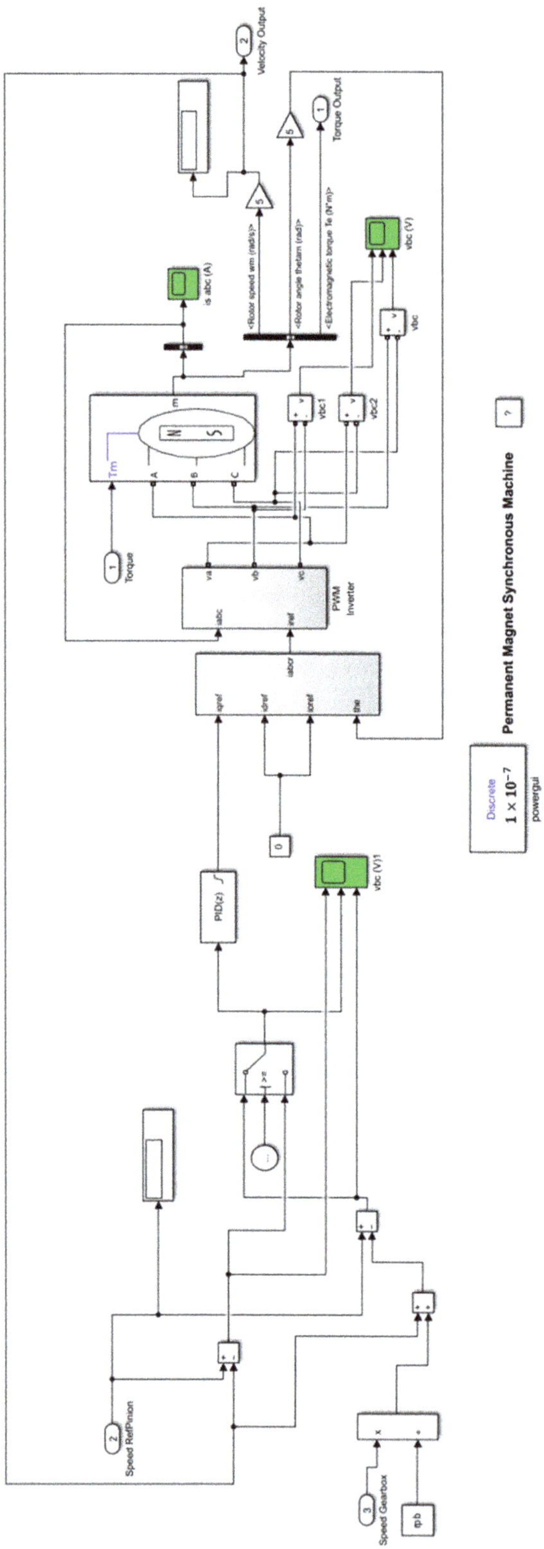


**Figure 6.** The Simulink model of an electric motor.

The permanent magnet synchronous machine block, shown in Figure 6, implements a three-phase or a five-phase permanent magnet synchronous machine. The stator windings are connected to an internal neutral point. The three-phase machine can have a sinusoidal or trapezoidal back EMF waveform. The rotor can be a round or salient pole for the sinusoidal machine. The rotor is round when the machine is trapezoidal. Preset models are available for the sinusoidal back EMF machine. The five-phase machine has a sinusoidal back EMF waveform and a round rotor. The permanent magnet synchronous machine block operates in either generator or motor mode. The mode of operation is dictated by the sign of the mechanical torque (positive for motor mode, negative for generator mode). The electrical and mechanical parts of the machine are each represented by a second-order state-space model. The sinusoidal model assumes that the flux established by the permanent magnets in the stator is sinusoidal, which implies that the electromotive forces are sinusoidal.

Equations (9)–(13) are expressed in the rotor reference frame using an extended Park transformation. All quantities in the rotor reference frame are referred to the stator [29–31].

$$\frac{d}{dt}i_{d1} = \frac{v_{d1}}{L} - \frac{R}{L}i_{d1} + \frac{L_q}{L}p\omega_m i_{q1} \tag{9}$$

$$\frac{d}{dt}i_{q1} = \frac{v_{q1}}{L} - \frac{R}{L}i_{q1} - \frac{L_d}{L}p\omega_m i_{d1} - \frac{\lambda p\omega_m}{L} \tag{10}$$

$$\frac{d}{dt}i_{d2} = \frac{v_{d2}}{L} - \frac{R}{L}i_{d2} \tag{11}$$

$$\frac{d}{dt}i_{q2} = \frac{v_{q2}}{L} - \frac{R}{L}i_{q2} \tag{12}$$

$$T_e = 2.5p\lambda i_{q1} \tag{13}$$

In the simulation framework developed in Simulink–MATLAB, the feedback signals are modeled as analog signals. A fixed discrete simulation time step of 0.1 μs (i.e., $1 \times 10^{-7}$ s) is used, corresponding to a system bandwidth of approximately 5 MHz. In this work, three different configurations are considered to evaluate the influence of the electrical subsystem on the dynamics of an electromechanical gear transmission system:

**(a) Gear-pair system without motor:** This configuration is used not only to isolate and evaluate the influence of the electric motor but also to validate the model against experimental data reported by Kahraman and Blankenship [32]. This validation step is essential to ensuring that the proposed Simulink–MATLAB model is both accurate and stable for analyzing the system's dynamic response.

**(b) Gear-pair system combined with motor (no coupling):** To incorporate the motor's effect, one approach is to connect the motor directly to the gear pair without including any feedback from the mechanical subsystem. In this case, only internal motor feedback (i.e., the motor's output speed) is used for control, as shown in Figure 7a. Although the input speed and torque applied to the gear transmission vary over time, they are not influenced by the gear transmission system, like the fundamental mesh frequency. This is because there is no feedback from the gear system to the motor, so the mechanical dynamics do not affect the motor's behavior.

**(c) Gear-pair system coupled with motor:** This configuration more closely reflects real-world conditions, as it includes feedback from the gear transmission. Specifically, the first derivative of the dynamic transmission error, denoted as $\dot{\lambda}$, is used to influence the motor's electrical behavior. This coupling introduces a two-way interaction between the mechanical and electrical parts of the system. The feedback from the gear transmission is shown in Figure 7b (blue path).

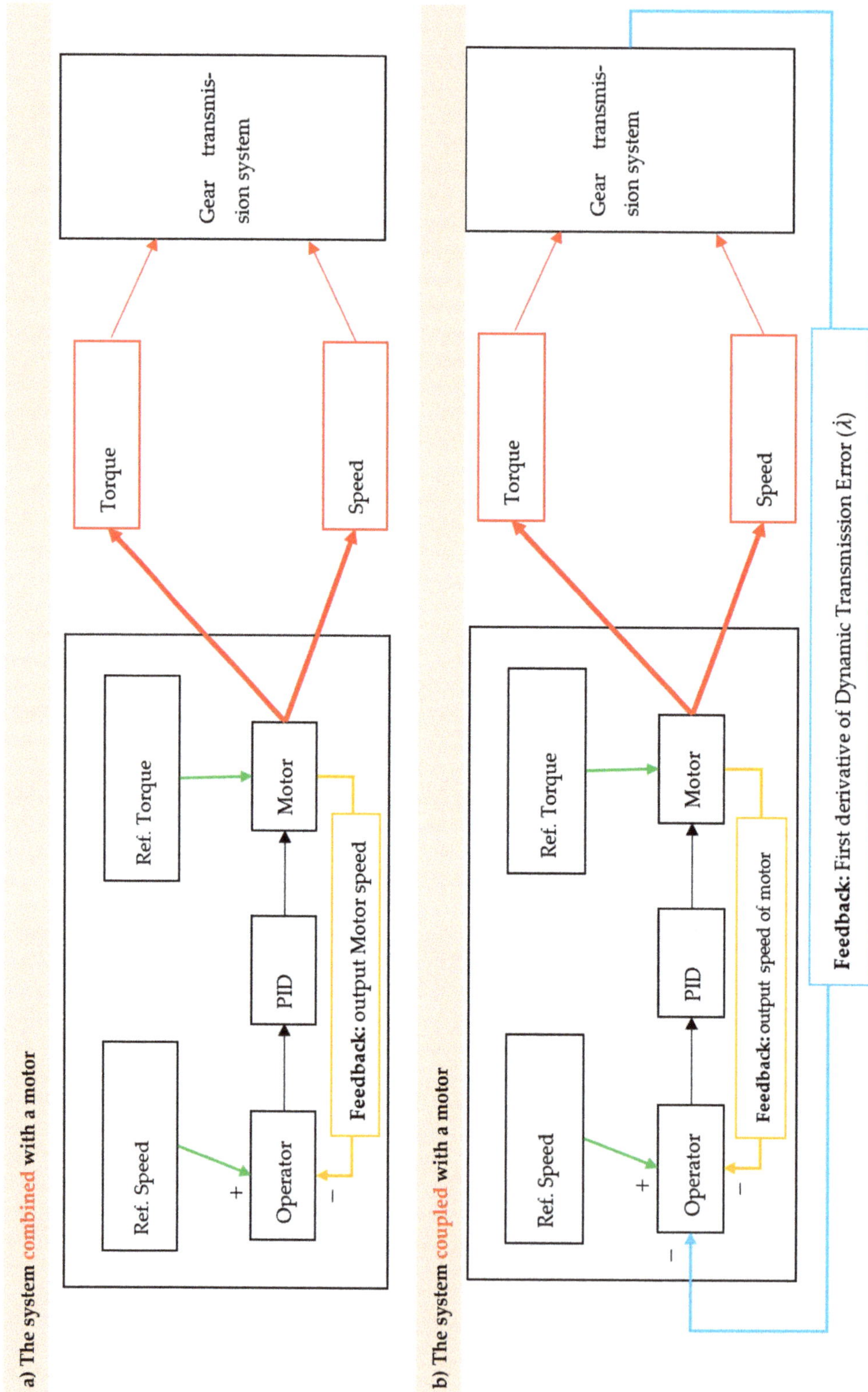


**Figure 7.** The proposed model for considering the effect of the motor.

## 3. Numerical Results

The nonlinear differential Equation (7) of the gear pair, which includes time-varying parameters, represents a non-smooth and non-autonomous dynamical system, after eliminating the rigid body motion; this equation, combined with the dynamic model of the electric motor, is solved numerically using Simulink–MATLAB. During each simulation, the transient is removed from the total response by cancelling the first 50 periods, obtaining the steady-state response. The total simulation time is set to 3.8 s, which means about 70 rounds of the pinion and 3500 meshing periods. The nominal torque and speed are 340 Nm and 115 rad/s, respectively, the fundamental shaft frequency is about 18.3 Hz, and the meshing frequency is 915 Hz, while the natural frequency of the system is 3051 Hz ($1.917 \times 10^4$ rad/s). The simulation assumes no manufacturing error nor tip or roof reliefs, i.e., $e(t) = 0$, and the mesh stiffness is simulated using seven harmonics in the Fourier series (the highest harmonic is 21,357 Hz).

### 3.1. Validation of the Nonlinear Gear Model

To assess the accuracy of the gear model, a comparison with experimental results is carried out. Kahraman et al. conducted a series of experiments on a setup of a gear pair where the backlash and teeth deflections were not negligible (parametric excitation) and input torque was present. One of these tests focused on a spur gear set with the geometrical and physical parameters presented in Table 1 [32,33]:

**Table 1.** Pinion and gear data.

| Tooth Parameters | Pinion | | Gear |
|---|---|---|---|
| No. of teeth | 50 | | 50 |
| Module [mm] | 3 | | 3 |
| Pressure angle [Deg] | 20 | | 20 |
| Base diameter [mm] | 140.95 | | 140.95 |
| Tooth thickness at pitch diameter [mm] | 4.64 | | 4.64 |
| Outer diameter [mm] | 156 | | 156 |
| Root diameter [mm] | 140.68 | | 140.68 |
| Face width [mm] | 20 | | 20 |
| Inertia [kg m$^2$] | 0.0074 | | 0.0074 |
| Backlash [mm] | | 0.1447 | |
| Center distance [mm] | | 150 | |
| Damping ratio [34] | | 0.01 | |

The amplitude–frequency diagrams and the root-mean-square (RMS) of the nondimensional dynamic transmission error are used to compare the experimental data [32] with the present numerical simulations. Figure 8 shows a good agreement between the numerical results and the experimental data. Both results, experimental data and Simulink results, clearly capture the key features of nonlinear dynamic behavior, including the softening behavior, primary and super-harmonic resonances, and jump phenomena. The overall shape and the frequency of the resonance are consistent across both cases, confirming the accuracy of the model. Minor discrepancies, such as higher amplitudes in the simulation, may be attributed to idealizations in damping or parameter estimation, but they do not detract from the overall close correspondence. These results confirm that the Simulink model provides a reliable representation of the system's nonlinear dynamic response.

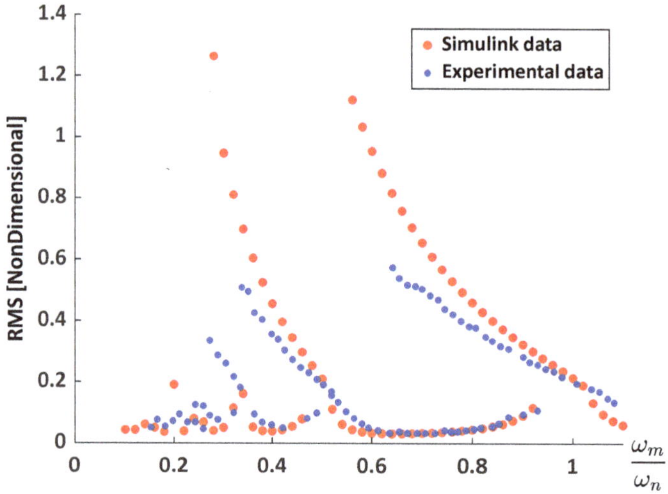


**Figure 8.** A comparison of the simulated model in Simulink and experimental data.

### *3.2. Dynamic Mesh Torque, Output Torque, and Speed of Motor*

To analyze the gear system coupled with the motor, during the simulation, the gear pair is coupled with the motor at $t = 0.2$ s over a total of 3.8 s. After that instant, the system receives the feedback from the gear pair. As mentioned, the nominal torque is 340 N·m, so as shown in Figure 9a, the output torque of the motor fluctuates around this value with about 14 N·m peak-to-peak oscillation, i.e., 2.06% fluctuation. Regarding the speed of the motor, it stabilizes at the nominal speed of 115 rad/s (18.3 Hz) after 0.2 s, see Figure 9b; after the transient, it continues to fluctuate about the nominal value. The amplitude of fluctuation for both systems, coupled and combined with the motor, is different. Mainly, these characteristics of the coupled system are higher than the combined one.

Regarding the motor speed, it takes some time to reach the nominal value of 115 rad/s (18.3 Hz), as illustrated in Figure 9b. Once stabilized, the speed fluctuates slightly around the nominal level. Notably, the amplitude of both torque and speed fluctuations differs between the coupled and combined systems. These fluctuations are generally more pronounced in the coupled configuration compared to the combined system without gear feedback. The repetitive dynamic load is the main cause of fatigue phenomena such as failure in the tooth root, pitting, and scoring on the tooth surface. Therefore, dynamic mesh torque (DMT) is a key parameter in manufacturing a gear transmission system. The DMT is evaluated by the following formula:

$$DMT_{pinion} = r_p\left[C_m\left(\dot{\lambda} - \dot{e}\right) + K_m f(\lambda - e)\right] \tag{14}$$

Figure 9c represents the DMT for the two cases: without and coupled with the motor. From the results, it seems the DMT of the coupled system has a higher frequency component than in the case without the motor.

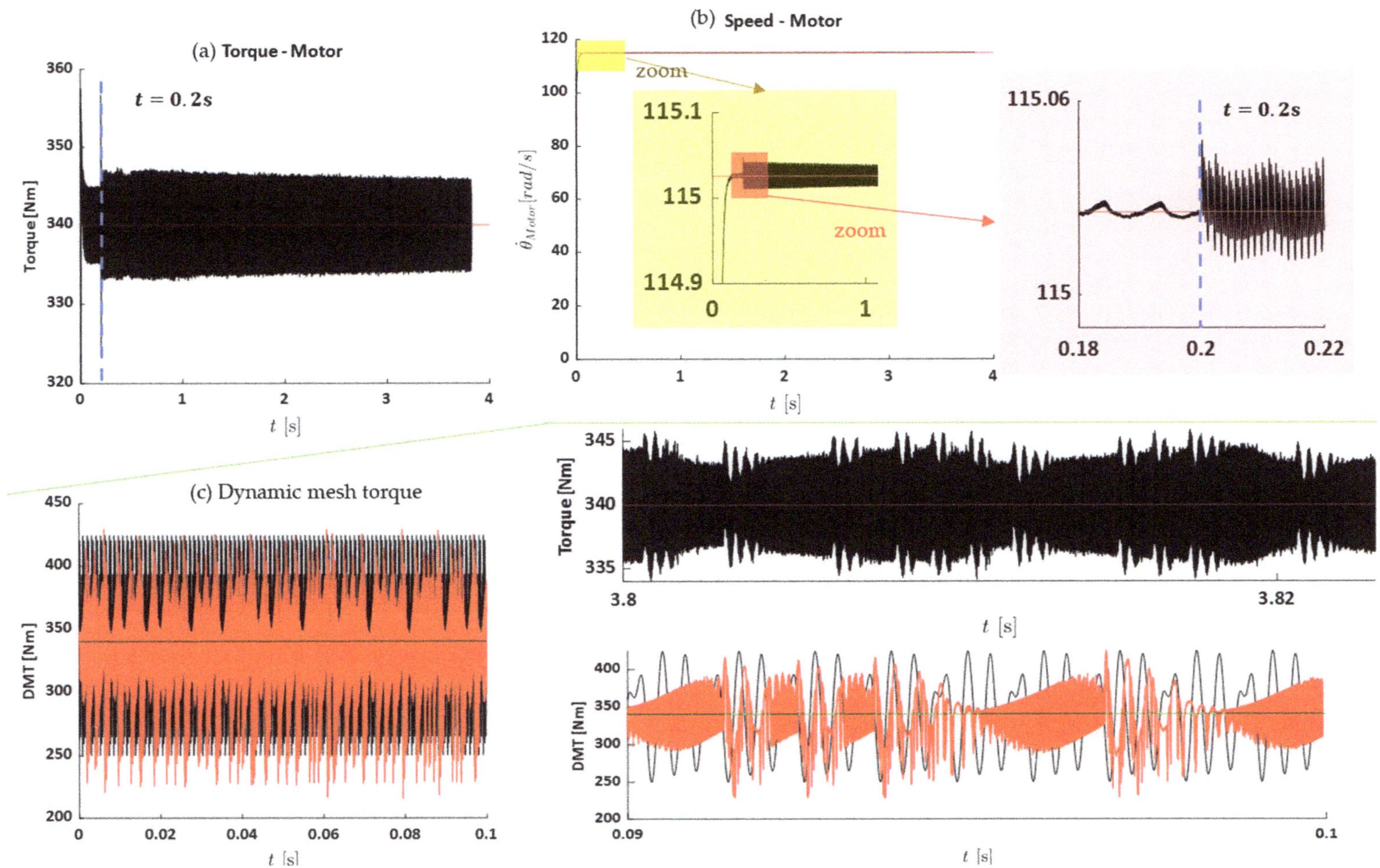


**Figure 9.** Output results of the motor: (**a**) torque, (**b**) speed, and (**c**) dynamic mesh torque for two cases: • coupled with the motor; • without the motor.

### 3.3. Dynamic Response of the Systems

The primary parameter used to assess the nonlinear vibrations of the system is the dynamic transmission error (DTE). Figure 10 shows the vibration levels experienced by gears for the three configurations: (i) coupled (red), (ii) combined (blue), and (iii) without the motor (black). Our findings indicate that connecting the electric motor to the gear transmission without incorporating feedback from the mechanical subsystem tends to overestimate the vibration response. This can potentially lead to incorrect system design or inappropriate component selection. The average dynamic transmission error for the three cases is as follows:

- **System without the motor:** $\lambda_{ave} = 0.085$ mm
- **Combined system:** $\lambda_{ave} = 0.082$ mm
- **Coupled system:** $\lambda_{ave} = 0.086$ mm

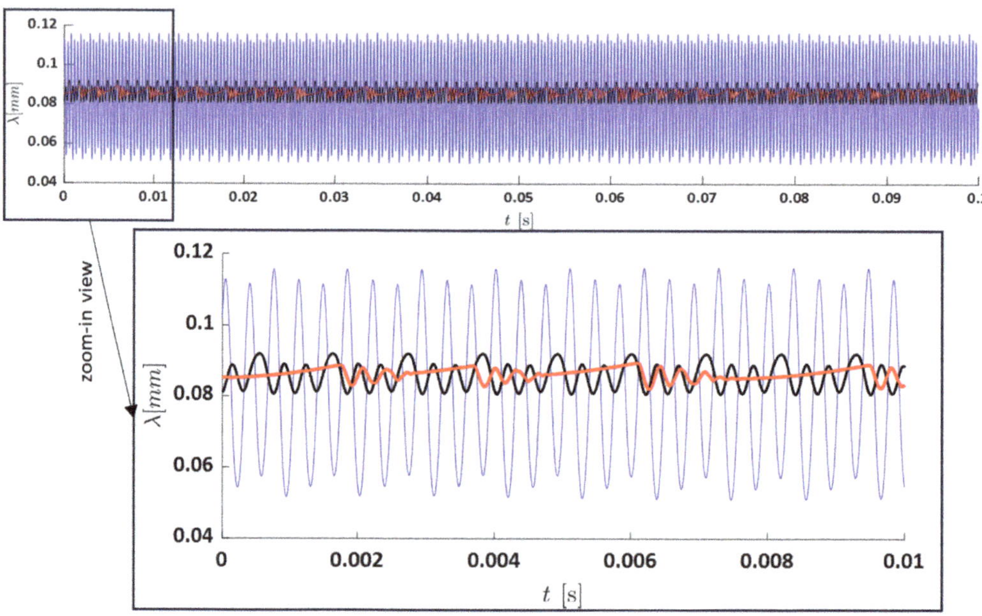


**Figure 10.** Dynamic transmission error, i.e., $\lambda$, for $T_n = 340$ N·m. (• red) Coupled with the motor; (• black) without the motor; (• blue) combined with the motor.

#### Dynamics of the Systems Under High and Low Levels of Torque

Here, the dynamics of the system under different levels of required torque are analyzed. In this part, two torque levels are considered: low and high, corresponding to half and double the nominal torque, respectively. At low torque (Figure 11), the effect of the motor, when combined with the gear transmission system, is negligible. However, when the motor is coupled with the gear transmission system, the differences become more pronounced. This suggests that the combined system does not fully account for the motor's effect on the dynamic response. In the second scenario, where the system operates under a high torque level, only two systems are compared: one with the motor coupled and one without, as shown in Figure 12.

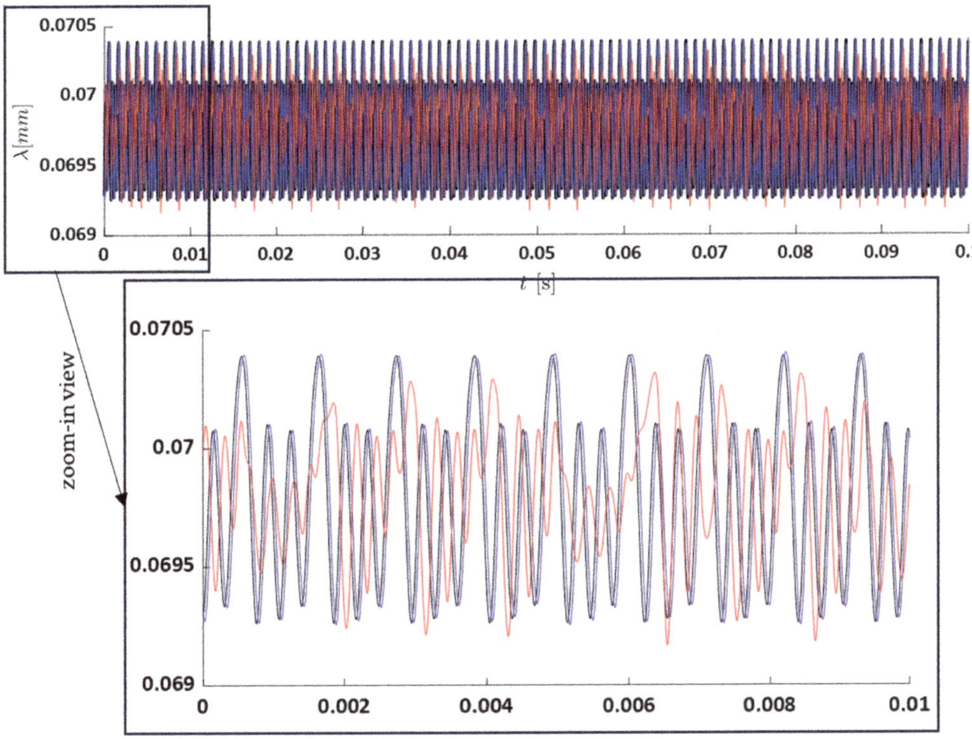


**Figure 11.** Dynamic transmission error, i.e., $\lambda$, for low-level torque: $T = 0.5 \times T_n$. (• red) Coupled with the motor; (• black) without the motor; (• blue) combined with the motor.

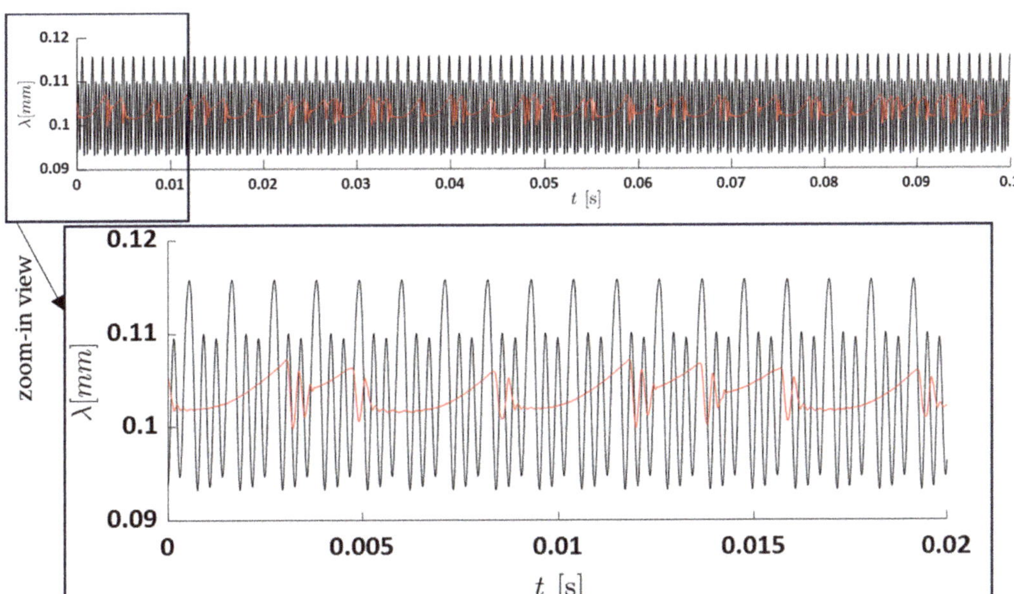


**Figure 12.** Dynamic transmission error, $\lambda$, for high-level torque: $T = 2 \times T_n$. (• red) Coupled with the motor; (• black) without the motor.

### 3.4. Spectral Analysis

Another insightful analysis of the system dynamics regards the spectrum, computed by using the fast Fourier transform (FFT). The spectra reveal a series of peaks at discrete frequencies, signifying the presence of periodic behavior with minor fluctuations over time (see Figure 13a,b). The strength of this periodic behavior is crucial; pronounced and distinct peaks denote a robust periodicity within the system. Conversely, if periodic behavior is weaker and less prominent, the spectrum will lack distinct peaks (see Figure 13c), resembling a continuous band with some degree of variability. This suggests a weaker or more intricate periodicity within the system. Notably, during the spectral analysis, distinct spikes appear at constant frequency intervals, denoted as $\Omega$, which correspond to beat frequencies, as shown in Table 2. These peaks are integer multiples of the fundamental mesh frequency, indicating that the observed frequencies are harmonically related to the gear meshing process. Consequently, the approximate period of the resulting modulation frequency can be estimated as: $T_i \approx 1/\Omega_i$.

**Table 2.** FFT spectrum data at $\omega_m = 115\ \frac{\text{rad}}{\text{s}} = 18.3$ Hz.

| $i$ | Combined | | Without | | Coupled | |
|---|---|---|---|---|---|---|
| | $\Omega_i$ [Hz] | $\frac{\Omega_i}{\omega_m}$ | $\Omega_i$ [Hz] | $\frac{\Omega_i}{\omega_m}$ | $\Omega_i$ [Hz] | $\frac{\Omega_i}{\omega_m}$ |
| 1 | 110 | 6 | 920 | 50 | 110 | 6 |
| 2 | 910 | 50 | 1830 | 100 | 370 | 20 |
| 3 | 1830 | 100 | 2741 | 150 | 910 | 50 |
| 4 | 2750 | 150 | 3661 | 200 | 2961 | 162 |
| 5 | 3660 | 200 | 4572 | 250 | 3221 | 176 |
| 6 | 4580 | 250 | 5492 | 500 | | |
| 7 | 5500 | 300 | 6403 | 550 | | |
| 8 | 6410 | 350 | 8244 | 450 | | |

Figure 14 shows the spectrum of the motor's output speed for three different systems. The system without the motor maintains a constant speed of 115 Hz. In contrast, the other two systems, i.e., coupled and combined systems, exhibit fluctuating speeds. The level of speed oscillation is higher in the system coupled with the motor (peak-to-peak is around 0.178 Hz) compared to the system combined with the motor (peak-to-peak is around 0.012 Hz). Interestingly, for the system combined with the motor, the spectrum reveals a single, prominent spike at 110 Hz, which is attributed to the shaft speed ($6 \times 18.3$ Hz), with no other significant spikes observed in the diagram (see Figure 14c, blue). In the system where the motor is coupled, in addition to the spike at 110 Hz, another spike appears at 915 Hz, corresponding to the mesh frequency (50 teeth $\times$ 18.3 Hz shaft speed). Furthermore, a sideband at 110 Hz is also present.

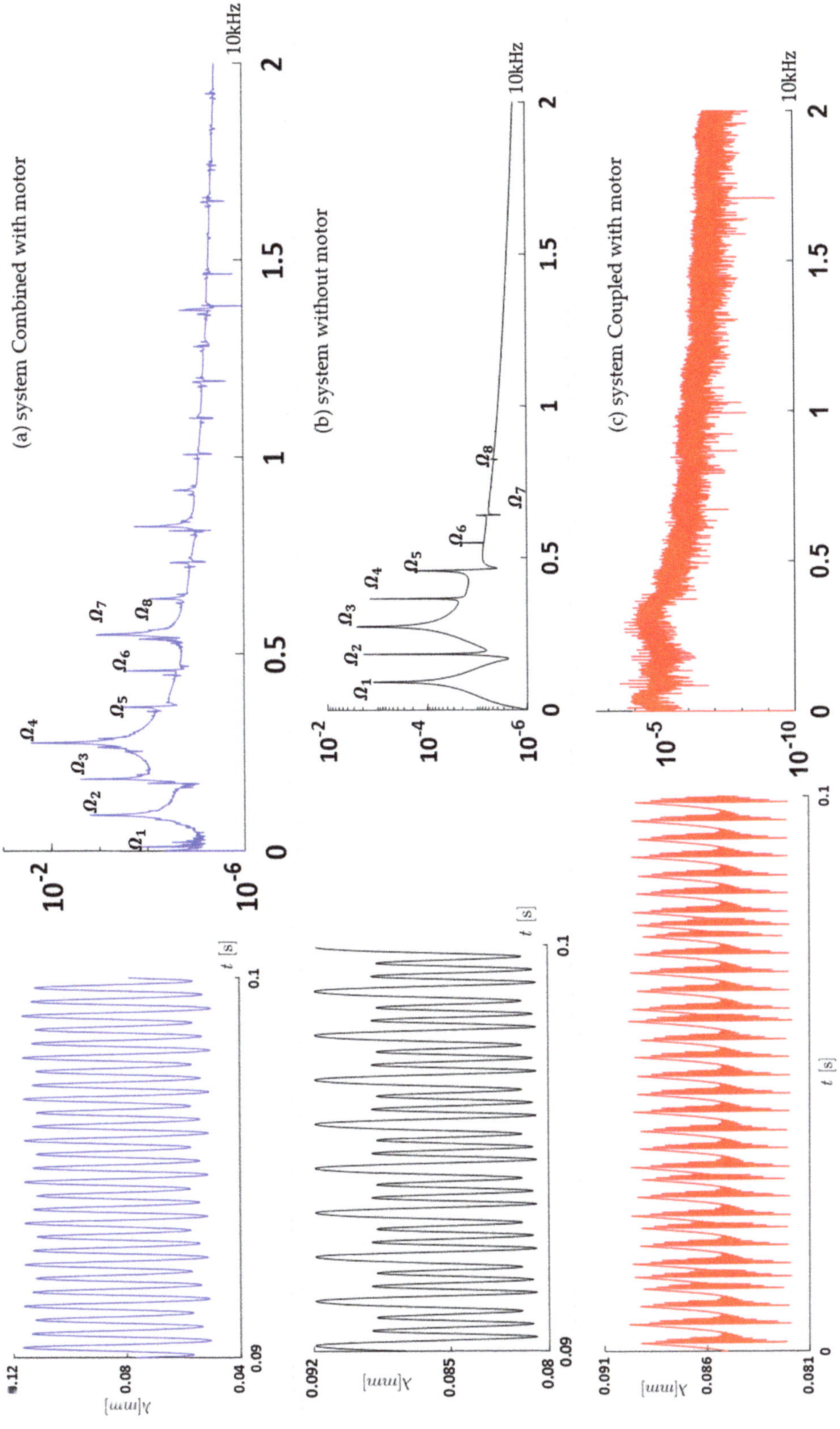


**Figure 13.** FFT spectrum analysis on dynamic transmission error: • coupled with the motor; • without the motor; • combined with the motor.

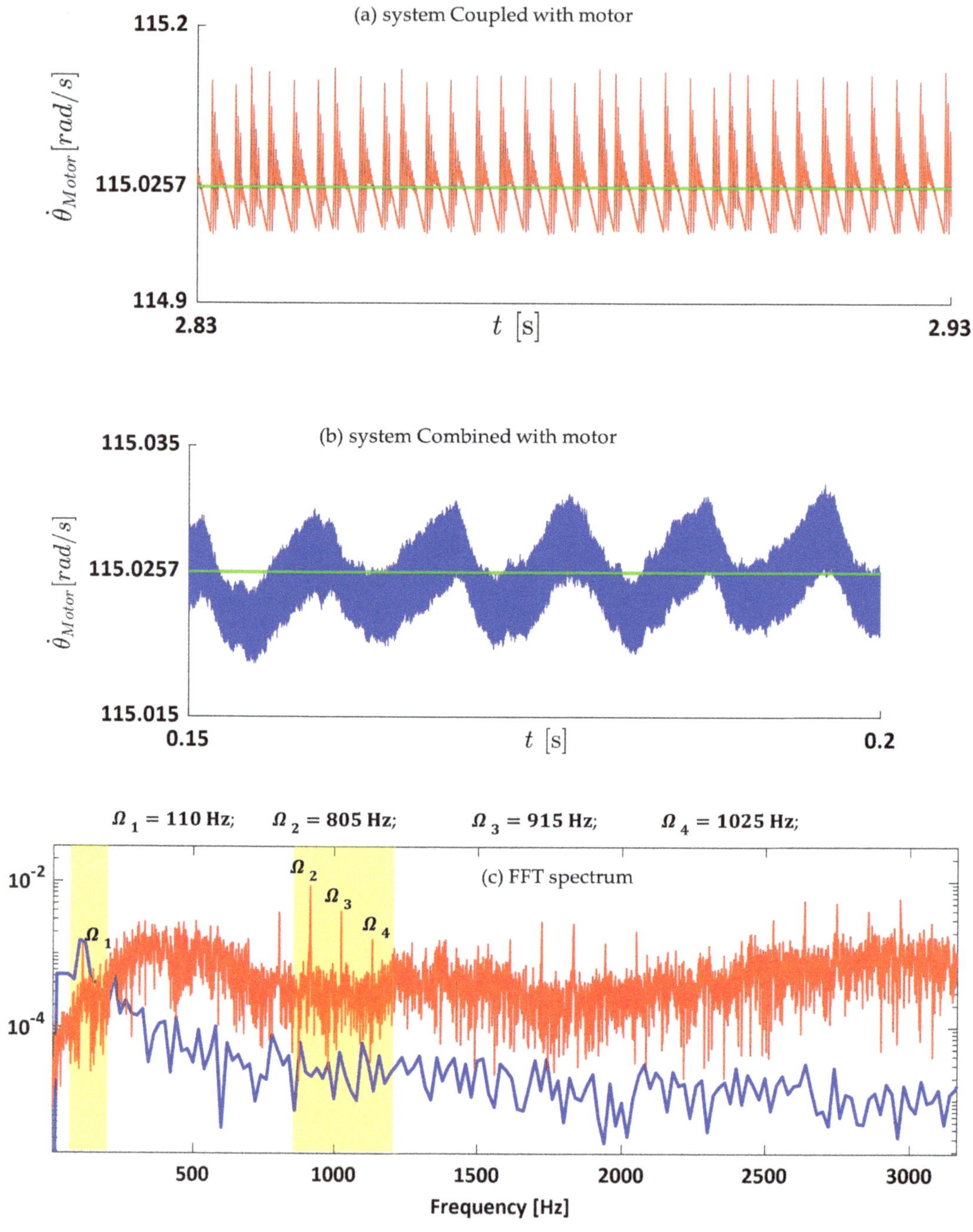


**Figure 14.** FFT spectrum analysis on output speed of the motor: • coupled with the motor; • combined with the motor.

## 4. Conclusions

This study has provided a detailed investigation into the dynamic interactions between a permanent magnet synchronous machine and a nonlinear spur gear transmission system. By analyzing three distinct models—namely, a standalone gear system, a combined system without dynamic feedback from the gear transmission system, and a fully coupled electromechanical system—we have demonstrated the influence of motor–gear coupling on system dynamics. The numerical simulations, validated against experimental data, confirm the reliability and accuracy of the proposed Simulink–MATLAB model in capturing the nonlinear behavior of gear transmissions. Importantly, the inclusion of mechanical feedback via the derivative of the dynamic transmission error significantly alters both the motor output characteristics and the overall vibrational response of the system. This coupled feedback loop introduces higher-frequency components in the dynamic mesh torque and amplifies oscillations in both torque and speed, particularly under nominal and high-load conditions.

Our findings highlight the limitations of traditional modeling approaches that neglect electromechanical feedback as they tend to underestimate or misrepresent dynamical behavior. Moreover, the spectral analysis reveals that dynamic coupling introduces harmonics and sidebands in the output motor speed, associated with gear meshing frequencies, which are absent in combined configurations. These insights are essential for the accurate design, control, and durability assessment of advanced electromechanical powertrains. Ignoring the bidirectional interaction between motor control and gear dynamics may result in suboptimal system performance or premature failure due to unaccounted fatigue mechanisms.

**Supplementary Materials:** The following supporting information, the Simulink Model, used to conduct this analysis, can be downloaded from the following link: https://www.mdpi.com/article/10.3390/vibration8030034/s1.

**Author Contributions:** Conceptualization, A.Z., F.P. and M.M.; methodology, A.Z., F.P. and M.M.; software, A.Z. and M.M.; validation, A.Z., F.P. and M.M.; formal analysis, A.Z., F.P. and M.M.; investigation, A.Z., F.P. and M.M.; resources, A.Z. and F.P.; data curation, A.Z. and M.M.; writing—original draft preparation, A.Z. and M.M.; writing—review and editing, A.Z., F.P. and M.M.; visualization, A.Z., F.P. and M.M.; supervision, A.Z. and F.P.; project administration, A.Z. and F.P.; funding acquisition, A.Z. and F.P. All authors have read and agreed to the published version of the manuscript.

**Funding:** The financial support was provided by Region Emilia Romagna PR FESR 2021–2027 Action 1.1.2, titled "THEORETIC", C.U.P.: E67G22000610001.

**Data Availability Statement:** The original data presented in the study are openly available in Mendeley data repository at Supplementary Materials https://doi.org/10.17632/5rjgkcyp7g.1.

**Conflicts of Interest:** The authors declare no conflicts of interest.

## List of Symbols

| | |
|---|---|
| $L$ | Armature inductance |
| $R$ | Resistance of the stator windings |
| $L_q, L_d$ | q-axis and d-axis inductances |
| $T_e$ | Electromagnetic torque |
| $i_{q1}, i_{d1}$ | $q1$-axis and $d1$-axis currents |
| $v_{q1}, v_{d1}$ | $q1$-axis and $d1$-axis voltages |
| $i_{q2}, i_{d2}$ | $q2$-axis and $d2$-axis currents |
| $v_{q2}, v_{d2}$ | $q2$-axis and $d2$-axis voltages |
| $\omega_m$ | Angular velocity of the rotor |

| | |
|---|---|
| $\lambda$ | Amplitude of the flux induced by the permanent magnets of the rotor in the stator phases |
| $p$ | Number of pole pairs |
| $I_g$, $I_p$ | Momentum inertia of the gear and pinion |
| $r_g$, $r_p$ | Pitch radius of gear and pinion, respectively |
| $\theta_p$, $\theta_g$ | Rotational degree of freedom for the pinion and gear |
| $T_m$ | Input torque |
| $T_l$ | Output torque |
| $\lambda$ | Dynamic transmission error |
| $e$ | Geometric transmission error |
| $K_m(t)$ | Mesh stiffness |
| $f(\lambda - e)$ | Backlash function |
| $n_s(t)$ | Input speed |
| $N_{\mathrm{p}}$ | Teeth number of pinion |
| $m_{\mathrm{eq}}$ | Equivalent mass |
| $T_{\mathrm{eq}}$ | Equivalent torque |
| $C_m$ | Damping |

## References


1. Qian, J.; Ji, C.; Pan, N.; Wu, J. Improved Sliding Mode Control for Permanent Magnet Synchronous Motor Speed Regulation System. *Appl. Sci.* **2018**, *8*, 2491. [CrossRef]
2. Xu, D.; Wang, T.; Liu, J.; Wei, H. Research on Optimal-Torque Control Method of Permanent Magnet Synchronous Motor. In Proceedings of the 2008 3rd IEEE Conference on Industrial Electronics and Applications, Singapore, 3–5 June 2008; pp. 1229–1233. Available online: https://ieeexplore.ieee.org/document/4582715 (accessed on 25 April 2025).
3. Zhang, G.; Yu, W.; Hua, W.; Cao, R.; Qiu, H.; Guo, A. The Design and Optimization of an Interior, Permanent Magnet Synchronous Machine Applied in an Electric Traction Vehicle Requiring a Low Torque Ripple. *Appl. Sci.* **2019**, *9*, 3634. [CrossRef]
4. Sreejeth, M.; Singh, M.; Kumar, P. Particle swarm optimisation in efficiency improvement of vector controlled surface mounted permanent magnet synchronous motor drive. *IET Power Electron.* **2015**, *8*, 760–769. [CrossRef]
5. Wang, S.C.; Nien, Y.C.; Huang, S.M. Multi-Objective Optimization Design and Analysis of V-Shape Permanent Magnet Synchronous Motor. *Energies* **2022**, *15*, 3496. [CrossRef]
6. Sun, X.; Hu, C.; Lei, G.; Guo, Y.; Zhu, J. State Feedback Control for a PM Hub Motor Based on Gray Wolf Optimization Algorithm. *IEEE Trans. Power Electron.* **2020**, *35*, 1136–1146. [CrossRef]
7. Shang, D.X.; Liu, Y.C.; Sun, F.J.; Zhang, H.Y. Study on DTC-SVM of PMSM Based on Propeller Load Characteristic. In Proceedings of the 2008 7th World Congress on Intelligent Control and Automation, Chongqing, China, 25–27 June 2008; pp. 6445–6449. Available online: https://ieeexplore.ieee.org/document/4593905 (accessed on 25 April 2025).
8. Molaie, M.; Deylaghian, S.; Iarriccio, G.; Samani, F.S.; Zippo, A.; Pellicano, F. Planet Load-Sharing and Phasing. *Machines* **2022**, *10*, 634. [CrossRef]
9. Jędryczka, C.; Danielczyk, D.; Szeląg, W. Torque Ripple Minimization of the Permanent Magnet Synchronous Machine by Modulation of the Phase Currents. *Sensors* **2020**, *20*, 2406. [CrossRef]
10. Dai, Y.; Lee, H.J. Torque Ripple and Electromagnetic Vibration Suppression of Fractional Slot Distributed Winding ISG Motors by Rotor Notching and Skewing. *Energies* **2024**, *17*, 4964. [CrossRef]
11. Hu, J.; Peng, T.; Jia, M.; Yang, Y.; Guan, Y. Study on Electromechanical Coupling Characteristics of an Integrated Electric Drive System for Electric Vehicle. *IEEE Access* **2019**, *7*, 166493–166508. [CrossRef]
12. Jiang, S.; Li, W.; Wang, Y.; Yang, X.; Xu, S. Study on electromechanical coupling torsional resonance characteristics of gear system driven by PMSM: A case on shearer semi-direct drive cutting transmission system. *Nonlinear Dyn.* **2021**, *104*, 1205–1225. [CrossRef]
13. Ge, S.; Hou, S.; Yao, M. Electromechanical Coupling Dynamic Characteristics of the Dual-Motor Electric Drive System of Hybrid Electric Vehicles. *Energies* **2023**, *16*, 3190. [CrossRef]
14. Guo, D.; Huang, D.; Ge, S.; Wang, Y.; Shen, Q. Vibration characteristics of electric drive system considering rotor-step skewing. *Proc. Inst. Mech. Eng. Part D* **2024**, *238*, 4142–4153. [CrossRef]
15. Xie, Y.; Lim, K.; Liu, H.; Zhan, Z.; Ren, X.; Li, X.; Zhou, R.; Gao, P.; Xiang, C. Modelling of electromechanical coupling dynamics for high-speed EHT system used in HEV and characteristics analysis. *Appl. Math. Model.* **2024**, *136*, 115614. [CrossRef]
16. Liu, Y.; Wang, Z.; Yang, J.; Zhang, Y.; Pu, W. Electromechanical dynamic behavior of gear systems caused by transient mixed lubrication of rubbing surface. *Tribol. Int.* **2025**, *204*, 110459. [CrossRef]

17. Dou, J.; Li, Z.; Yao, H.; Ding, M.; Wei, G. Torsional vibration suppression and electromechanical coupling characteristics of electric drive system considering misalignment. *Appl. Math. Mech.* **2024**, *45*, 1987–2010. [CrossRef]
18. Li, H.; Guo, Y.; Xu, Q. PMSM Torque Ripple Suppression Method Based on SMA-Optimized ILC. *Sensors* **2023**, *23*, 9317. [CrossRef] [PubMed]
19. Müller, K.; Wanke, A.; Burkhardt, Y.; Gersem, H.D. Evaluation of Torque Ripple and Tooth Forces of a Skewed PMSM by 2D and 3D FE Simulations. *arXiv* **2025**. Available online: http://arxiv.org/abs/2503.16279 (accessed on 30 April 2025).
20. Bai, W.; Qin, D.; Wang, Y.; Lim, T.C. Dynamic characteristic of electromechanical coupling effects in motor-gear system. *J. Sound Vib.* **2018**, *423*, 50–64. [CrossRef]
21. Yi, Y.; Qin, D.; Liu, C. Investigation of electromechanical coupling vibration characteristics of an electric drive multistage gear system. *Mech. Mach. Theory* **2018**, *121*, 446–459. [CrossRef]
22. Ge, S.; Qiu, L.; Zhang, Z.; Wang, H.; Hu, M. Electromechanical coupling dynamic characteristics of electric drive system for electric vehicle. *Nonlinear Dyn.* **2024**, *112*, 6101–6136. [CrossRef]
23. Chen, X.; Yuan, S.; Peng, Z. Nonlinear vibration for PMSM used in HEV considering mechanical and magnetic coupling effects. *Nonlinear Dyn.* **2015**, *80*, 541–552. [CrossRef]
24. Molaie, M.; Samani, F.S.; Zippo, A.; Pellicano, F. Spiral Bevel Gears: Nonlinear dynamic model based on accurate static stiffness evaluation. *J. Sound Vib.* **2023**, *544*, 117395. [CrossRef]
25. Molaie, M.; Samani, F.S.; Zippo, A.; Iarriccio, G.; Pellicano, F. Spiral bevel gears: Bifurcation and chaos analyses of pure torsional system. *Chaos Solitons Fractals* **2023**, *177*, 114179. [CrossRef]
26. Anuradha, G.; Sagi, R.P.; Shakya, P.; Sekhar, A.S. Influence of Geometric Parameters on the Dynamic Performance of Spiral Bevel Gear. *J. Vib. Eng. Technol.* **2024**, *12*, 9097–9111. [CrossRef]
27. Molaie, M.; Zippo, A.; Pellicano, F. Chaotic dynamics of spiral bevel gears. *Int. J. Non-Linear Mech.* **2025**, *175*, 105098. [CrossRef]
28. Molaie, M.; Samani, F.S.; Pellicano, F. Spiral Bevel Gears Nonlinear Vibration Having Radial and Axial Misalignments Effects. *Vibration* **2021**, *4*, 666–678. [CrossRef]
29. Grenier, D.; Dessaint, L.A.; Akhrif, O.; Bonnassieux, Y.; Le Pioufle, B. Experimental nonlinear torque control of a permanent-magnet synchronous motor using saliency. *IEEE Trans. Ind. Electron.* **1997**, *44*, 680–687. [CrossRef]
30. Toliyat, H.A. Analysis and Simulation of Multi-Phase Variable Speed Induction Motor Drives Under Asymmetrical Connections. In Proceedings of the Applied Power Electronics Conference APEC '96, San Jose, CA, USA, 3–7 March 1996; Volume 2, pp. 586–592. Available online: https://ieeexplore.ieee.org/abstract/document/500500 (accessed on 25 April 2025).
31. Baudart, F.; Labrique, F.; Matagne, E.; Telteu, D.; Alexandre, P. Control Under Normal and Fault Tolerant Operation of Multiphase SMPM Synchronous Machines with Mechanically and Magnetically Decoupled Phases. In Proceedings of the 2009 International Conference on Power Engineering, Energy and Electrical Drives, Lisbon, Portugal, 18–20 March 2009; pp. 461–466. Available online: https://ieeexplore.ieee.org/abstract/document/4915200 (accessed on 25 April 2025).
32. Kahraman, A.; Blankenship, G.W. Experiments on Nonlinear Dynamic Behavior of an Oscillator with Clearance and Periodically Time-Varying Parameters. *J. Appl. Mech.* **1997**, *64*, 217–226. [CrossRef]
33. Bonori, G.; Pellicano, F. Non-smooth dynamics of spur gears with manufacturing errors. *J. Sound Vib.* **2007**, *306*, 271–283. [CrossRef]
34. Blankenship, G.W.; Kahraman, A. Steady state forced response of a mechanical oscillator with combined parametric excitation and clearance type non-linearity. *J. Sound Vib.* **1995**, *185*, 743–765. [CrossRef]

vibration

MDPI

*Article*

# Dynamics of Steel Gas Pipelines: Finite Element Simulation of Damaged Sections Reinforced with Composite Linings

**Arman Moldagaliyev [1], Nurlan Zhangabay [2,*], Marco Bonopera [3,*], Raimberdi Ristavletov [1], Ulzhan Ibraimova [2] and Shairbek Yeshimbetov [4]**

[1] Department of Mechanics and Mechanical Engineering, Mukhtar Auezov South Kazakhstan University, Tauke Khan Av., 5, Shymkent 160012, Kazakhstan; arm_mold81@mail.ru (A.M.); r.raimberdi12@rambler.ru (R.R.)
[2] Department of Architecture and Urban Planning, Mukhtar Auezov South Kazakhstan University, Tauke Khan Av., 5, Shymkent 160012, Kazakhstan; ibraimova_uljan@mail.ru
[3] Department of Civil Engineering, College of Engineering, National Taiwan University, No. 1, Sec. 4, Roosevelt Rd., Taipei 10617, Taiwan
[4] Department of Scientific and Innovative Work, Peoples' Friendship University Named After Academician A. Kuatbekov, Tole bi Str., 32B1, Shymkent 160011, Kazakhstan; s.yeshimbetov111@mail.ru
* Correspondence: nurlan.zhanabay777@mail.ru (N.Z.); marco.bonopera@unife.it (M.B.)

**Abstract:** Based on a reduced model of a linear section of a steel gas pipeline between four supports and with a crack-like through defect, ANSYS FE software is used in this study to develop numerical approaches regarding three key parameters of a composite bandage in the form of a circular lining: the type of composite material and the length and thickness of the composite lining. The approach for assessing the static strength of a damaged section of a steel pipeline with a composite lining that is subjected to internal pressure allows for the determination of the optimal thickness of the composite lining itself, which is equal to the indicator "50.0% to 62.5%" of the pipe thickness. Furthermore, the approach for assessing the dynamic strength and analyzing the possible destruction of the reinforced damaged section of a pipeline experiencing an increase in internal pressure allows for the determination of the optimal length of the composite lining, which, in turn, should be at least 241.2 mm. This work also considers cases when there is no internal pressure and the steel pipeline is subjected to critical pressure. It is found that the frequency spectrum of pipeline oscillations without a composite lining is higher than that with a composite lining. The difference between the corresponding dynamic oscillations increases with the thickness or the length of the composite lining. In the absence of internal pressure, all frequencies of the steel pipeline with a crack closed by a composite lining are paired. This pairing is disrupted when the pipeline is subjected to critical internal pressure, and the difference between its oscillation frequency spectrum without and with a composite lining increases. In this case, the oscillation modes significantly differ from those of the same pipeline structure when unloaded. The results ensure the optimal stress distribution in the defect area of a steel pipeline wall and improve the reliability and safety of pipelines under seismic actions. The approach for increasing dynamic strength and eliminating defects can be applied to pipelines with a large diameter regardless of the causes and geometric dimensions of the defects. Moreover, this approach to increasing the strength can be used by various industries and/or institutes which work on the design of new, earthquake-resistant, reinforced pipelines.

**Keywords:** composite lining; finite element; numerical simulation; oscillation frequency; oscillation mode shape; pressure; steel gas pipeline

## 1. Introduction

One of the main tasks in an oil and gas complex is to increase the requirements for ensuring the safe operation of pipelines and reducing the risk of accidents. The need to consider these aspects is due to the threats of large-scale fires, explosions, soil and water pollution, and leaks of petroleum products, in the context of which it is objectively impossible to completely eliminate the possibility of pipeline destruction. A pipeline is a complex piece of engineering infrastructure, the reliable operation of which depends on a number of factors and, above all, is associated with ensuring the strength and crack resistance of welded pipes as well as ensuring resistance to destruction. The annual growth of gas production and the increment in its specific weight in the fuel and energy balance require the construction of steel gas pipelines, an increase in their throughput, an increase in the capacity of gas pipelines by constructing ones with large diameters, and an increase in the working pressure [1]. Hence, there is a need to reconstruct them and pay great attention to maintaining them in working and technically suitable conditions since incidents which occur on such infrastructure are associated with important material costs, fire and explosion hazards, risks of environmental pollution, and danger to human life [2–6]. On the national scale in the Republic of Kazakhstan, this problem is highly relevant. With large reserves of natural gas, oil, and gas condensate and an extensive network of pipelines, Kazakhstan has a serious impact on the development of the worldwide energy market [7]. In this regard, a necessary condition for stabilizing and boosting the economy, ensuring the integrity of the country, and improving the standard of living of the population is the creation of a dynamically developing, sustainably functioning, and balanced pipeline transport system.

The causes of destruction and approaches to increasing the strength properties of major steel pipelines are quite popular issues to study, in fact, and have been analyzed by several scientists for a long time. For example, Zhao et al. [8] conducted a review regarding dents on pipelines, but this work alone is not complete since there are many other factors in the destruction of gas pipelines to be considered [9–11]. It is known that during long-term operations, dents can form on pipelines, which, in turn, give impetus to the appearance of cracks [12,13]. In the works of Shirazi et al. [14] and Zhangabay et al. [15,16], imperfections in the form of cracks were investigated. Various types of cracks and their effects on the bearing capacity of pipelines were assessed, and it was specifically stated that the presence of cracks in a pipeline increases the possibility of its destruction [15–19]. As measures to strengthen pipelines in the works of Alabtah et al. [19] and Zhu et al. [20], it was demonstrated that the use of composite linings made of fiber-reinforced polymers improves the rigidity, strength, pressure resistance, and durability of pipes. Bandaging damaged sections of oil pipelines has widely been utilized as a measure to increase the operation life of oil pipelines in seismically hazardous areas. In particular, composite linings have increasingly been utilized because of their high strength, durability, and resistance to corrosion and mechanical loads. When designing a bandage for damaged sections of pipelines, the optimal parameters of the composite lining, i.e., length, thickness, and material properties, should be estimated according to computational studies [21,22]. Lim et al. [23], Dong [24], and Kong et al. [25] presented results from the successful use of composites to strengthen pipelines in which the authors proposed an effective method for repairing steel pipes using carbon fiber. The effect of defect sizes and the mechanical properties of putty on the rupture characteristics of repaired pipes were investigated using rupture tests and finite element (FE) analyses. The outcome of the study showed the effectiveness of the carbon fiber in repairing defects on a steel pipe. Since carbon fiber has shown its effectiveness, the scientific hypothesis of this work served to develop the proposed study on the use of carbon fiber in estimating the natural oscillation frequencies of a steel gas pipeline when amplified by the proposed material in the form of a composite bandage. Nordhagen et al. [26] also achieved the same positive effect but, conversely, chose steel wire as a reinforcement measure. Furthermore, preliminary works of a typical section of a main pipeline between four supports with damage in the form of thinning of the original pipe thickness and through cracks showed that the presence of a carbon fiber lining no thicker than half the pipe thickness completely compensated for the stress concentration in the damage area [27]. Such studies were performed numerically using ANSYS 19.2 software. Also, 30 changes in the spectrum of natural frequencies of damaged sections of a pipeline reinforced with a composite lining were analyzed depending on the size and location of the lining itself. In short, it was found that a pipeline reinforced with a composite lining has a higher spectrum of oscillation frequencies compared to an unreinforced one [22]. Thus, the application of composite linings can be adopted as a way to tune out damaged sections from critical low frequencies. Furthermore, it was found that the oscillation modes of a pipeline section with damage and secured with a composite lining partly differ from the oscillation modes of the same section without a composite lining. As a result, when designing composite linings for damaged sections of a pipeline, it is advisable to conduct computational studies, which should include the following: (1) assessment of the stress–strain state of the damaged section reinforced with a composite lining at the operating internal pressure; (2) assessment of the dynamic strength and analysis of the possible destruction of the reinforced damaged section when the internal pressure increases until reaching critical pressure; (3) determination of the influence of the composite lining on the spectrum of the natural frequencies and oscillation modes of the bandaged section of the pipeline between the supports.

The outcomes from comprehensive studies on main pipelines can generally complement and enrich the data known on their operations under seismic impacts. Experiments and modeling serve to create approaches for designing pipelines with composite linings, eliminating their defects, and increasing their dynamic strength in seismic areas. In addition, their scientific effect is to create the foundations for the design of main pipelines with composite linings, increase their seismic resistance, and eliminate accidental product leakage. The difference that characterizes the design solution proposed in this work is the chance of regulating the dynamic properties of a pipeline by selecting the dimensions of its composite lining in order to detach the corresponding dominant (resonant) frequencies of ground vibrations. Furthermore, the developed approach for increasing the dynamic strength of pipelines, the methodology of calculation, and the recommendations for the design of earthquake-resistant pipelines can be adopted to improve the design codes of oil and petroleum pipelines.

## 2. Materials and Methods

### *2.1. Static Assessment of a Reinforced Gas Pipeline*

The minimum thickness of a composite lining is determined using the solution of the elastic–plastic deformation of a section of a hardened steel pipe between supports with a longitudinal through crack and subjected to the maximum internal pressure. The goal is to find the minimum thickness of the composite lining which compensates for the effect of the maximum allowable internal pressure in the pipeline with a through-crack-like defect. Moreover, this work is conducted using an FE model of a linear section of a main pipeline, which makes it possible to use the approach presented below to calculate the optimal thickness of the composite linings. The FE model was designed in and realized by the ANSYS 19.2/Workbench program complex. Numerical analysis of the static strength was performed in the calculation module "Static Structural". The problem was solved using a three-dimensional (3D) formulation. The condition of the automatic selection of FE types was fulfilled. Specifically, the FE model of the linear section of the pipeline between supports was assumed as follows: one extreme support implements the condition of the rigid sealing of the pipe; two internal supports are freely movable; and the second extreme one implements the condition of sealing the pipe with a chance of movement in the longitudinal direction (Figure 1). Accordingly, the conditions were realized on the outermost supports, "Fixed Support" and "Displacement", with "z = Free" and with the coordinate axis Oz directed along the investigated pipe section. The contact interaction between the internal supports and the pipe was modeled as "Frictionless". The dimensions of the FE model in the longitudinal direction are shown in Figure 1 (in millimeters). The diameter of the pipe section is equal to 300 mm, whilst the thickness of its wall is 0.8 mm. The corresponding material is of C245 steel grade with a yield strength of 240 MPa and a tensile strength of 360 MPa. The inner surface of the pipeline is subjected to uniformly distributed stationary pressure. As preliminary studies have demonstrated [27], such pressure causes a practically uniform stress state in an undamaged pipe on a linear section of the pipeline between the extreme supports. Together, freely movable supports do not significantly influence the axisymmetric nature of the distribution of the stresses in the axial section of the pipeline, and consequently, they can be neglected [27]. At the same time, a part of the structure of the actual length between the two supports was additionally modeled. The pipe is attached to four supports by four frames. The lower ends of the supports are rigidly fixed. The fastening of the upper ends of the supports with frames is modeled by one-piece "Bonded" joints. The frame at the support that implements the condition of rigid sealing of the pipe is rigidly fixed. Its contact with the pipe is modeled as "Bonded". The contact between the pipe and the two central, freely movable supports is modeled as "Frictionless". Moreover, the frame of the longitudinally movable support is fixed, so that movement in the direction normal to the pipe surface is impossible. Also, the model takes into account the weight of the structure [27].

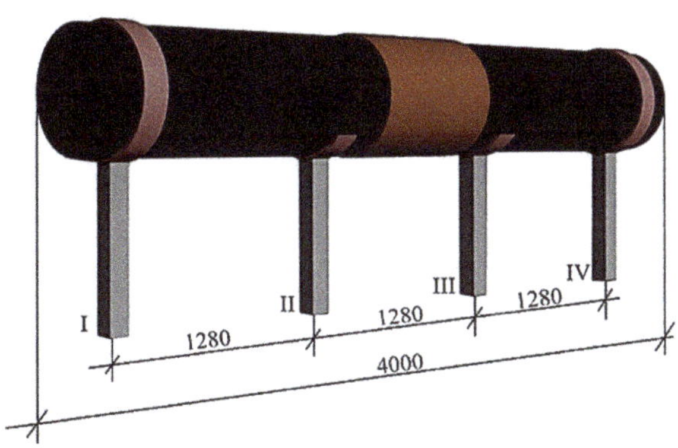


**Figure 1.** Representation of the linear section of the steel gas pipeline [27,28] (unit: millimeter).

The critical pressure is calculated by using the FE method. Its value is equal to the internal pressure, which causes a stress state in the pipeline between the supports which, in turn, is equal to the yield strength of the steel material. For this FE model of the pipeline (Figure 1), made of the C245 steel grade with a yield strength of 240 MPa and a pipe wall thickness of 0.8 mm, the critical pressure $P_{cr}$ = 6.426 kPa. Figure 2 shows the stresses in the pipeline subjected to stationary uniform internal critical pressure. As can be seen from the modeling, the stresses are close to the yield strength. In this case, the edge effects, caused by the condition of fixing the structure, are localized at the edges of the pipeline and do not affect the stress state in its central part. Therefore, the subsequent FE analyses are performed on a model with a crack and a composite lining which, in turn, are both located in the central part of the pipeline.

**Figure 2.** Stresses along the FE model of the steel gas pipeline without defects and corresponding to an internal pressure equal to 6.426 kPa (unit: kilopascal).

The damage on the investigated section of the pipeline in the form of a through longitudinal crack is simulated by ANSYS 19.2 [29]. Preliminary studies proved that such damage is the most dangerous because it can cause avalanche destruction of the pipeline [29,30]. Crack-like defects of various widths are simulated. Then, the section of the pipeline with a crack is reinforced with a circular lining made by a composite material and having the following orthotropic properties: elastic moduli $E_{xx}$ = 35 GPa, $E_{\varphi\varphi}$ = 35 GPa, $E_{zz}$ = 8 GPa; shear moduli $G_{x\varphi}$ = 6 GPa, $G_{\varphi z}$ = 30 GPa, $G_{xz}$ = 30 GPa; Poisson's ratios $\nu_{x\varphi}$ = 0.01, $\nu_{\varphi z}$ = 0.09, $\nu_{xz}$ = 0.09; and density $\rho$ = 1477 kg/m$^3$. These values were obtained experimentally for the carbon fiber composite based on the prepreg SIGRAPREG C U200-0/NF-E310/30% [31,32]. This composite material has been successfully used as face skins, including in structures with imperfections [32,33]. The composite wrap satisfies Hooke's law [27,28]. Notably, the strength properties of the composite lining significantly overcome those of the pipeline material, and the length of the composite lining significantly exceeds the crack length.

The thickness of the composite lining is estimated which will compensate for the pressure in the area of the through-crack-like defect. Specifically, it is assumed that the composite lining's thickness should bear stresses on its outer surface in the area of the through crack that do not exceed the tensile strength value.

The effectiveness of the repair technology for steel pipes with corrosion and crack-like defects directly depends on the choice of three key parameters of the composite linings. These are the type of composite material, the length of the lining, and the thickness of the lining. Changing the composite material and the geometric parameters of the lining can reduce the weight of the bonded pipe, minimize the cost of its repair, and increase the durability and the strength of the reinforced pipe section with defects.

Optimization of the composite lining in order to achieve the nominal strength of the bonded pipe with a minimum weight of the lining and a maximum savings in composite material should be performed by numerical simulation. In this case, as a rule, the choice of the composite material is executed in advance. Therefore, the purpose of optimization is to determine the minimum length and minimum thickness of the composite lining that will ensure that the static and dynamic strengths of the bonded section of pipeline with a defect between supports stays at the strength level of this section of the pipeline without a defect.

The basic principles of the gradient descent method are used. This allows us to find independent local extremes of the functions of the minimum thickness and the minimum length of the composite lining [30].

For the minimum thickness of the lining $x_k$, $k = 1, 2, 3 \ldots$, it is possible to write

$$x_{k+1} = x_k - \alpha \nabla f(x_k), \tag{1}$$

where $\alpha$ is the step in finding the minimum thickness and $\nabla f(x_k)$ is the gradient of the function at a point $x_k$. For the minimum length of the lining $y_k$, $k = 1, 2, 3 \ldots$, it is then possible to write

$$y_{k+1} = y_k - \beta \nabla g(y_k), \tag{2}$$

where $\beta$ is the step in finding the minimum length and $\nabla g(y_k)$ is the gradient of the function at a point $y_k$. The initial data for Equations (1) and (2) are determined by FE analysis in Ansys, while the geometric parameters of the composite lining are adjusted according to the numerical simulation results.

The problem was solved using a 3D formulation. The FE model consists of two 3D bodies: a steel pipe and a composite lining (arrow 2). The contact between the parts of the pipeline provides us with equal displacements and stresses at the contact points. The composite lining is located at the center of the pipeline, overlapping the crack. In addition, the composite lining is mounted on the pipeline without any tension. The pipeline's inner surface and the area of the composite lining corresponding to the thickness of the through crack are subjected to stationary internal pressure. In this study, the maximum allowable internal pressure in the pipe, namely, the critical pressure $f$ = 6.426 kPa, is used as the stationary internal pressure. The edges of the pipeline are fixed. On the right edge, displacements in the longitudinal direction are allowed, whilst in the radial direction, displacements are equal to "zero". On the left edge, a rigid fixation is simulated. Such a fixation permits us to minimize the influence of the edge effects on the calculations of the stress–strain state of the structure. According to the results, a displacement and a stress field of the reinforced composite lining of the steel pipeline having a longitudinal crack are formed. During calculations, the crack's width varies. It is set by the arc length of the outer radius of the pipe in degrees. The optimal thickness of the composite lining is then determined according to the FE results. When forming the FE model, the fact that the pad is made by an orthotropic material was assumed. Specifically, to properly account for its non-isotopic properties, the orientation of the axes in the FE model were set for the composite lining, i.e., corresponding to the mechanical parameters of the material, as shown in Figure 3. The $O_x$ axis is shown in red and directed along the cylindrical shell, the $O_y$ axis is shown in green and directed tangentially to the cylindrical shell, and the $O_z$ axis is depicted in blue and directed normally to the cylindrical shell.

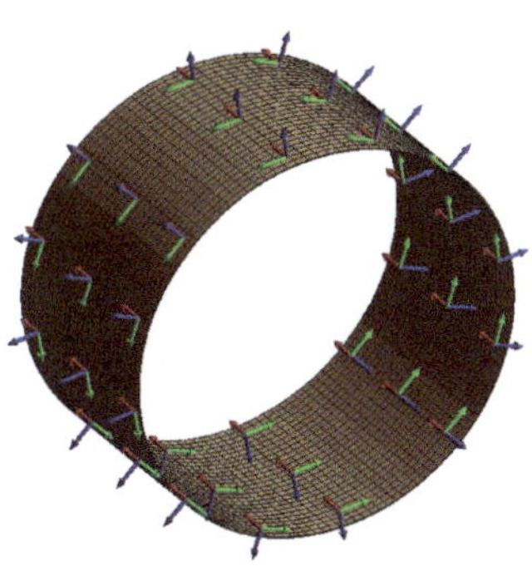

**Figure 3.** Orientation of the axes $O_x$, $O_y$, and $O_z$ in the orthotropic composite lining [27,28].

Specifically, the convergence of the FE model was achieved by the standard method of reducing the size of the FE by varying the "Element Size" parameter. Notably, the central fragment of the FE breakdown of the pipe with the composite lining is shown in Figure 4.

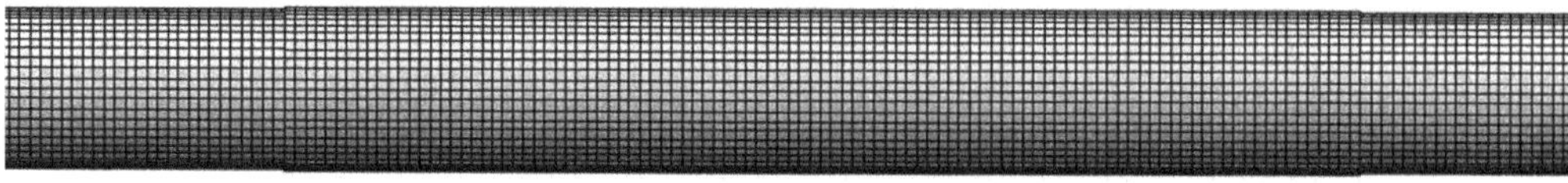

**Figure 4.** Fragment of the FE model of the steel gas pipeline with the composite lining.

The convergence of the FE model was verified using a test example that accounted for prestresses induced by critical pressure. The FE calculations were performed for cracks with thicknesses of 0.1° and 1.6° of the outer diameter of the steel pipeline; a 0.4 mm thick lining was modeled. The results of the convergence analysis are presented in Table 1. In detail, the table shows the stresses on the composite lining's outer diameter, at the central point of the pipeline, and opposite to the crack. The difference in values between the parameter sizes of 4.0 mm and 2.0 mm does not exceed the percentage of 2.2%, and the difference between 2.0 mm and 1.0 mm does not exceed the percentage of 0.5%. Such results allow us to declare that the FE model with an "Element Size" parameter not greater than 2.0 mm satisfies the requirements for the convergence of the FE solution for the stresses. The results presented below were gained with an "Element Size" parameter equal to 1.0 mm. In this case, the FE model contains $1.13 \times 10^6$ elements and $7.37 \times 10^6$ nodes.

**Table 1.** Convergence analysis of the FE model: stresses at a given point depending on the "Element Size" parameter, in MPa (megapascals).

| Crack Width, Degrees | Element Size Parameter, mm | | |
|---|---|---|---|
| | 4 | 2 | 1 |
| 0.1 | 23.4 | 22.9 | 22.8 |
| 1.6 | 475.6 | 474.9 | 474.7 |

## 2.2. *Dynamic Assessment of a Reinforced Gas Pipeline*

Numerical analysis of the dynamic strength was executed using the "Explicit Dynamics" calculation module. The previously developed FE model was supplemented with the "Plastic Strain Failure" module. This allowed the simulation of the structural failure due to the plastic deformation of the material which, in turn, was modeled using the "Bilinear Isotropic Hardening" approach. The minimum length of the composite lining was determined using the solution of the dynamic elastic–plastic deformation of a section of a hardened steel pipeline with a longitudinal through crack in which the pressure increases from zero to a critical value. Such pressure growth usually leads to the formation of through cracks, and cracks in the longitudinal direction can cause avalanche failure of the pipe section between supports [15–17]. The composite lining absorbs a significant amount of energy during internal pressure growth due to stress dissipation throughout the structure. The fibers in the composite act as a barrier against rapid crack propagation, reducing the rate of crack growth. Thus, composites with directional reinforcement or a laminated structure can have residual stresses that reduce the likelihood of rapid crack propagation [34]. An additional purpose is to determine the minimum length of the composite lining that prevents the growth of a longitudinal through crack when the pressure in the pipeline increases until a critical value is reached and that overlaps with the existing through defect. In this case, the composite lining's thickness is considered as optimal and selected on the basis of the findings that are described in Section 2.1. The dynamic problem, as well the static one, is solved for an FE model of the linear section of the steel pipeline (Figure 1), which allows the use of the approach presented below to estimate the composite lining's optimal length. Note that the results obtained by numerical modeling should be used as initial data for subsequent experimental studies. Damage to the assumed section of the pipeline is modeled in the form of a through longitudinal crack with a width equal to 0.40, which corresponds to 1.0 mm. According to preliminary studies, it can be stated that such a crack, when the internal pressure increases until a critical value is reached, increases in the longitudinal direction caused by the local destruction of the material at the crack tips and at its edges [29,30]. Thus, the deformation pattern of a steel pipeline without and with a long composite lining is analyzed. The dynamic problem of the longitudinal crack propagation in a pipeline without a composite lining is solved. Specifically, the initial lining's length is determined, which is equal to the sum of the maximum crack length and two sections per bevel, as described in Moldagaliyev et al. [27,28]. In addition, the dynamic problem of the longitudinal crack propagation in a pipeline with a composite lining four times longer than the crack length is solved. High-stress-concentration areas in the pipeline are identified, and the initial composite lining's length is also adjusted. Then, verification is executed for a composite lining of the selected length.

The results of the works for a carbon fiber composite lining and its mechanical properties, which were presented in Moldagaliyev et al. [27,28], are reported below. The choice of a carbon fiber composite lining is due to the high specific strength of this material, which allows it to significantly increase the load-bearing capacity of a pipe with a lining without a significant increase in its weight. Its high fatigue resistance should also be noted, which is essential for the operation of gas pipelines in seismically dangerous areas. Modeling of anisotropic properties of composite material is carried out on the basis of orthotropic models and is in good agreement with experimental data [31,32]. An important issue is the question of modeling the contact interaction between the carbon fiber composite lining and the steel pipe. Depending on the bonding method, the contact bond is specified with the possibility of considering sliding, adhesion, or debonding at critical loads. In this study, this bond was modeled in ANSYS-19.2 as a "Bonded" rigid contact. The composite lining's thickness was selected according to the approach described in Section 2 and is equal to 50% of the pipe thickness—0.4 mm. Such a choice is supported by numerical studies, which showed that a smaller thickness does not provide sufficient strength for the bandaged section of the pipe, while a larger thickness does not give a significant increase in strength but increases the mass of the structure [27,28]. In the FE model, as well as in the static model, of the dynamic deformation of a steel pipeline with a composite lining in the crack area, the internal pressure effect is considered, as is that of the pressure on the lining from the opening crack edges. Therefore, the stresses in the composite lining in the crack area may exceed the results previously obtained. In this regard, the final thickness of the composite lining, if necessary, should be adjusted according to the outcomes of the dynamic analyses.

Specifically, the FE model consists of two 3D bodies: a steel pipeline and a composite lining. The crack and the composite lining are located at the center of the pipeline. The lining is mounted on the pipeline without any tension. At the initial time, the full contact is simulated between the contact surfaces of both elements, which may be disrupted during the deformation and the local destruction. Furthermore, the model is constructed from FEs and supports the plastic flow of the steel pipeline and the destruction of the pipe and lining material. To ensure the convergence of the solution, a uniform grid similar to that of the static FE model is adopted (Figure 3) along with the selection of the number of FEs by the thickness of the pipe and composite lining. In this case, the linear size in the FE model does not overcome the value of 2.0 mm, which guarantees the convergence of the FE solution. The edges of the steel pipeline are fixed: longitudinal displacements are allowed on the right edge, whilst the radial displacements are "zero"; rigid fixation of the pipeline is simulated on the left edge. In addition, the inner surface of the pipeline, the area of the composite lining on the through crack, and the crack edges are subjected to non-stationary internal pressure. On the basis of the obtained results, dynamic fields of displacements, stresses, and plastic deformations of the composite lining on the pipeline with a longitudinal crack are formed. The composite lining's length was varied, whilst its optimal length was found according to the FE calculations.

### 2.3. Assessing the Influence of a Composite Lining on the Spectrum of Natural Frequencies and Oscillation Modes

The spectrum of the natural frequencies and oscillation modes of the FE model of a linear section of a steel pipeline are determined. A pipeline section fixed between supports is considered (Figure 1). It is modeled so that the left support implements the condition of rigid pipe sealing. Conversely, the right support implements the condition of pipe sealing with the chance of displacement in the longitudinal direction. Two internal, freely movable supports are also modeled, which, in turn, limit the displacement of two contact areas of the pipeline with the support along the vertical direction. A further goal is to analyze the effect of a composite lining of a given length and thickness on the oscillations of a linear section of a steel pipeline. The cases of a not-loaded pipeline with internal pressure and that of a pipeline subjected to the critical internal pressure of 6.426 kPa are considered. The steel pipeline is modeled with a longitudinal through crack. Specifically, a crack with a width of 0.4°, which corresponds to 1.0 mm, is assumed. The crack length is chosen equal to 200 mm. Moreover, the crack is located at the central part of the steel pipeline, i.e., between the freely movable supports. A circular carbon fiber composite lining, whose mechanical properties were presented in Dong [24] and Kong et al. [25], is modeled. The composite lining's thickness is 37.5%, 50%, and 62.5% of the pipe thickness, which is equal to 0.3, 0.4, and 0.5 mm, respectively. Furthermore, the composite lining's length is 10%, 15%, and 20% of the pipeline length, which is equal to 400, 600, and 800 mm, respectively. The edges of the composite lining are symmetrically positioned with respect to the crack center in the longitudinal direction. Conversely, prestresses caused by internal pressure are modeled in the "*Static Structural*" block. This part of the FE model is specifically illustrated in Section 2.1.

Additionally, the static analysis block includes the boundary conditions which simulate the effect of the two freely movable supports: "*Displacement 2*" and "*Displacement 3*". According to the achieved results, the displacement fields and the equivalent stresses are gained. They determine the natural frequencies and oscillation modes of the stressed pipeline within the "*Modal*" block. For each calculation case, the first 20 frequencies and oscillation modes of the pipe section with a longitudinal through crack and reinforced with a composite lining are obtained. Moreover, for comparing the influence of the composite lining on the pipeline oscillation, frequencies, and oscillation forms of a given pipe section without a crack and without the lining are determined.

This work is an intermediate contribution which was developed in the form of modeling reduced steel gas pipelines with composite linings. Its numerical findings are correlated with the experimental outcomes on reduced models of a steel gas pipeline where a steel wire was adopted as a prestressing tool to prevent the crack development [16]. Further experiments regarding steel gas pipelines reinforced with composite linings will be combined with the aforementioned results [27,28,31].

## 3. Results and Discussion

### *3.1. FE Modeling of the Thickness of a Composite Lining*

The crack width varies during the calculation process. It is set by the arc length of the outer radius of the pipe in degrees. The optimal thickness of the lining is determined based on the calculation results. Table 2 shows the analysis of the stress state of a steel pipeline's section with a composite lining on an area with a crack of a width equal to 0.1°; 0.2°; 0.4°; 0.8°; and 1.6°, respectively. Note that the arc length of 0.4° corresponds to a crack of a width of 1.0 mm and with an accuracy of hundredths. The composite linings of thicknesses of 0.3, 0.4, and 0.5 mm are investigated. Figure 5 shows the location of control point no. 1, arranged on the outer surface of the composite lining and above the crack.

**Table 2.** Stresses in the composite lining and at control point no. 1 (Figure 5).

| Crack Width, Degrees | Composite Lining's Thickness, mm/% of the Steel Pipeline's Thickness | | |
|---|---|---|---|
| | **0.5/62.5** | **0.4/50.0** | **0.3/37.5** |
| 0.1 | 21.8 | 22.9 | 24.3 |
| 0.2 | 66.3 | 118.6 | 179.9 |
| 0.4 | 172.9 | 257.0 | 377.2 |
| 0.8 | 262.8 | 359.1 | 528.3 |
| 1.6 | 351.8 | 474.9 | 680.1 |

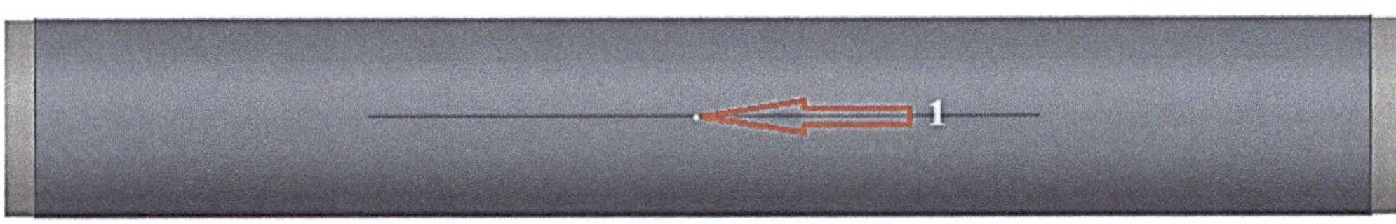


**Figure 5.** Control point no. 1 for the stress-state analysis of a bandaged section of a steel pipeline with a defect.

The stresses at control point no. 1 show that the composite linings with a thickness of 37.5% of the pipe thickness have stresses in the through-crack area lower than the tensile strength limit of 490 MPa [25] and corresponding to the crack widths of 0.1°, 0.2°, and 0.4° only. Thus, the selection of the composite lining with a thickness lower than 37.5% of the pipe thickness is inappropriate. Linings with values of 50% and 62.5% of the thickness compensate for the internal pressure in a pipeline subjected to the critical load even in the area of a through-crack-like defect for all the cases considered. However, for linings with 50% thickness of the pipe thickness, the stresses in the area of the through defect reach 96.9% of the tensile strength. And for linings with 62.5% thickness of the pipe thickness, the percentage is 71.7% only. Consequently, the selection of a lining with 62.5% thickness is preferable. Therefore, the distribution of the stress state along the steel pipeline is the same for all the cases under investigation. As an example, Figure 6 illustrates the distribution of the stress state on the section of the pipeline between the supports for the case with a crack having a width of 0.1° and a composite lining having a thickness of 0.4 mm.

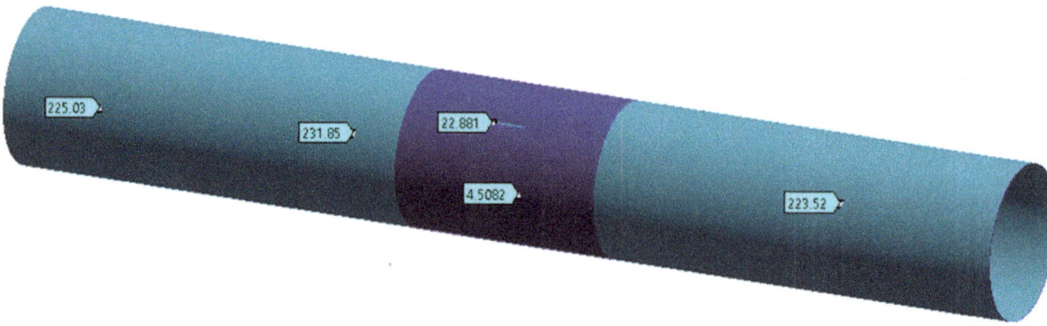


(a)

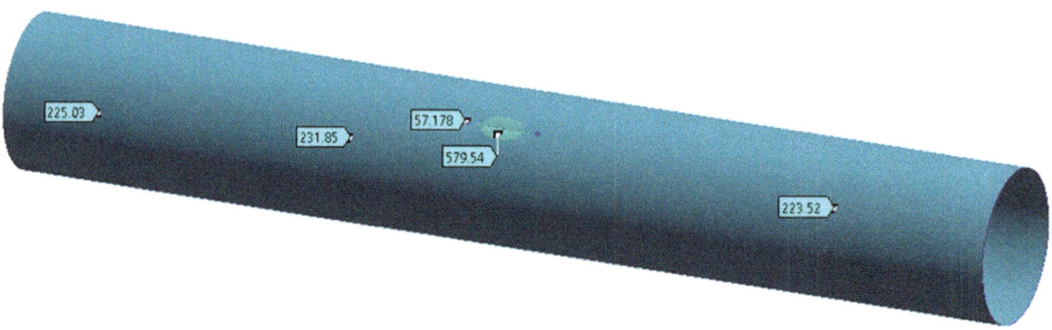


(b)

**Figure 6.** Von Mises stresses for a crack having a width of 0.1° and a 0.4 mm thick composite lining: (**a**) steel pipe with the composite lining; (**b**) steel pipe with a crack under the composite lining (unit: megapascal).

The results depicted in Figure 6 demonstrate that the effect of through defects of small thickness on the pipeline strength is fully compensated for by a composite lining with a thickness of 50% of the pipe thickness. While in the pipeline's area close to the through crack, the stresses exceed the tensile strength, in the composite lining, the maximum stresses do not overcome 4.7% of the tensile strength. Figure 7 shows the stress-state distribution along the pipeline fixed between the supports for the case of a crack having a width of 1.6° and a 0.5 mm thick composite lining. As the crack width increases, the overall stress state in both the pipe and the composite lining increases significantly. In this case, the process is local. At the crack edges, the stress level seriously exceeds the tensile strength of the steel material, whilst in the remaining parts of the pipeline, the stress state remains within the steel's elastic behavior. Conversely, in the composite lining, a stress concentration area is formed above the through defect. In width, it slightly exceeds the crack width. And with distance from the defect area, the stresses in the composite lining quickly decrease to values which do not overcome 10% of the tensile strength. As a result, carbon fiber composite linings with a thickness of the indicator "50.0% to 62.5%" of the steel pipe thickness compensate for the impact of the internal critical load in a pipeline with through-crack-like defects.

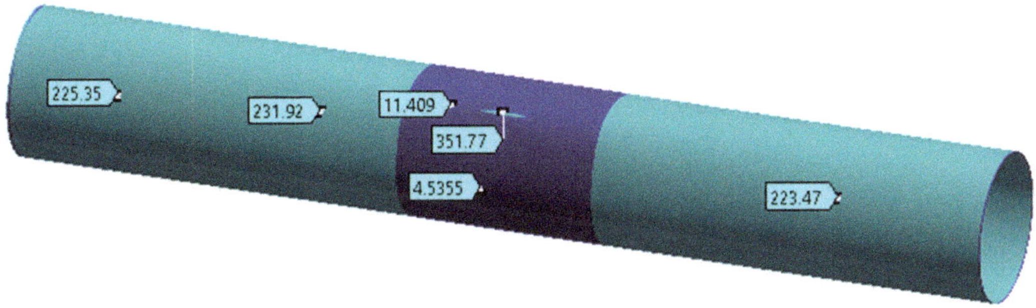


(**a**)

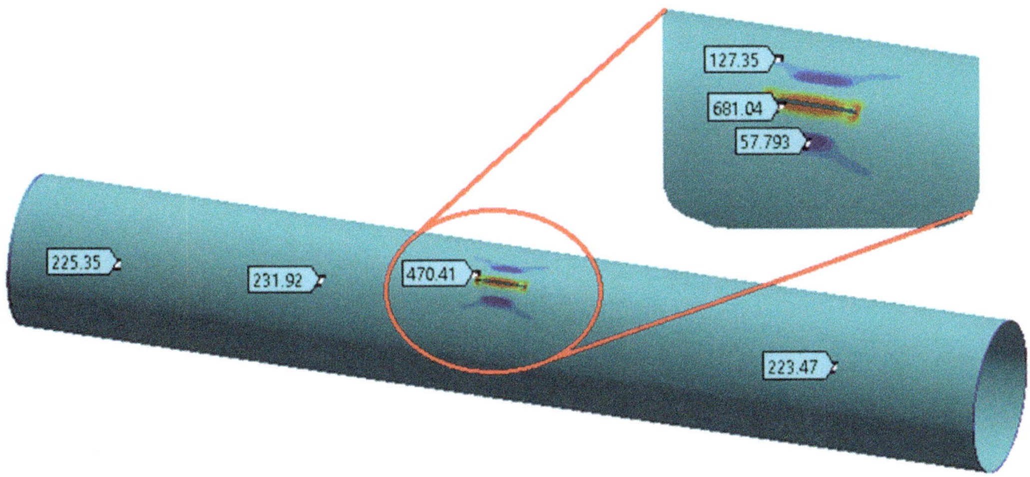


(**b**)

**Figure 7.** Von Mises stresses for a crack having a width of 1.6° and a 0.5 mm thick composite lining: (**a**) steel pipe with the composite lining; (**b**) steel pipe with a crack under the composite lining (unit: megapascal).

### *3.2. FE Modeling of the Length of a Composite Lining*

The initial maximum size of the circular composite lining is estimated as follows. The destruction of a pipeline section with a longitudinal crack and without a composite lining subjected to a non-stationary pressure (which increases until a critical value is reached) is investigated. As shown in the works conducted by Moldagaliyev et al. [27] and Zhangabay et al. [31], non-stationary pressure causes crack growth along the pipeline and the opening of its walls. The local destruction of the steel material at the crack tip is caused by large plastic deformations [35–38]. In order to avoid the mutual influence of deformation processes at the opposite crack tips, a crack of 200 mm in length is considered. Figure 8 displays the plastic deformations in a pipeline with a crack of a width of 4°. Note that plastic deformations are concentrated at the tips of the crack and at the initial time of their occurrence (Figure 8a). In this case, the longitudinal coordinate of the left tip of the crack corresponds to a value of 1900 mm (arrow no. 1), whilst the right tip corresponds to a value of 2100 mm (arrow no. 2).

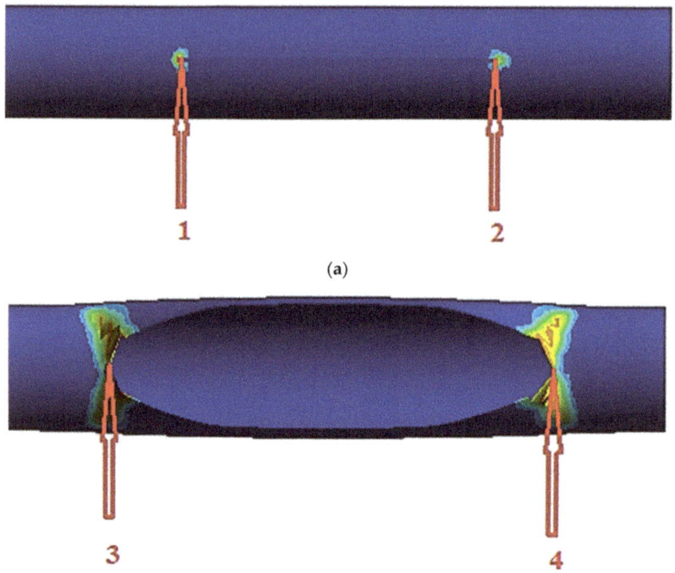


**Figure 8.** Plastic deformations in a steel pipeline: (**a**) at the initial time of the crack growth; (**b**) at the time of stopping the crack growth in the longitudinal direction.

Over time, the crack grows in the longitudinal direction, and its edges open simultaneously (Figure 8b). The symmetry of the crack development is observed along both longitudinal and radial directions. By the time the crack growth in the longitudinal direction stops, the left coordinate has a value equal to 1880 mm (arrow no. 3), while the right one has a value of 2120 mm (arrow no. 4). The initial values of the coordinates correspond to the values of 1900 mm for the left and 2100 mm for the right. Therefore, the crack elongation from each end is equal to 20 mm, which, specifically, is 25 times greater than the pipe thickness. The von Mises equivalent stresses at the time of the crack growth cessation in the longitudinal direction are shown in Figure 9.

**Figure 9.** Von Mises equivalent stresses in a steel pipeline and at the time of the crack growth cessation in the longitudinal direction (unit: megapascal).

The stress state of the open crack allows us to declare that there is no stress concentration at the crack tips. Indeed, the stress values in the pipeline do not reach the yield strength of the steel equal to 240 MPa. Thus, there are no reasons for further crack growth in the longitudinal direction. Therefore, regarding the steel pipeline under investigation, the initial length of the composite lining should not be lower than (3) [39]

$$l_{cw} = s_p + 2(s_{tl} + s_{ol}), \tag{3}$$

$$(l_{cw})_{max} = 200 \text{ mm} + 2(0.6 \text{ mm} + 20 \text{ mm}) = 241.2 \text{ mm}.$$

where $s_p$ is the length of the defect along the pipe; $s_{ol}$ is the overlap length corresponding to the distance to which the composite lining covers the defect; and $s_{tl}$ is the length of the projection of the bevel edge of the lining onto the pipe axis. A preliminary numerical simulation of the bandage is aimed at choosing the optimal amount ($s_{tl}$ + $s_{ol}$).

Preliminary studies have shown that the thickness of the carbon fiber composite lining does not generally exceed half the thickness of the steel pipe. In our case, this thickness is equal to 0.4 mm. Consequently, the length of the bevel edge projection of the lining onto the pipe axis was chosen as one and a half times the maximum lining thickness. Due to its small value, this parameter does not significantly impact the results and can be adjusted if necessary.

The effect of the composite lining on the crack development when the internal pressure increases to a critical value was additionally studied. In order to avoid the influence of the edge effects, the composite lining's length was taken to be significantly greater than the maximum dimensions of the crack. Specifically, it was taken to be equal to 800 mm and with a crack of 200 mm. Figure 10 displays the von Mises equivalent stresses along a pipeline with a composite lining at steady critical pressure. Considering the stress distribution, it is possible to state that the composite lining permits us to restore the functionality of the damaged pipeline. Consequently, the maximum stresses in the concentration areas do not exceed the value of 232 MPa, thus not reaching the yield strength of 240 MPa. Simultaneously, the stress at the crack tip is half of the yield strength, i.e., equal to 119 MPa only. As a result, there are no conditions for the crack growth in the longitudinal direction. Figure 11 illustrates the plastic deformations along the pipeline under the composite lining and subjected to critical pressure. They are only concentrated in the localized area of the crack tip and do not propagate in the longitudinal direction.

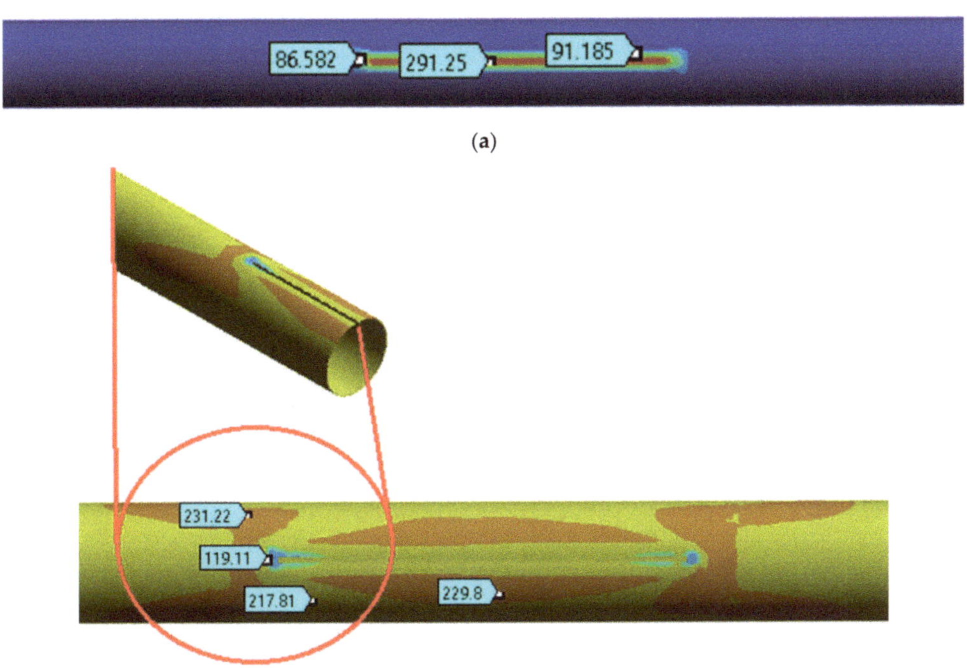


**Figure 10.** Von Mises equivalent stresses at critical pressure: (**a**) in a steel pipeline with a long composite lining; (**b**) in a steel pipeline with a crack under the composite lining (unit: megapascal).

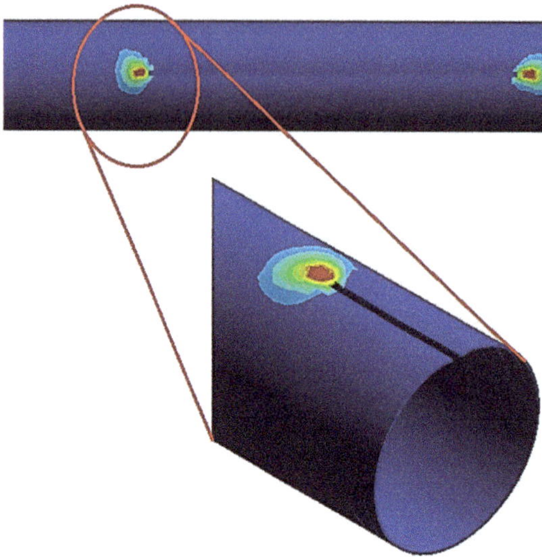

**Figure 11.** Plastic deformations in a steel pipeline under the composite lining and subjected to critical pressure.

The coordinates of the crack tip correspond to the values of 1900 mm for the left and 2100 mm for the right, i.e., both at the beginning of the deformation process and at the end. This fact indicates the absence of the crack growth in the longitudinal direction.

The stresses in the composite lining and corresponding to the through-crack area do not overcome the value of 292 MPa. In detail, they are far from the tensile strength limit, which is equal to 490 MPa, and quickly decrease to a minimum value with distance from the through crack. The stress distribution shows that the stresses, close to the yield strength of the steel, do not only occur near the crack edges but, additionally, along the pipeline and at the left and right of the crack. The length of each stress concentration is approximately half of the initial crack's length. Based on this fact, the length of the composite lining on a longitudinal crack is recommended to be twice as long as the crack length. Moreover, a verification of the deformation of the steel pipeline with a through crack of length $l$ and reinforced with a circular composite lining of length $2l$ is conducted. Figure 12 depicts the von Mises equivalent stresses in the pipeline with the composite lining and subjected to critical pressure.

**Figure 12.** Von Mises equivalent stresses in a steel pipeline with an overlay having an optimal length and subjected to critical pressure (unit: megapascal).

The maximum stress level in the pipeline and outside the composite lining does not exceed 86% of the steel yield strength. Conversely, the maximum stress in the composite lining does not overcome 60% of the tensile strength of the composite lining itself. Figure 13 shows the von Mises equivalent stresses in the pipeline under the composite lining and subjected to critical pressure.

**Figure 13.** Von Mises equivalent stresses in a steel pipeline with a composite lining having an optimal length and subjected to critical pressure (unit: megapascal).

The stresses in the entire area which is under the composite lining do not reach the steel yield strength. Thus, there is no crack growth in the longitudinal direction. Consequently, the presence of the composite lining stops the crack from developing under the influence of the internal pressure. Figure 14 displays the plastic deformations in the pipeline under the composite lining and subjected to critical pressure. Specifically, they are only localized in the area of the crack tip and do not propagate in the longitudinal direction. Such outcomes confirm that the optimal length of the composite lining should be twice that of the crack length.

**Figure 14.** Plastic deformations in a steel pipeline under the composite lining having an optimal length and subjected to critical pressure.

### 3.3. *Comparative Analysis of the FE Modeling Results*

A comparative analysis of the obtained results described in the previous sections with a bandaged longitudinal through crack and a similarly designed steel pipeline without defects is presented as follows.

#### 3.3.1. Case 1: No Internal Pressure on the Steel Gas Pipeline

Tables 3–5 depict the FE results for the first 20 frequencies of oscillations of a steel pipeline without and with a composite lining of 37.5%, 50%, and 62.5% of the pipe thickness, respectively. The composite lining's length is equal to 10% of the pipeline length. The findings prove that in the absence of internal pressure, all oscillation frequencies for both pipelines without and with a composite lining on a longitudinal through crack are paired. This indicates that for the pipeline under investigation, a 1.0 mm wide crack does not significantly affect the frequency spectrum. Indeed, the frequency spectrum of oscillations of a pipeline without a composite lining is higher than that of a pipeline with a composite lining. The difference between the corresponding frequencies increases with an increment in the composite lining's thickness. This phenomenon is explained by the fact that the presence of an overlay leads to a decrease in the frequency spectrum. Moreover, an increase in the thickness of the overlay creates an additional resistance due to the mass of the overlay for frequency values in the normal state (without a lining).

**Table 3.** Oscillation frequencies of an unloaded steel pipeline: composite lining's thickness—37.5% of the pipeline thickness; composite lining's length—10% of the pipeline length.

| Frequency No. | Oscillation Frequencies, in Hz | | Difference in Values, % |
|---|---|---|---|
| | Pipeline without Composite Lining | Pipeline with Composite Lining | |
| 1 | 29.29 | 29.13 | 0.6 |
| 2 | 29.29 | 29.13 | 0.6 |
| 3 | 34.84 | 34.66 | 0.5 |
| 4 | 34.84 | 34.66 | 0.5 |
| 5 | 49.96 | 49.63 | 0.7 |
| 6 | 49.96 | 49.63 | 0.7 |
| 7 | 52.21 | 51.92 | 0.6 |
| 8 | 52.21 | 51.92 | 0.6 |
| 9 | 54.84 | 54.78 | 0.1 |
| 10 | 54.84 | 54.78 | 0.1 |
| 11 | 59.32 | 59.26 | 0.1 |
| 12 | 59.32 | 59.26 | 0.1 |
| 13 | 75.50 | 75.06 | 0.6 |
| 14 | 75.50 | 75.06 | 0.6 |
| 15 | 77.36 | 77.26 | 0.1 |
| 16 | 77.36 | 77.26 | 0.1 |
| 17 | 78.28 | 78.21 | 0.1 |
| 18 | 78.28 | 78.21 | 0.1 |
| 19 | 79.28 | 78.87 | 0.5 |
| 20 | 79.28 | 78.87 | 0.5 |

**Table 4.** Oscillation frequencies of an unloaded steel pipeline: composite lining's thickness—50% of the pipeline thickness; composite lining's length—10% of the pipeline length.

| Frequency No. | Oscillation Frequencies, in Hz | | Difference in Values, % |
|---|---|---|---|
| | Pipeline without Composite Lining | Pipeline with Composite Lining | |
| 1 | 29.29 | 29.06 | 0.8 |
| 2 | 29.29 | 29.06 | 0,8 |
| 3 | 34.84 | 34.61 | 0.7 |
| 4 | 34.84 | 34.61 | 0.7 |
| 5 | 49.96 | 49.51 | 0.9 |
| 6 | 49.96 | 49.51 | 0.9 |
| 7 | 52.21 | 51.88 | 0.6 |
| 8 | 52.21 | 51.88 | 0.6 |
| 9 | 54.84 | 54.76 | 0.1 |
| 10 | 54.84 | 54.76 | 0.1 |
| 11 | 59.32 | 59.25 | 0.1 |
| 12 | 59.32 | 59.25 | 0.1 |
| 13 | 75.50 | 75.02 | 0.6 |
| 14 | 75.50 | 75.02 | 0.6 |
| 15 | 77.36 | 77.24 | 0.1 |
| 16 | 77.36 | 77.24 | 0.1 |
| 17 | 78.28 | 78.21 | 0.1 |
| 18 | 78.28 | 78.21 | 0.1 |
| 19 | 79.28 | 78.81 | 0.6 |
| 20 | 79.28 | 78.81 | 0.6 |

**Table 5.** Oscillation frequencies of an unloaded steel pipeline: composite lining's thickness—62.5% of the pipeline thickness; composite lining's length—10% of the pipeline length.

| Frequency No. | Oscillation Frequencies, in Hz | | Difference in Values, % |
|---|---|---|---|
| | Pipeline without Composite Lining | Pipeline with Composite Lining | |
| 1 | 29.29 | 29.01 | 1.0 |
| 2 | 29.29 | 29.01 | 1.0 |
| 3 | 34.84 | 34.58 | 0.8 |
| 4 | 34.84 | 34.58 | 0.8 |
| 5 | 49.96 | 49.40 | 1.1 |
| 6 | 49.96 | 49.40 | 1.1 |
| 7 | 52.21 | 51.84 | 0.7 |
| 8 | 52.21 | 51.84 | 0.7 |
| 9 | 54.84 | 54.75 | 0.2 |
| 10 | 54.84 | 54.75 | 0.2 |
| 11 | 59.32 | 59.24 | 0.1 |
| 12 | 59.32 | 59.24 | 0.1 |
| 13 | 75.50 | 74.96 | 0.7 |
| 14 | 75.50 | 74.96 | 0.7 |
| 15 | 77.36 | 77.21 | 0.2 |
| 16 | 77.36 | 77.21 | 0.2 |
| 17 | 78.28 | 78.21 | 0.1 |
| 18 | 78.28 | 78.21 | 0.1 |
| 19 | 79.28 | 78.72 | 0.7 |
| 20 | 79.28 | 78.72 | 0.7 |

Tables 6 and 7 illustrate the FE results for the first 20 frequencies of oscillations of a steel pipeline without and with a composite lining of 15% and 20% of the pipeline length, respectively. The composite lining's thickness is 50% of the pipe thickness. All oscillation frequencies retain their pairing, associated with the axial symmetry of the pipeline with a circular composite lining and the mechanical properties of the composite lining itself. In this case, the difference between the corresponding frequencies of the pipeline without and with a composite lining increase with an increment in the composite lining's length. Specifically, such an increment leads to a decrease in the frequency spectrum values (Table 8), as does an increase in the composite lining's thickness (Table 9). This phenomenon, as well as in the case of thickness, is explained by the fact that the presence of an overlay leads to a decrease in the frequency spectrum. Also, an increase in the length of the overlay creates additional resistance due to the mass of the overlay for frequency values in the normal state.

**Table 6.** Oscillation frequencies of an unloaded steel pipeline: composite lining's thickness—50% of the pipeline thickness; composite lining's length—15% of the pipeline length.

| Frequency No. | Oscillation Frequencies, in Hz | | Difference in Values, % |
|---|---|---|---|
| | Pipeline without Composite Lining | Pipeline with Composite Lining | |
| 1 | 29.29 | 28.94 | 1.2 |
| 2 | 29.29 | 28.94 | 1.2 |
| 3 | 34.84 | 34.50 | 1.0 |
| 4 | 34.84 | 34.50 | 1.0 |
| 5 | 49.96 | 49.31 | 1.3 |
| 6 | 49.96 | 49.31 | 1.3 |
| 7 | 52.21 | 51.72 | 0.9 |
| 8 | 52.21 | 51.72 | 0.9 |
| 9 | 54.84 | 54.71 | 0.2 |

Table 6. *Cont.*

| Frequency No. | Oscillation Frequencies, in Hz | | Difference in Values, % |
|---|---|---|---|
| | Pipeline without Composite Lining | Pipeline with Composite Lining | |
| 10 | 54.84 | 54.71 | 0.2 |
| 11 | 59.32 | 59.20 | 0.2 |
| 12 | 59.32 | 59.20 | 0.2 |
| 13 | 75.50 | 74.78 | 1.0 |
| 14 | 75.50 | 74.78 | 1.0 |
| 15 | 77.36 | 77.15 | 0.3 |
| 16 | 77.36 | 77.15 | 0.3 |
| 17 | 78.28 | 78.15 | 0.2 |
| 18 | 78.28 | 78.15 | 0.2 |
| 19 | 79.28 | 78.65 | 0.8 |
| 20 | 79.28 | 78.65 | 0.8 |

**Table 7.** Oscillation frequencies of an unloaded steel pipeline: composite lining's thickness—50% of the pipeline thickness; composite lining's length—20% of the pipeline length.

| Frequency No. | Oscillation Frequencies, in Hz | | Difference in Values, % |
|---|---|---|---|
| | Pipeline without Composite Lining | Pipeline with Composite Lining | |
| 1 | 29.29 | 28.83 | 1.6 |
| 2 | 29.29 | 28.83 | 1.6 |
| 3 | 34.84 | 34.40 | 1.3 |
| 4 | 34.84 | 34.40 | 1.3 |
| 5 | 49.96 | 49.10 | 1.8 |
| 6 | 49.96 | 49.10 | 1.8 |
| 7 | 52.21 | 51.57 | 1.2 |
| 8 | 52.21 | 51.57 | 1.2 |
| 9 | 54.84 | 54.62 | 0.4 |
| 10 | 54.84 | 54.62 | 0.4 |
| 11 | 59.32 | 59.12 | 0.3 |
| 12 | 59.32 | 59.12 | 0.3 |
| 13 | 75.50 | 74.56 | 1.3 |
| 14 | 75.50 | 74.56 | 1.3 |
| 15 | 77.36 | 77.02 | 0.4 |
| 16 | 77.36 | 77.02 | 0.4 |
| 17 | 78.28 | 78.06 | 0.3 |
| 18 | 78.28 | 78.06 | 0.3 |
| 19 | 79.28 | 78.54 | 0.9 |
| 20 | 79.28 | 78.54 | 0.9 |

**Table 8.** Oscillation frequencies of an unloaded steel pipeline with a composite lining of different lengths; the thickness of the composite lining is 50% of the pipeline thickness.

| Frequency No. | Oscillation Frequencies, in Hz | | |
|---|---|---|---|
| | 10% of Pipeline Length | 15% of Pipeline Length | 20% of Pipeline Length |
| 1 | 29.06 | 28.94 | 28.83 |
| 2 | 29.06 | 28.94 | 28.83 |
| 3 | 34.61 | 34.50 | 34.40 |
| 4 | 34.61 | 34.50 | 34.40 |
| 5 | 49.51 | 49.31 | 49.10 |
| 6 | 49.51 | 49.31 | 49.10 |

**Table 8.** *Cont.*

| Frequency No. | Oscillation Frequencies, in Hz | | |
|---|---|---|---|
| | 10% of Pipeline Length | 15% of Pipeline Length | 20% of Pipeline Length |
| 7 | 51.88 | 51.72 | 51.57 |
| 8 | 51.88 | 51.72 | 51.57 |
| 9 | 54.76 | 54.71 | 54.62 |
| 10 | 54.76 | 54.71 | 54.62 |
| 11 | 59.25 | 59.20 | 59.12 |
| 12 | 59.25 | 59.20 | 59.12 |
| 13 | 75.02 | 74.78 | 74.56 |
| 14 | 75.02 | 74.78 | 74.56 |
| 15 | 77.24 | 77.15 | 77.02 |
| 16 | 77.24 | 77.15 | 77.02 |
| 17 | 78.21 | 78.15 | 78.06 |
| 18 | 78.21 | 78.15 | 78.06 |
| 19 | 78.81 | 78.65 | 78.54 |
| 20 | 78.81 | 78.65 | 78.54 |

**Table 9.** Oscillation frequencies of an unloaded steel pipeline with a composite lining of different thicknesses; the length of the composite lining is 10% of the pipeline length.

| Frequency No. | Oscillation Frequencies, in Hz | | |
|---|---|---|---|
| | 37.5% of Pipeline Length | 50% of Pipeline Length | 62.5% of Pipeline Length |
| 1 | 29.13 | 29.06 | 29.01 |
| 2 | 29.13 | 29.06 | 29.01 |
| 3 | 34.66 | 34.61 | 34.58 |
| 4 | 34.66 | 34.61 | 34.58 |
| 5 | 49.63 | 49.51 | 49.40 |
| 6 | 49.63 | 49.51 | 49.40 |
| 7 | 51.92 | 51.88 | 51.84 |
| 8 | 51.92 | 51.88 | 51.84 |
| 9 | 54.78 | 54.76 | 54.75 |
| 10 | 54.78 | 54.76 | 54.75 |
| 11 | 59.26 | 59.25 | 59.24 |
| 12 | 59.26 | 59.25 | 59.24 |
| 13 | 75.06 | 75.02 | 74.96 |
| 14 | 75.06 | 75.02 | 74.96 |
| 15 | 77.26 | 77.24 | 77.21 |
| 16 | 77.26 | 77.24 | 77.21 |
| 17 | 78.21 | 78.21 | 78.21 |
| 18 | 78.21 | 78.21 | 78.21 |
| 19 | 78.87 | 78.81 | 78.72 |
| 20 | 78.87 | 78.81 | 78.72 |

The oscillation modes for all the parameters of the composite lining are the same. They are depicted in Figure 15. The first through fourth modes have no nodes in the longitudinal direction. Conversely, the fifth and sixth modes have one node each in the longitudinal direction. For higher modes of natural oscillations, the number of nodes in the longitudinal direction increases uniformly. Furthermore, the first and fifth modes have six nodes each in the circumferential direction; the second and sixth modes have eight; the third mode has four; and the fourth mode has ten. Consequently, at initial dynamic frequencies, oscillations occur in the circumferential direction. As the frequency increases, oscillations in the axial direction are added to the oscillations in the circumferential direction. It can be stated that combined natural oscillation modes predominate.

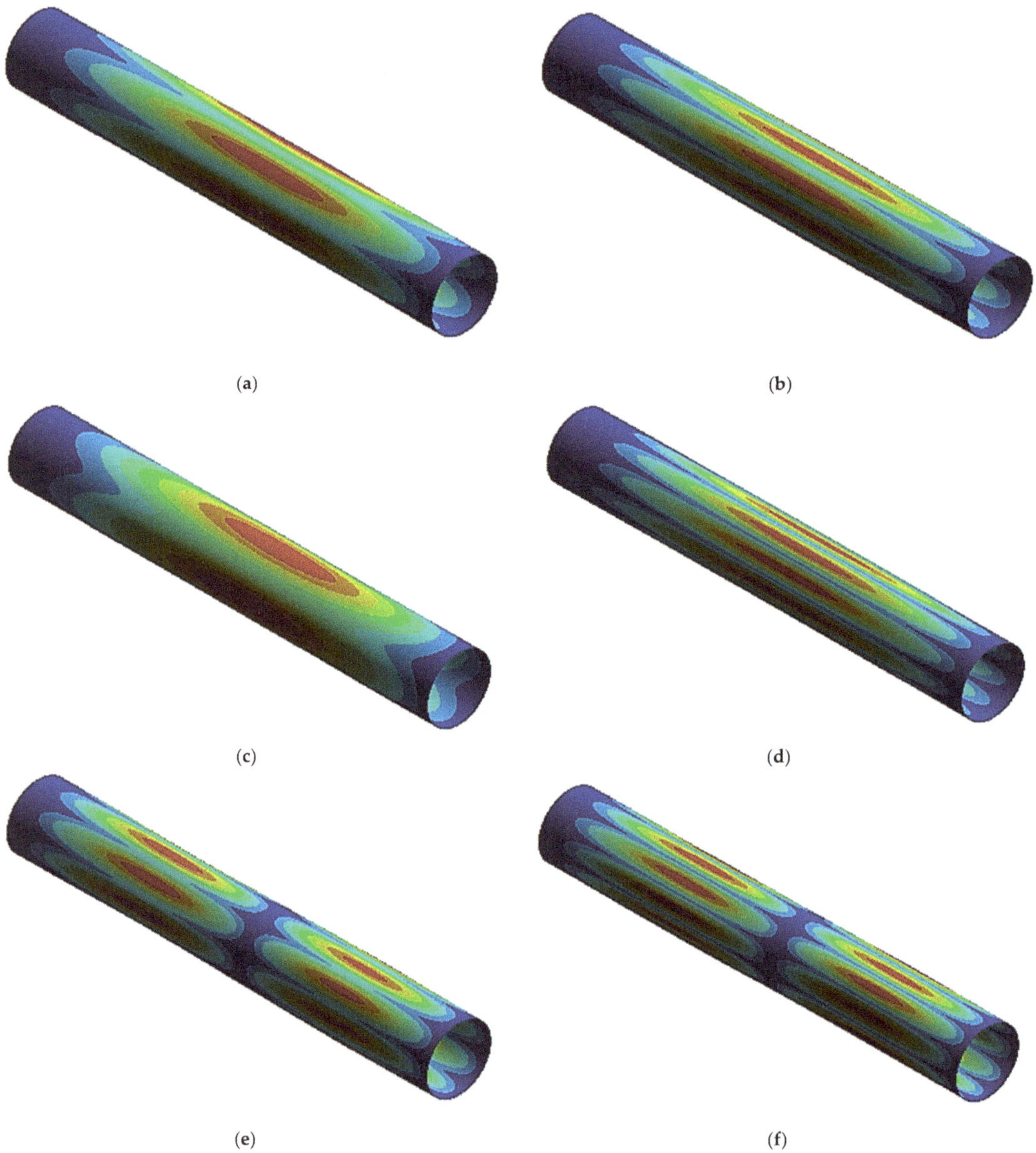

**Figure 15.** Oscillation modes of an unloaded steel pipeline with a composite lining: (**a**) first; (**b**) second; (**c**) third; (**d**) fourth; (**e**) fifth; (**f**) sixth mode.

### 3.3.2. Case 2: Steel Gas Pipeline Subjected to Critical Internal Pressure

Tables 10–12 present the FE results for the first 20 frequencies of pipeline oscillations without and with a composite lining of 37.5%, 50%, and 62.5% of the pipeline thickness, respectively. The composite lining's length is 10% of the pipeline length. In detail, the results demonstrate that the frequency pairing is violated for a steel pipeline stressed by internal pressure and with a crack closed by a composite lining. This is explained by the violation of the axisymmetry in the pipeline area with a 1.0 mm wide crack and subjected to internal critical pressure. In this case, the frequency spectrum of pipeline oscillations without a composite lining is higher than that of pipeline oscillations with a composite lining. In addition, the difference between the corresponding oscillation frequencies increases with an increment in the composite lining's thickness. This is explained by the creation of additional resistance due to the mass of the pad for frequency values in the normal state (without a lining).

**Table 10.** Oscillation frequencies of a steel pipeline subjected to critical pressure: composite lining's thickness—37.5% of the pipeline thickness; composite lining's length—10% of the pipeline length.

| Frequency No. | Oscillation Frequencies, in Hz | | Difference in Values, % |
|---|---|---|---|
| | Pipeline without Composite Lining | Pipeline with Composite Lining | |
| 1 | 137.42 | 136.31 | 0.8 |
| 2 | 137.42 | 136.34 | 0.8 |
| 3 | 151.84 | 150.57 | 0.8 |
| 4 | 151.84 | 150.58 | 0.8 |
| 5 | 207.07 | 206.71 | 0.2 |
| 6 | 207.07 | 206.81 | 0.1 |
| 7 | 251.02 | 248.79 | 0.9 |
| 8 | 251.02 | 248.81 | 0.9 |
| 9 | 260.46 | 260.10 | 0.1 |
| 10 | 260.46 | 260.11 | 0.1 |
| 11 | 290.92 | 289.19 | 0.6 |
| 12 | 290.92 | 289.20 | 0.6 |
| 13 | 315.45 | 314.28 | 0.4 |
| 14 | 322.38 | 320.40 | 0.6 |
| 15 | 322.38 | 320.40 | 0.6 |
| 16 | 349.45 | 347.13 | 0.7 |
| 17 | 349.45 | 347.17 | 0.7 |
| 18 | 351.06 | 348.66 | 0.7 |
| 19 | 351.06 | 348.92 | 0.6 |
| 20 | 353.33 | 353.00 | 0.1 |

**Table 11.** Oscillation frequencies of a steel pipeline subjected to critical pressure: composite lining's thickness—50% of the pipeline thickness; composite lining's length—10% of the pipeline length.

| Frequency No. | Oscillation Frequencies, in Hz | | Difference in Values, % |
|---|---|---|---|
| | Pipeline without Composite Lining | Pipeline with Composite Lining | |
| 1 | 137.42 | 136.17 | 0.9 |
| 2 | 137.42 | 136.19 | 0.9 |
| 3 | 151.84 | 150.39 | 1.0 |
| 4 | 151.84 | 150.4 | 1.0 |
| 5 | 207.07 | 206.68 | 0.2 |
| 6 | 207.07 | 206.77 | 0.1 |
| 7 | 251.02 | 248.44 | 1.0 |
| 8 | 251.02 | 248.46 | 1.0 |
| 9 | 260.46 | 260.06 | 0.2 |
| 10 | 260.46 | 260.07 | 0.1 |
| 11 | 290.92 | 288.97 | 0.7 |
| 12 | 290.92 | 288.98 | 0.7 |
| 13 | 315.45 | 314.11 | 0.4 |

**Table 11.** *Cont.*

| Frequency No. | Oscillation Frequencies, in Hz | | Difference in Values, % |
|---|---|---|---|
| | Pipeline without Composite Lining | Pipeline with Composite Lining | |
| 14 | 322.38 | 320.14 | 0.7 |
| 15 | 322.38 | 320.14 | 0.7 |
| 16 | 349.45 | 346.42 | 0.9 |
| 17 | 349.45 | 346.46 | 0.9 |
| 18 | 351.06 | 348.59 | 0.7 |
| 19 | 351.06 | 348.85 | 0.6 |
| 20 | 353.33 | 352.97 | 0.1 |

**Table 12.** Oscillation frequencies of a steel pipeline subjected to critical pressure: composite lining's thickness—62.5% of the pipeline thickness; composite lining's length—10% of the pipeline length.

| Frequency No. | Oscillation Frequencies, in Hz | | Difference in Values, % |
|---|---|---|---|
| | Pipeline without Composite lining | Pipeline with Composite Lining | |
| 1 | 137.42 | 135.87 | 1.1 |
| 2 | 137.42 | 135.89 | 1.1 |
| 3 | 151.84 | 150.03 | 1.2 |
| 4 | 151.84 | 150.04 | 1.2 |
| 5 | 207.07 | 206.62 | 0.2 |
| 6 | 207.07 | 206.7 | 0.2 |
| 7 | 251.02 | 247.75 | 1.3 |
| 8 | 251.02 | 247.77 | 1.3 |
| 9 | 260.46 | 259.98 | 0.2 |
| 10 | 260.46 | 259.99 | 0.2 |
| 11 | 290.92 | 288.53 | 0.8 |
| 12 | 290.92 | 288.55 | 0.8 |
| 13 | 315.45 | 313.78 | 0.5 |
| 14 | 322.38 | 319.62 | 0.9 |
| 15 | 322.38 | 319.62 | 0.9 |
| 16 | 349.45 | 345.00 | 1.3 |
| 17 | 349.45 | 345.04 | 1.3 |
| 18 | 351.06 | 348.46 | 0.7 |
| 19 | 351.06 | 348.71 | 0.7 |
| 20 | 353.33 | 352.92 | 0.1 |

Tables 13 and 14 depict the outcomes for the first 20 frequencies of pipeline oscillations without and with a composite lining equal, respectively, to 15% and 20% of the pipeline length. The composite lining's thickness is 50% of the pipeline thickness. The difference between the corresponding frequencies of the pipeline without and with a composite lining increase with the increment in the composite lining's length, both for the paired and unpaired frequencies. Increasing the composite lining's length and thickness leads to a decrement in the frequency spectrum values of the pipeline due to the additional weight of the lining (Tables 15 and 16).

**Table 13.** Oscillation frequencies of a steel pipeline subjected to critical pressure: composite lining's thickness—50% of the pipeline thickness; composite lining's length—15% of the pipeline length.

| Frequency No. | Oscillation Frequencies, in Hz | | Difference in Values, % |
|---|---|---|---|
| | Pipeline without Composite Lining | Pipeline with Composite Lining | |
| 1 | 137.42 | 135.61 | 1.3 |
| 2 | 137.42 | 135.61 | 1.3 |
| 3 | 151.84 | 149.7 | 1.4 |
| 4 | 151.84 | 149.7 | 1.4 |
| 5 | 207.07 | 206.53 | 0.3 |
| 6 | 207.07 | 206.53 | 0.3 |
| 7 | 251.02 | 247.14 | 1.6 |
| 8 | 251.02 | 247.14 | 1.6 |
| 9 | 260.46 | 259.78 | 0.3 |
| 10 | 260.46 | 259.78 | 0.3 |
| 11 | 290.92 | 288.35 | 0.9 |
| 12 | 290.92 | 288.35 | 0.9 |
| 13 | 315.45 | 313.45 | 0.6 |
| 14 | 322.38 | 319.41 | 0.9 |
| 15 | 322.38 | 319.41 | 0.9 |
| 16 | 349.45 | 343.98 | 1.6 |
| 17 | 349.45 | 343.98 | 1.6 |
| 18 | 351.06 | 348.06 | 0.9 |
| 19 | 351.06 | 348.06 | 0.9 |
| 20 | 353.33 | 352.66 | 0.2 |

**Table 14.** Oscillation frequencies of a steel pipeline subjected to critical pressure: composite lining's thickness—50% of the pipeline thickness; composite lining's length—20% of the pipeline length.

| Frequency No. | Oscillation Frequencies, in Hz | | Difference in Values, % |
|---|---|---|---|
| | Pipeline without Composite Lining | Pipeline with Composite Lining | |
| 1 | 137.42 | 135.06 | 1.7 |
| 2 | 137.42 | 135.06 | 1.7 |
| 3 | 151.84 | 149.05 | 1.9 |
| 4 | 151.84 | 149.05 | 1.9 |
| 5 | 207.07 | 206.18 | 0.4 |
| 6 | 207.07 | 206.18 | 0.4 |
| 7 | 251.02 | 245.93 | 2.1 |
| 8 | 251.02 | 245.93 | 2.1 |
| 9 | 260.46 | 259.34 | 0.4 |
| 10 | 260.46 | 259.34 | 0.4 |
| 11 | 290.92 | 287.97 | 1.0 |
| 12 | 290.92 | 287.97 | 1.0 |
| 13 | 315.45 | 312.8 | 0.8 |
| 14 | 322.38 | 318.92 | 1.1 |
| 15 | 322.38 | 318.92 | 1.1 |
| 16 | 349.45 | 341.95 | 2.2 |
| 17 | 349.45 | 341.95 | 2.2 |
| 18 | 351.06 | 346.91 | 1.2 |
| 19 | 351.06 | 346.91 | 1.2 |
| 20 | 353.33 | 352.06 | 0.4 |

**Table 15.** Oscillation frequencies of a steel pipeline subjected to critical pressure and with a composite lining of different lengths; the composite lining's thickness is 50% of the pipeline thickness.

| Frequency No. | Oscillation Frequencies, in Hz | | |
|---|---|---|---|
| | 10% of Pipeline Length | 15% of Pipeline Length | 20% of Pipeline Length |
| 1 | 136.17 | 135.61 | 135.06 |
| 2 | 136.19 | 135.61 | 135.06 |
| 3 | 150.39 | 149.70 | 149.05 |
| 4 | 150.40 | 149.70 | 149.05 |
| 5 | 206.68 | 206.53 | 206.18 |
| 6 | 206.77 | 206.53 | 206.18 |
| 7 | 248.44 | 247.14 | 245.93 |
| 8 | 248.46 | 247.14 | 245.93 |
| 9 | 260.06 | 259.78 | 259.34 |
| 10 | 260.07 | 259.78 | 259.34 |
| 11 | 288.97 | 288.35 | 287.97 |
| 12 | 288.98 | 288.35 | 287.97 |
| 13 | 314.11 | 313.45 | 312.8 |
| 14 | 320.14 | 319.41 | 318.92 |
| 15 | 320.14 | 319.41 | 318.92 |
| 16 | 346.42 | 343.98 | 341.95 |
| 17 | 346.46 | 343.98 | 341.95 |
| 18 | 348.59 | 348.06 | 346.91 |
| 19 | 348.85 | 348.06 | 346.91 |
| 20 | 352.97 | 352.66 | 352.06 |

**Table 16.** Oscillation frequencies of a steel pipeline subjected to critical pressure and with a composite lining of different lengths; the composite lining's length is 10% of the pipeline length.

| Frequency No. | Oscillation Frequencies, in Hz | | |
|---|---|---|---|
| | 37.5% of Pipeline Length | 50% of Pipeline Length | 62.5% of Pipeline Length |
| 1 | 136.31 | 136.17 | 135.87 |
| 2 | 136.34 | 136.19 | 135.89 |
| 3 | 150.57 | 150.39 | 150.03 |
| 4 | 150.58 | 150.40 | 150.04 |
| 5 | 206.71 | 206.68 | 206.62 |
| 6 | 206.81 | 206.77 | 206.7 |
| 7 | 248.79 | 248.44 | 247.75 |
| 8 | 248.81 | 248.46 | 247.77 |
| 9 | 260.10 | 260.06 | 259.98 |
| 10 | 260.11 | 260.07 | 259.99 |
| 11 | 289.19 | 288.97 | 288.53 |
| 12 | 289.20 | 288.98 | 288.55 |
| 13 | 314.28 | 314.11 | 313.78 |
| 14 | 320.40 | 320.14 | 319.62 |
| 15 | 320.40 | 320.14 | 319.62 |
| 16 | 347.13 | 346.42 | 345.00 |
| 17 | 347.17 | 346.46 | 345.04 |
| 18 | 348.66 | 348.59 | 348.46 |
| 19 | 348.92 | 348.85 | 348.71 |
| 20 | 353.00 | 352.97 | 352.92 |

It can be noted that the oscillation modes of the loaded steel pipeline (Figure 16) differ significantly from the oscillation modes of the unloaded one (Figure 15). They are the same for all the cases which have been assumed. In detail, the first mode of oscillation is a radial natural mode; the second and third modes of oscillation have four nodes in the circumferential direction; the second mode of oscillation has no nodes in the axial direction; and the third one has one node. Furthermore, the fourth and fifth oscillation modes are similar to the second and third ones, respectively. They have six nodes in the axial direction. The sixth mode with six nodes in the axial direction already has two nodes in the axial direction. In addition, the seventh oscillation mode corresponds to the "tension–compression" oscillations.

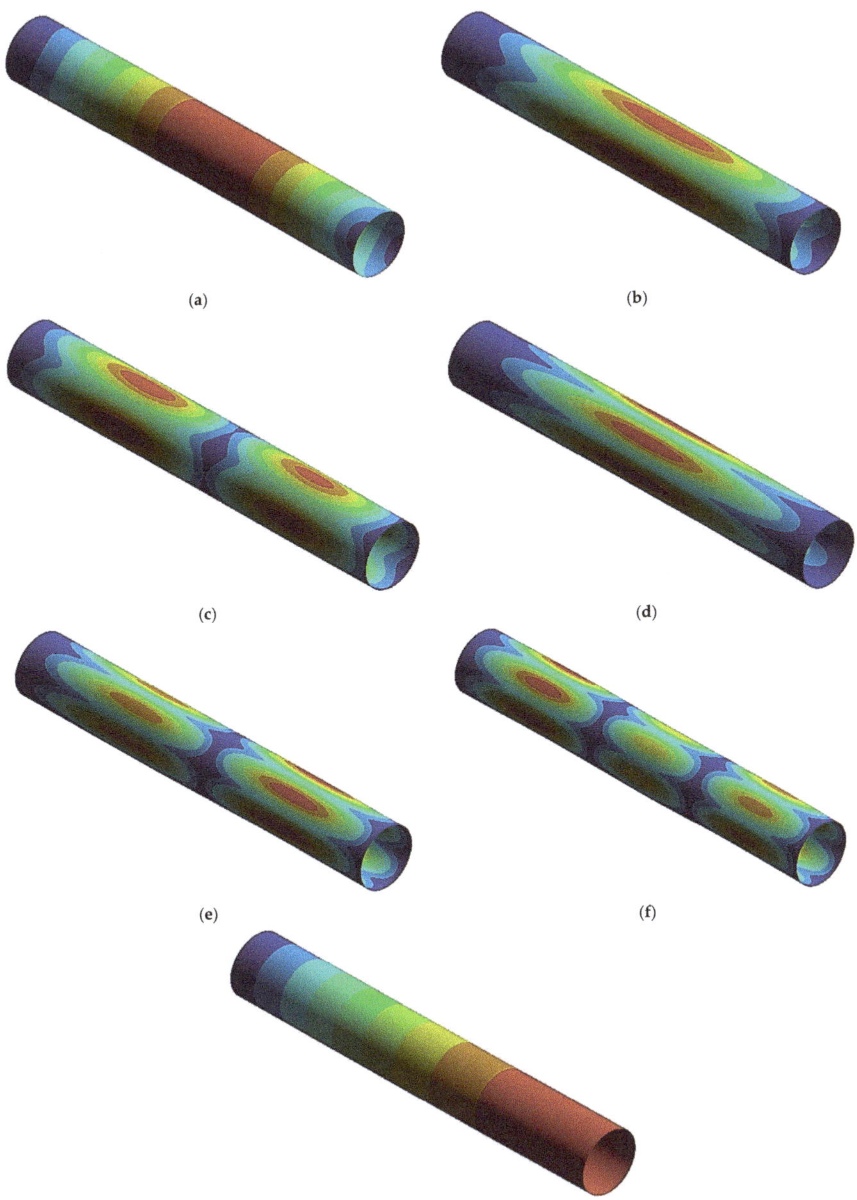

**Figure 16.** Oscillation modes of a steel pipeline subjected to internal pressure and with a crack closed by a composite lining: (**a**) first; (**b**) second; (**c**) third; (**d**) fourth; (**e**) fifth; (**f**) sixth; (**g**) seventh mode.

## 4. Discussion

This work is a continuation of studies on increasing the reliability of steel gas pipelines [27,28], in which reduced FE modeling was simulated to compare the obtained results with those from experimental simulations. These findings properly correlate with the numerical ones of full-scale gas pipelines [26], which indicates the reliability of the research as a whole. Thus, on the basis of a common FE model of the linear section of a steel pipeline arranged between four supports and with a crack-like through defect, numerical modeling of three key parameters of a bandage in the form of circular composite linings were developed (Figure 1) [27,28]: the type of composite material and the length and thickness of the composite lining. These approaches are advisable to be used in the design of composite linings for damaged sections of steel pipelines to reduce the costs of repair work and ensure the reliability of the operation of bandaged sections. FE modeling was performed using ANSYS 19.2 [29] according to the implemented models included in the unified calculation project "Program for calculating the linear section of main pipelines with composite linings for analyzing the static and dynamic stress–strain state, possible destruction and oscillations in the ANSYS software". We also note the variety of previous works in the field on strengthening gas pipelines with different types of profiles, including steel [26] and composite materials [19,21,23–25]. However, the aforementioned contributions mainly deal with problems related to the issues of reinforcement caused by different types of defects. Yet, the investigation with changes in the thickness analysis of composite linings has not been executed with sufficient depth, which requires additional investigation. In addition, the task of analyzing natural frequencies using composite overlays has not been performed except in the previous works conducted by the authors [15,16]. Previous work has also proposed defining the material of the composite lining as an elastic carbon-fiber-reinforced composite material, fiberglass, or any other material with elastic orthotropic mechanical properties [27,28]. The developed approach for assessing the static strength of a damaged section of a steel pipeline with a composite lining and subjected to internal pressure allows the determination of the optimal thickness of the composite lining itself (Table 2, Figures 6 and 7). Specifically, the results show that composite linings made of carbon fiber and with a thickness of the indicator "50.0% to 62.5%" of the pipeline thickness compensate for the internal critical pressure effect in the pipeline itself and with through-crack-like defects. The approach for assessing the dynamic strength and analyzing the possible destruction of a strengthened damaged section of a steel pipeline subjected to an internal pressure increment also permits determining the optimal length of the composite lining, which is at least equal to 241.2 mm (Figure 9).

When designing a steel gas pipeline for operation in seismically hazardous areas, its spectrum of natural frequencies and oscillation modes is analyzed in order to tune it out from the resonant frequencies. Therefore, when bandaging damaged sections, it is necessary to assume the degree of influence of the composite lining on the spectrum of natural frequencies and oscillation modes of the pipeline. To analyze this influence, a numerical approach was implemented for assessing the composite lining effect on the spectrum of the natural frequencies and oscillation modes of the bandaged section of the pipeline fixed between supports. Using an FE model of the linear section of a steel pipeline between four supports as an example, a comparative analysis between the FE results for a pipeline with a bandaged longitudinal through crack and those for a similarly designed pipeline without defects was performed. The first case considered no internal pressure, while the second considered the pipeline subjected to critical internal pressure. In the first case, the FE results of the first 20 frequencies of pipeline oscillations without and with a composite lining assumed a thickness of 37.5%, 50%, and 62.5% of the pipeline thickness, respectively. The composite lining's length was equal to 10% of the pipeline length. Moreover, the FE results proved that in the absence of internal pressure, all oscillation frequencies are paired for pipelines both without and with a composite lining on a longitudinal through crack. This indicates that for the steel pipeline under investigation, a 1.0 mm wide crack does not seriously affect the frequency spectrum. Indeed, the frequency spectrum of pipeline oscillations without a composite lining is higher than that of the pipeline with a composite one. The difference between the corresponding frequencies increases with the increment of the composite lining's thickness. The FE results for the first 20 oscillation frequencies of a pipeline without and with a composite lining consider the percentages of 15% and 20% of the pipeline length, respectively. The composite lining's thickness is 50% of the pipeline thickness. All oscillation frequencies retain their pairing associated with the axial symmetry of the pipeline with a circular lining and the mechanical properties of the composite lining itself (Tables 3–7). In this case, the difference between the corresponding pipeline's frequencies without and with a composite lining increases with an increment in the composite lining's length. In the absence of internal pressure, all oscillation frequencies for a pipeline with a crack closed by a composite lining are paired. In the second case, this pairing is violated when the pipeline is subjected to internal critical pressure. Furthermore, the difference between the oscillation frequency spectrum of the pipeline without a composite lining and that of the pipeline with a composite lining increase (Tables 10–16). In short, the oscillation modes of the loaded steel pipeline (Figure 16) differ significantly from the oscillation modes of the unloaded one (Figure 15). They are equal for all the cases which were considered. Notably, as a disadvantage of this work, the FE analyses do not take into account the friction between the steel pipeline and the corresponding bandage. However, this assumption does not significantly affect the aforementioned outcomes, but the influence of the friction will be considered in additional investigations.

The results of this contribution fully confirm the hypothesis of the theoretical study of natural gas pipelines presented by Kong et al. [25], which indicates the use of the obtained values as a correlation of experimental findings. Together, the usefulness of this work at the national [40,41] and international [42–46] scale is justified by the lack of design codes for increasing the bearing capacity and decreasing the wear of gas pipelines in the Republic of Kazakhstan, where it is more than 70% [47].

## 5. Conclusions

According to an FE model of the linear section of a steel gas pipeline arranged between four supports and with a crack-like through defect, numerical modeling of three key parameters of a bandage in the form of circular composite linings was implemented: the type of composite material and the length and thickness of the composite lining. An approach for assessing the static strength of a damaged section of a steel pipeline with a composite lining subjected to internal pressure was developed that makes it possible to estimate the optimal thickness of the composite lining itself. An approach for assessing the dynamic strength and analyzing the possible destruction of a strengthened damaged section of the same pipeline with an increment in internal pressure was also implemented, which, in turn, makes it possible to determine the composite lining's optimal length. An analysis of the spectrum of the natural frequencies and oscillation modes of the pipeline was performed in order to tune it out from the resonant frequencies for seismically hazardous areas. In fact, using an FE model of the linear section of a steel pipeline fixed between four supports, a comparative analysis of the results of a supported pipeline with a bandaged longitudinal through crack and a similarly designed pipeline without defects was executed. The cases were considered of when there is no internal pressure and when the pipeline is subjected to critical internal pressure. Thus, it was found that the frequency spectrum of the pipeline oscillations without a composite lining is higher than that of the pipeline oscillations with a composite lining. The difference between the corresponding frequencies increases when increasing the thickness or length of the composite lining. In the absence of internal pressure, all oscillation frequencies for a steel pipeline with a crack and closed by a composite lining are paired. This pairing is violated when the pipeline is exposed to internal critical pressure, and the difference between the frequency spectrum of pipeline oscillations without and with a composite lining increases. The oscillation modes of the loaded steel pipeline largely differ from the oscillation modes of the same structure when it is unloaded. In relation to the above conclusions, the proposed approaches, through the application of the composite linings, can be adopted as a methodology for tuning out damaged areas from critical low frequencies.

**Author Contributions:** Conceptualization, A.M., N.Z. and M.B.; methodology, A.M. and U.I.; investigation, A.M.; data curation, A.M., R.R. and S.Y.; writing—original draft preparation, A.M. and S.Y.; writing—review and editing, A.M., N.Z. and M.B.; supervision, A.M. and M.B.; project administration, A.M., R.R. and U.I.; funding acquisition, A.M. All authors have read and agreed to the published version of the manuscript.

**Funding:** This research was funded by the Science Committee of the Ministry of Education and Science of the Republic of Kazakhstan (Grant No. AP14872527).

**Data Availability Statement:** Data is contained within the manuscript.

**Acknowledgments:** The authors would like to thank the A. Pidhornyi Institute of Mechanical Engineering Problems of the National Academy of Sciences of Ukraine and the National Science and Technology Council (NSTC) of Taiwan for providing advisory assistance to this work.

**Conflicts of Interest:** The authors declare that they have no known competing financial interests or personal relationships that could have appeared to influence the work reported in this manuscript.

## References


1. The Global Gas Market. Available online: https://fin-plan.org/blog/investitsii/mirovoy-rynok-gaza/ (accessed on 2 September 2024).
2. This Day in History Information Portal. Available online: https://www.history.com/this-day-in-history/natural-gas-explosion-kills-schoolchildren-in-texas (accessed on 2 September 2024).
3. The New York Times. Scores Dead in Kenyan Pipeline Fire. Available online: https://www.nytimes.com/2011/09/13/world/africa/13kenya.html (accessed on 2 September 2024).
4. BBC News Information Portal. Taiwan Gas Blasts in Kaohsiung Kill at Least 25. Available online: https://www.bbc.com/news/world-asia-28594693 (accessed on 2 September 2024).
5. History. Sewers Explode in Guadalajara, Mexico, Killing Hundreds. 13 November 2009. Available online: https://www.history.com/this-day-in-history/sewers-explode-in-guadalajara (accessed on 23 October 2024).
6. The Telegraph. Burned to Death in Kenya Pipeline Fire. 2011. Available online: https://www.telegraph.co.uk/news/worldnews/africaandindianocean/kenya/ (accessed on 23 October 2024).
7. KazMunayGas. Available online: https://www.kmg.kz/ru (accessed on 23 October 2024).
8. Zhao, J.; Lv, Y.R.; Cheng, Y.F. Standards and methods for dent assessment and failure prediction of pipelines: A critical review. *Pet. Sci.* **2022**, *19*, 3029–3045. [CrossRef]
9. Zhangabay, N.; Ibraimova, U.; Suleimenov, U.; Moldagaliyev, A.; Buganova, S.; Jumabayev, A.; Kolesnikov, A.; Tursunkululy, T.; Zhiyenkulkyzy, D.; Khalelova, A.; et al. Factors affecting extended avalanche destructions on long-distance gas pipe lines: Review. *Case Stud. Constr. Mater.* **2023**, *19*, e02376. [CrossRef]
10. Shang, J.; Gao, R.; Xing, B.; Wei, H.; Hua, Z. Enhanced hydrogen degradation of two pipeline steels by increasing inert gas pressure in hydrogen-containing mixtures: Experimental and theoretical insights. *Corros. Sci.* **2024**, *240*, 112466. [CrossRef]
11. Röthig, M.; Hoschke, J.; Tapia, C.; Venezuela, J.; Atrens, A. A review of gas phase inhibition of gaseous hydrogen embrittlement in pipeline steels. *Int. J. Hydrog. Energy* **2024**, *60*, 1239–1265. [CrossRef]
12. Zhao, P.; Shuai, J.; Sun, M.; Lv, Z.; Xu, K.; Wang, Y. Burst pressure of thin-walled pipes with dent and gouge defects. *Thin-Walled Struct.* **2021**, *159*, 107213. [CrossRef]
13. He, Z.; Zhou, W. Fatigue reliability analysis of dented pipelines. *J. Pipeline Sci. Eng.* **2021**, *1*, 290–297. [CrossRef]

14. Shirazi, H.; Eadie, R.; Chen, W. A review on current understanding of pipeline circumferential stress corrosion cracking in near-neutral PH environment. *Eng. Fail. Anal.* **2023**, *148*, 107215. [CrossRef]
15. Zhangabay, N.; Ibraimova, U.; Ainabekov, A.; Buganova, S.; Moldagaliev, A. Finite-Element Modeling of the Temperature Effect on Extended Avalanche Damage of Gas Main Pipelines. *Materials* **2024**, *17*, 1963. [CrossRef]
16. Zhangabay, N.; Ibraimova, U.; Tursunkululy, T.; Buganova, S.; Moldagaliev, A.; Duissenbekov, B. Experimental Study of Avalanche Damage Protection Methods for Main Steel Gas Pipelines. *Materials* **2024**, *17*, 3171. [CrossRef]
17. Niazi, H.; Eadie, R.; Hao, Z. High pH stress corrosion cracking initiation and crack evolution in buried steel pipelines: A review. *Eng. Fail. Anal.* **2020**, *120*, 105013. [CrossRef]
18. Nikiforov, D.A.; Abdullin, N.V. Increasing the bearing capacity of insulation coatings of underground pipelines. *Constr. Repair Gas Oil Pipelines Gas Oil Storage Facil.* **2020**, *4*, 57–60. [CrossRef]
19. Alabtah, F.G.; Mahdi, E.; Eliyan, F.F. The use of fiber reinforced polymeric composites in pipelines: A review. *Compos. Struct.* **2021**, *276*, 114595. [CrossRef]
20. Zhu, Y.; Liu, G.; Cui, Z.; Yang, H.; Liu, F.; Jiang, B.; Chen, L. Effect and mechanism of ionic liquid-polymer composite coating on enhancing hydrogen embrittlement resistance of X80 pipeline steel for hydrogen blended natural gas transportation. *Int. J. Hydrog. Energy* **2024**, *80*, 1305–1316. [CrossRef]
21. Dumitrescu, A.; Minescu, M.; Dinita, A.; Lambrescu, I. Corrosion Repair of Pipelines Using Modern Composite Materials Systems: A Numerical Performance Evaluation. *Energies* **2021**, *14*, 615. [CrossRef]
22. Shafaee Fallah, A.; Sadeghian, M.; Golmakani, M.E. Experimental and Numerical Study on the Strength of Repaired Steel Pipes with Composite Patches under Internal Pressure. *Mech. Adv. Compos. Struct.* **2023**, *10*, 437–448. [CrossRef]
23. Lim, K.S.; Azraai, S.N.A.; Noor, N.M.; Yahaya, N. An Overview of Corroded Pipe Repair Techniques Using Composite Materials. *Int. J. Mater. Metall. Eng.* **2016**, *10*, 19–25. [CrossRef]
24. Dong, C. Flexural behavior of carbon and glass reinforced hybrid composite pipes. *Compos. Part C Open Access* **2021**, *4*, 100090. [CrossRef]
25. Kong, D.; Huang, X.; Xin, M.; Xian, G. Effects of defect dimensions and putty properties on the burst performances of steel pipes wrapped with CFRP composites. *Int. J. Press. Vessel. Pip.* **2020**, *186*, 104139. [CrossRef]
26. Nordhagen, H.O.; Kragset, S.; Berstad, T.; Morin, A.; Dorum, C.; Munkejord, S.T. A new coupled fluid-structure modelling methodology for running ductile fracture. *Comput. Struct.* **2017**, *94–95*, 13–21. [CrossRef]
27. Moldagaliyev, A.; Zhangabay, N.; Suleimenov, U.; Avramov, K.; Raimberdiyev, T.; Chernobryvko, M.; Umbitaliyev, A.; Jumabayev, A.; Yeshimbetov, S. Deformation features of trunk pipelines with composite linings under static loads. *East. Eur. J. Enterp. Technol.* **2023**, *5*, 34–42. [CrossRef]
28. Moldagaliyev, A.; Zhangabay, N.; Bonopera, M.; Raimberdiyev, T.; Yeshimbetov, S.; Galymzhan, S.; Anarbayev, Y. Finite-Element Analysis of Oscillations in Damaged Pipeline Sections Reinforced with a Composite Material. *Model. Simul. Eng.* **2024**, *15*, 2827002. [CrossRef]
29. ANSYS Products 19.2. Available online: https://getintopc.com/softwares/ansys-products-19-2-free-download-6384058/ (accessed on 6 September 2024).
30. Kiwiel, K. Convergence and efficiency of subgradient methods for quasiconvex minimization. *Math. Program.* **2001**, *90*, 1–25. [CrossRef]
31. Uspensky, B.; Derevianko, I.; Avramov, K.; Polishchuk, O.; Salenko, A. Experimental and numerical study on fatigue of sandwich plates with honeycomb core manufactured by fused deposition modeling. *Appl. Compos. Mater.* **2022**, *29*, 2033–2061. [CrossRef]
32. Chernobryvko, M.; Avramov, K.; Pierre, C.; Derevianko, I.; Maksymenko-Sheiko, K.; Uspensky, B. Influence of cell breakings on the buckling of composite sandwich shells with FDM additively manufactured honeycomb cores. *Proc. Inst. Mech. Eng. Part G J. Aerosp. Eng.* **2024**. [CrossRef]
33. Chernobryvko, M.; Avramov, K.; Pierre, C. Buckling of Cylindrical Sandwich Panels with Imperfections in Honeycomb Cores Manufactured by Fused Deposition Modeling. In *Advances in Mechanical and Power Engineering II. CAMPE 2023*; Altenbach, H., Gao, X.W., Syngellakis, S., Cheng, A.H.D., Lampart, P., Tkachuk, A., Eds.; Lecture Notes in Mechanical Engineering; Springer: Cham, Switzerland, 2025. [CrossRef]
34. Zhang, P.; Heyne, M.A.; To, A.C. Biomimetic Staggered Composites with Highly Enhanced Energy Dissipation: Design, Modeling, and Test. *arXiv* **2015**. [CrossRef]
35. Zhangabay, N.; Ibraimova, U.; Bonopera, M.; Suleimenov, U.; Avramov, K.; Chernobryvko, M.; Utelbayeva, A.; Uspenskyi, B. Finite-Element Modeling of the Dynamic Behavior of a Crack-like Defect in an Internally Pressurized Thin-Walled Steel Cylinder. *Appl. Sci.* **2024**, *14*, 1790. [CrossRef]
36. Sen, T.; Reddy, H.N.J. Strengthening of RC beams in flexure using natural jute fibre—Textile reinforced composite system and its comparative study with CFRP and GFRP strengthening systems. *Int. J. Sustain. Built Environ.* **2013**, *2*, 41–55. [CrossRef]

37. Altenbach, H.; Breslavsky, D.V.; Chernobryvko, M.V.; Senko, A.; Tatarinovaal, O. Fast Fracture of Conic Shell Under the Action of Belt Explosive Charge. In *Advances in Mechanical and Power Engineering;* Lecture Notes in Mechanical Engineering; Springer: Cham, Switzerland, 2022; pp. 366–376. [CrossRef]
38. Chernobryvko, M.; Kruszka, L.; Vorobiev, Y. Thermo-Elastic-Plastic Constitutive Model for Numerical Analysis of Metallic Structures under Local Impulsive Loadings. *Applied Mech. Mater.* **2014**, *566*, 493–498. [CrossRef]
39. Williamson, T.D. *RES-Q Wrap Design & Installation of RES-QTM Composite Wrap on Pipelines;* T.D. Williamson, S.A.: Tulsa, OK, USA, 2012; Available online: https://scholar.google.com/scholar_lookup?title=RES-Q+Wrap+Design+&+Installation+of+RES-QTM+Composite+Wrap+on+Pipelines&author=Williamson,+T.D.&publication_year=2012 (accessed on 20 January 2025).
40. SP RK EN 1993-4-3-2007-2011: Design of Steel Structures. Ch.4-3 Pipelines. 2016. Available online: https://online.zakon.kz/Document/?doc_id=34586480 (accessed on 2 September 2024).
41. Sanitary Rules of the Republic of Kazakhstan EN 1998-4:2006/2012: Seismic Design. Part 4. Bunkers, Reservoirs and Pipelines. 2012. Available online: https://online.zakon.kz/Document/?doc_id=37105813&doc_id2=37807474#activate_doc=2&pos=1;-0.0999908447265625&pos2=3;-100.09999084472656 (accessed on 2 September 2024).
42. API (American Petroleum Institute). *API Specification 5L*, 46th ed.; API: Washington DC, USA, 2018; Available online: https://buy-pipe.com/home/structure/item_214/955b42b7590d39be6f4d268afcd0a015.pdf (accessed on 2 September 2024).
43. Eurocode 8: Design of Structures for Earthquake Resistance. Part 4: Silos, Tanks, and Pipelines. 2006. Available online: https://www.phd.eng.br/wp-content/uploads/2014/12/en.1998.4.2006.pdf (accessed on 2 September 2023).
44. Eurocode 3: Design of Steel Structures. Part 4: Silos, Tanks, and Pipelines. 2005. Available online: https://www.phd.eng.br/wp-content/uploads/2015/12/en.1993.1.8.2005-1.pdf (accessed on 2 September 2024).
45. *ANSI/ASVE B 31G;* Manual for Determining the Remaining Strength of Corroded Pipelines. ASME: New York, NY, USA, 1984. Available online: https://law.resource.org/pub/us/cfr/ibr/002/asme.b31g.1991.pdf (accessed on 2 September 2024).
46. *ANSI/ASVE B31.8-73;* Gas Transmission and Distribution, Piping Systems. ASME: New York, NY, USA, 2004. Available online: https://law.resource.org/pub/us/cfr/ibr/002/asme.b31.8.2003.pdf (accessed on 2 September 2024).
47. Zona, K.Z. Iznos Kazaxstanskix Gazoprovodov Sostavlyaet Bolee-70%. Available online: https://zonakz.net/2021/03/12/iznos-kazaxstanskix-gazoprovodov-sostavlyaet-bolee-70/ (accessed on 2 September 2024).

*vibration*

MDPI

*Article*

# Methodology for Designing Vibration Devices with Asymmetric Oscillations and a Given Value of the Asymmetry of the Driving Force

**Mihail D. Gerasimov [1], Nickolai S. Lubimyi [1,*], Andrey A. Polshin [1], Boris S. Chetverikov [1] and Anastasia Chetverikova [2]**

[1] Department of Hoisting-and-Transport and Road Machines, Belgorod State Technological University named after V.G. Shukhova, st. Kostyukov 46, 308012 Belgorod, Russia; mail_mihail@mail.ru (M.D.G.); info@polshin.ru (A.A.P.); await_rescue@mail.ru (B.S.C.)

[2] Faculty of Foreign Languages, Institute of Pedagogy, The National Research University—Belgorod State University, st. Pobeda, 85, 308015 Belgorod, Russia; aanastasia.chetverikova@yandex.ru

* Correspondence: nslubim@bk.ru; Tel.: +7-951-136-40-17

**Abstract:** In mechanical engineering, the building industry, and many other branches of industry, vibration machines are widely used, in which circular and directed oscillations predominate in the form of movement of the working equipment. This article examines methods for generating asymmetric oscillations, which are estimated by a numerical parameter, namely by the coefficient of asymmetry of the magnitude of the driving force when changing the direction of action in a directed motion within each period of oscillations. It is shown that for generating asymmetric mechanical vibrations, vibration devices are used, consisting of vibrators of directed vibrations, called stages. These stages form the total asymmetric driving force. The behavior of the total driving force of asymmetric vibrations and the working equipment of the vibration machine are described by analytical equations, which represent certain laws of motion of the mechanical system. This article presents a numerical analysis of methods for obtaining laws of motion for a two-stage, three-stage, and four-stage vibration device with asymmetric oscillations. An analysis of the methodology for obtaining a generalized law of motion for a vibration device with asymmetric oscillations is performed based on the application of polyharmonic oscillation synthesis methods. It is shown that the method of forming the total driving force of a vibration device based on the coefficients of the terms of the Fourier series has limited capabilities. This article develops, substantiates, and presents a generalized method for calculating and designing a vibration device with asymmetric oscillations by the value of the total driving force and a given value of the asymmetry coefficient in a wide range of rational designs of vibration machines. The proposed method is accompanied by a numerical example for a vibration device with an asymmetry coefficient of the total driving force equal to 10.

**Keywords:** vibration machines; vibration device; asymmetric vibrations; driving force; summation of vibrations; asymmetry coefficient of the driving force

## 1. Introduction

Vibrational processes are part of oscillatory processes, which are reflected in the fundamental works of scientists from different countries [1–6].

Theoretical developments that form the science of vibrational processes and technologies are based on the general laws of oscillation theory [7–12], which, in turn, are the theoretical basis for many areas of research and their practical use [13–16].

In accordance with the widespread use of vibration technologies and vibration machines in a number of industrial production of materials, works, and services, methods for their design are described to one degree or another in specialized technical literature [10,11,17]. A special place in the development of methods for general and special design and calculation of parameters of vibration machines is occupied by the works of Andronov A.A., Ashavskiy A.M., Bauman V.A., Blekhman I.I., Mandelstam L.I. Strelkov S.P.

In works [17–19], fundamental requirements and criteria for the development of [17] laws of change and control of the driving force of a vibration machine depending on the area of application and its functional tasks are formed in a sufficiently complete manner for the prospective design of vibration machines. In order to implement the requirements for the development of design methods and calculation of technological parameters of vibration machines, formulated in [17,18,20] and in a number of other specialized studies [21–24], when solving new technical problems, their formalization and additional analysis of the kinetics of processes occurring under the conditions of superimposed vibration effects are required each time [25,26].

Problems of this kind include, in particular, the development of a methodology for designing vibration devices with asymmetric oscillations or with an asymmetric driving force. Considering that this scientific direction is still in its infancy, some features of terminology and definitions should be noted. In this article, the terms 'asymmetric driving force' and 'asymmetric oscillations' are considered equivalent, as well as 'asymmetry coefficient of the driving force magnitude' and 'dynamic coefficient of the oscillatory system'.

The issues of parameter asymmetry in oscillatory processes are common topics in scientific publications. In the works [26–28], some classification features of asymmetry in oscillatory, and therefore, in vibrational processes are given, namely:

- geometric asymmetry,
- asymmetry of frictional properties,
- force asymmetry,
- temporal asymmetry of excitation,
- frequency asymmetry,
- functional asymmetry,
- kinematic asymmetry,
- structural (constructive) asymmetry,
- gradient asymmetry,
- wave asymmetry,
- initial asymmetry, i.e., associated with the initial conditions of motion.

Without going into a detailed description of each type of classification feature, it can be noted that the selected type of asymmetry can be closely related to another type of asymmetry or have conditions for the transition of one into another. The use of one or another type of asymmetry in the scientific or technical sphere is associated with specific conditions and priorities of the authors. In this work, force asymmetry is considered a complex parameter that can be assessed and characterized by a numerical value, namely the coefficient of asymmetry of the driving force ($k_{as}$) or the coefficient of dynamism of the oscillatory system ($K_d$ or $k_d$) [29,30]. Asymmetric force is considered a force ($F$) of directed action along some straight line with a variable magnitude, whose magnitude in one direction significantly, or by some number of times, exceeds the magnitude of the component acting in the opposite direction while one of the components, for example, the larger one, can be taken as positive ($+F$), and in the opposite one, negative ($-F$). The magnitude of the asymmetry coefficient of the driving force ($k_{as}$) or the dynamic coefficient of the oscillatory system ($K_d$ or $k_d$) can be calculated from the expressions:

$$k_{as} = K_d = k_d = -\frac{+F}{-F} = \frac{F}{|-F|} = \frac{F_{uw}}{F_{is}} \tag{1}$$

where $F_{uw}$ is the magnitude of the driving force acting in the direction of performing 'useful work'; $F_{is}$ is the magnitude of the driving force acting in the direction of performing 'idle stroke', in the opposite direction; and (+$F$) and ($-F$), respectively, are the components of the total value of the driving force, acting alternately in opposite directions.

The novelty of the study is that for the first time, an equation of the total driving force is given, ensuring the obtaining of any coefficient (for technical and technological tasks) of asymmetry of the driving force, for example, more than 10. This is achieved by the fact that on the basis of the analysis of a number of works in the field of development of methods for calculating and designing vibration machines, the authors introduce a new term in relation to the form of oscillations of the working equipment of vibration machines which is asymmetric oscillations. This type of oscillation has a numerical characteristic, namely, the asymmetry coefficient of the driving force. This work currently lacks a general methodology for designing and calculating vibration devices with asymmetric oscillations, and the available individual examples for vibration devices with low values of the asymmetry coefficient of the driving force, for example, in the range of 2–4, are of an exclusively local nature for specific conditions. The novelty is that the use of the method of decomposition of piecewise smooth or monotone functions into a Fourier series for creating a general law for designing and calculating vibration devices with asymmetric oscillations is not applicable. Since it does not allow for the obtaining of a coefficient of asymmetry of the driving force with a greater value, for example, more than four.

The purpose of this article is to develop an engineering methodology for designing vibration devices with asymmetric oscillations based on the initial parameters: the magnitude of the total driving force acting in the direction of performing useful work and the asymmetry coefficient of the magnitude of the total driving force.

## 2. Materials and Methods

This article uses methods of analytical, numerical, and comparative research, logically closely related to classical methods of vibration theory and real methods of calculation and design of technological vibration machines, with their parameters and conditions of the application area, such as vibration displacement, vibration sorting, vibration compaction, vibration driving of piles and others. To evaluate the existing patterns of designing vibration devices with asymmetric oscillations, numerical methods for estimating the coefficient of asymmetry of the magnitude of the driving force and the obtained values of its asymmetry coefficient are used. To evaluate the possibility of creating a general methodology for designing vibration devices with asymmetric oscillations and a generalized law for the formation of asymmetric oscillations, classical laws of analytical mathematics are used, i.e., expansion of piecewise smooth and continuous functions in a Fourier series. When creating a mathematical model of a vibration device with asymmetric oscillations, the d'Alembert principle was used to obtain the equilibrium equations of the systems under consideration. Theorems on the properties of optimal Maxwell–Fejer polynomials, which specify the optimal Maxwell–Fejer impulse, were used as a method for assessing the created general methodology for designing and calculating vibration devices with asymmetric oscillations. When deriving the general equation for the value of the total driving force, the search method by simplex ($S^2$-method) was used in problems of constructing a regular simplex for a given base point and scale factor.

## 3. Results

Vibratory machines occupy an important place in a number of technological processes, for example, in operations of sorting and separating various materials [31] into fractions, inter-operational transportation [32,33], compaction and molding [31,34,35], immersion of structural elements into the ground and their extraction from the ground [22,36,37], cutting [38,39], and destruction and digging [40,41]. In the building industry, vibration machines are used to produce building materials and perform construction and road construction work. In vibration machines, the driving force of the working equipment is the 'driving force'. The driving force serves to periodically move the structural elements of the machines with a certain amplitude $A$ and frequency $f$ and is characterized by the magnitude, modulus or amplitude value, direction or, line of action, point of application, and the nature of the behavior or position of the line of action in time [42–44].

Among the various types of vibration machines, a class of vibration inertial machines stands out, which the driving force of the vibration device is generated by one or several shafts rotating at an angular velocity $\omega$ (rad/s); with eccentric weight fixed to them, with a mass of $m$ (kg), the center of gravity of which is shifted relative to the axis of rotation by a certain value, $r$ (m), called eccentricity, and is described by the law of change [29,35].

$$F = m \cdot r \cdot \omega^2 \cdot cos(\omega t + \varphi \big) \tag{2}$$

where $F = m \cdot r \cdot \omega^2$ is the amplitude value of the driving force, H; $t$ is the current time, s; $\varphi$ is the initial phase of oscillations, deg, rad; $m$ is imbalance mass, kg; $r$ is imbalance eccentricity, m; and $\omega$ is the angular velocity of rotation of the unbalanced shaft, rad/s.

In the case where mechanical vibrations are generated by a rotating driving force ($F$), such vibrations are called circular or elliptical, depending on the ratio of resistance forces along the coordinate axes.

The magnitude of the driving force for a driving force of directional action, without asymmetry, i.e., when $k_{as} = K_d = 1.0$, acting in the direction of performing useful work ($F_{uw}$), is equal in magnitude to its action in the opposite direction, i.e., in the direction of performing an idle stroke ($F_{is}$).

However, the equality of the components of the total driving force in the direction of performing useful work and in the direction of performing idle stroke during vibration displacement and vibration transport of materials, during vibration immersion and vibration extraction of structural elements into and from the soil, as well as during vibration compaction is undesirable. In such a case, the following relationship must be fulfilled: $F_{uw} \gg F_{is}$, and accordingly, $k_{as} = K_d \gg 1.0$.

This problem can be solved by asymmetric oscillations, which allow obtaining a significant difference in the magnitude of the driving force acting in the direction of performing useful work, $F_{uw} = +F$, and in the direction of performing idle stroke, $|F_{is}| = \frac{F_{uw}}{k_{as}}$. In this case, the ratio of forces ($F_{uw}$ and $F_{is}$) is estimated by the asymmetry coefficient of the driving force, $k_{as}$, or the dynamic coefficient of the oscillatory system (1).

Further improvement of vibratory machines is connected with the task of developing the methodology and principles of designing vibration devices with asymmetric oscillations. However, at present this direction is still in the state of development. In accordance with this, this article is aimed at developing and forming the methodology and principles of designing vibration devices with asymmetric oscillations. At the same time, the objective of this article is to analyze existing methods for designing asymmetric oscillations of vibration machines and comparative studies of the results of their use.

At present, there are some theoretical provisions [9,11,14,25,27,45] that allow to consider them as prerequisites for the formation of a methodology for designing vibration devices with asymmetric oscillations. The existing theoretical provisions do not yet have a generally accepted analytical completeness, but they have some general conclusions.

The first conclusion is based on the fact that asymmetric oscillations can be obtained by adding several directed oscillations. As a result of adding the driving forces of the directed oscillations, a total directed driving force is created that is asymmetric in magnitude in opposite directions. The second conclusion is that the rotation frequencies of the unbalanced shafts of each subsequent directed oscillation, starting with the second one, are multiples of the first, Table 1.

**Table 1.** Ratio of angular frequencies of vibrators with directional oscillations in a single vibration device with asymmetric oscillations.

| Characteristics of Oscillations | First Stage | Second Stage | Third Stage | | *n*th Stage |
|---|---|---|---|---|---|
| Frequency of single-directed oscillations | $\omega_1$ | $\omega_2$ | $\omega_3$ | ... | $\omega_n$ |
| The ratio of the frequencies of single-directed oscillations to the frequency of the first stage | $\omega_1$ | $2\omega_1$ | $3\omega_1$ | ... | $n\omega_1$ |

Attempts to use asymmetric vibrations in the building industry were made as early as the middle of the last century [31], although the term 'asymmetric vibrations' had not yet been introduced.

The vibration device [31] consists of four unbalanced shafts, installed two on top of each other, forming two pairs, each of which generates directed vibrations with equal driving forces ($F$). The vibration frequency of one pair of unbalanced shafts is 400 rpm, forming the first stage of the vibration device, and the rotation frequency of the second pair of unbalanced shafts is 800 rpm, forming the second stage of the vibration device.

Using the vibration parameters from the provisions of [31], it is possible to calculate the magnitude of the total driving force as the sum of two directed oscillations, forming a vibration device with two vibration stages.

$$F_{sum2} = F_1 + F_2 = m_1 r_1 \omega_1^2 cos\omega_1 t + 0.5 m_1 r_1 (2\omega_1)^2 cos(2\omega_1 t + \pi) \quad (3)$$

where $F_1$ and $F_2$ are, respectively, the driving force of the vibration device of the first and second stages of directed oscillations, kN; $m_1$ is the mass of the eccentric weight of the first stage of directed oscillation, kg; $r_1$ is the eccentricity of the eccentric weight of the first stage of directed oscillation, m; and $t$ is the current time within one period of oscillations, s.

The following parameter values are accepted according to the recommendations [27]: $m_1$ = 10, $m_2$ = 5 (kg), $r_1$ = 0.02, $r_2$ = 0.01 (m), $\omega_1$ = 41.9, and $\omega_2 = 2\omega_1$ = 83.8 (rad/s). The static moment of the imbalances of the two-stage vibration device ($M_{st2}$) of the first term in Equation (3) is $M_{1st2} = m_1 \cdot r_1 = 0.2$ (kg·m), and of the second term in (3) is $M_{2st2} = m_2 \cdot r_2 = 0.05$ (kg·m), and their ratio.

$$M_{1st2} : M_{2st2} = 100\% : 25\% \quad (4)$$

In this case, the oscillation frequency is $f$ = 6.67 Hz, and the oscillation period is $T$ = 0.15 s. Calculation of forces based on 20 values of the oscillation period shows that the value of the total driving force with the law of its change according to Equation (3) in the direction of performing useful work ($F_{uw}$) is $F_{sum2} = F_{uw} = -701.4$ kN, and in the direction of performing idle stroke is $F_{is} = 394.5$ kN. Hence, for a two-stage vibration device with asymmetric oscillations (3), the asymmetry coefficient will be: $k_{as2} = \frac{|-701.4|}{394.5} = 1.78$.

The graph of the change in the components and the total value of the driving force, according to the recommendations [27] based on the calculations, is shown in Figure 1. The direction of the total driving force ($F_{sum2}$) is toward the negative values of the ordinate axis.

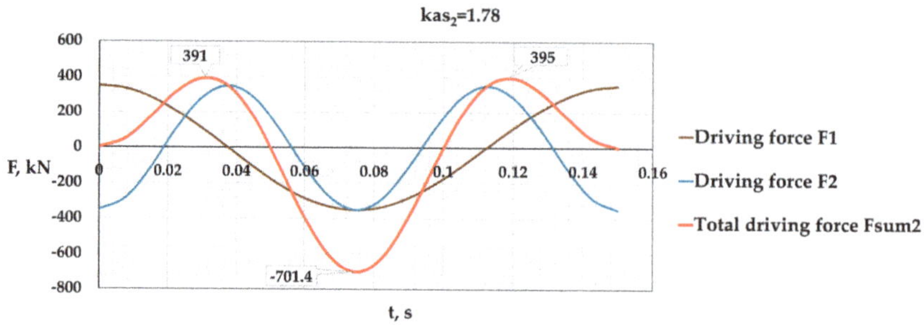


**Figure 1.** Adding two ($F_1$ and $F_2$) directed oscillations to obtain the total oscillation by initial conditions [31]. The asymmetry coefficient of the driving force is $k_{as2} = k_{d2} = 1.78$, ($F_{sum2}$).

In the graph, Figure 1, the direction of useful work performed by the driving force ($F_{sum2}$) is taken in the direction of negative values of the ordinate axis, for example, for driving piles.

On the one hand, the method of this technical solution [31] consists of doubling the angular frequency of rotation of the second stage eccentric weights in relation to the first. On the other hand, the mass and eccentricity of the second stage eccentric weights are halved, so that the amplitude values of the magnitude of the driving force of the first and second stages remain equal, $|F_1| = |F_2| = 350.7$ kN, and the sum is $|F_1| + |F_2| = 701.4$.

It should be noted that the largest number of publications and patents related to asymmetric oscillations of any parameter are associated with two-stage vibration devices, the asymmetry coefficient ($k_{as2}$) of which tends to two: $k_{as2} \rightarrow 2.0$. Where the index ($as2$) characterizes the asymmetry of the parameter of a two-stage vibration device generating a total driving force by two coordinated directional harmonic oscillations with multiple frequencies.

Another example of the development of a methodology for designing a vibration device with asymmetric oscillations is the work [46]. In this work, the term asymmetric oscillations also does not appear. In [46], an equation is given for the total value of the driving force of forced oscillations, consisting of three components $y(t) = F_{sum3}$:

$$y(t) = 4.5sin\frac{\omega}{2}t + 3sin\left(\omega t - \frac{\pi}{2}\right) + 1.5sin(\frac{3}{2}\omega t - \pi) \tag{5}$$

where 4.5, 3, 1.5 are the amplitudes of harmonics in the spectrum of subharmonic oscillations [46]; $\varphi_i = 0, -\frac{\pi}{2}, -\pi$ constitute the initial phase angles; and $\omega$ is the angular velocity, rad/s.

To carry out a comparative calculation, it was assumed that $\omega$ = 104.7 rad/s, and the magnitude of the amplitude value of the function $y(t) = F_{sum3}$ = 9.0 [47].

The authors of [46] draw attention to the fact that Equation (5) is used to tune the resonant oscillations of a vibration machine with a piecewise linear characteristic of elastic connections and, under these specific conditions, is the law of change of the function $y(t)$. Using the parameters related to the vibration device [46], the calculation of the value of the total driving force has been performed, the graph of which is shown in Figure 2. The direction of action of the value of the total driving force ($F_{sum3}$) is chosen similarly to [46], upward, toward the positive values of the ordinate axis.

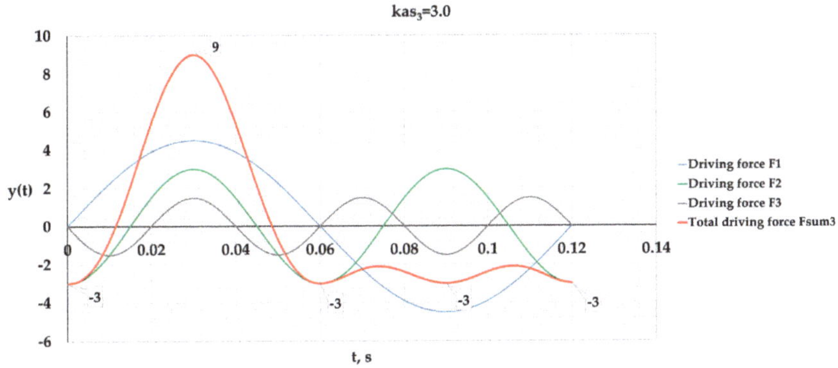


**Figure 2.** Addition of three directional oscillations ($F_1$ and $F_2$, $F_3$) to obtain a total oscillation ($F_{sum3}$) with asymmetry of the driving force $k_{as3} = k_{d3} = 3.0$, [46].

As can be seen from Figure 2, the total driving force ($F_{sum3}$) is the sum of three separate synchronized harmonic oscillations, that is, three stages of the vibration device with multiple frequencies, is asymmetric in magnitude, with amplitude values in the direction of performing useful work ($F_{sum3} = 9.0$) and in the direction of performing idle stroke ($F_{is3} = 3.0$) with an asymmetry coefficient: $k_{as3} = k_{d3} = \frac{9.0}{3.0} = 3.0$. It can be written that the amplitude value of the function.

$$y(t) = F_{sum3} = F_{13} + F_{23} + F_{33}, \tag{6}$$

where $F_{13}, F_{23}, F_{33}$ are, respectively, the force of the first, second, and third components of the total driving force at three stages of the vibration device. Equation (6) can be written as follows:

$$y(t) = F_{sum3} = \lambda_1 \cdot F_{sum3} + \lambda_2 \cdot F_{sum3} + \lambda_3 \cdot F_{sum3} \tag{7}$$

where $\lambda_1$, $\lambda_2$, $\lambda_3$ are, respectively, the coefficients that determine the ratio of the component forces in fractions of the value of the total driving force. In the case of Equation (5), it can be written as follows:

$$F_{sum3} = 0.5 \cdot F_{sum3} + 0.33 \cdot F_{sum3} + 0.17 \cdot F_{sum3} \tag{8}$$

In this case, the value of the static moment of the unbalance of the three-stage vibration device ($M_{st3}$) is, for the first stage, $M_{1st3} = m_1 \cdot r_1 = 1.64$ (kg·m), for the second stage, $M_{2st3} = m_2 \cdot r_2 = 0.274$ (kg·m), and for the third stage, $M_{3st3} = m_3 \cdot r_3 = 0.061$ (kg·m). Their ratio is as follows:

$$M_{1st3} : M_{2st3} : M_{3st3} = 100\% : 16.67\% : 3.71\% \tag{9}$$

The next stage of development and improvement of the methodology for designing and calculating vibration devices with asymmetric oscillations can be attributed to the works [48,49].

In these works, the definition of 'asymmetric' oscillations also does not appear. At the same time, oscillations generated by vibration devices with similar parameters form oscillations with an asymmetric value of the total driving force. For a vibration device with four vibration stages, it is recommended to adopt the ratio of static moments of eccentric weight, expressed as a percentage:

$$M_1 : M_2 : M_3 : M_4 = 100 : 18.72 : 5.6 : 1.38 \tag{10}$$

The recommendations given in [49] for selecting the parameters of the vibration device make it possible to calculate the output parameters and plot a graph of the change in the components and the total driving force within the oscillation period, as shown in Figure 3. The magnitude of the total driving force ($F_{sum4}$) is calculated in the direction of the negative values of the ordinate axis, downwards.

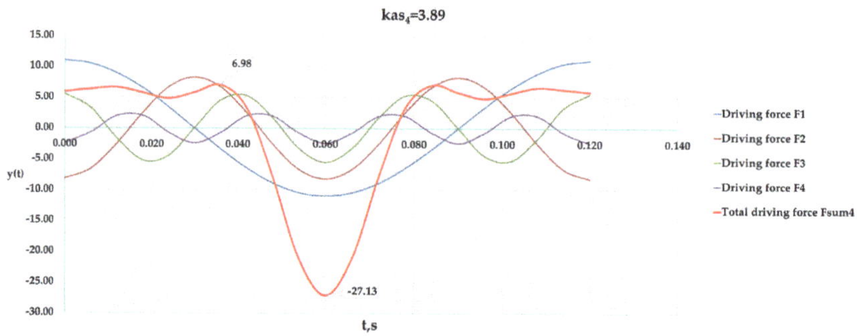


**Figure 3.** Addition of four directional oscillations ($F_1$ and $F_2$, $F_3$, $F_4$) to obtain a total oscillation with asymmetry of the driving force $k_d = 4.0$, ($k_{as4} = k_{d4} = 3.89$, ($F_{sum4}$)).

Equation (10) can be considered as a method for designing a vibration device with an asymmetry coefficient of the driving force equal to four.

From the graph, Figure 3, it is evident that the magnitude of the total driving force is asymmetrical, since the force acting in the direction of performing useful work exceeds the force acting in the direction of performing idle stroke by the value of the asymmetry coefficient, which is equal to $k_{ac} = k_d = -\frac{-27.13}{6.98} = 3.89$.

The design methodology for a vibration device with asymmetric oscillations according to [49] for a given amplitude value of the total driving force is carried out in several steps, or stages, for example, for a four-stage vibration device.

At the first stage, the amplitude value of the total driving force and the asymmetry coefficient equal to $k_{as} = k_d = 4.0$. are determined.

At the second stage, the angular speed of rotation of the unbalanced shafts of the first and subsequent stages are assigned: $\omega_1$; $\omega_2 = 2\omega_1$; $\omega_3 = 3\omega_1$; $\omega_4 = 4\omega_1$.

At the third stage, the parameters of the static moment, $M_{1st}$, of the unbalance of the first stage are selected based on its mass, $m_1$, and eccentricity, $m_1$: $M_{1st} = m_1 r_1$.

At the fourth stage, according to relation (10), the static moments of the unbalances of the second, third, and fourth stages, $M_2, M_3, M_4$, are determined.

At the fifth stage, according to the obtained values of the static moments, the masses of the unbalances of all stages, $m_i$, and their eccentricities, $r_i$, are determined.

At the sixth stage, the selected parameters are checked for the value of the total driving force, $F_{sum4}$.

Attempts to create vibration devices with an asymmetry coefficient greater than $k_{ac} \geq 4.0$ are not presented in [49].

The design methodology for a vibration device with asymmetric oscillations, which follows from [49], is artificially cumbersome for engineering calculations, since the static moments of the imbalances are intermediate parameters between the initial conditions and the value of the total driving force.

From the analysis of the methods of calculation and design of vibration devices with asymmetric oscillations, it is evident that in the considered sources of information related to mechanical oscillations, the term 'asymmetric' oscillations or vibration devices with 'asymmetric' oscillations are not actually used. The works considered above, which by the nature of the problems can be attributed to works connected with asymmetric oscillations of vibration machines, do not contain a unified methodology either from the point of view of the formulation of terminology or from the point of view of the formulation of the initial statement of problems and the sequence of their solution. Thus, there is a need to distinguish asymmetric oscillations in mechanical systems into an independent class and the need to obtain a general equation of the total driving force with a given asymmetry coefficient.

An important stage in the development of a methodology for designing vibration devices with asymmetric oscillations is associated with work, devoted to obtaining the parameters of real vibration devices using coefficients, obtained by expanding functions into a Fourier series [31].

The main theoretical basis is the works [31,40], in which the question of the formation of asymmetric oscillations has been determined using the coefficients of the Fourier series obtained by decomposing piecewise smooth or continuous monotone functions. It was assumed that, by successively accepting the coefficients of the terms under the summation sign as multipliers to the corresponding component forces, it is possible to establish a rational relationship of such components of elementary forces, which in sum makes it possible to obtain an asymmetric value of the total driving force.

The conducted studies on the expansion of piecewise smooth and monotone functions in a Fourier series in a wide range of possible variants established that the method of using the coefficients of the Fourier series has certain limitations [49,50]. These limitations are in the fact that the method allows obtaining the total driving force with the value of the asymmetry coefficient within the limits $k_{ac} = k_d = 1 \ldots 4$. The numerical analysis of the use of coefficients, terms of the Fourier series, is considered using the example of a function:

$$y = cos^{2m}(x/2), \tag{11}$$

and the corresponding series:

$$y(x) = \frac{(2m)!}{2^{2m}(m!)^2} + \frac{1}{2^{2m-1}} \sum_{k=0}^{m-1} \frac{(2m)!}{k! \cdot (2m-k)!} cos(m-k)x \tag{12}$$

Let us consider the problem of changing the value of the asymmetry coefficient of the total oscillation obtained by using, respectively, 1. . .6 terms of the Fourier series in the expansion of the function $y = cos^{12}\frac{x}{2}$.

There is the following series:

$$cos^{12}\frac{x}{2} = 0.226 + 0.387 \cdot cosx + 0.242 \cdot cos2x + 0.107 \cdot cos3x + 0.032 \cdot cos4x + 0.006 \cdot cos5x + 0.0005 \cdot cos6x \tag{13}$$

where the variable $x = \omega t$; $\omega$ is the angular velocity of the first unbalanced shaft; $\omega$ = 52.3 rad/s is taken; the oscillation frequency is $f$ = 8.33 Hz; and the oscillation period is $T$ = 0.12 s. As a result, a sequence of the value of the asymmetry coefficient of the total driving force ($k_{asn}$) is obtained, depending on the number ($n$) of terms of the series with the corresponding coefficients used, as shown in Figure 4.

It is obvious that for the first two or three components of the vibration device stages, the value of the asymmetry coefficient increases, practically, by one. At four stages, the growth of the asymmetry coefficient decreases, and at five and six stages, the growth of the asymmetry coefficient stops. This happens because the subsequent numerical values of the Fourier series coefficients decrease, tend to zero and become insignificant.

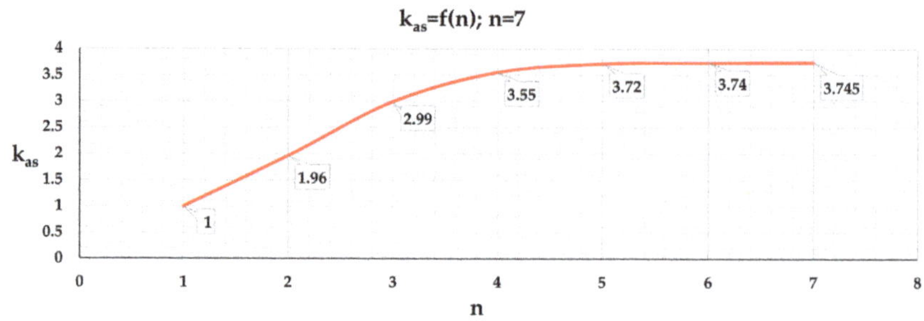


**Figure 4.** Graph of the change in the asymmetry coefficient ($k_{asn}$) of the total driving force of the vibration device, formed on the basis of the terms of the Fourier series of the function $= cos^{12}\frac{x}{2}$ for the first six terms ($n = 1\ldots6$).

The main parameter determining the effect of asymmetric mechanical vibrations of a vibration device with asymmetric vibrations is the value of the total driving force obtained by adding two or more symmetric driving forces directed along a common straight line. Such components of driving forces are generated by separate vibrators called vibration stages. Thus, the value of the total asymmetric driving force, $F_{sum}$, is the sum of several elementary symmetric forces, $F_i$, directed along a common straight line, i.e.,:

$$F_{sum} = \sum_{i=1}^{n} F_i = F_1 + F_2 + \ldots + F_n \tag{14}$$

where $i$ is the serial number of the driving force, the number of the stage of the vibration device with asymmetric oscillations; $i = 1, 2, \ldots, n$; $n$ is the number of received components, the number of stages of the vibration device with asymmetric oscillations; $n_i$ is the $i$-th stage of the vibration device with asymmetric oscillations, consisting of $n$ stages; and $F_i$ are forces that make up the total driving force.

A vibration device with asymmetric oscillations can be formed on the basis of a sequential connection into a single mechanism of paired vibrators with circular oscillations, each pair of which forms a stage of the vibration device, or on the basis of planetary vibrators with directional oscillations.

The following problems are of scientific interest:

- The procedure for determining the rational number of stages, and therefore the rational coefficient of asymmetry of the driving force of a vibration device with asymmetric oscillations, for the use of asymmetric oscillations in specific conditions [37];
- What ratio of driving forces of each stage in the value of the total driving force is rational;
- How to achieve the greatest coefficient of asymmetry of the total driving force and, therefore, the coefficient of dynamism of a mechanism with asymmetric oscillations;

A vibration mechanism is considered, consisting of $n$ pairs of vibrators with circular oscillations, each equivalent pair, the stage of which generates a directed driving force corresponding to its parameters, as shown in Figure 5.

In Figure 5, $\varphi_k$ is the angle of deviation of the $k$-th eccentric weight from the horizontal. The following designations are used for the $k$-th vibrator where $m_k$ is the mass of the eccentric weight, $R_k$ is the radius of the guide circle, $\varphi_{k0}$ is the initial angle of deviation of the eccentric weight (initial phase), and $\omega_k$ is the angular velocity of rotation of the eccentric weight.

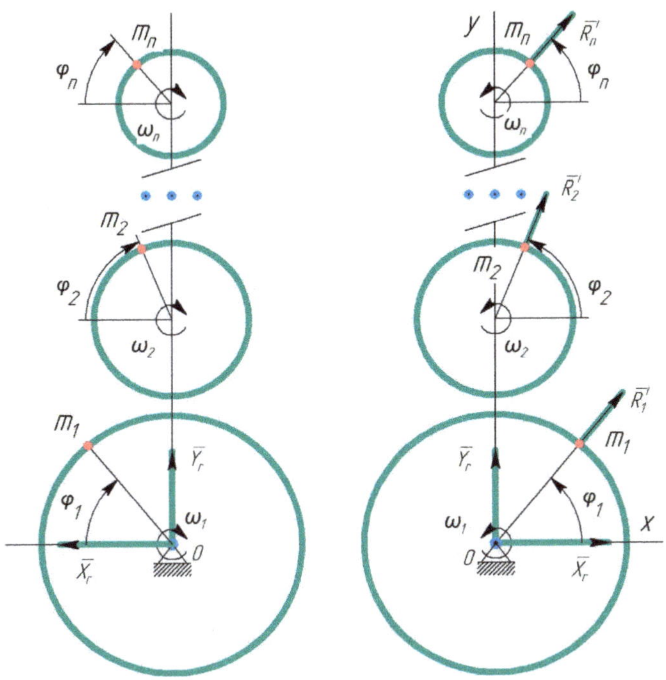

**Figure 5.** A diagram of a vibration mechanism consisting of $n$ pairs of unbalanced vibrators, each pair of which forms a vibration block, a stage that generates directed oscillations.

Let us introduce the inertial forces $\overline{R}_k^i$, applied to the eccentric weight, and the components of the support reaction $\overline{X}_r$ and $\overline{Y}_r$. The total horizontal component of the support reaction of the mechanism, due to the mirror arrangement of the vibrators, is equal to zero. Neglecting the weight of the guide circles and eccentric weight due to their smallness in comparison with the inertial forces (at sufficiently large values of the angular velocities $\omega_k$), based on d'Alembert's principle [22,37], the following equilibrium equation in projection onto the $O_y$ axis is obtained:

$$Y_r + 2\sum_{k=1}^{n} \overline{R}_k^i sin\varphi_k = 0, \tag{15}$$

where

$$\overline{R}_k^i = m_k R_k \omega_k^2,\ \varphi_k = \omega_k t + \varphi_{k0}. \tag{16}$$

Thus, the value of the vertical component of the support reaction can be calculated using the following equation:

$$Y_r = -2\sum_{k=1}^{n} \overline{R}_k^i sin\varphi_k, \tag{17}$$

or

$$Y_r = \sum_{k=1}^{n} \left(-2m_k R_k \omega_k^2\right) sin(\omega_k t + \varphi_{k0}), \tag{18}$$

Considering further only pairs of vibrators, the angular velocities of rotation of the eccentric weight that are related by the relation:

$$\omega_k = k\omega_1 \tag{19}$$

where $\omega_1$ is the angular velocity of rotation of the pair of the first stage of eccentric weight, Equation (8) is written as shown below:

$$Y_r = \sum_{k=1}^{n} a_k sin(k\varphi + \varphi_{k0}), \omega_k = k\omega_1 \quad (20)$$

where $\varphi = \omega_1 t$, and parameters $a_k$ can take any values (due to the choice of the value of the radii of the guide circles $R_k$ and the mass of the eccentric weight $m_k$).

The minimum absolute value of $Y_r$, as a function of the angle $\varphi$, denoted by $Y_{ri}$, is the dynamic force acting in the direction of idle stroke. The maximum absolute value of $Y_r$, denoted by $Y_{ra}$, is the immersion force acting in the direction of performing useful work.

The ratio of these forces $\left(-\frac{Y_{ri}}{-Y_{ra}}\right)$ is less than one.

$$k_d = -\frac{Y_{ri}}{-Y_{ra}} \leq 1.0 \quad (21)$$

If the inverse ratio $\left(-\frac{-Y_{ra}}{Y_{ri}}\right)$ is used, then its value is greater than one.

$$k_d = -\frac{-Y_{ra}}{Y_{ri}} \geq 1.0 \quad (22)$$

It is called the coefficient of dynamism of the vibration system or the coefficient of asymmetry ($k_{as} = k_d$) of the driving force of the vibration mechanism (the minus sign in Equations (21) and (22) is introduced due to the fact that one of these quantities is negative, opposite in the direction of action to the other).

The task is to achieve the maximum value of the dynamic coefficient of the vibration mechanism $k_d$ at a given immersion force $Y_{ra}$, which is further denoted as $A$, by combining several pairs of planetary vibrators and varying their parameters.

As follows from Equation (20), the minimum value of the function is achieved in the case when for all $k$:

$$sin(k\varphi + \varphi_{k0}) = -1 \quad (23)$$

The maximum immersion force $A$ in this case is determined with the equation:

$$A = \sum_{k=1}^{n} a_k \quad (24)$$

Condition (23) is satisfied if:

$$k\varphi + \varphi_{k0} = \frac{3\pi}{2} + 2\pi m \quad (25)$$

where $m$ can take any values.

From Equation (25) for the value of the initial phase $\varphi_{k0}$, the following equation is obtained:

$$\varphi_{k0} = \frac{\pi}{2}(k-1) \quad (26)$$

As an illustration, Figure 6 shows support reactions of each of the seven pairs of vibrators with coefficients $a_k$ equal to one and initial phases $\varphi_{k_0}$ equal to zero. The maximum immersion force of the first pair of vibrators is achieved at $\varphi$ = 270°. At this angle, the immersion forces of the second and fourth pairs of vibrators are equal to zero, and the third pair of vibrators at this angle has the maximum value of the lifting force.

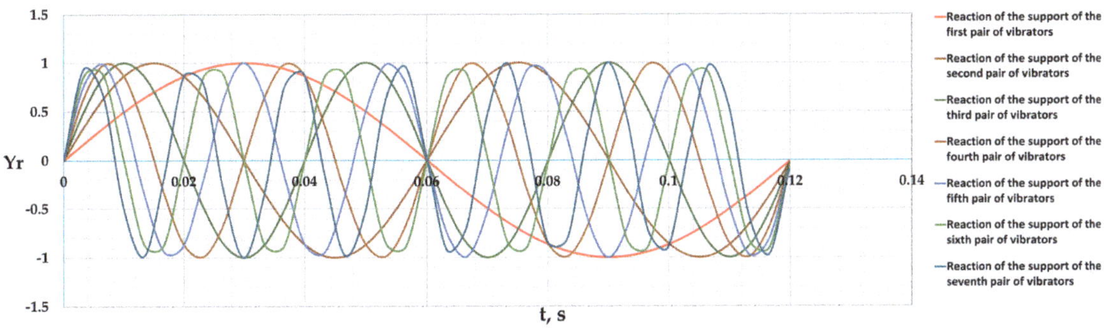


**Figure 6.** Support reactions of each of the seven pairs of vibrators with the value $a_k = 1.0$ and initial phases $\varphi_{k0} = 0$.

The sum of support reactions of all seven pairs of vibrators is shown in Figure 7. The amplitude value of the immersion force and the lifting force are equal to each other and are $Y_{ra} = Y_{ri}$ = 5.3, and the dynamic coefficient is 1, $k_d = 1.0$.

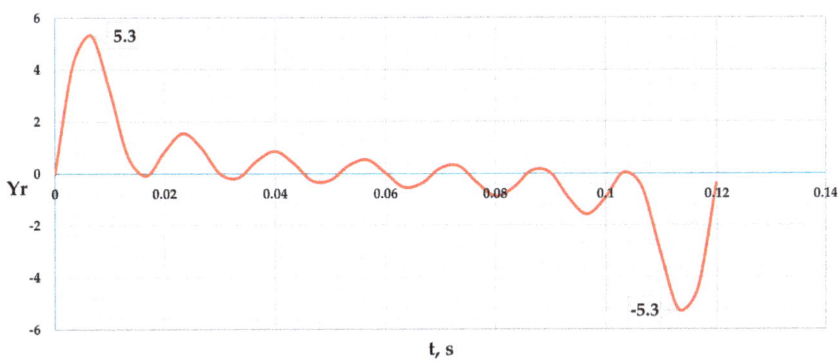


**Figure 7.** The sum of support reactions of all seven pairs of vibrators with the value $a_k = 1.0$ and initial phases $\varphi_{k0} = 0$.

Figure 8 shows the support reactions of each of the five pairs of vibrators with coefficients $a_k$ equal to one and initial phases $\varphi_{k0}$ calculated using Equation (26). The maximum values of the immersion forces of all pairs of vibrators are achieved at $\varphi$ = 270°.

The sum of support reactions of all five pairs of vibrators is shown in Figure 9. The immersion force in this case is equal to 7.002, the lifting force is equal to 2.119, and the dynamic coefficient is equal to $k_d$ = 3.3.

It is evident from the graph in Figure 9 that in the direction of useful work, the vibration device with asymmetric oscillations generates a total driving force of $F_{uw} = |6.99|$ kN. In the direction of idle stroke, a total value is equal to $F_{is} = 1.93$ kN. The asymmetry coefficient of the total driving force is $k_{as} = \frac{6.99}{1.93} = 3.62$. On the graph of the obtained function $Y_r = f(\varphi)$, there are two values that practically reduce the asymmetry coefficient from 7.0 to 3.62.

In the works [40], a problem is formulated related to the useful use of a directing or polyharmonic pulse. It is assumed that the mathematical model of a polyharmonic pulse is a trigonometric polynomial:

$$f_n(t, \lambda) = \sum_{k=0}^{n} \lambda_k cos(kt),\ t \in [0, \pi],$$
$$\lambda_k = (\lambda_0,\ \lambda_1,\ \ldots,\ \lambda_n) \tag{27}$$

and related functionality:

$$K_n(\lambda) = \frac{masf_n(t,\lambda)_t}{minf(t,\lambda)_t} \tag{28}$$

called the 'asymmetry coefficient'.

In this case, the problem of maximizing the functional $K_n(\lambda)$ over variations of $\lambda$ is set:

$$K_n(\lambda) \to \sup_\lambda, \tag{29}$$

under the following conditions:

$$\int_0^{\pi} f_n(t,\lambda)dt = 0,\ \sum_{i=0}^{n} \lambda_i = c > 0 \tag{30}$$

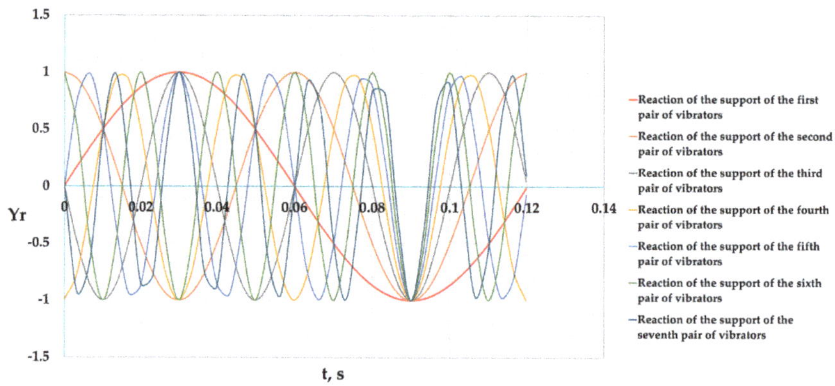


**Figure 8.** Support reactions of each of the seven pairs of vibrators with coefficients $a_{ki} = 1.0$ and initial phases $\varphi_{k0} = \frac{\pi}{2}(k-1)$, calculated using Equation (26).

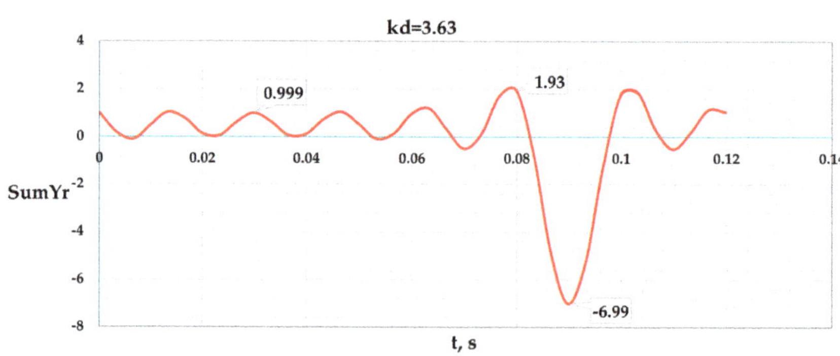


**Figure 9.** The sum of support reactions of all seven pairs of vibrators with coefficients $a_k = 1.0$ and initial phases $\varphi_{k0} = \frac{\pi}{2}(k-1)$, calculated using Equation (26).

As a result, in [40], the following is given:

**Theorem 1.** *A polynomial [40] is optimal if, up to a constant factor, it has the form of a Fejér sum*

$$f_n(t) = \sum_{k=1}^{n} (n+1-k)cos(kt) \tag{31}$$

In this case, there is equality:

$$\max K_n(\lambda)_\lambda = n \tag{32}$$

In addition, in [40], theorems on the properties of the optimal Maxwell–Fejer polynomials defining the optimal (in the sense of the asymmetry coefficient) impulse—the Maxwell–Fejer impulse have been formulated and proven. The graph of the Maxwell–Fejer impulse, which is important for engineering and technical problems, is shown in Figure 10. It is noted that a characteristic property of the Maxwell–Fejer impulse is the location of the minima at the same level.

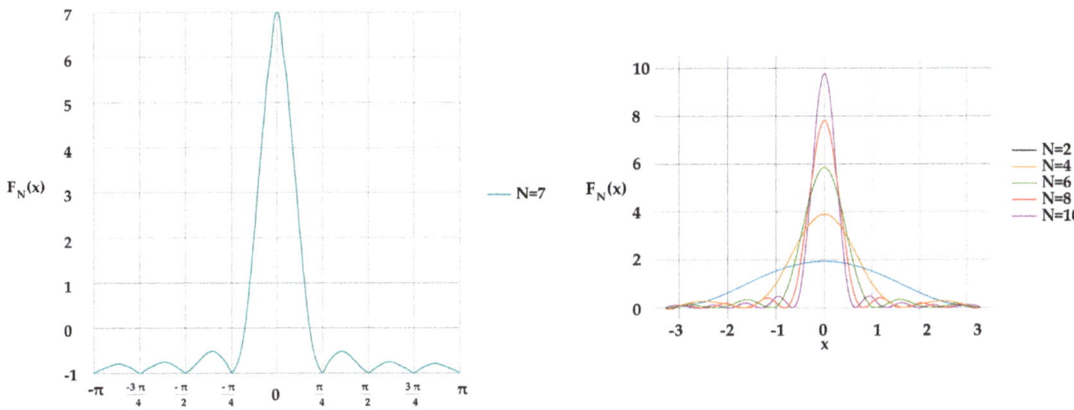


(a) (b)

**Figure 10.** Fejer kernel, (**a**) is for N = 7, (**b**) is for several Fejer kernels [39–41].

It is obvious that the behavior of the graphs in Figures 9 and 10, reflecting the general trends of the calculation and design methods of vibration devices with asymmetric oscillations and the optimal value of the functional ($\max K_n(\lambda)_\lambda = \max k_d(\lambda)_\lambda = n$), do not ensure complete coincidence of the results.

The research must return to the previously calculated vertical support reaction $Y_r$ (20) $Y_r = \sum_{k=1}^{n} a_k sin(k\varphi + \varphi_{k0})$, where $\varphi = \omega_1 t$. Thus, only pairs of vibrators are considered, angular velocities of rotation of the guide circles, which are related by the relation (26), $\varphi_{k0} = \frac{\pi}{2}(k-1)$. The conditions when the minimum value of Equation (30) is achieved, the maximum immersion force, $A$, which is determined with Equation (24), are considered.

For two pairs of vibrators in [31,37,40], it is theoretically proven that the maximum dynamic coefficient of the vibration mechanism $k_d = 2$ for a given immersion force $A$ is achieved with coefficients:

$$\begin{cases} a_1 = 2m_1R_1\omega^2 = \frac{2}{3}A \\ a_2 = 8m_2R_2\omega^2 = \frac{1}{3}A \end{cases} \tag{33}$$

For a vibration mechanism consisting of more than two pairs of planetary vibrators, the determination of the optimal values of the parameters $a_k$, at which the maximum dynamic coefficient $k_d$ is achieved (for a given immersion force $A$), was carried out numerically. The simplex search method ($S^2$-method) proposed in [47,50] was used. This method has no relation to the simplex method of linear programming, and the similarity of the names is accidental.

The implementation of the optimum search algorithm consists of two types of calculations: the construction of a regular simplex for a given base point and scale factor and the calculation of the coordinates of the reflected point.

The calculation of the coordinates of the vertices of the simplices for a given initial (base) point $a^{(0)}$ is performed using the following equations [51]:

$$a^{(i)} = \begin{cases} a_j^{(0)} + \delta_1, \ if\ j \neq i, \\ a_j^{(0)} + \delta_2, if\ j = i, \end{cases} \tag{34}$$

where the increments $\delta_1$ and $\delta_2$ depend on the dimension $N$ of the problem and are determined with the equations:

$$\delta_1 = \left( \frac{\sqrt{N+1} + N - 1}{N\sqrt{2}} \right) \alpha \tag{35}$$

$$\delta_2 = \left( \frac{\sqrt{N+1} - 1}{N\sqrt{2}} \right) \alpha \tag{36}$$

where $\alpha$ is the scale factor (reduction coefficient).

Calculations of the second type are related to the reflection of a point relative to the center of gravity of the simplex. If $a^{(j)}$ is the point to be reflected, then the center of gravity of the remaining points of the simplex $a_c$ is calculated using the following equation:

$$a_n^{(j)} = 2a_c - a^{(j)} \tag{37}$$

For calculations of a three-pair unbalanced vibrator, the immersion force $A = 6$ has been set, and the point with coordinates $a_1 = 1.0,\ a_2 = 1.0$ has been chosen as the base point. The coefficient $a_3$, in accordance with (24), has been calculated using the following equation:

$$a_3 = A - a_1 - a_2 \tag{38}$$

Since in the problem under consideration $N = 2$ (two variable parameters are $a_1$ and $a_2$), and the scale factor $\alpha$ has been chosen equal to one, the increments in this case are $\delta_1 = 0.9659,\ \delta_2 = 0.2588$. As the objective function, since the method used allows calculating the minimum of the function, the function $1/k_d$ has been used.

The calculation of the objective function values has been performed as follows. For the given values of the coefficients $a_1$ and $a_2$ and the coefficient $a_3$ calculated using Equation (38), the values of the function $Y_r(\varphi)$ have been calculated when the angle $\varphi$ changed from 0° to 360° with a step of 1°, the minimum $Y_{ri}$ and maximum $Y_{ra}$ values of the function have been determined, and the dynamic coefficient (1) has been calculated

$$k_d = -\frac{Y_{ri}}{Y_{ra}}$$

and then the value $1/k_d$ has been used.

Figure 11 graphically presents the calculated simplices in the process of determining the optimal values of the parameters $a_1$ and $a_2$. Table 2 shows the coordinates of the vertices of the simplices. Figure 12 shows the values of the parameters at the vertices of the simplices and the corresponding values of the dynamic coefficient $k_d$.

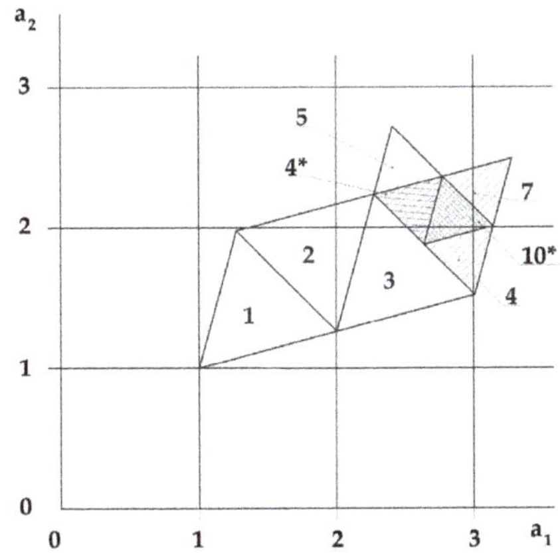


**Figure 11.** Calculation of simplexes in the process of determining the optimal values of parameters $a_1$ and $a_2$.

**Table 2.** Coordinates of the vertices of the simplices.

| Vertices | 1 | 2 | 3 | 4 | 5 | 6 | 7 | 8 | 9 | 10 |
|---|---|---|---|---|---|---|---|---|---|---|
| $a_1$ | 1.00 | 1.26 | 1.97 | 2.22 | 2.93 | 3.19 | 2.35 | 2.71 | 2.58 | 3.06 |
| $a_2$ | 1.00 | 1.97 | 1.26 | 2.22 | 1.52 | 2.48 | 2.71 | 2.35 | 1.87 | 2.00 |
| $< a_3 >$ | 4.00 | 2.78 | 2.78 | 1.55 | 1.55 | 0.33 | 0.94 | 0.94 | 1.55 | 0.94 |
| $k_d$ | 1.47 | 1.72 | 1.72 | 2.42 | 2.02 | 2.18 | 2.21 | 2.55 | 2.66 | 2.99 |

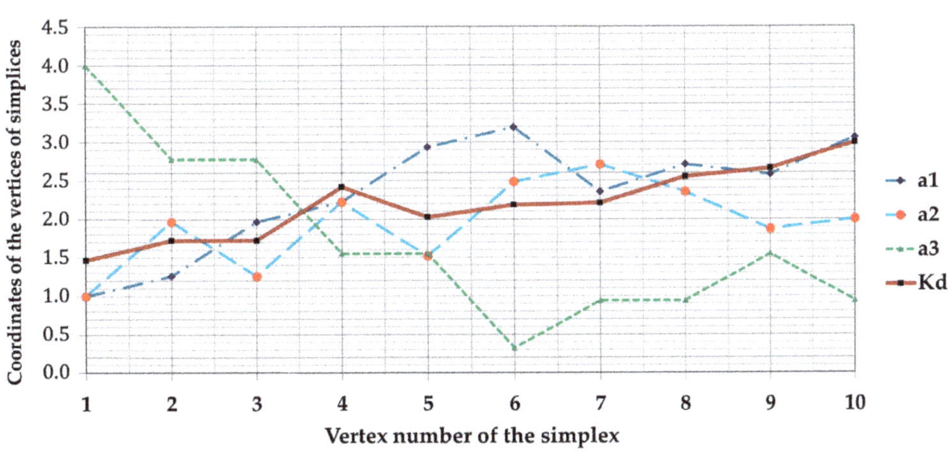


**Figure 12.** Values of parameters at the vertices of simplexes and the corresponding values of the dynamic coefficient $k_d$.

On the simplex '4', there was a coincidence in the minimum point, since the value of the objective function $1/k_d$ at the following iteration steps with a change of simplex vertices (they are not shown in the figures and in the table) turned out to be greater than the values at the vertices of the simplex of the previous iteration. The value of the reduction coefficient $\alpha$ was reduced to 0.5. Vertex '4' was taken as the base point, as having the minimum value of the objective function in the simplex. In Figure 11, it is marked with the symbol (*).

At the vertex '10' of the simplex '7', also marked with the symbol (*), $a_1 = 3.06$, $a_2 = 2.00$ (marked with a dot in Figure 11), the minimum of the objective function $1/k_d$ has been reached. The construction of further simplices, even with a decrease in the value of the reduction coefficient $a$, did not lead to a decrease in the objective function.

The results of the calculations led to the following conclusions: for a given immersion force $A$, the maximum value of the dynamic coefficient $1/k_d$ is achieved with the following values of the coefficients $a_k$:

$$\begin{cases} a_1 = \frac{1}{2}A = \frac{3}{6}A \\ a_2 = \frac{1}{3}A = \frac{2}{6}A \\ \quad a_3 = \frac{1}{6}A \end{cases} \tag{39}$$

Figure 12 and (39) show the results of calculating the support reaction $Y_r(\varphi)$ and the contribution of each of the three stages of unbalanced vibration blocks with directional vibrations for the above values of the input parameters.

Thus, for a given total immersion force $A$, the maximum possible dynamic coefficient for a two-stage unbalanced vibration mechanism with asymmetric oscillations is equal to two and is achieved with coefficients (33). For a three-stage unbalanced vibration mechanism with asymmetric oscillations, it is equal to three and is achieved with coefficients (39). For a four-stage unbalanced vibration mechanism with asymmetric oscillations, the dynamic coefficient can be equal to four and is achieved with coefficients (40)

$$\begin{pmatrix} a_1 = \frac{4}{10}A, \\ a_2 = \frac{3}{10}A \\ a_3 = \frac{2}{10}A \\ a_4 = \frac{1}{10}A \end{pmatrix} \tag{40}$$

Based on the results obtained and presented in this article, for a vibration mechanism consisting of $n$ vibration stages, for example, $n$ planetary vibrators with directional oscillations or $n$ pairs of vibrators with circular oscillations, each pair generating directional driving forces, it follows that in order to obtain a total asymmetric oscillation, the following conditions must be met:

1. The maximum value of the total driving force acting in the direction of performing useful work is achieved when (26):

$$\varphi_{k0} = \frac{\pi}{2}(k-1)$$

in this case, the magnitude of the total driving force acting in the direction of performing useful work, for example, immersion, is determined with Equation (24):

$$A = \sum_{k=1}^{n} a_k$$

2. The maximum value of the asymmetry coefficient of the total driving force, or the dynamic coefficient of the oscillatory system ($k_{as}$, $k_d$), is equal to $n$ and is achieved at the following values of the coefficients of a number of vibration stages

$$a_i = \frac{n+1-i}{\sum_{k=1}^{n} k} \cdot A \tag{41}$$

where $k_d$ is the dynamic coefficient of the oscillatory system (the asymmetry coefficient of the total driving force, $k_{as}$); $k$ is the serial number of the vibration stage generating the directed oscillations (the directed driving force); $n$ is the number of vibration stages in the vibration device with asymmetric oscillations; $i$ is the serial number of the term of the total driving force included in the total driving force of the vibration device with asymmetric oscillations; $A$ is the specified value of the total driving force acting in the direction of performing useful work; and $a_i$ is the part of the value of the total driving force that falls on the $i$-th stage of the vibration device. When considering Equation (24), the sum of the values of $a_i$ is always: $\sum_{i=1}^{n} a_i = 1.0$.

Using Equation (41), $a_k = (a_1, a_2, \ldots, a_n)$, the general equation of oscillations of a vibration device asymmetric with a total driving force can be written in the form:

$$F_{sum} = \sum_{i=1}^{n} a_i \cdot F_i = a_1 \cdot F_{sum} + a_2 \cdot F_{sum} + \cdots + a_n \cdot F_{sum} \quad (42)$$

where $a_i$ are the coefficients of the terms of the series of the total driving force of a vibration device with asymmetric oscillations when using the value $A$ = 1.0 in Equation (42), and then $F_{sum} = 1.0$.

## 4. Discussion

The behavior of the total driving force $F_{sum} = A$ for an arbitrary value of its magnitude and a given asymmetry coefficient, using the equations obtained (24,26,41,42) as an example, have been considered. For comparison, the number of stages of the vibration device to be seven, $n$ = 7.0, has been taken. The rotation frequency of the eccentric weight shafts, rpm, is, respectively, 500, 1000, 1500, 2000, 2500, 3000, and 3500. The total value of the driving force $F_{sum}$ = 10 kN. The given asymmetry coefficient of the driving force is equal to $k_{as} = k_d = 7.0$. The graph of the change in $F_{sum} = f(\varphi)$ has the form shown in Figure 13.

As the calculation result and the behavior of the graph in Figure 13d show, the obtained method for designing vibration devices with asymmetric oscillations fully corresponds to the optimality condition of the Maxwell–Feuillere polynomial, which assigns the optimal (in the sense of the asymmetry coefficient) impulse to the functional (28).

When using Equations (24), (26), (41) and (42) in designing a vibration device with asymmetric oscillations and a given coefficient of asymmetry of the total driving force, there is no need to calculate and plot graphs in Figure 13a–c, since the calculation provides for obtaining graph Figure 13d directly.

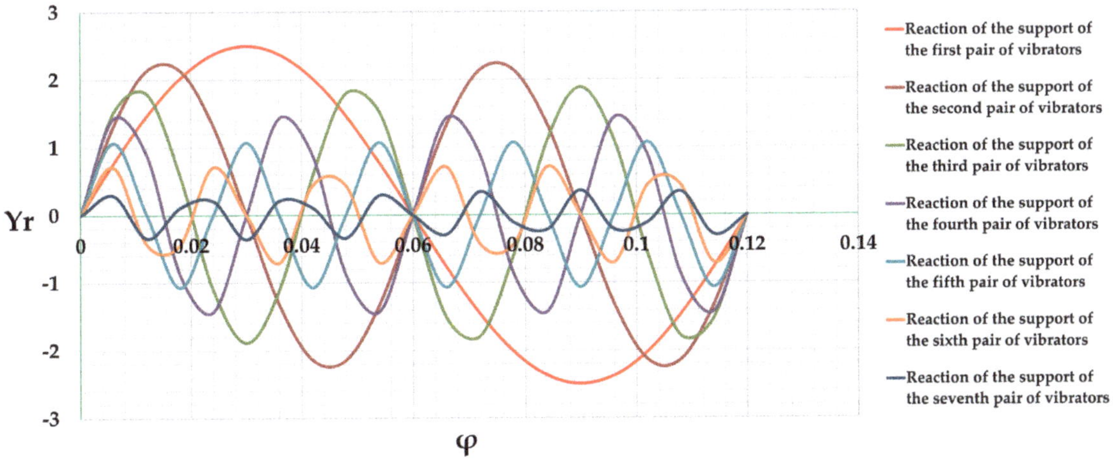


(a)

**Figure 13.** *Cont.*

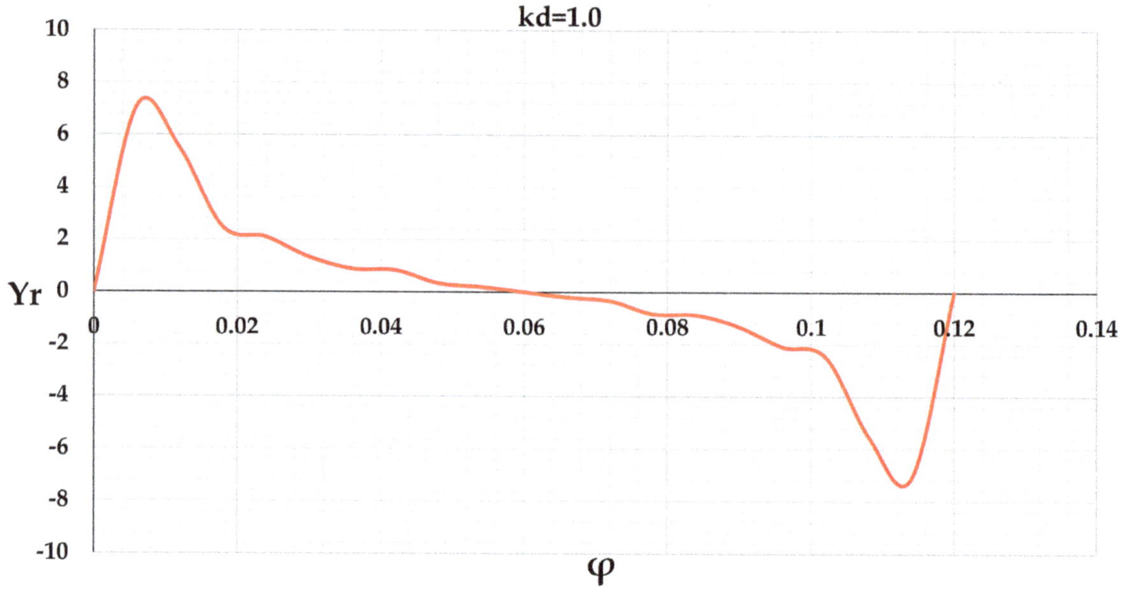


(b)

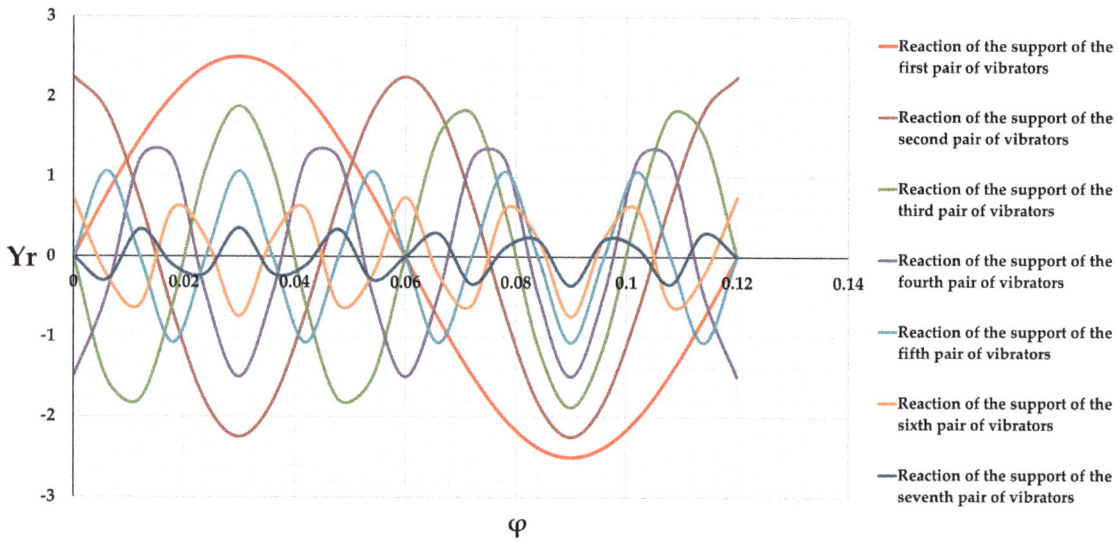


(c)

**Figure 13.** *Cont.*

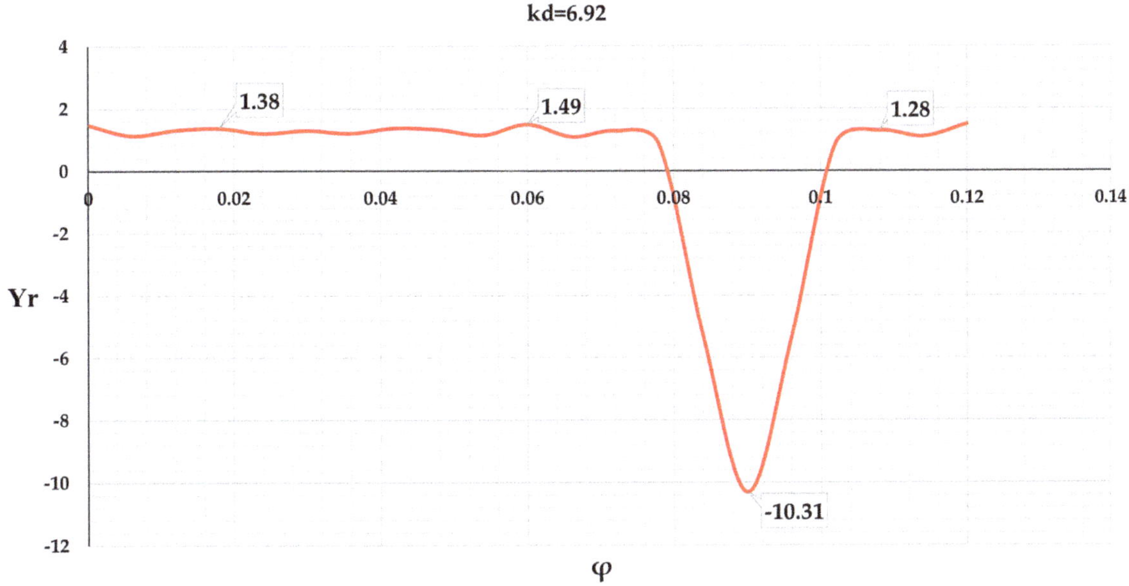


(**d**)

**Figure 13.** Calculation graphs of the magnitude of the driving forces of a vibration device with asymmetric oscillations with the total driving force $F_{sum}$ = 10 kN and the specified asymmetry coefficient $k_{as}$ = 7.0. (**a**) is support reactions of each of the seven pairs of vibrators with the value $F_{sum}$ = 1 kN and the initial phases $\varphi_{k0} = 0$; (**b**) is the sum of the support reactions of all seven pairs of vibrators with the initial phases $\varphi_{k0} = 0$; (**c**) is the support reactions of each of the seven pairs of vibrators with the initial phases $\varphi_{k0} = \frac{\pi}{2}(k-1)$, (**d**) is the sum of the support reactions of all seven pairs of vibrators with the initial phases $\varphi_{k0} = \frac{\pi}{2}(k-1)$.

## 5. Conclusions

The design methodology, in this case, comes down to the following actions.

1. Determination of the coefficients $a_i$ for the terms of a series of seven members of the total driving force of a vibration device with asymmetric oscillations when used in Equation (24): $A$ = 1.0;
2. Determination of the terms of forces of a series of seven members from the expression: $a_i \cdot F_{sum}$;
3. Checking the obtained value of the total driving force using Equation (42);
4. Assignment of the mass and eccentricity of the eccentric weight of each stage using known engineering methods based on the limitations associated with the dimensions of the specific area of application of the product.

$$m_i r_i \omega_i^2 = a_i \cdot F_{sum} \quad (43)$$

where $m_i$ is the total mass of the eccentric weight of the $i$-th stage of the vibration device; $r_i$ is the eccentricity of the eccentric weight of the $i$-th stage of the vibration device; and $\omega_i$ is the angular velocity of the unbalanced shaft of the $i$-th stage, related to the angular velocity of rotation of the unbalanced shaft of the first stage by a multiple ratio.

As a comparative example, a numerical example of the initial data is given: $F_{sum}$ = 10 kN is the total value of the driving force with asymmetric oscillations and an asymmetry coefficient $k_{as}$ = 10.

1. Calculation (41) of coefficients $a_i$ for components of the total driving force $F_{sum} = 1$ kN is performed;
2. Rotation frequencies of unbalanced shafts of ten stages of the vibration device with asymmetric oscillations are assigned, $\omega_i = n \cdot \omega_1$;
3. Calculation of the value of components ($F_i$) of the total driving force $F_i = 10$ kN is calculated;
4. Based on design considerations and process kinetics features, masses ($m_i$) and eccentricity ($r_i$) of imbalances of the corresponding stages of the vibration device are assigned;
5. A general table of parameters is compiled, as shown in Table 3;
6. Calculation and control graphs of changes in the driving force value within the oscillation period are performed, as shown in Figure 14.

**Table 3.** The method of determining the parameters of a vibration device with asymmetric oscillations with the value of the total driving force $F_{sum} = 10$ kN and the asymmetry coefficient $k_{as}$ =10.

| Parameter Name | Unit of Measurement | Vibration Device Stage | | | | | | | | | | Sum $a_i$ |
|---|---|---|---|---|---|---|---|---|---|---|---|---|
| | | 1 | 2 | 3 | 4 | 5 | 6 | 7 | 8 | 9 | 10 | |
| Coefficient $a_i$ at $F_{sum}$= 1 kN | - | 0.18 | 0.16 | 0.146 | 0.127 | 0.11 | 0.091 | 0.073 | 0.055 | 0.036 | 0.018 | 1.0 |
| Magnitude of the component $F_i$ at $F_{sum}$ = 10 kN | kN | 1.82 | 1.64 | 1.46 | 1.27 | 1.09 | 0.91 | 0.73 | 0.55 | 0.36 | 0.18 | 10 |
| Angular velocity of imbalance of the $i$-th stage | $s^{-1}$ | 52.36 | 104.72 | 157.08 | 209.44 | 261.8 | 314.16 | 366.52 | 418.88 | 471.24 | 523.6 | - |
| Moment of imbalance, $m_i \cdot r_i$ | kg·m | 3.14 | 4.19 | 3.14 | 2.09 | 2.62 | 3.14 | 3.67 | 2.09 | 1.41 | 1.05 | - |
| Mass imbalance, $m_i$ | kg | 5.52 | 1.86 | 1.47 | 1.45 | 0.8 | 0.46 | 0.27 | 0.32 | 0.27 | 0.17 | - |
| Eccentricity, $r_i$ | m | 0.06 | 0.04 | 0.02 | 0.01 | 0.01 | 0.01 | .0.01 | 0.005 | 0.003 | 0.002 | - |

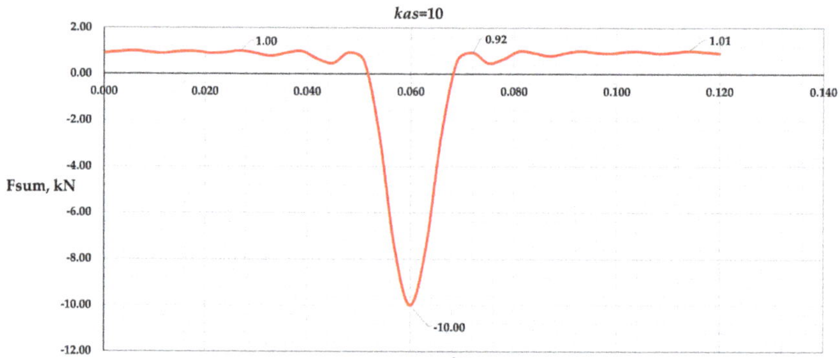


**Figure 14.** Graph of change in total driving force $F_{sum} = f(t)$.

Based on the calculation results, a graph of the total driving force $F_{sum} = f(t)$ is constructed within one oscillation period ($T$ = 0.12 s).

Thus, the magnitude of the driving force $F_{sum}$ = 10 kN is obtained, acting in the direction of performing useful work with the asymmetry coefficient $k_{as}$ = 10, which is generated by ten stages of the vibration device with directed oscillations with matched oscillation frequencies. The magnitude of the total driving force in the direction of the idle stroke is 1 kN.

In accordance with the stated objective of the research, a methodology and technique for calculating and designing the parameters of vibration devices with asymmetric oscillations for a given value of the total driving force and a given asymmetry coefficient have been developed.

A general equation for the total driving force of a vibration device with asymmetric oscillations, ensuring any asymmetry coefficient specified from the point of view of technological feasibility, is obtained and proposed.

The proposed method of calculating and designing vibration devices with asymmetric oscillations using a generalized equation of the total driving force allows for the conversion of existing models of vibration machines from uniform oscillations to asymmetric ones, thereby increasing their operating efficiency with improved specific parameters.

The authors have developed and created a test bench, shown in Figure 15, consisting of 6 pairs of steps. The bench allows for the replacement of pairs of imbalances, changing and selecting the corresponding moments of inertia, masses of imbalances, eccentricity, and angular velocities. It allows for the obtaining of an asymmetric driving force with a given asymmetry coefficient according to the developed methodology. For the presented test bench, consisting of 6 steps, the asymmetry coefficient in the range of 2.0...6.0 is realized.

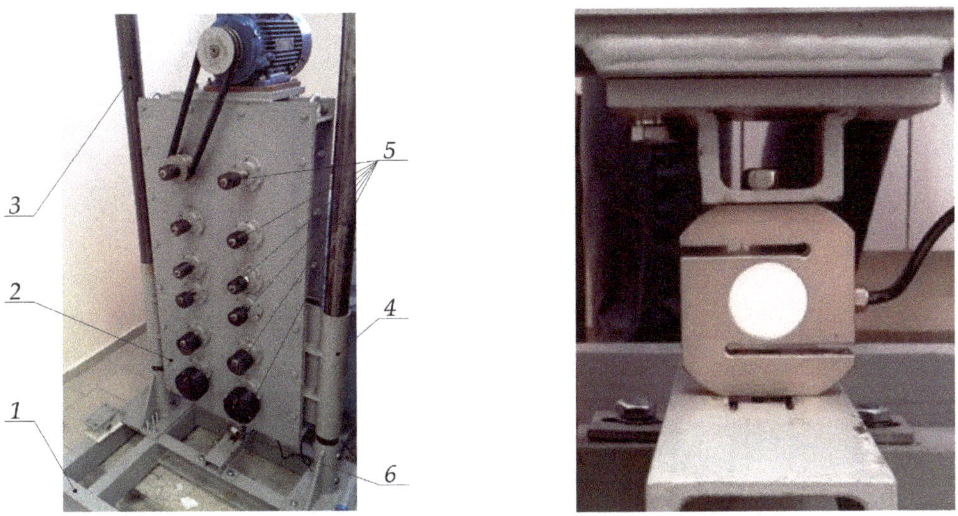


(a) (b)

**Figure 15.** (**a**) The test bench with synchronized pairs of imbalances for multi-stage production of asymmetric oscillations: 1 is a base, 2 is a body, 3 is a guide rod, 4 is the sliders, 5 is the imbalances, 6 is a driving force sensor. (**b**) A driving force magnitude recording sensor.

**Author Contributions:** M.D.G.: Conceptualization, methodology; N.S.L.: data curation, writing—original draft preparation; A.A.P.: supervision, resources; B.S.C.: visualization, investigation, validation; A.C.: software. All authors have read and agreed to the published version of the manuscript.

**Funding:** The study was supported by the Russian Science Foundation Grant No. 23-79-10022, https://rscf.ru/project/23-79-10022/.

**Data Availability Statement:** The data presented in this study are available upon request from the respective author. The data are not publicly available as they are also part of ongoing research.

**Acknowledgments:** The authors are grateful to the team of the High Technology Center (HTC) of BSTU and V.G. Shukhov for assistance in providing design capacities.

**Conflicts of Interest:** The authors declare no conflicts of interest.

## References


1. Olukan, T.; Farshid, I. *Oscillations: Theory and Applications in AFM*; IOP Publishing: Bristol, UK, 2024; pp. 1–21. [CrossRef]
2. Yaïr, H.; Santos, S.; Elsherbiny, L.; Chiesa, M. Competition between transient oscillations and early stochasticity in exponentially growing populations. *Phys. Rev. Res.* **2024**, *6*, 033320. [CrossRef]
3. Chen, L.; Chen, J.; Tan, C.; Duan, P.; Liu, X.; Liu, M.; Fa, M.; Feng, J.; Zhang, X.; Zheng, B. The effect of channel size and wall erosion on the low-frequency oscillation in Hall thruster. *J. Phys. D Appl. Phys.* **2024**, *58*, 015203. [CrossRef]
4. Ginoux, J.-M. From Nonlinear Oscillations to Chaos Theory. In *The Foundations of Chaos Revisited: From Poincaré to Recent Advancements*; Springer: Berlin/Heidelberg, Germany, 2016; pp. 27–47. [CrossRef]
5. Gatti, P.L. *Applied Structural and Mechanical Vibrations: Theory and Methods*, 2nd ed.; CRC Press: Boca Raton, FL, USA, 2014; pp. 1–635. [CrossRef]
6. Ouali, M.; Magraoui, R. Contribution to Conditional Maintenance by Vibration Analysis of Rotating Machine Mechanical Failures and Proposed Solutions. *Rev. Gestão Soc.Ambient.* **2024**, *18*, e08541. [CrossRef]
7. Fatih, P.; İsmail, E.; Cihan, M. Vibration analysis of half rail vehicle model with 10 degrees of freedom based on mechanical-electrical analogy theory. *Int. J. Heavy Veh. Syst.* **2021**, *28*, 521–541. [CrossRef]
8. Polidor, B.; Nicusor, D.; Cornelia, D. Dynamic Performances of Technological Vibrating Machines. *Symmetry* **2022**, *14*, 539. [CrossRef]
9. Hordieiev, A.; Stariy, A.; Urbaniuk, E.; Hordieiev, A. Design of a vibrating machine module for pollution cleaning and a general method for calculating its elements. *Her. Khmelnytskyi Natl. Univ. Tech. Sci.* **2022**, *315*, 40–46. [CrossRef]
10. Lyan, I.; Panovko, G.; Shokhin, A. Creation and verification of spatial mathematical model of vibrating machine with two self-synchronizing unbalanced exciters. *J. Vibroeng.* **2021**, *23*, 1524–1534. [CrossRef]
11. Lyan, I.; Panovko, G.; Shokhin, A. Modeling of a vibrating machine with spatial oscillations of the working body. *IOP Conf. Ser. Mater. Sci. Eng.* **2021**, *1129*, 012038. [CrossRef]
12. Zhao, C.; Zhao, Q.; Gong, Z.; Wen, B. Synchronization of Two Self-Synchronous Vibrating Machines on an Isolation Frame. *Shock Vib.* **2011**, *18*, 73–90. [CrossRef]
13. Gurskyi, V.; Kuzio, I.; Krot, P.; Zimroz, R. Energy-Saving Inertial Drive for Dual-Frequency Excitation of Vibrating Machines. *Energies* **2021**, *14*, 71. [CrossRef]
14. Tian, Y.; Tu, C. Statistical Prediction and Analysis on Parameters of Vibrating Sinking Pipe Gravel Pile Machines. *J. Phys. Conf. Ser.* **2021**, *1744*, 022131. [CrossRef]
15. Jatsun, S.; Zaharov, I. Optimization of the Vibrating Technological Machines. In Proceedings of the 6th World Congresses of Structural and Multidisciplinary Optimization, Rio de Janeiro, Brazil, 30 May–3 June 2005; pp. 1–10.
16. Wu, F.; Zhang, Z. Abnormal Detection of Vibrating Sinking Pipe Gravel Pile Machines Based On Support Vector Machine Model. *J. Phys. Conf. Ser.* **2021**, *1744*, 022132. [CrossRef]
17. Astashev, V.; Krupenin, V.L. Quantitative evaluations of the effectiveness of vibrating and vibroimpact machines. *IOP Conf. Ser. Mater. Sci. Eng.* **2020**, *709*, 033041. [CrossRef]
18. Srikanth, V.; Kuznetsov, A. Symmetry-breaking of turbulent flow due to asymmetric vortex shedding in periodic porous media. *arXiv* **2024**, arXiv:2407.12955. [CrossRef]
19. Jiaguo, Y.; Ying, P. Research on Design of Vibration Platform of Molded Case Circuit Breaker and Reliability of Platform Frame. *Am. J. Eng. Technol. Manag.* **2018**, *3*, 35–41. [CrossRef]
20. Zhang, J.; Wang, X. Optimal Design and Experimental Study of Piezoelectric Vibrator With Multi-Order Bending Mode and Multi-Operating Frequency. *IEEE Access* **2023**, *99*, 45605–45617. [CrossRef]
21. Gurskyi, V.; Krot, P.; Korendiy, V.; Zimroz, R. Dynamic Analysis of an Enhanced Multi-Frequency Inertial Exciter for Industrial Vibrating Machines. *Machines* **2022**, *10*, 130. [CrossRef]
22. Xu, Y.; Luo, A. Paired asymmetric periodic oscillations in a pair of first-order asymmetric nonlinear circuit systems. *Mech. Syst. Signal Process.* **2022**, *171*, 108810. [CrossRef]
23. Fedyushkin, A. The Influence of Controlled Vibration Effects on Fluid Flow in Technological and Engineering Processes. In *Proceedings of the 10th International Conference on Civil Engineering*; Springer: Berlin/Heidelberg, Germany, 2024; pp. 656–664. [CrossRef]
24. Mousavi, S.; Fakhari, V. Design and development of closed-loop controllers for trajectory tracking of a planar vibration-driven robot. *J. Vib. Control* **2024**, 10775463241276026. [CrossRef]
25. Amiri, N.; Sohrabi, K.; Eftekharian, G.; Fakhari, V. Optimization and control of an energy-efficient vibration-driven robot. *J. Vib. Control* **2023**, *30*, 2184–2199. [CrossRef]
26. Hanns, W.T.; Roberta, L.K.; Bert, U.; Ralph, H. Good Vibrations: Asymmetric Vibrations for Directional Haptic Cues. In Proceedings of the World Haptics 2009—Third Joint EuroHaptics conference and Symposium on Haptic Interfaces for Virtual Environment and Teleoperator Systems, Salt Lake City, UT, USA, 18–20 March 2009; pp. 285–589. [CrossRef]

27. He, W.; Tang, B.; Jia, S. Design, simulation and development of a broadband electromagnetic angular vibrator. *Int. J. Appl. Electromagn. Mech.* **2018**, *57*, 295–307. [CrossRef]
28. Gouskov, A.; Panovko, G.; Shokhin, A. To the Issue of Control Resonant Oscillations of a Vibrating Machine with Two Self-synchronizing Inertial Exciters. In *Proceedings of the 14th International Conference on Vibration Problems*; ICOVP 2019, Lecture Notes in Mechanical Engineering; Springer: Singapore, 2020. [CrossRef]
29. Du, Y.; Dang, P.; Liu, Z.; Qiao, C. Performance improvement of a low-frequency vibration generator by using iterative learning control. *Rev. Sci. Instrum.* **2022**, *93*, 105005. [CrossRef] [PubMed]
30. Fanghella, P.; Bruzzone, L.; Ellero, S. Dynamic Balancing of a Nutating Planetary Bevel Gear Train. *Mech. Mach. Sci.* **2015**, *25*, 23–33. [CrossRef]
31. Gerasimov, M.D.; Glagolev, S.N.; Mkrtychev, O.V. Determination of The Driving Force's Asymmetry Factor and The Vibrostand's Work's Analysis. *Int. J. Appl. Eng. Res.* **2015**, *10*, 45181–45185.
32. Kidokoro, J.; Mian, S.; Hoque, L.; Yu, P.; Okimura, K.; Nakanishi, T. Observation of transient aspects of self-sustained oscillations and the role of parallel capacitance in VO2-based planar devices. *J. Appl. Phys.* **2024**, *136*, 065303. [CrossRef]
33. Jazar, R.; Marzbani, H. Mechanical Vibrations Modeling. In *Vehicle Vibrations*; Springer: Cham, Switzerland, 2024; pp. 55–78. [CrossRef]
34. Noda, S.; Igari, R.; Endo, K.; Yamanobe, K.; Takayama, T. Switching backdrivability of a planetary gear by vibration: Design parameter setting and excited force estimation. *ROBOMECH J.* **2023**, *10*, 25. [CrossRef]
35. Yamamoto, T.; Ishida, Y.; Kazuo, A. On the Subharmonic Oscillations of Unsymmetrical Shafts. *Trans. Jpn. Soc. Mech. Eng.* **1978**, *44*, 1934–1942. [CrossRef]
36. Lubimyi, N.S.; Chepchurov, M.; Polshin, A.A.; Gerasimov, M.D.; Chetverikov, B.S.; Chetverikova, A.; Tikhonov, A.A.; Maltsev, A. Reducing the Cost of 3D Metal Printing Using Selective Laser Melting (SLM) Technology in the Manufacture of a Drill Body by Reinforcing Thin-Walled Shell Forms with Metal-Polymers. *J. Manuf. Mater. Process.* **2024**, *8*, 44. [CrossRef]
37. Gerasimov, M.; Lubimyi, N.; Yakovlev, E. Comparative analysis of the parameters of generators with circular and asymmetric oscillations. *IOP Conf. Ser. Mater. Sci. Eng.* **2020**, *945*, 012003. [CrossRef]
38. Petro, K.; Roman, S.; Oleksii, S.; Serhii, B. Vertical vibrating moveable device. *Vib. Eng. Technol.* **2022**, 29–35. [CrossRef]
39. Tyuremnov, I.; Shorokhov, D. Vibrating roller with compacted soil interaction modelling. *Russ. Automob. Highw. Ind. J.* **2024**, *21*, 202–216. [CrossRef]
40. Kostin, V.; Kostin, D.; Sapronov, Y. Maxwell-Fejer polynomials and optimization of polyharmonic impulse. *Dokl. Math.* **2012**, *86*, 512–514. [CrossRef]
41. Nessmann, A. Polyharmonic Functions in the Quarter Plane. *Electron. J. Comb.* **2024**, *31*, 1–63. [CrossRef] [PubMed]
42. Luan, T.; Khatir, S.; Thi, T.; Baets, B. Exponential-trigonometric optimization algorithm for solving complicated engineering problems. *Comput. Methods Appl. Mech. Eng.* **2024**, *432*, 117411. [CrossRef]
43. Andrukhiv, A.; Sokil, M.; Sokil, B.; Fedushko, S.; Syerov, Y.; Karovič, V.; Klynina, T. Influence of Impulse Disturbances on Oscillations of Nonlinearly Elastic Bodies. *Mathematics* **2021**, *9*, 819. [CrossRef]
44. Aoyama, Y.; Lehmamnn, P.; Theodorou, E. Second Order Stein-Variational Dynamic Optimization. *arXiv* **2024**, arXiv:2409.04644. [CrossRef]
45. Bai, J.; Zeng, Z. The Euler-Lagrange equations of nabla derivatives for variational approach to optimization problems on time scales. *Adv. Contin. Discret. Models* **2024**, *2024*, 33. [CrossRef]
46. Tsifansky, S.L.; Beresnevich, V.I. Method of Tuning a Vibration Machine with a Piecewise Linear Characteristic of Elastic Connections to Resonant Oscillations. 15 November 1987. (SU). Prior. SU 3910845/24-28. SU Patent 1,351,696, 17 June 1985. Available online: https://yandex.ru/patents/doc/SU1351696A1_19871115 (accessed on 12 September 2024).
47. Gerasimov, M.D.; Vorobyov, N.D.; Gerasimov, D.M. Mathematical model of a two-stage vibration mechanism. *Bull. BSTU Named After V.G. Shukhov* **2017**, *2*, 195–201. [CrossRef]
48. Albrecht, K.; Christian, H. Vibration Generator. 28 September 2010. (DE). Prior. US 2009/0243410. U.S. Patent 7,804,211, 1 October 2009.
49. Gerasimov, M.D. Addition of oscillations in vibration exciters. *Bull. BSTU Named After V.G. Shukhov* **2016**, *3*, 116–121.
50. Gerasimov, M.D. Method for Generating Directed Inertial Asymmetric Oscillations of Working Equipment of Vibration Machines. 12 January 2021. (RU). Prior. RU 2020121504. RU Patent 2,740,282, 5 August 2020.
51. Korn, G.; Korn, T. *Handbook of Mathematics for Scientists and Engineers*; The Main Editorial Board of Physical and Mathematical Literature; "Science": New York, NY, USA, 1973; pp. 1–832.

vibration

MDPI

Article

# Analysis of Displacement Transmissibility and Bifurcation Behavior in Nonlinear Systems with Friction and Nonlinear Spring

Deog Jae Hur [1] and Sung Chul Hong [2,*]

[1] AI & Mechanical Center, Institute for Advanced Engineering, Youngin-si 17180, Republic of Korea; djhur@iae.re.kr

[2] Departments of Mechanical & Automotive Engineering, Halla University, Wonju-si 26404, Republic of Korea

* Correspondence: schong@halla.ac.kr

**Abstract:** In this paper, a nonlinear vibration system with friction and linear and nonlinear springs is modeled and analyzed. The analysis examined how the combination of nonlinear variables affects the displacement of the system using the slowly varying amplitude and phase (SVAP) method. The break-loose frequency at which relative motion begins was obtained as a function of the friction ratio, and it was found that the displacement transmissibility differed depending on the change in design parameters. The displacement transmissibility response showed a unique phenomenon in which bifurcation occurred in the front resonant branch before the maximum response point when the linear damping coefficient was small and the friction coefficient was large, and the displacement transfer curve was separated at a specific parameter value. This phenomenon can be divided into three parameter zones considering the bifurcation pattern and stability of the displacement transmissibility curve. In addition, a 3-D spatial zone of dimensionless parameters was presented, which can predict stability during the design process, along with the drawing method and procedure. This can be conveniently utilized in the process of setting the parameters of the isolators considering the stability of the response during the design. In the analysis and design process of vibration isolators with friction damping, this study has important implications for practical applications.

**Keywords:** nonlinear vibration isolator (NVI); the slowly varying amplitude and phase (SVAP); displacement transmissibility (DTR); trajectories of limit point bifurcation (TLPB); tracking diagram (TD)

## 1. Introduction

The need for vibration isolation to protect devices from vibration due to earthquakes or environmental disturbances is greatly increasing with the development and precision of high-tech industries. Structural vibrations are widely present in aerospace, machinery, optics, submarines, architectural engineering, etc., and can sometimes affect the precision, safety, and reliability of objects, leading to unexpected results. Therefore, research on vibration isolation devices with nonlinearity to enhance precision in device protection is being conducted in various fields [1–5]. In addition, to enhance vibration isolation, the designed isolators consist of various types with springs with restoring forces and dampers that dissipate energy, with these elements frequently exhibiting nonlinear characteristics [6–13]. Vibration isolation is primarily achieved using materials such as rubber mounts, metal springs, air springs, and wire-mesh/cable systems [14]. Exhibiting minimal damping and a significantly high transmissibility at resonance, metal springs are used alongside viscous dampers to enhance vibration isolation performance. Generating friction due to contact between wires under load, wire-mesh vibration isolation systems exhibit significant behavior, leading to the presence of both static and dynamic nonlinearities in stiffness [13]. In addition, dampers utilizing piston friction in cylinder-type air springs and friction in magnetic springs are also employed [15]. Friction dampers, being easy to install and maintain, generate friction without being significantly influenced by load magnitude or inherent frequency, allowing for various designs based on surface shape or material [16]. The isolator, composed of these elements, can be briefly illustrated as a nonlinear system consisting of mass, a spring with nonlinear characteristics, and viscous and friction dampers, as shown in Figure 1.

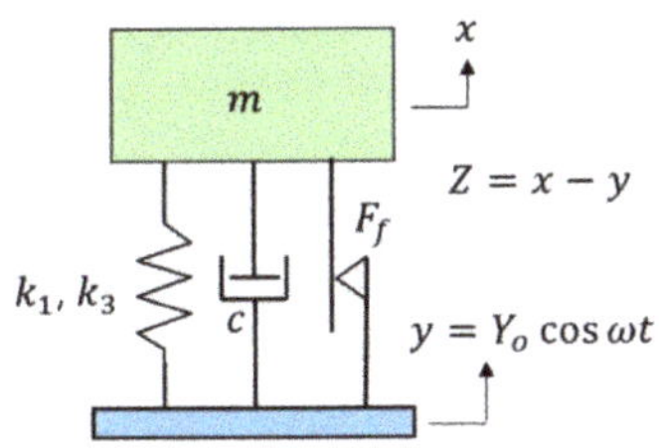


**Figure 1.** Base excited nonlinear vibration system including friction damping.

The analysis methods for nonlinear vibration systems apply approximate techniques such as the harmonic balance method (HBM) [17–19], the averaging method [18–20], the multiple scale method [21], etc. With the study of these methods, research on the characteristics of nonlinear systems with friction and quasi-zero stiffness as vibration isolation is actively being conducted [22,23]. Brfennan et al. [24] analyzed the jump-up and jump-down phenomena through the theoretical framework of a damped Duffing oscillator. In their analysis of the isolator, Carrella et al. [25] transformed the force–displacement characteristics into those of a hardening Duffing oscillator and applied the approximate harmonic balance method to derive analytical expressions for force and displacement transmissibility. Additionally, Barquist et al. [26] presented analytical calculations of the response of a Duffing oscillator to low-frequency variations in resonant frequency and damping. Liu et al. [27], proposing a mechanical model for a Duffing-type isolator subjected to excitation, aimed to establish criteria for jump avoidance under both base and forced excitation conditions. Murata et al. [28] analyzed the motion of diaphragm air springs using the Helmholtz–Duffing equation, also investigating the effects of jump phenomena and hysteresis through bifurcation studies, which highlighted the relationship between excitation frequency and amplitude that causes discontinuous changes in motion amplitude. Kovacic et al. [29] studied the nonlinear characteristics of a system with quasi-zero stiffness, which operates without linear stiffness terms and exhibits hardening behavior, focusing on the response of an asymmetric Duffing oscillator at its first resonance. Shi et al. [30] derived the frequency response relationships of nonlinear systems using averaging methods and elliptic functions. In their study on friction damping, Mario et al. [31] analytically examined a linear oscillator with dry friction, obtaining displacement transmissibility under base excitation and exploring the characteristics of friction damping. Wang et al. [32] conducted a study on the vibrations, stability, and bifurcation mechanisms of nonlinear systems associated with dry friction dampers for supercritical transmission shafts. However, they did not clarify the interrelationships among the variables for design implementation, and the absence of linear stiffness has been shown to pose challenges in analyzing the system's characteristics.

Uzdin et al. [33] conducted a study on a seismic isolation system, which comprises support elements connected in series with elastic and friction components, modeling the friction element as a dry friction damper. The analysis revealed that, to avoid hazardous resonant vibrations, the friction force and damping must exceed 10% of the structural weight, which he proposed as a critical design parameter. Benacchio et al. [34] investigated the effects of the nonlinear parameter of dry friction on the dynamic behavior of oscillators. The equations of motion were formulated according to the Mathieu–Duffing equation, using the harmonic balance method and variable amplitude method to obtain solutions. The study demonstrated that an isolator incorporating dry friction exhibits bifurcations of isolated periodic solutions. Zucca and Ferhatoglu [35,36] argued that non-uniqueness in simulating friction damping can lead to convergence problems in numerical methods that directly compute periodic limit states. Thus, he interpreted the condition of ignoring the non-uniqueness of residual traction by setting the time average of the slip displacement to zero. Ferhatoglu et al. [37] focused on the first-order resonance with clear mode separation based on nonlinear modes. First, the boundaries of nonlinear modal properties, especially modal frequency that depend on amplitude, damping ratio, and displacement, are calculated and interpreted as closed-form expressions governing the amplitude–frequency curve for harmonic excitation. Starossek [38] considered a hardening model of the Duffing equation combining Coulomb and viscous damping, observing abnormal jumps in a system subjected to base excitation. Ravindra and Mallik [39] considered the equations of a system with a hardening-type third-order nonlinear spring that combines Coulomb and viscous damping, observing abnormal jumps due to base excitation. This provided a relationship between the frequency and amplitude of excitation, leading to a discontinuous change in the amplitude of motion. Yu and Zang [40] replaced dry friction with viscous damping to obtain the displacement transmission rate of systems with dry friction. However, an analysis of non-resonance with nonlinear stiffness has not been implemented. Huang et al. [41] introduced mass and stiffness as new parameters to improve the diversity of the equations of motion that include nonlinear damping, analyzing them using the averaging method. Yu et al. [42] approximated the equations of a system with dry friction and third-order stiffness to derive the transfer function, further investigating the effects of each parameter on non-resonance, and then validated the theory through empirical experiments. However, they did not approach the dimensional analysis of the equations, introducing parameters of different dimensions, which complicates the overall variable relationship analysis. Additionally, there was no three-dimensional analysis of the dimensionless coefficients of viscous damping, friction damping, and nonlinear spring constants for implementation in design.

Therefore, in this study, in order to examine the change in the stability of the response according to the elements that make up the system, the following was conducted. First, a mathematical model is derived for a single-degree-of-freedom base excitation vibration model characterized by nonlinear properties, including stiffness with first- and third-order terms, along with dry friction. To examine the general relationships between the variables in the mathematical model, dimensionless parameters are introduced, and the response characteristics are derived using the SVAP method. The analysis is conducted through numerical methods based on theoretical governing equations. The discussion of the results aims to investigate the bifurcation phenomena in the solutions according to the nonlinear values of friction and stiffness in the transmissibility characteristics, as well as to study stability. Additionally, to utilize this information as design criteria while considering system stability, the interrelationships among viscous damping, stiffness nonlinearity, and friction damping will be examined, and the system's transmissibility and stability will be analyzed.

## 2. Equation of Motion for Base Exited System

The single-degree-of-freedom vibration system model with friction damping, viscous damping, and nonlinear stiffness is shown in Figure 1. In this vibration system, $c$ denotes the viscous damping coefficient, $F_f$ represents the friction force generated by the friction damper, which is modeled using the Coulomb friction model, $k_1$ is the stiffness coefficient, and $k_3$ is the coefficient of cubic nonlinear stiffness. When the base is excited with the displacement $y$, mass $m$ moves with the displacement $x$, and $Z$ represents the relative displacement of the two displacements.

To investigate the isolation characteristics of the vibrating system shown in Figure 1, the transmissibility is investigated when the base is harmonically excited with $Y_o cos\omega t$. The equation of motion for the vibration isolator in Figure 1 is expressed as follows:

$$m\frac{d^2Z}{dt^2}+c\frac{dZ}{dt}+k_1Z+k_3Z^3+F_f\,sgn\left(\frac{dZ}{dt}\right)=m\Omega^2Y_ocos(\omega t) \quad (1)$$

To express Equation (1) in dimensionless form, one defines the following non-dimensional parameters:

$$z = \frac{Z}{Y_o}, \quad \omega_n = \sqrt{\frac{k_1}{m}}, \quad \tau = \omega_n t, \quad \Omega = \frac{\Omega}{\omega_n}, \\ \zeta = \frac{c}{2m\omega_n}, \quad \mu = \frac{k_3 Y_o^2}{k_1}, \quad \eta = \frac{F_f}{2k_1 Y_o}. \tag{2}$$

Using these dimensionless parameters (2), Equation (1) is rewritten as the following non-dimensional equation:

$$\ddot{z} + 2\zeta\dot{z} + z + \mu z^3 + 2\eta\, sgn(\dot{z}) = \Omega^2 cos(\Omega\tau) \tag{3}$$

where $z$ is the dimensionless relative displacement, $\omega_n$ is the natural frequency, $\zeta$ is the damping ratio, $\mu$ is the coefficient of cubic nonlinearity, $\eta$ is the friction ratio, $\Omega$ is the excitation frequency ratio, $\tau$ is the dimensionless time, the "dots" denote derivatives with respect to $\tau$, i.e., $\dot{z} = dz/d\tau$ and $\ddot{z} = d^2z/d\tau^2$.

## 3. Response Analysis of Nonlinear Isolation System for Base Excited

To solve the nonlinear Equation (3), the SVAP method [18,19] is used. This method serves as a first approximation of the averaging method and is also referred to as the method of first-order averaging. The important feature of this approach is that it not only determines the steady-state periodic solution but also allows for the determination of the system's transient behavior as a limit cycle periodic solution.

The steady-state solution of Equation (3) and its time derivative form are given by Equations (4) and (5), respectively.

$$z = u(\tau)cos(\Omega\tau + \phi(\tau)) \tag{4}$$

$$\dot{z} = -u(\tau)\Omega cos(\Omega\tau + \phi(\tau)) \tag{5}$$

where $u$ and $\phi$ are functions of $\tau$. To determine these solutions, they must satisfy certain condition. The condition can be obtained by differentiating Equation (4) with respect to $\tau$ and comparing it with Equation (5), as follows:

$$\dot{u}\ cos(\Omega\tau + \phi) - \dot{\phi}u\ sin(\Omega\tau + \phi) = 0 \tag{6}$$

Differentiating Equation (5) with respect to time, and then substituting it along with Equations (4) and (5) into Equation (3) and simplifying, yields the following equations:

$$\dot{u}\ sin(\Omega\tau + \phi) + \dot{\phi}u\ cos(\Omega\tau + \phi) = \left(-u\Omega + \frac{u}{\Omega} + \frac{3\mu}{4\Omega}u^3\right)cos(\Omega\tau + \phi) \\ + \frac{\mu}{4\Omega}u^3\ cos(3\Omega\tau + 3\phi) - 2\zeta u\ sin(\Omega\tau + \phi) + 2\frac{\eta}{\Omega}sgn(\dot{u}) - \Omega\ cos\ \Omega\tau \tag{7}$$

Solving Equations (6) and (7) simultaneously for $\dot{u}$ and $\dot{\phi}$ and subsequently carrying out averaging over the period, $T = 2\pi/\Omega$, according to the SVAP method, yields the following first-order differential equation:

$$\dot{u} = -\frac{1}{2\Omega}\left(2\zeta\Omega u + \frac{8}{\pi}\eta + \Omega^2 sin(\phi)\right) \tag{8}$$

$$\dot{\phi} = \frac{1}{2\Omega u}\left((1 - \Omega^2)u + \frac{3}{4}\mu u^3 - \Omega^2\ cos(\phi)\right) \tag{9}$$

The averaged Equations (8) and (9) form an autonomous system, as time $\tau$ is not explicitly present in the right-hand side. These steady-state responses, denoted as $u_s$ and

$\phi_s$, are expressed as the following nonlinear algebraic equations by substituting $\dot{u} = \dot{\phi} = 0$, $u = u_s$ and $\phi = \phi_s$ into Equations (8) and (9).

$$2\zeta\Omega u_s + \frac{8}{\pi}\eta + \Omega^2 sin(\phi_s) = 0 \tag{10}$$

$$\left(1 - \Omega^2\right)u_s + \frac{3}{4}\mu u_s^3 - \Omega^2 cos(\phi_s) = 0 \tag{11}$$

By eliminating $\phi_s$ from Equations (10) and (11), the sixth-order equation for the stationary response $u_s$ can be obtained as follows.

$$a_6 u_s^6 + a_4 u_s^4 + a_2 u_s^2 + a_1 u_s + a_0 = 0 \tag{12}$$

where

$$\begin{aligned} a_6 = \left(\tfrac{3}{4}\mu\right)^2, \quad & a_4 = \tfrac{3}{2}\mu\left(1 - \Omega^2\right), \quad a_2 = \Omega^4 + 2\left(2\zeta^2 - 1\right)\Omega^2 + 1, \\ & a_1 = \tfrac{32}{\pi}\zeta\eta\Omega, \quad a_0 = \left(\tfrac{8}{\pi}\eta\right)^2 - \Omega^4. \end{aligned} \tag{13}$$

Using the solution of Equation (12), the phase $\phi_s$ have the form

$$\phi_s = \tan^{-1}\left(\frac{-2\zeta\Omega u_s - \frac{8}{\pi}\eta}{(1 - \Omega^2)u_s + \frac{3}{4}\mu u_s^3}\right). \tag{14}$$

In the case where $u_s = 0$ in Equation (12), there is no relative displacement. Therefore, the constant term $a_0$ in Equation (12) must be zero, i.e.,

$$\frac{8}{\pi}\eta - \Omega_c^2 = 0 \tag{15}$$

In Equation (15), $\Omega_c$ represents the critical frequency at which relative motion occurs and means the sticking to sliding transition frequency. This is commonly referred to as the starting frequency of motion [13], also known as the onset frequency of slip or the break-loose frequency [33]. If the applied frequency is less than $\Omega_c$, relative motion remains in a locked state, and if the applied frequency exceeds $\Omega_c$, relative motion will occur. Figure 2 illustrates the relationship described by Equation (15), showing the slip and sticking regions in the $\Omega - \eta$ plane and the threshold line. In Figure 2, the friction ratio corresponding to the resonant frequency, $\Omega = \Omega_n = 1$, is $\pi/8$. Therefore, if $\eta > \pi/8 \simeq 0.393$, relative displacement does not occur even at the resonant frequency.

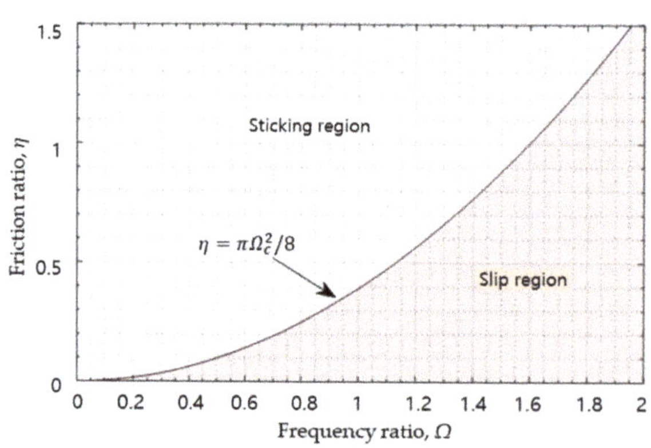


**Figure 2.** Sticking and slip regions in $\Omega - \eta$ plane and the separating threshold line.

The stability of the solutions obtained from Equations (10) and (11) can be obtained from the eigenvalue analysis of the Jacobian matrices of Equations (8) and (9). If the real value of the eigenvalue is positive, it is unstable, and if it is negative, the solution is stable. The Jacobian matrix for stability analysis is as follows.

$$J(u_s, \phi_s) = \begin{bmatrix} -\zeta & -\frac{\Omega}{2}cos(\phi_s) \\ \frac{1}{4\Omega u_s^2}\left(3\mu u_s^3 + 2\Omega^2 \cos\left(\phi_s\right)\right) & \frac{\Omega}{2u_s}sin(\phi_s) \end{bmatrix} \tag{16}$$

The vibration transmission characteristics can be identified through the displacement transmissibility of absolute displacement, which can be expressed as the ratio of the absolute displacement amplitude of the mass and the base excitation amplitude, and the absolute displacement response of the mass is represented as the sum of the relative displacement response and the base excitation displacement as follows:

$$x = u_s Y_0\, cos(\Omega\tau + \phi_s) + Y_0\, cos(\Omega\tau). \tag{17}$$

Therefore, the absolute displacement transmissibility $T_r$ and the phase $\psi$ of the absolute displacement due to the base excitation are as follows:

$$T_r = \frac{|x|}{Y_0} = \sqrt{1 + 2u_s \cos\left(\phi_s\right) + u_s^2} \tag{18}$$

$$\psi = \tan^{-1}\frac{-u_s\, sin(\phi_s)}{1 + u_s\, cos(\phi_s)} \tag{19}$$

## 4. Results and Discussion

The displacement responses for the dimensionless parameters ($\zeta$, $\eta$, $\mu$) defined in Equation (2) are obtained from Equations (12) and (14). Based on these results, the absolute displacement transmissibility (DTR) can be obtained using Equation (18). The stability of these solutions is assessed using Equation (16). In this study, the DTR characteristics according to changes in friction damping, viscous damping, and nonlinear stiffness will be examined.

### 4.1. System with Zero Nonlinear Spring Coefficient ($\mu = 0$)

In the case where the coefficient of nonlinear stiffness $k_3$ in the equation of motion Equation (1) is zero, the DTR can be obtained by numerical analysis, and the result can be obtained, as shown in Figure 3. As shown in Figure 3a,b, as the excitation frequency ratio increases, the frequency at which DTR begins to become greater than one increases as the friction ratio $\eta$ increases. This is because the break-loose frequency increases as the friction ratio increases, so there is no relative motion until the excitation frequency reaches the break-loose frequency, maintaining $T_r = 1$. Also, all DTR curves always pass the point of $T_r = 1$ when $\Omega = \sqrt{2}$. Since this is a case that satisfies $u_s = -2\, cos\phi_s$ in Equation (18), by substituting this into Equation (11), confirms that $\Omega = \sqrt{2}$. When $\eta < \pi/8$ ($\Omega_c < 1$), resonance occurs at $\Omega = 1$. For instance, as shown in Figure 3a, when $\eta = 0$ or $\eta = 0.2$ ($\eta < \pi/8$) with $\zeta = 0$, the DTR becomes infinite at $\Omega = 1$ in both cases. However, in Figure 3b, where $\zeta \neq 0$, the DTR is finite due to the effect of viscous damping. When $\eta > \pi/8$ ($\Omega_c > 1$), the DTR rapidly decreases, and when $\eta \geq \pi/4$ ($\Omega_c \geq \sqrt{2}$), the DTR becomes less than or equal to one at all excitation frequencies $\left(\Omega \geq \sqrt{2}\right)$, resulting in vibration isolation.

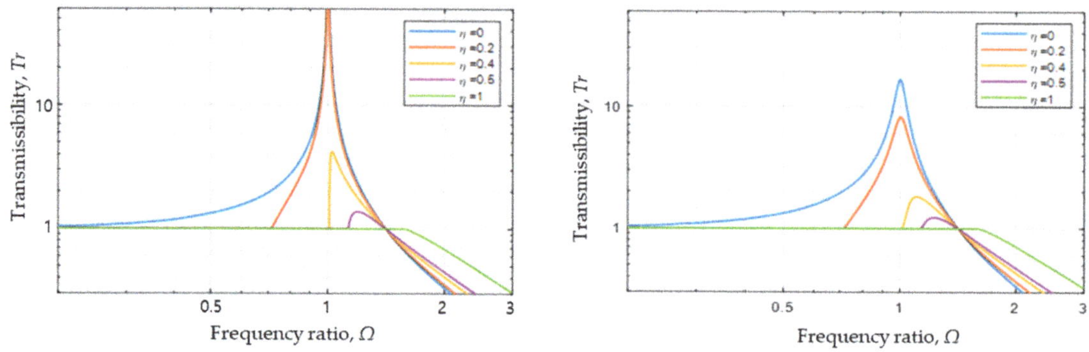


(a) $\zeta = 0.00$ (b) $\zeta = 0.03$

**Figure 3.** Absolute displacement transmissibility of linear spring system ($\mu = 0$). (**a**) Transmissibility curve due to frequency ratio at $\zeta$ = 0.00 for $\eta$ = 0, 0.2, 0.4, 0.5, 1.0, (**b**) Transmissibility curve due to frequency ratio at $\zeta$ = 0.03 for $\eta$ = 0, 0.2, 0.4, 0.5, 1.0.

### 4.2. System with Positive Nonlinear Spring Coefficient ($\mu > 0$)

For the system where the linear and nonlinear spring constants are non-zero, the TRs of the equation of motion (1) were numerically analyzed with respect to changes in the damping ratio $\zeta$ and friction ratio $\eta$. The results are shown in Figure 4. Depending on the frequency ratio, the system has either one or three solutions for the DTR. The stability of these solutions is evaluated using Equation (16), and unstable solutions are represented by dashed lines. A limit point (LP) bifurcation occurs when the system transitions from one solution to three, or from three solutions to one, and this LP bifurcation is marked with circular symbols. When the excitation frequency is lower than the break-loose frequency, no relative motion occurs due to the friction damper, and the DTR remains at one. As shown in Figure 4a, when the friction ratio $\eta$ increases, the break-loose frequency also increases. With a further increase in the break-loose frequency, the front resonant branch of the DTR curve bifurcates into one unstable solution and two stable solutions, forming a concave shape. As the values continue to rise, the front resonant branch and rear resonant branch meet at a single point, indicating that the solution possesses multiple roots. Subsequently, the DTR curve separates into a host branch and a sub-branch. The results of this numerical analysis are presented in Figure 4. In Figure 4a–c, examining the cases where $\eta$ is 0.6 and 1, it can be seen that as the damping ratio $\zeta$ increases, the separation of the DTR curve also occurs. From the perspective of the damping ratio, considering $\eta = 0.6$, it is noted that as $\zeta$ increases to 0, 0.04, and 0.06, the DTR curve separates into the host branch and sub-branch. In Figure 4d, when the damping ratio is large (e.g., $\zeta = 0.2$), it is noted that even as the friction ratio increases, the front resonant branch maintains stable solutions without bifurcation, while the rear resonant branch shows bifurcation or stable solutions depending on the value of the damping ratio.

In the analysis of Figure 4, DTR can be classified into three types based on the system design parameters, as shown in Figure 5. These types include Type I, where the front resonant branch of the DTR curve undergoes bifurcation; Type II, where the TR curve separates into a host branch and a sub-branch; and Type III, where there is no bifurcation in the front resonant branch of the DTR curve.

Type I, illustrated in Figure 5a, features a concave or narrow-waist shape due to the LP bifurcation points A and B, where the corresponding frequencies satisfy $\Omega_A > \Omega_B$. This type also exhibits abnormal jump characteristics. When the excitation frequency is swept upward from point S, it follows the S-A path and jumps up at point A, moving along the upper resonant branch to point E. Conversely, when sweeping the frequency downward from point F, it follows the F-C path, jumps up at point C, and subsequently moves along the upper resonant branch to point B before jumping down and reaching point S. This results in a double jump phenomenon.

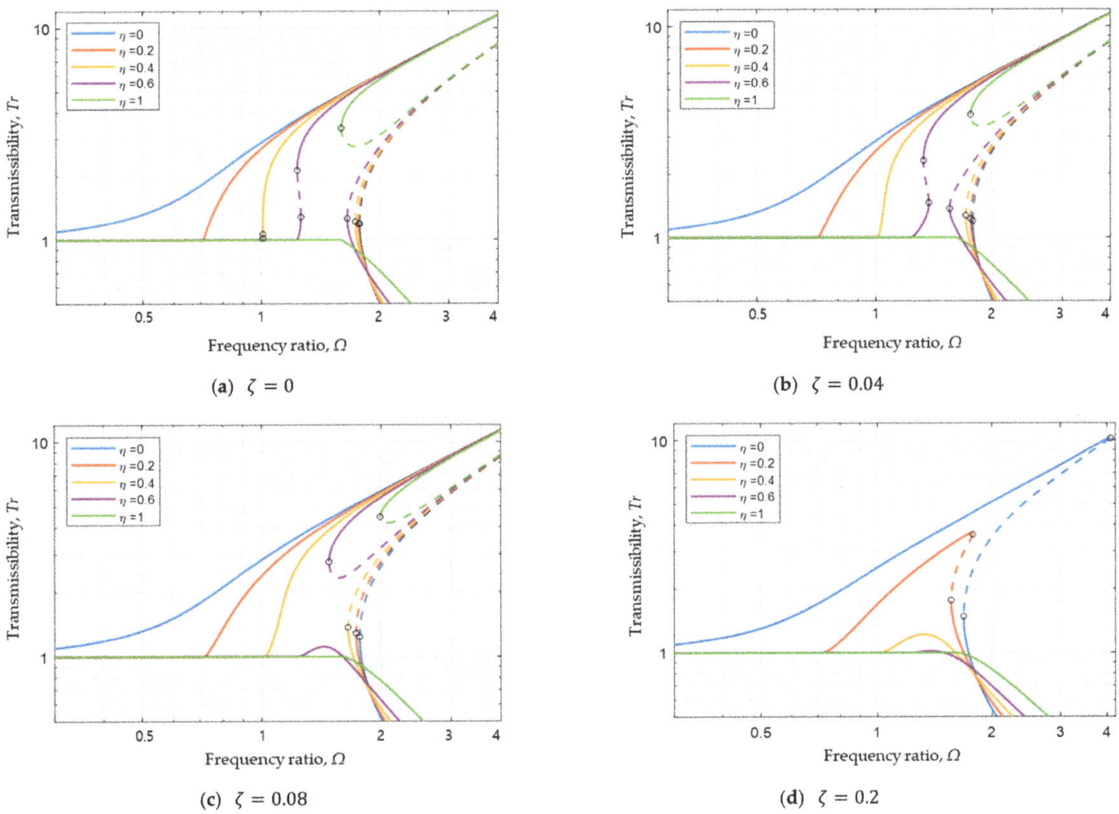


(**a**) $\zeta = 0$ (**b**) $\zeta = 0.04$ (**c**) $\zeta = 0.08$ (**d**) $\zeta = 0.2$

**Figure 4.** Absolute displacement transmissibility of a nonlinear system with respect to friction and frequency ratio for a spring stiffness ratio of $\mu = 0.2$. The dashed lines denote the unstable branch regions, and the circle markers indicate the locations of the LP bifurcations. (**a**) Transmissibility curve due to frequency ratio at $\zeta = 0.00$, (**b**) Transmissibility curve due to frequency ratio at $\zeta = 0.04$, (**c**) Transmissibility curve due to frequency ratio at $\zeta = 0.08$, (**d**) Transmissibility curve due to frequency ratio at $\zeta = 0.2$ for $\eta = 0, 0.2, 0.4, 0.6, 1.0$.

Type II, depicted in Figure 5b, exhibits DTR characteristics that are separated into a host branch and a sub-branch. In the case of a downward sweep from point E of the sub-branch, a jump down occurs at point D along the upper sub-branch, although this is a very specific scenario and has a low probability of occurrence in practical engineering. Upward or downward sweeps from points S or F lie on the stable host branch along the S-F path. However, the separated sub-branch has a large transmissibility characteristic, and due to its distinct properties, it may not be identified in analyses using the shooting method [43]. Therefore, caution should be exercised during the analysis.

Type III, shown in Figure 5c, is characterized by the absence of bifurcation in the front resonant branch of the DTR curve. If the viscous damping is significant, the rear resonant branch also exhibits stable solutions.

Consequently, it is essential to investigate the relationships among the design parameters: the damping ratio $\zeta$, the friction ratio $\eta$, and the coefficient of cubic nonlinearity $\mu$, as bifurcation phenomena related to instability arise based on these relationships. Understanding these relationships is critical for the vibration isolation design of nonlinear systems.

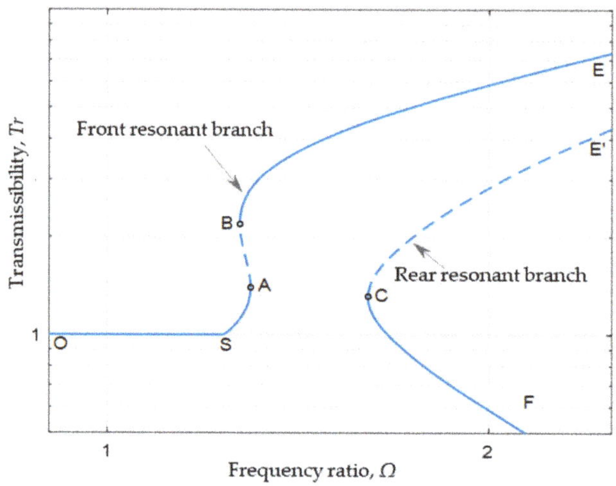


(a) Type I

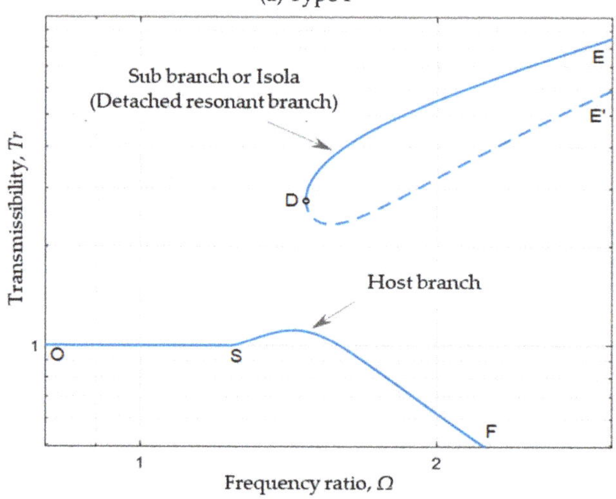


(b) Type II

**Figure 5.** *Cont.*

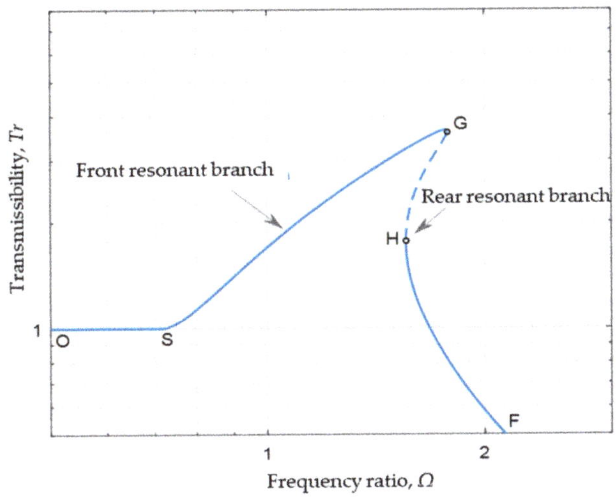


(c) Type III

**Figure 5.** Three types of displacement transmissibility responses that occur depending on system design parameters; the dotted line is an unstable region; and the end point of the dotted line is the LP Bifurcation point. (**a**) Type I with a concave or narrow-waist shape due to the LP bifurcation points ($\zeta = 0.02$, $\mu = 0.2$, $\eta = 0.6$), (**b**) Type II separated into a host branch and a sub-branch ($\zeta = 0.08$, $\mu = 0.2$, $\eta = 0.6$), (**c**) Type III without bifurcation in front resonant branch curve ($\zeta = 0.2$, $\mu = 0.2$, $\eta = 0.2$).

The LP bifurcations of the DTR curves vary according to the design parameters: friction ratio, damping ratio, and coefficient of cubic nonlinearity. By tracking the changes in these bifurcations, one can understand the characteristics of the DTR curve. First, fixing $\zeta = 0.02$ and $\eta = 0.2$, and calculating the bifurcation points of the DTR curves for several friction ratio values, the results shown in Figure 6a are obtained. In this figure, points A and C merge into point D ($\eta = 0.6565$) as $\eta$ increases. Beyond this point, the bifurcations disappear. Meanwhile, point B follows a path from point E to point F as $\eta$ increases. Figure 6b shows the trajectory of the bifurcation points in the $\Omega - \eta$ plane, obtained by incrementally changing the friction ratio $\eta$. The points corresponding to the bifurcations in Figure 6a are marked with the same labels in Figure 6b. The kink point on the trajectory curve is labeled as $K_p$, the intersection with the horizontal line is labeled as $R_p$, and the critical point is denoted as $C_p$. As seen in Figure 6b, bifurcation points A and C move closer together as $\eta$ increases and eventually converge at the critical point $C_p$, which corresponds to point D ($\eta = \eta_D$). Conversely, as $\eta$ decreases, bifurcation points B and A approach each other, forming a kink point $K_p$, and the unstable region disappears.

Thus, when $\eta_{K_p} < \eta < \eta_{C_p}$, the DTR curve exhibits a Type I form with two bifurcations, leading to unstable solutions on the front resonant branch. When $\eta > \eta_{C_p}$, the DTR curve separates into a host branch and a sub-branch with one bifurcation, forming a Type II shape. Finally, when $\eta < \eta_{K_p}$, the DTR curve has no unstable solutions on the front resonant branch, resulting in a Type III shape. Consequently, by using the trajectory curve as shown in Figure 6b, the frequency response characteristics can be effectively identified according to the variation in $\eta$.

The following analyzes the changes in bifurcation points with respect to the damping ratio $\zeta$. DTR curves for various $\zeta$ values were calculated, and the bifurcation points of each DTR curve were tracked and plotted on the $\Omega - \eta$ plane, as shown in Figure 7. First, for $\zeta = 0$, the kink point $K_p$ is located at the coordinates $(\Omega, \eta)$ = (1, 0.393) on the boundary line between stick and slip, which can be verified by the break-loose frequency relation (15).

The critical point $C_p$ is located at the coordinates (1.492, 0.695). As $\zeta$ increases to 0.02 and 0.08, the range of friction ratios between $K_p$ and $C_p$ on the trajectory curve, i.e., from $\eta_{K_p}$ to $\eta_{C_p}$, decreases. Consequently, as viscous damping increases, the range of friction ratios for which the DTR curve takes the form of Type I diminishes. Additionally, since the friction ratio at the critical point $C_p$, $\eta_{C_p}$, decreases, the range of friction ratios forming the Type II curve increases. When $\zeta$ = 0.15, $C_p$ and $R_p$ disappear, leaving only $K_p$. At this point, $K_p$ is located at the coordinates (1.4417, 0.3835). If $\eta > \eta_{K_p} = 0.3835$, the DTR curve exhibits Type II characteristics, and if $\eta < \eta_{K_p}$, it exhibits Type III characteristics. Similarly, when $\zeta$ = 0.2, only $K_p$ exists at the coordinates (1.4977, 0.2405), and since $\eta$ is always less than $\eta_{K_p}$, the DTR curve shows Type III characteristics.

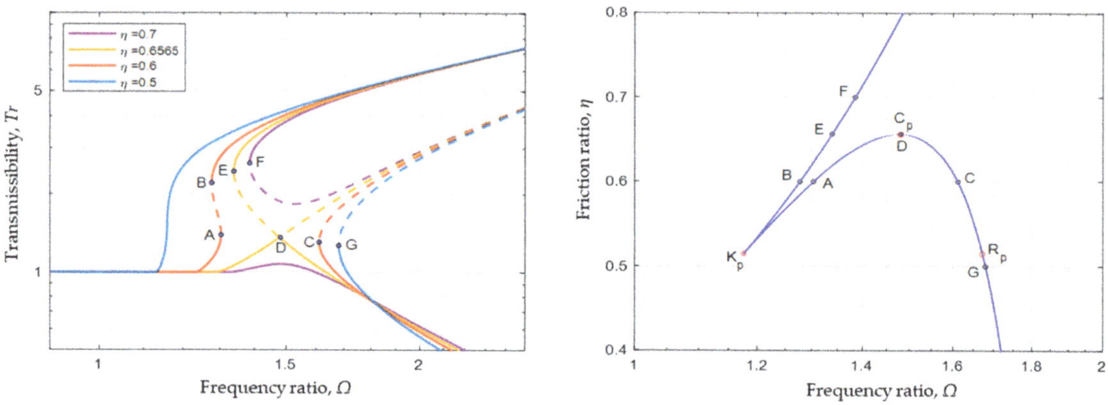


(a) DTR curves and LPs (b) Trajectory of LPs

**Figure 6.** Transition of TR curves and the trajectory of LP bifurcations in the $\Omega - \eta$ plane, with $\mu = 0.2$ and $\zeta = 0.02$. (**a**) Transmissibility curve due to frequency ratio for η = 0.7, 0.6565, 0.6, 0.5, 0.4, 0.5, 1.0 and Limit Point, (**b**) Trajectory of Limit Points due to friction ratio at the left figure.

By tracking the changes in the bifurcation points $C_p$, $K_p$, and $R_p$ along the bifurcation path in Figure 7 and reconstructing them on the $\zeta - \eta$ plane using their $\zeta$ and $\eta$ values, the tracking diagram shown in Figure 8 is obtained. In Figure 8, the $R_p$ line partially overlaps with the $K_p$ line, and the $C_p$ line decreases as $\zeta$ increases, eventually merging with the $K_p$ and $R_p$ lines and vanishing at $\zeta \simeq 0.13$. These two lines divide the $\zeta - \eta$ plane into three regions. Analyzing the DTR curve characteristics in the three regions of Figure 8, in Zone I, where $\eta_{K_p} < \eta < \eta_{C_p}$, the DTR curve exhibits Type I characteristics with bifurcations in both the front and rear resonant branches. Zone III, where $\eta < \eta_{K_p}$ shows no bifurcation in the front resonant branch but bifurcation in the rear resonant branch, displaying Type II and Type III characteristics. In Zone II, where $\eta > \eta_{C_p}$ and $\eta > \eta_{K_p}$, the DTR curve splits into host and sub-branches, with the host branch having no bifurcation, corresponding to Type II and Type III characteristics. Zone II can be considered a parameter region where the system's response is stable.

The tracking diagram in Figure 8, obtained by tracing significant points such as $K_p$ and $C_p$ along the bifurcation curves, provides valuable information for selecting design parameters, $\eta$ and $\zeta$, to stabilize the system, offering highly useful guidance for the design process.

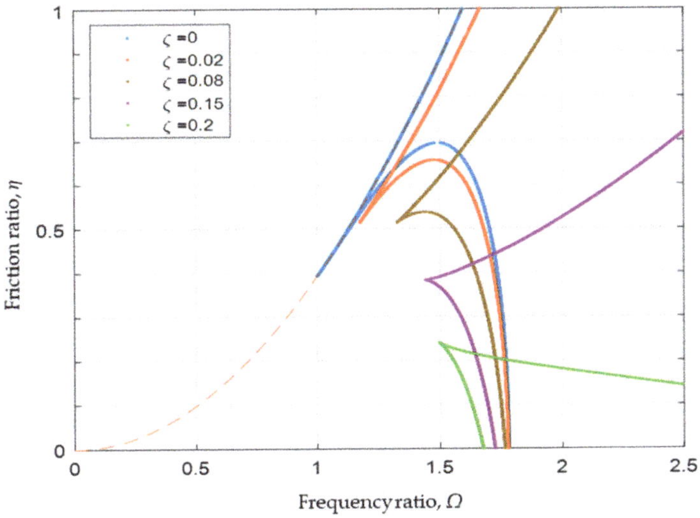


**Figure 7.** Trajectories of LP Bifurcations according to various damping ratios $\zeta$ for $\mu = 0.2$ in the $\Omega - \eta$ plane. The dashed line means the boundary between stick and slip.

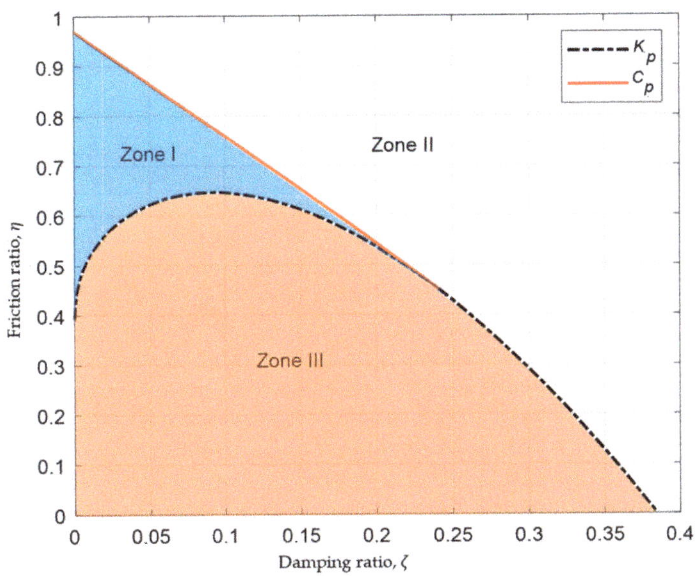


**Figure 8.** Tracking diagram of $C_p$, $K_p$, and $R_p$ points in $\zeta - \eta$ plane with $\mu = 0.2$ and regions delineated by them. (Sky blue is zone I, white is zone II, coral is zone III).

The following analyzes the system's response characteristics with changes in the value of the coefficient of cubic nonlinearity $\mu$, presenting the tracking diagram of $C_p$ and $K_p$ at $\mu = 0.5$ in Figure 9. As seen in Figure 9, this diagram resembles Figure 8 but shows enlarged Zone I and Zone III. Notably, as the nonlinear stiffness increases, Zone I expands, which requires careful consideration when selecting the design parameters for the system. Figure 10 illustrates the effects of the nonlinear stiffness coefficient by adding cases for $\mu$ = 0.5 and 1, represented in a three-dimensional $\zeta - \eta - \mu$ space. As shown in Figure 10, it consists of three distinct spatial regions, and the DTR characteristics in each region are extensions of the results already discussed in Figure 8. Therefore, the results of Figure 10 can be used as useful material for setting the design parameters for vibration isolation, namely, the friction ratio, damping ratio, and the coefficient of cubic nonlinearity, according to specific objectives.

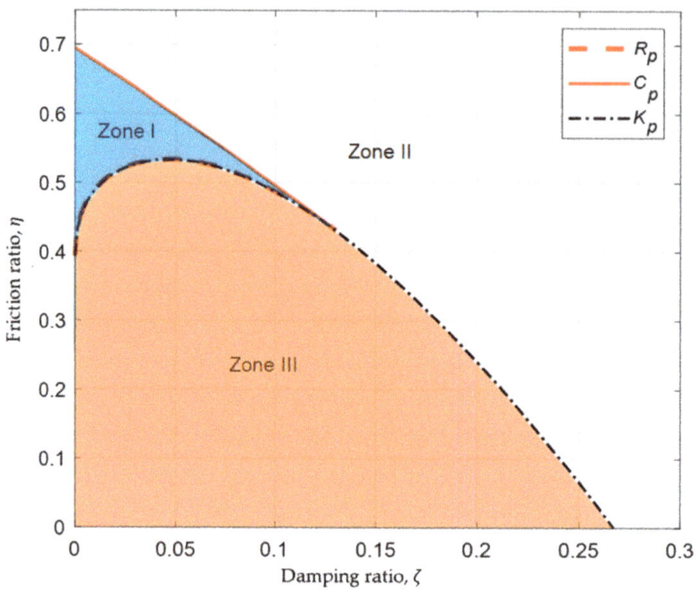


**Figure 9.** Tracking diagram of $C_p$ and $K_p$ points in $\zeta - \eta$ plane with $\mu = 0.5$ and regions delineated by them. (Sky blue is zone I, white is zone II, coral is zone III).

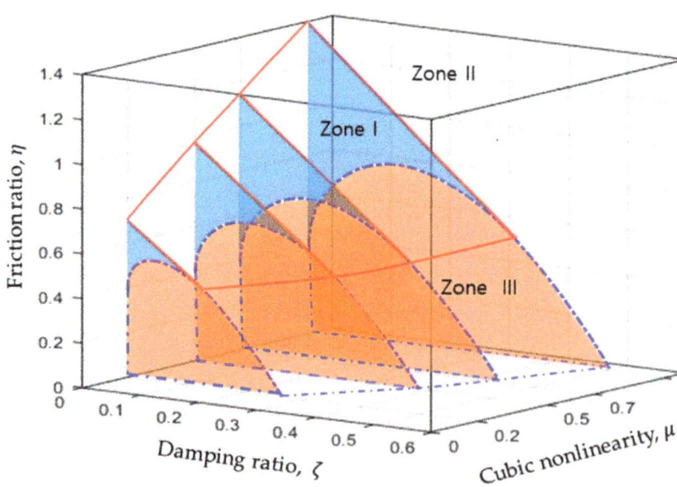


**Figure 10.** Tracking diagram of $C_p$ and $K_p$ points in $\zeta - \mu - \eta$ space and regions delineated by them. (Zone I is a volume where the sky blue colored areas are connected in the nonlinear stiffness ratio $\mu$ direction, zone II is a volume where the white-colored areas are connected, and zone III is a volume where the coral-colored areas are connected).

## 5. Conclusions

The response characteristics of the nonlinear isolation system with friction damping, viscous damping, and nonlinear stiffness, attributed to the base excitation, were analyzed using the SVAP method, and the results are as follows.

The break-loose frequency where the relative motion of the system occurs can be predicted by deriving the functional relationship of the friction damping ratio. In a nonlinear system with friction damping, the break-loose frequency increases as the friction coefficient increases, and the bifurcation phenomenon of the solution occurs in the front resonant branch. On the other hand, when the friction and viscous damping ratio increases, the bifurcation phenomenon of the solution disappears in the front resonant branch, and the bifurcation of the solution also disappears in the rear resonant branch. These phenomena can be classified into three types. The three types include configurations where unstable points (saddle points) exist in certain intervals of the DTR curve due to variations in design parameters, configurations with separated shapes of the DTR curve, and configurations where no separation phenomenon occurs. By tracking the bifurcation and its trajectory with respect to the design parameters, and following the key points $K_p$ and $C_p$, the system's parameter space could be divided into three zones on the $\eta - \zeta$ plane, allowing the specific type of DTR curve in each zone to be identified. Furthermore, by incorporating the nonlinear stiffness $\mu$, the analysis was extended into a three-dimensional space of $\zeta$, $\eta$, and $\mu$, confirming that as $\mu$ increases, the number of unstable regions also increases. The presentation of the three-dimensional space can serve as a valuable resource for easily determining stability and can be utilized for selecting design parameters. In other words, this is expected to serve as a resource for analyzing and selecting the interrelationships of design parameters such as the friction ratio $\eta$, damping ratio $\zeta$, and nonlinear stiffness coefficient $\mu$, which are key design variables in vibration isolation systems to ensure stability. Furthermore, by presenting a methodology for analyzing the response characteristics using the SVAP method, it is anticipated that this approach can be widely applied to single-degree-of-freedom nonlinear vibration isolation systems with friction and 1st- and 3rd-order nonlinear springs. Future research may explore additional nonlinear effects or integrate different damping mechanisms to expand upon this study.

**Author Contributions:** All authors, having participated in the discussions leading to the expression of this idea and agreeing on the content, made equal scientific contributions. S.C.H., focusing on the derivation of the equations of motion, numerical programming and simulation, and result analysis, and D.J.H., concentrating on partial numerical interpretation, result analysis, and manuscript construction, contributed to the work. The writing of the paper was improved through mutual reviews, with all authors having read the final manuscript and approved its accuracy. All authors have read and agreed to the published version of the manuscript.

**Funding:** This work was supported by a Korea Planning & Evaluation Institute of Industrial Technology (KEIT) grant funded by the Korean government, MOTIE (No. 20018493).

**Institutional Review Board Statement:** Not applicable.

**Informed Consent Statement:** Not applicable.

**Data Availability Statement:** Data are contained within the article.

**Conflicts of Interest:** The authors declare no conflicts of interest.

## Nomenclature

| | |
|---|---|
| $m$ | mass of system |
| $c$ | viscous damping coefficient |
| $k_1$ | stiffness coefficient |
| $k_3$ | coefficient of cubic nonlinear stiffness |
| $F_f$ | friction force |
| $x$ | displacement of mass |

| | |
|---|---|
| $y$ | displacement of base |
| $Z$ | relative displacement of the mass with respect to the base |
| $\omega$ | excitation frequency |
| $Y_o$ | amplitude of base displacement |
| $z$ | dimensionless relative displacement |
| $\omega_n$ | natural frequency |
| $\tau$ | dimensionless time |
| $\Omega$ | frequency ratio |
| $\zeta$ | damping ratio |
| $\eta$ | friction ratio |
| $\mu$ | coefficient of cubic nonlinearity |
| $u$ | amplitude of relative displacement response |
| $\phi$ | phase of relative displacement response |
| $u_s$ | steady-state amplitude of relative displacement response |
| $\phi_s$ | steady-state phase of relative displacement response |
| $\Omega_c$ | break-loose frequency, starting frequency of motion, or critical frequency |
| $T_r$ | absolute displacement transmissibility (ADTR) |
| SVAP | slowly varying amplitude and phase |
| DTR | displacement transmissibility |
| LP | Limit Point |
| $K_p$ | kink point |
| $C_p$ | critical point |
| $R_p$ | intersection point |

## References


1. Liu, C.; Jing, X.; Daley, S.; Li, F. Recent advances in micro-vibration isolation. *Mech. Syst. Signal Process.* **2015**, *56–57*, 55–80. [CrossRef]
2. Yan, B.; Yu, N.; Wu, C. A state-of-the-art review on low-frequency nonlinear vibration isolation with electromagnetic mechanisms. *Appl. Math. Mech.-Engl.* **2022**, *43*, 1045–1062. [CrossRef]
3. Santhosh, B.; Narayanan, S.; Padmanabhan, C. Nonlinear Dynamics of Shrouded Turbine Blade System with Impact and Friction. *Appl. Mech. Mater.* **2014**, *76*, 81–92. [CrossRef]
4. Le, T.D.; Ahn, K.K. A vibration isolation system in low frequency excitation region using negative stiffness structure for vehicle seat. *J. Sound Vib.* **2011**, *330*, 6311–6335. [CrossRef]
5. Le, T.D.; Ahn, K.K. Experimental investigation of a vibration isolation system using negative stiffness structure. *Int. J. Mech. Sci.* **2013**, *70*, 99–112. [CrossRef]
6. Lang, Z.Q.; Jing, X.J.; Billings, S.A.; Tomlinson, G.R.; Peng, Z.K. Theoretical study of the effects of nonlinear viscous damping on vibration isolation of sdof systems. *J. Sound Vib.* **2009**, *323*, 352–365. [CrossRef]
7. Yuvaraju, B.A.G.; Srinivas, J.; Nanda, B.K. Nonlinear dynamics of friction-induced regenerative chatter in internal turning with process damping forces. *J. Sound Vib.* **2023**, *544*, 117386. [CrossRef]
8. Ruzicka, J.E.; Derby, T.F. *Influence of Damping in Vibration Isolation, SVM-7*; Shock and Vibration Information Center: Washington DC, USA, 1971.
9. Ibrahim, R.A. Recent advances in nonlinear passive vibration isolators. *J. Sound Vib.* **2008**, *314*, 371–452. [CrossRef]
10. Ledezma-Ramírez, D.F.; Tapia-González, P.E.; Ferguson, N.; Brennan, M.; Tang, B. Recent advances in shock vibration isolation: An overview and future possibilities. *Appl. Mech. Rev.* **2019**, *71*, 060802. [CrossRef]
11. Rivin, E.I. *Passive Vibration Isolation*; ASME Press: New York, NY, USA, 2003.
12. Hong, S.C.; Hur, D.J. Dynamic behavior of a simple rolling seismic isolator with a position restoring device. *Appl. Sci.* **2018**, *8*, 1910. [CrossRef]
13. Hur, D.J.; Hong, S.C. Analysis of an isolation system with vertical spring-viscous dampers in horizontal and vertical ground motion. *Appl. Sci.* **2020**, *10*, 1141. [CrossRef]
14. Shin, Y.H.; Lee, J.H.; Jung, B.C.; Moon, S.J. Design of passive vibration isolation element by wire mesh material for developing a hybrid mount. *Trans. Korean Soc. Noise Vib. Eng.* **2020**, *30*, 75–81. [CrossRef]
15. Alrajhi, J.; Alhaifi, K.; Alardhi, M.; Alhaifi, N.; Alazmi, J.; Khalfan, A.; Alkhulaifi, K. The numerical analysis of single degree of freedom vibration system with non-linearity. *J. Mech. Eng. Autom.* **2023**, *12*, 6–19.
16. Min, K.W.; Kim, H.S. Performance based design of friction dampers for seismically excited structures. *J. Earthq. Eng. Soc. Korea* **2003**, *7*, 17–24.
17. Krack, M.; Gross, J. *Harmonic Balance for Nonlinear Vibration Problems*; Springer Nature: Cham, Switzerland, 2019.
18. Mickens, R.E. *Truly Nonlinear Oscillations, Harmonic Balance, Parameter Expansions, Iteration, and Averaging Methods*; World Scientific Publishing Company: Danvers, MA, USA, 2010.
19. Hagedorn, P. *Non-Linear Oscillations*; Clarendon Press: Oxford, UK, 1981.

20. Sanders, J.A.; Verhulst, F.; Murdock, J. *Averaging Methods in Nonlinear Dynamical Systems*, 2nd ed.; Springer: New York, NY, USA, 2007.
21. Nayfeh, A.H.; Mook, D.T. *Nonlinear Oscillations*; John Wiley & Sons: Hoboken, NJ, USA, 1995.
22. Shahraeeni, M.; Sorokin, V.; Mace, B.; Ilanko, S. Effect of damping nonlinearity on the dynamics and performance of a quasi-zero-stiffness vibration isolator. *J. Sound Vib.* **2022**, *526*, 116822. [CrossRef]
23. Ma, Z.; Zhou, R.; Yang, Q. Recent advances in quasi-zero stiffness vibration isolation systems: An overview and future possibilities. *Machines* **2022**, *10*, 813. [CrossRef]
24. Brennan, M.J.; Kovacic, I.; Carrella, A.; Waters, T.P. On the jump-up and jump-down frequencies of the Duffing oscillator. *J. Sound Vib.* **2008**, *318*, 1250–1261. [CrossRef]
25. Carrella, A.; Brennan, M.J.; Waters, T.P.; Lopes, V. Force and displacement transmissibility of a nonlinear isolator with high-static-low-dynamic-stiffness. *Int. J. Mech. Sci.* **2012**, *55*, 22–29. [CrossRef]
26. Barquist, C.S.; Jiang, W.G.; Gunther, K.; Lee, Y. Determining the source of phase noise: Response of a driven Duffing oscillator to low-frequency damping and resonance frequency fluctuations. *Phys. D Nonlinear Phenom.* **2021**, *427*, 13299. [CrossRef]
27. Liu, S.; Peng, G.; Li, Z.; Li, W. Analytical jump-avoidance criteria of Duffing-type vibration isolation systems under base and force excitations based on concave-convex property. *J. Vib. Control* **2023**, *29*, 4082–4092. [CrossRef]
28. Murata, A.; Kume, Y.; Hashimoto, F. Application of catastrophe theory to forced vibration of a diaphragm air spring. *J. Sound Vib.* **1987**, *112*, 31–44. [CrossRef]
29. Kovacic, I.; Brennan, M.J.; Lineton, B. On the resonance response of an asymmetric Duffing oscillator. *Int. J. Non-Linear Mech.* **2008**, *43*, 858–867. [CrossRef]
30. Shi, B.; Yang, J.; Li, T. Enhancing Vibration Isolation Performance by Exploiting Novel Spring-Bar Mechanism. *Appl. Sci.* **2021**, *11*, 8852. [CrossRef]
31. Marino, L.; Cicirello, A.; Hills, D.A. Displacement transmissibility of a Coulomb friction oscillator subject to jointed base-wall motion. *Nonlinear Dyn.* **2019**, *98*, 2595–2612. [CrossRef]
32. Wang, D.; Song, L.; Zhu, R.; Cao, P. Nonlinear dynamics and stability analysis of dry friction damper for supercritical transmission shaft. *Nonlinear Dyn.* **2022**, *110*, 3135–3149. [CrossRef]
33. Uzdin, A.M.; Kuznetsova, I.O.; Frese, M.; Nazarova, S.S.; Nazarov, A.A. Analysis of the behaviour of seismic isolated structure on bearings connected to the structure with a dry friction damper. *AIP Conf. Proc.* **2023**, *2612*, 040012.
34. Benacchio, S.; Giraud-Audine, C.; Thomas, O. Effect of dry friction on a parametric nonlinear oscillator. *Nonlinear Dyn.* **2022**, *108*, 1005–1026. [CrossRef]
35. Zucca, S.; Firrone, C.M. Nonlinear dynamics of mechanical systems with friction contacts: Coupled static and dynamic multi-harmonic balance method and multiple solutions. *J. Sound Vib.* **2014**, *333*, 916–926. [CrossRef]
36. Ferhatoglu, E.; Zucca, S. Determination of periodic response limits among multiple solutions for mechanical systems with wedge dampers. *J. Sound Vib.* **2021**, *494*, 115900. [CrossRef]
37. Ferhatoglu, E.; Groß, J.; Krack, M. Frequency response variability in friction-damped structures due to non-unique residual tractions: Obtaining conservative bounds using a nonlinear-mode-based approach. *Mech. Syst. Signal Process.* **2013**, *201*, 110651. [CrossRef]
38. Starossek, U. Exact analytical solutions for forced cubic restoring force oscillator. *Nonlinear Dyn.* **2016**, *83*, 2349–2359. [CrossRef]
39. Ravindra, B.; Mallik, A.K. Hard Duffing-type Vibration Isolator with Combined Coulomb and Viscous Damping. *Int. J. Non-Linear Mech.* **1993**, *28*, 427–440. [CrossRef]
40. Yu, H.; Zhang, L. Free resonance analysis of bilinear hysteresis dry friction damper. *J. Vib. Shock* **2016**, *3*, 92–95.
41. Huang, X.; Sun, J.; Hua, H.; Zhang, Z. The isolation performance of vibration systems with general velocity-displacement-dependent nonlinear damping under base excitation: Numerical and experimental study. *Nonlinear Dyn.* **2016**, *85*, 777–796. [CrossRef]
42. Yu, H.; Xu, Y.; Sun, X. Analysis of the non-resonance of nonlinear vibration isolation system with dry friction. *J. Mech. Sci. Technol.* **2018**, *32*, 1489–1497. [CrossRef]
43. Dhooge, A.; Govaerts, W.; Kuznetsov, Y.A. MATCONT: A Matlab Package for Numerical Bifurcation Analysis of ODEs. *ACM Trans. Math. Softw.* **2003**, *29*, 141–164. [CrossRef]

*vibration*

MDPI

*Article*

# Unidirectional Flow Through Time-Dependent Cross-Sectional Areas of a Compliant Tube and a Valve: A Nonlinear Model

**Christos Manopoulos [1,*], Sokrates Tsangaris [1], Christina Georgantopoulou [2] and Dimitrios Mathioulakis [2]**

[1] Laboratory of Biofluid Mechanics & Biomedical Technology, Fluids Section, School of Mechanical Engineering, National Technical University of Athens, Heroon Polytechniou 9, Zografos, 15780 Athens, Greece; sgt@fluid.mech.ntua.gr

[2] Mechanical Department, School of Engineering, Bahrain Polytechnic, Isa Town P.O. Box 33349, Bahrain; christina@polytechnic.bh (C.G.); dimitrios.mathioulakis@polytechnic.bh (D.M.)

* Correspondence: manopoul@fluid.mech.ntua.gr or manopoul@central.ntua.gr

**Abstract:** This work investigates the conditions for net flow generation by a straight tube with a cross-sectional area harmonically varying in time that connects two tanks—a problem that is mainly found in the design of impedance pumps. By assuming a quasi-one-dimensional flow and applying continuity and momentum equations, a first-order differential equation with respect to the flow rate is derived and presented for the first time, including a nonlinear term that is responsible for net flow rate generation. Namely, the net flow rate is found to be nonzero (as is the nonlinear term) if the cross-sectional areas of the two tanks are unequal and one of them is smaller than that of the straight tube. In this case, the flow is directed from the smaller to the larger tank and the net flow rate increases with the frequency of the tube's cross-sectional area variation. In contrast, when the tanks' cross-sections are equal, the net flow is generated only if a valve is installed, e.g., at one end of the tube, due to the large asymmetries imposed in the hydraulic losses with respect to the tube mid-length. Compared with constant valve opening, the net flow rate is augmented significantly if the valve opening is time-dependent. By employing the same equation, the flow rate of an intra-aortic counter-pulsating balloon pump is also examined, in which the valve (representing the aortic valve) opens during the shrinkage of the tube, and it is shown that the net flow rate increases with the frequency and amplitude of the tube's cross-sectional area. Conclusively, the harmonic oscillation in time of a tube's wall can cause unidirectional flow only if asymmetric losses are present with respect to its mid-length.

**Keywords:** impedance pump; net flow rate; nonlinear flow model; time-dependent cross-sectional area; local losses; intra-aortic balloon pump

## 1. Introduction

Although the majority of pumps (e.g., centrifugal or axial flow pumps) essentially consist of a number of rotating blades—the kinetic energy of which is transferred to the conveyed fluid—there are also some other types of pumps that add energy to fluids by means of the moving walls of flexible conduits or membranes. In the second category of pumps, there are two main types: peristaltic and impedance pumps. In peristaltic pumps, a flexible tube is pressed by a rotating roller so that a moving stenosis is formed, traveling along the tube. As a result, the cross-sectional area of the tube is a well-defined function of axial distance and time. In contrast, an impedance pump consists mainly of a flexible tube, a part of which is periodically compressed and decompressed by a pincher with a given amplitude and frequency. Especially in micro-scale applications, impedance pumps cause the motion of fluid with a membrane that oscillates over a cavity by an external energy source, e.g., piezoelectric actuators, electromagnetic forcing, etc. [1]. Compared with peristaltic pumps, the net flow rate of impedance pumps is harder to predict since this is mainly affected by the reflection of longitudinal pressure waves at sites of the hydraulic loop, where there is an abrupt change in the tube's material, like at the connection of a soft tube with a stiff one. As a result, the time-dependent streamwise pressure distribution is quite complex, which, in turn, deforms the tube walls and accordingly influences the pump flow rate. In contrast to peristaltic pumps, in which the flow rate is a linear function of the frequency of the rollers, in impedance pumps, the flow rate is a nonlinear function of the pincher frequency [2]. More specifically, the net flow rate maximizes at various pinching frequencies as a result of a resonance phenomenon, namely, when these frequencies coincide with the natural frequencies of the hydraulic installation. Moreover, it is interesting to note that with the increase in pinching frequency, the sign of the net flow rate also changes, alternating from positive to negative. The above natural frequencies mainly depend on (a) the tube material properties (e.g., Young's modulus, Poisson's ratio, and wall thickness), which affect the speed of the traveling pressure disturbances, and (b) the length of the tube. It is thus the complex interaction between the fluid and the deformable tube wall that makes the problem difficult to solve. The above has been examined in multiple publications in the open literature in the last seventy years, some of which are mentioned here [3–12]. Nevertheless, there are still open questions that need to be answered regarding a more efficient design for this type of pump.

In micro impedance pumps, the whole flexible area moves in a predetermined way over a cavity. In order to increase the net flow rate, taper-type tube segments are normally introduced at the connection of the cavity with the microchannels so that the flow is turned into a unidirectional flow despite the harmonic motion in time of the membrane [13]. Essentially, the taper-tube segments cause an asymmetry to the hydraulic losses with respect to the membrane location so that although the fluid goes back and forth (induced by the membrane motion), its displacement is nonzero during each period. In [14], various configurations of piezoelectric impedance pumps are presented, and in [15], several flow rectifiers applied in micropumps—as active and passive valves—are discussed. In a recent effort by this group, a time-dependent passive valve was introduced in a valveless pump circuit, enhancing the net flow rate several times, especially at small amplitudes of the pump's pincher [16].

Another interesting impedance type of pump used in hemodynamics is the so-called intra-aortic balloon pump, which is a type of mechanical support for the failing heart that was introduced in clinical practice in the 1960s [17]. It consists of a balloon of cylindrical shape inserted into the human descending aorta, which inflates during diastole and deflates during systole, affecting the aortic pressure waveform and the blood flow rate. In other words, the cross-section of the free passage of blood in the descending aorta varies in time, as it is increased during systole and decreased during diastole. The basic parameters involved in this pump are the volume of the balloon and the timing of its inflation–deflation with respect to the aortic valve opening–closing. Numerical flow simulations of this pump, including the arterial system, have been presented in the literature based on lumped flow models with electrical circuit analogies, including resistance, inductance, and capacitance elements [18–20].

In the present numerical work, a quasi-one-dimensional flow model is employed in the study of a pumping system that consists of a straight tube of a predetermined cross-sectional area harmonically varying in time with or without a valve at one of its two ends, of a constant or time-varying opening. The tube is connected to two tanks of arbitrary cross-sections. The aim of this work is the investigation of the conditions for net flow rate generation in impedance pumps. Based on continuity and momentum equations, a first-order nonlinear differential equation of the flow rate is derived, which is presented and solved here for the first time, which is one of the novel features of this work. Most importantly, it is shown that its nonlinear term is responsible for net flow rate generation, being nonzero only if asymmetries in the hydraulic losses exist with respect to the mid-length of the tube. These losses are due to either abrupt changes in the fluid conduit cross-sectional area or, more importantly, the presence of a valve. Practically, the examined hydraulic system resembles an impedance pump in which the tube's walls or a membrane are oscillating in a known, deterministic way, not being affected by the fluid forces. In other words, the procedure followed does not take into account the case where the shape of the tube changes due to its interaction with the fluid, which occurs in some impedance pumps. Moreover, the case of a valve of a time-dependent opening installed in the system, besides its potential application in impedance pumps, is also examined in the context of an intra-aortic balloon pump for various heart rates and inflated diameters of the balloon, assuming that the valve plays the role of the aortic valve. The findings of this work could be used in the design of impedance pumps and intra-aortic balloon pumps.

## 2. Materials and Methods

This work numerically examines the flow between two open tanks, 1 and 2, connected by a straight tube of length $L$, with a cross-sectional area $A(t)$ harmonically varying in time with a cyclic frequency $\omega$, including, in the general case, a time-dependent constriction at the right end of the tube $A_{02}(t)$ (Figure 1).

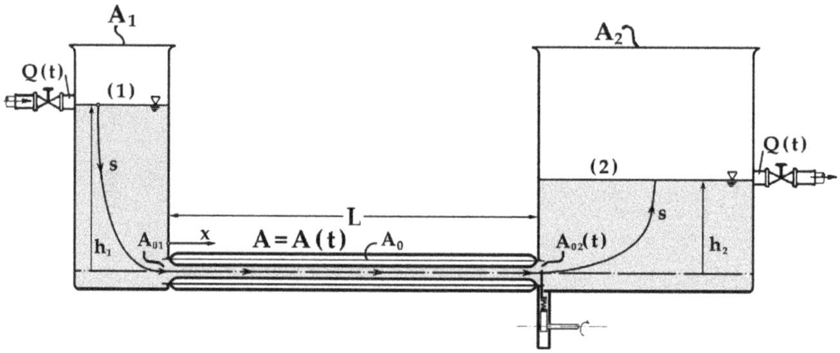


**Figure 1.** Hydraulic installation.

The cross-sectional areas of the tanks are denoted as $A_1$ and $A_2$, and their elevations are $h_1$ and $h_2$, respectively. The general expression of the continuity equation is:

$$\frac{\partial \rho}{\partial t} + div(\rho \vec{v}) = 0 \tag{1}$$

where $\rho$ is the fluid density and $\vec{v}$ is the fluid velocity vector. By employing a quasi-one-dimensional flow model (taking into account the radial displacement of the tube wall) and assuming that the flow is incompressible, after some manipulation, the above equation is written as follows:

$$\frac{\partial A}{\partial t} + \frac{\partial (A v_x)}{\partial x} = 0 \tag{2}$$

where $v_x$ is the axial (average in space) fluid velocity in a tube's cross-section.

By assuming that the flow is laminar, the momentum equation along an arbitrary direction s is given by:

$$\rho \frac{\partial v_s}{\partial t} + \frac{\rho}{2} \frac{\partial (v_s^2)}{\partial s} + \frac{\partial p}{\partial s} + \rho g \frac{dh}{ds} + 8\pi \rho \nu \frac{v_s}{A} = 0 \tag{3}$$

where $v_s$ is the fluid velocity along $s$, $p$ is the static pressure, $h$ is the elevation, and $\nu$ is the fluid's kinematic viscosity. In Equation (3), the fluid inertia is counterbalanced by the pressure, gravity, and viscosity forces. In particular, the last term of (3) is the known viscous term of the Poiseuille flow. More details about the derivation of Equation (3) can be found in [21].

In case a tube's cross-sectional area changes stepwise, like at the junctions of the tube with the tanks, the hydraulic losses depend on the ratio of the two areas and the direction of the flow. Namely, if the flow is directed from a smaller to a larger cross-section ('case 1'), the losses are higher (due to flow separation) than in the opposite direction ('case 2'). Based mainly on experimental observations, the hydraulic losses are expressed as a function of the dynamic pressure at the smaller cross-sectional area multiplied by a coefficient. The loss coefficient at tank 1's tube junction is designated by the Greek letter '$\xi$', whereas, at tank 2's tube junction, it is by the Greek letter '$\zeta$'. Each of the two coefficients has an index numbered '1' if the flow is from smaller to larger cross-sections and has '2' when the flow is in the opposite direction. Moreover, the expressions of the coefficients depend on whether the flow is directed from the tube to the tank or vice versa. The coefficients $\xi_1$ and $\xi_2$ are given by the following expressions (4)–(7) in which the cross-sectional area of tank 1 has been nondimensionalized by the tube time-averaged cross-sectional area $A_0$, namely $\widetilde{A}_1 = A_1/A_0$ [22]:

(a) If $A_1 > A_0$ and the flow direction is from the tube to tank 1 ('case 1'), the coefficient of the hydraulic losses $\xi_1$ is:

$$\xi_1 = \left(1 - \frac{1}{\widetilde{A}_1}\right)^2 \tag{4}$$

(b) If $A_1 > A_0$ and the flow direction is from tank 1 to the tube, namely towards a reduced area ('case 2'), the coefficient of the hydraulic losses $\xi_2$ is:

$$\xi_2 = 0.587 + \frac{0.395}{\sqrt{\widetilde{A}_1}} - \frac{4.538}{\widetilde{A}_1} + \frac{14.243}{\sqrt{\widetilde{A}_1^3}} - \frac{19.222}{\widetilde{A}_1^2} + \frac{8.54}{\sqrt{\widetilde{A}_1^5}} \tag{5}$$

(c) If $A_1 < A_0$ and the flow direction is from tank 1 to the tube ('case 1'), the coefficient of the hydraulic losses $\xi_1$ is:

$$\xi_1 = (1 - \widetilde{A}_1)^2 \tag{6}$$

(d) If $A_1 < A_0$ and the flow direction is from the tube to tank 1 ('case 2'), the coefficient of the hydraulic losses $\xi_2$ is:

$$\xi_2 = 0.587 + 0.395\sqrt{\widetilde{A}_1} - 4.538\widetilde{A}_1 + 14.243\sqrt{\widetilde{A}_1^3} - 19.222A_1^2 + 8.54\sqrt{\widetilde{A}_1^5} \tag{7}$$

Similarly, the hydraulic coefficient at the junction of the tube with tank 2, denoted by '$\zeta$' with an index of 1 or 2, is expressed in the same way as the coefficient $\xi$, in which $\widetilde{A}_2$ is used instead of $\widetilde{A}_1$.

A spatial integration of the momentum and continuity equations, taking into account the hydraulic losses at the two tank-tube junctions, ends up with the following first-order nonlinear differential equation with respect to the flow rate $Q$ at tank 1's tube junction:

$$B(t)\dot{Q} + C(t)Q + D(t)Q^2 + E(t) = 0 \tag{8}$$

where the coefficients $B(t)$, $C(t)$, $D(t)$, and $E(t)$ are:

$$B(t) = \frac{h_1}{A_1} + \frac{h_2}{A_2} + \frac{L}{A(t)}, \tag{9}$$

$$C(t) = 8\pi v\left(\frac{h_1}{A_1^2} + \frac{h_2}{A_2^2} + \frac{L}{[A(t)]^2}\right) - \left(\frac{\pm\zeta_{1,2}}{A_{02}^2} + \frac{1}{A_2^2}\right)\frac{dV(L,t)}{dt} + \frac{\partial}{\partial t}\left(\frac{L}{A(t)}\right), \tag{10}$$

$$D(t) = \frac{1}{2}\left(\frac{1}{A_2^2} - \frac{1}{A_1^2} \pm \frac{\xi_{1,2}}{A_{01}^2} \pm \frac{\zeta_{1,2}}{A_{02}^2}\right), \tag{11}$$

$$\begin{aligned} E(t) = {} & g(h_2 - h_1) - \frac{h_2}{A_2}\frac{d^2V(L,t)}{dt^2} + \frac{1}{2}\left(\frac{\pm\zeta_{1,2}}{A_{02}^2} + \frac{1}{A_2^2}\right)\left[\frac{dV(L,t)}{dt}\right]^2 - 8\pi v\frac{h_2}{A_2^2}\frac{dV(L,t)}{dt} \\ & - \frac{L}{A(t)}\frac{d^2V(L,t)}{dt^2} - \frac{\partial}{\partial t}\left(\frac{L}{A(t)}\right)\frac{dV(L,t)}{dt} - 8\pi v\frac{L}{[A(t)]^2}\frac{dV(L,t)}{dt}, \end{aligned} \tag{12}$$

In the above expressions, $A_{01}$ and $A_{02}$ refer to the cross-sectional areas at the tube-tank junctions, and $V(L,t)$ is the time-dependent volume of the tube. In the expressions (10) to (12), the positive sign in the coefficients $\xi_i$ and $\zeta_i$ is used when the flow direction is from tank 1 to tank 2. Briefly, Equation (8) was derived according to the following procedure: a spatial integration of the momentum

Equation (3) was applied in three cases, namely between the free surface of tank 1 and its junction with the tube, between the free surface of tank 2 and its junction at the other end of the tube, and finally, between the two ends of the tube. Moreover, based on spatial integration of the continuity equation, an expression of the fluid speed along the tube was derived, which is a function of the tube's cross-sectional area and the flow rate at one end of the tube. More details regarding the derivation of the above expressions are in [23].

In order to generalize the results, the following non-dimensional numbers are used,

$$\widetilde{t} = t\omega,\ \widetilde{A}_i = A_i/A_0,\ \widetilde{Q} = Q/(A_0\omega L)\ and\ W = R_0\sqrt{\omega/\nu}, \tag{13}$$

where $A_i$ refers to $A_{01}$, $A_{02}$, and $A_b$; $R_0$ is the tube's internal radius corresponding to the tube's undeformed cross-sectional area $A_0$; and $\omega$ is the cyclic frequency of the tube's oscillating area $A$ varying with an amplitude $A_b$, namely

$$\widetilde{A} = 1 + \widetilde{A}_b cos\widetilde{t} \tag{14}$$

The non-dimensional number $W$ in (13) is the known Womersley number. If the tank elevations are $h_1 = h_2 = 0$, the coefficients $B(t)$ to $E(t)$ are simplified and are given in non-dimensional form in (15), in which $V(L,t)$ being the total tube internal volume has been replaced by the product $A(t)\cdot L$:

$$\widetilde{B}(\widetilde{t}) = \frac{1}{\widetilde{A}},$$

$$\widetilde{C}(\widetilde{t}) = \frac{8}{W^2}\frac{1}{\widetilde{A}^2} - \left(\frac{\pm\zeta_{1,2}}{\widetilde{A}_{02}^2} + \frac{1}{\widetilde{A}_2^2} + \frac{1}{\widetilde{A}^2}\right)\frac{d\widetilde{A}}{d\widetilde{t}},$$

$$\widetilde{D}(\widetilde{t}) = \frac{1}{2}\left(\frac{1}{\widetilde{A}_2^2} - \frac{1}{\widetilde{A}_1^2} \pm \frac{\xi_{1,2}}{\widetilde{A}_{01}^2} \pm \frac{\zeta_{1,2}}{\widetilde{A}_{02}^2}\right),$$

$$\widetilde{E}(\widetilde{t}) = -\frac{1}{2\widetilde{A}}\frac{d^2\widetilde{A}}{d\widetilde{t}^2} + \frac{1}{2}\left(\frac{\pm\zeta_{1,2}}{\widetilde{A}_{02}^2} + \frac{1}{\widetilde{A}_2^2} + \frac{1}{\widetilde{A}^2}\right)\left(\frac{d\widetilde{A}}{d\widetilde{t}}\right)^2 - \frac{8}{W^2}\frac{1}{2\widetilde{A}^2}\frac{d\widetilde{A}}{d\widetilde{t}}, \tag{15}$$

The nonlinear differential Equation (8), in its non-dimensional form, is solved numerically with respect to $\widetilde{Q}$ by using a Runge–Kutta of fourth order, as it is explained in Appendix A, with a non-dimensional time interval of $10^{-3}$ to secure a small truncation error (of the order of $10^{-12}$). It has to be noted that due to the complex coefficients of Equation (8), which is the known Riccati differential equation, no analytical solution is available. By knowing $\widetilde{Q}(\widetilde{t})$, the net flow rate was calculated based on three periods of oscillation after the solution reached a steady state. Since the hydraulic loss coefficients at the tank-tube junctions $\xi_{1,2}$ and $\zeta_{1,2}$ take relatively low values, namely between 0 and 1, it was decided to include a gate valve in the hydraulic circuit, installed at the tank 2-tube junction, aiming to enhance the asymmetries in the hydraulic installation by increasing the flow resistance at one of the tube ends. Moreover, by including the gate valve, it was also possible to simulate the operation of an intra-aortic balloon pump by assuming that the valve plays the role of the aortic valve. According to Figure 2, the loss coefficient $\zeta$ of a gate valve takes a value as high as 400 when its opening $h$ is 6.6% of the tube diameter $d$ or 8.4% of the tube cross-sectional area [24].

Due to the induced flow asymmetries, the gate valve can further increase the net flow rate, as it is shown below in the Section 3. In order to estimate the valve's hydraulic losses, the spatially average fluid velocity at the gate opening is calculated based on the local flow rate and its area $A_{02}$, which is given by:

$$A_{02} = \left[\pi - 2\left(cos^{-1}\mu - \mu\sqrt{1-\mu^2}\right)\right]\frac{d^2}{4},\ \ 0 < \mu < 1 \tag{16}$$

where $\mu = h/d$ (see Figure 2). The derivation of (16) was based on simple geometrical formulas by computing the area between two equal circles (note that the gate valve seat area equals the pipe area) of diameter $d$, the centers of which are at a distance $h$ apart (see Figure 2).

Another option was also examined, namely when the gate valve opening varies in time, which could be used in an impedance pump to control its flow rate or to simulate the opening–closing of the aortic valve in the study of an intra-aortic balloon pump. In this case, the following expression of $A_{02}$ was arbitrarily used:

$$\widetilde{A}_{02} = \widetilde{A}_{02,min} + \left(1 - \widetilde{A}_{02,min}\right)cos^{40}\left[sin^4(\widetilde{t}/2)\right], \tag{17}$$

where $\widetilde{A}_{02,min}$ is the minimum dimensionless area of the gate valve opening. Expression (17) was devised based on the fact that the aortic valve opens and closes in an almost stepwise manner. By knowing the area $A_{02}$ at each time instant from Equation (17), the ratio $\mu$ was calculated from (16), and consequently, the hydraulic loss coefficient $\zeta$ of the valve was determined from the graph of Figure 2.

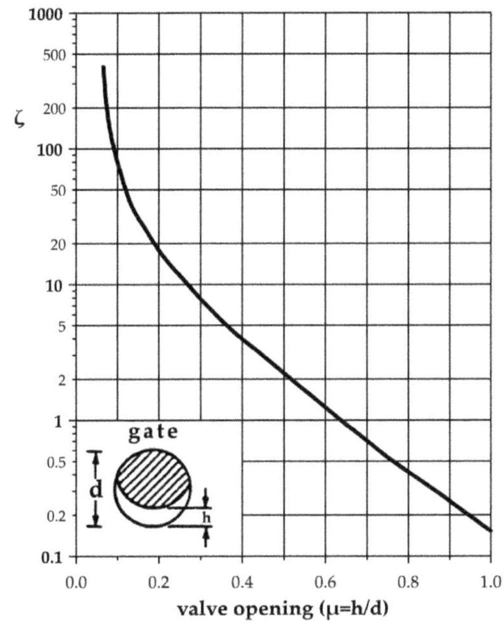


**Figure 2.** Gate valve loss coefficient [24] (p. 273).

In the next paragraph, the net flow rate is examined in the following three cases: (a) Two tanks at zero height are connected by a straight horizontal tube of uniformly oscillating walls, with no valve present. In this case, the net flow rate is calculated for various ratios of the tanks' cross-sectional areas $\widetilde{A}_1/\widetilde{A}_2$ for a certain Womersley number, as well as for a range of Womersley numbers and given ratios $\widetilde{A}_1/\widetilde{A}_2$. (b) Aside from the oscillating tube, the influence of a gate valve is also studied, where it is installed at tube one's end, the opening of which is either constant or time-dependent. The time-dependent opening of the valve is shown to be far more efficient in net flow rate generation compared with the constant opening. (c) The flow rate of an intra-aortic balloon pump is examined for various heart rates and balloon-inflated diameters (by varying the $A_b$ parameter), considering that the valve plays the role of a leaky aortic valve.

## 3. Results and Discussion

### 3.1. No Valve Is Used

If the gate valve is absent and the two tanks have equal cross-sectional areas ($A_1 = A_2$), the coefficient $D(t)$ of the nonlinear term of Equation (8) becomes zero, as well as the net flow rate. Since the tube area varies harmonically in time and there are no flow asymmetries with respect to the mid-length of the tube, the fluid moves back and forth without being transferred from one tank to the other. However, when $A_1$ differs from $A_2$, the hydraulic loss coefficients $\xi_{1,2}$ and $\zeta_{1,2}$ are different, inducing flow asymmetries so that the flow becomes, on average, unidirectional, directed from the smaller to the larger tank. Figure 3 shows the net flow rate as a function of the ratio of the cross-sectional areas of the tanks $\widetilde{A}_1/\widetilde{A}_2$ for various values of $\widetilde{A}_2$, by keeping constant the frequency of the tube oscillation ($W = 5$) with a 10% tube area amplitude ($\widetilde{A}_b = 0.1$). It is known that for small $W$ values (e.g., $W < 1$), the flow in a tube is quasi-steady. However, by progressively increasing $W$, the unsteadiness of the flow becomes dominant, changing the parabolic velocity profile to a flatter one [25]. In the present case, the spatial change in the velocity in each cross-section is not taken into account since the flow is assumed to be one-dimensional, and a spatially average velocity is considered in each tube cross-section.

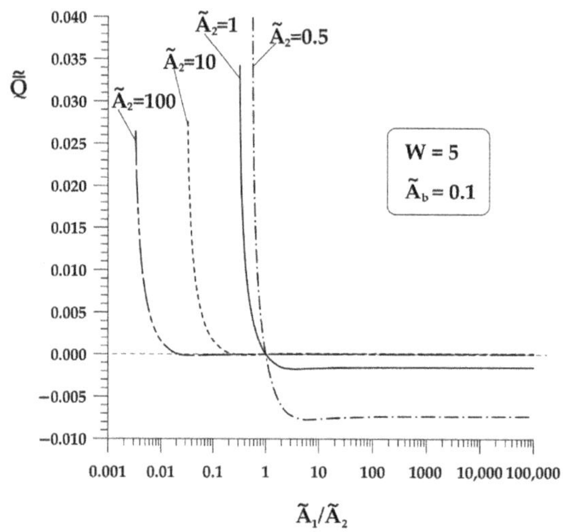


**Figure 3.** Net flow rate versus tank cross-sectional areas. $W = 5$, $\widetilde{A}_b = 0.1$.

All curves in Figure 3 go through the point $\widetilde{A}_1/\widetilde{A}_2 = 1$ and zero net flow rate, which is the case of no flow asymmetries in the system. In contrast, the net flow rate increases exponentially when one of the two tanks' cross-sectional areas takes smaller values than that of the tube; for example, in the case $A_1 = 0.326\ A_0$ for which the non-dimensional net flow rate takes a value of 0.028 (see Figure 3, $\widetilde{A}_2 = 1$ and $\widetilde{A}_1/\widetilde{A}_2 = 0.326$). However, when both tanks have larger cross-sections than that of the tube, the net flow rate is again zero. This finding could be applied to impedance micro pumps, in which the size of the cavity where the membrane oscillates has to be larger than at least one of the micro-tubes that is connected to the cavity.

When the frequency of the tube walls' oscillation increases, the net flow rate also increases. Figure 4 shows the variations in the net flow rate with the Womersley number when one or both tanks have smaller cross-sections than the tube's cross-section. Moreover, the net flow rate takes nonzero values when the Womersley number exceeds 0.5, and it increases exponentially for $W > 5$. Since W is inversely proportionate to the square root of the fluid kinematic viscosity, the net flow rate of water will be higher than that of blood and much higher than of air. The opposite signs of the two curves of Figure 4 denote opposite flow directions, with the positive to be directed from tank 1 to tank 2. Therefore, like in Figure 3, the flow is directed from the smaller to the larger tank.

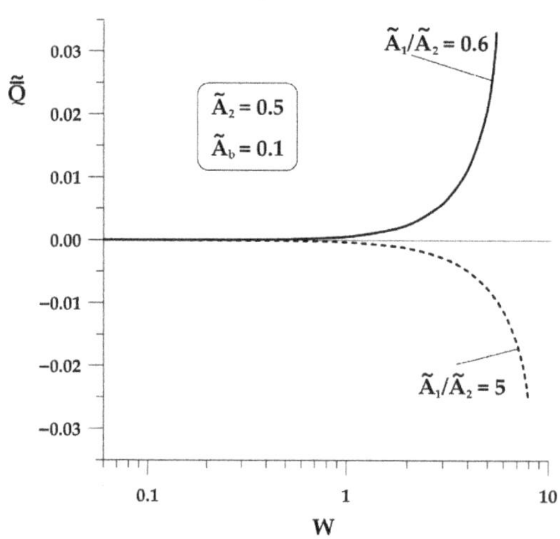


**Figure 4.** Net flow rate versus Womersley number.

### 3.2. A Gate Valve Is Used

#### 3.2.1. Constant Valve Opening

When the gate valve is introduced at the connection of the tube with tank 2, the pumping effect of the oscillating tube is enhanced, so if the tanks have equal cross-sectional areas, a net flow rate is established. As it is shown in Figure 5 for equal tanks, $\widetilde{A}_b = 0.1$ and $W = 5$, the flow is directed towards tank 2 and maximizes when the gate valve loss coefficient is $\zeta = 6.5$, corresponding to a valve opening of 37.6% of the mean tube cross-section. Consequently, in an impedance pump, the use of a valve in the hydraulic system could boost the net flow rate when the opening of the valve is suitable. In practice, instead of valves, taper tubes are used as flow rectifiers, with hydraulic losses strongly dependent on their geometry (width-taper angle) and Reynolds number. According to [26], the performance of such devices drops or even vanishes at $Re$ below 50, which seems to be remedied by using taper tubes with flexible covers so that, in this case, they work efficiently, even for $Re < 10$ [27].

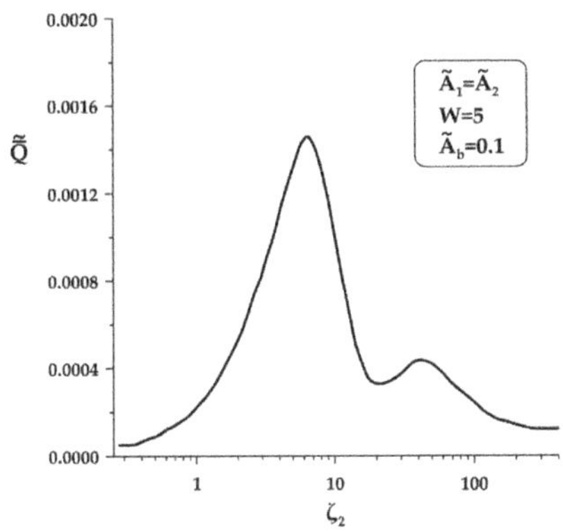


**Figure 5.** Net flow rate versus valve loss coefficient. Valve opening constant.

#### 3.2.2. Time-Dependent Valve Opening

In this paragraph, the opening of the valve is considered time-dependent. The valve can be part of the hydraulic loop of an impedance pump as a means to control its flow rate, or in the case of an intra-aortic balloon pump, it can simulate the operation of an aortic valve. Among several scenarios regarding the timing between the valve opening and the tube cross-sectional area variation, the case of the balloon pump was followed. Namely, the opening of the valve coincides with an increase in the tube cross-sectional area since, at the same time, the balloon shrinks. Figure 6 shows the variations in the tube cross-sectional area $\widetilde{A}(\widetilde{t})$ along with that of the valve opening $\widetilde{A}_{02}$, following Equation (17), with $A_{02,min}$ being equal to 8.4 % of $A_0$ (corresponding to $\zeta \approx 400$) and $\widetilde{A}_b = 0.1$.

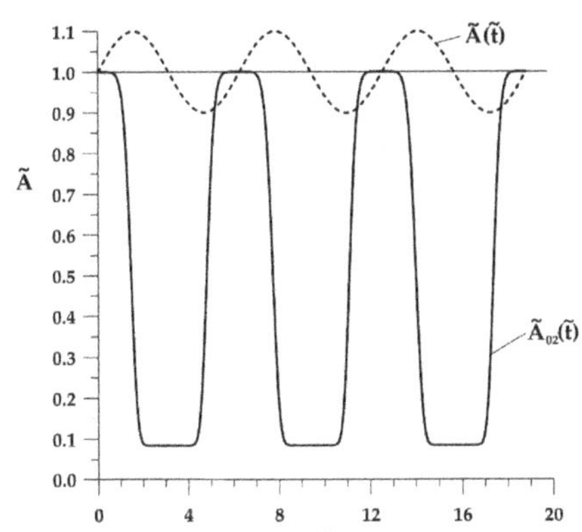


**Figure 6.** Tube (dashed line) and gate valve (continuous line) cross-sectional areas versus non-dimensional time. $\widetilde{A}_{02,min} = 0.084$.

By varying the minimum area of the valve and keeping the frequency of the tube and valve oscillation constant ($W = 5$), the graph in Figure 7 was obtained for equal tank cross-sections. It is clear from Figure 7 that the net flow rate is about 10 times larger compared to the case where the valve opening is constant, and the maximum value of the net flow rate occurs for $\widetilde{A}_{02,\min} = 25.3\%$. Moreover, it is interesting to note that the maximization of the net flow rate does not require that the valve closes completely. A similar phenomenon is detected in [16], in which an unsteady stenosis increased the net flow rate by 19 times.

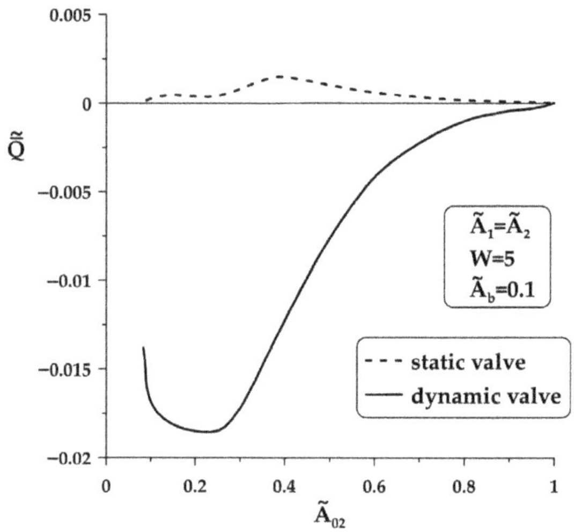


**Figure 7.** Net flow rate versus valve minimum cross-sectional area. Static and dynamic cases.

By increasing the frequency of the straight tube pulsation and its cross-sectional area amplitude $\widetilde{A}_b$, the non-dimensional net flow rate increases up to a Womersley value of 5 and then it remains constant (see Figure 8). The fact that the non-dimensional net flow rate is independent of $W$ when $W$ takes high values is attributed to the $W^{-2}$ terms in Equation (8), which tend toward zero when $W$ goes to infinity. Therefore, the net flow rate $\overline{Q}$ at high frequencies increases linearly with the frequency, according to the definition of the non-dimensional flow rate $\widetilde{Q} = Q/(A_0\omega L)$, namely, $\overline{Q} = A_0\omega L\widetilde{\overline{Q}}$.

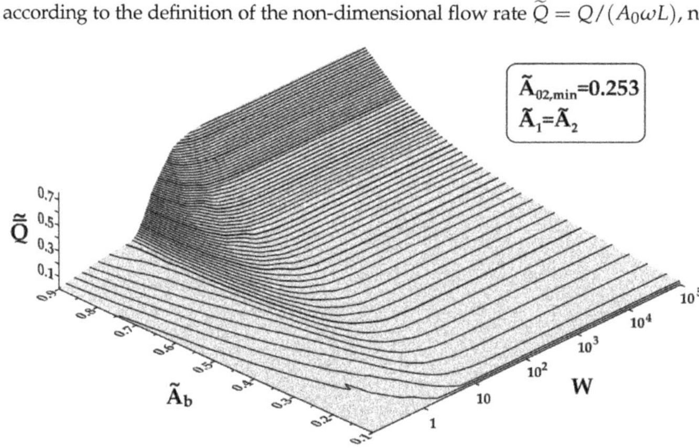


**Figure 8.** Dimensionless net flow rate versus $\widetilde{A}_b$ and $W$.

### 3.2.3. Intra-Aortic Balloon Pump

For the case of the aortic balloon pump, the following assumptions were made: blood is the flowing medium of hematocrit 40 so that its kinematic viscosity is 4.0072 cSt [28]; the heart rate varies from 70 to 120 beats/min, or the cyclic frequency $\omega$ changes from 7.3 up to 12.5 rad/s; the aorta internal diameter is 20 mm; the two tanks have the same diameter as that of the aorta; the length of the balloon is 0.26 m, based on [29]; $A_b$ varies from 6 to 24.5% of $A_0$; and the minimum opening of the valve is 8.4% of $A_0$. It should be noted that in balloon pumps, the inflated balloon's diameter normally does not exceed 80% of the descending thoracic aorta diameter [30], or the $A_b$ maximum value is close to 30% of $A_0$. Moreover, the present model simulates cases of leaky aortic valve leaflets, where the area of the valve that remains open, according to [31], is 25% of the cross-sectional area of the left ventricular outflow tract (for mild cases).

Figure 9 shows the net flow rate iso-contours as a function of both the inflation of the balloon—which is related to $A_b$—and the heart rate. According to Figure 9, the generated net flow rate takes a maximum value close to 1 L/min for a frequency of 2 Hz and $A_b$ = 24% of $A_0$. Nevertheless, it has to be remembered that in real applications, the arterial tree plays a significant role in the pump's performance, such as its compliance, which has been shown to reduce the balloon efficiency when this increases [32]. Of course, the procedure, although relatively simplistic, is able to reproduce the basic features of the pump.

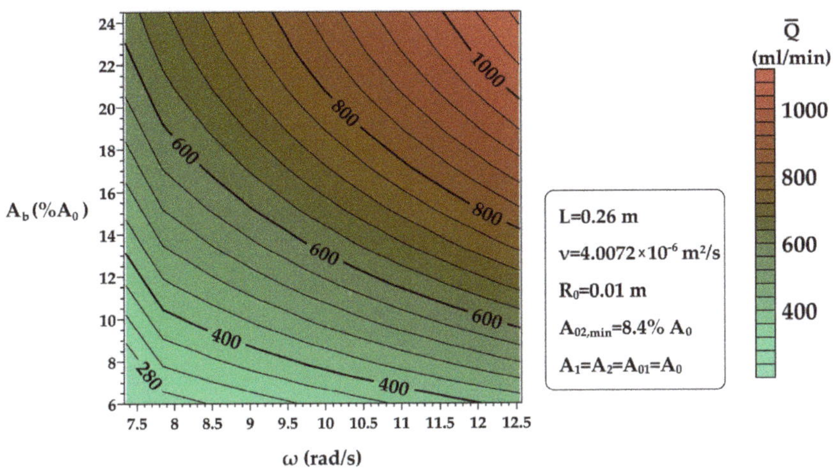


**Figure 9.** Net flow rate versus frequency and tube oscillation amplitude.

## 4. Conclusions

This work numerically examines, by employing a quasi-one-dimensional flow model, the conditions for which the net flow rate is generated by the harmonic variations in time of the cross-sectional area of a tube connecting two tanks. A first-order nonlinear differential equation of the flow rate was derived, showing that its nonlinear term is responsible for the net flow rate generation, being associated with asymmetries in the hydraulic losses with respect to the mid-length of the tube. The net flow rate is nonzero if the tank cross-sectional areas are not equal; it takes higher values if at least one of the tanks has a smaller area than that of the tube, is directed from the smaller to the larger tank, and increases with the frequency of the tube oscillation. When the tanks' cross-sections are equal, the net flow is null. In this case, if a valve is installed, the flow becomes unidirectional, and it is further increased if the valve opening is time-dependent, maximizing for a particular minimum valve opening. By simulating the operation of an intra-aortic balloon pump so that the valve opening coincides with the shrinkage of the tube, the net flow rate increases with the heart rate and the inflation of the balloon.

**Author Contributions:** Conceptualization, S.T. and C.M.; methodology, S.T. and C.M.; software, C.M.; validation, C.M.; formal analysis, C.M.; investigation, C.M. and D.M.; resources, C.M., C.G. and D.M.; data curation, C.M.; writing—original draft preparation, C.M. and D.M.; writing—review and editing, C.M., S.T., C.G. and D.M.; visualization, C.M.; supervision, S.T.; project administration, C.M. All authors have read and agreed to the published version of the manuscript.

**Funding:** This research received no funding.

**Data Availability Statement:** Data are contained within the article.

**Conflicts of Interest:** The authors declare no conflicts of interest.

## Appendix A

*Numerical Method for Solving the Nonlinear Ordinary Differential Equation*

To solve the ordinary differential Equation (8), a FORTRAN program implementing the classical fourth-order Runge–Kutta method is used as a single-step time marching scheme. In this method, the time derivative of the unknown dimensionless flow rate function is isolated as follows:

$$\frac{d\widetilde{Q}}{d\widetilde{t}} = \dot{\widetilde{Q}} = -\frac{\widetilde{C}(\widetilde{t})\cdot\widetilde{Q} + \widetilde{D}(\widetilde{t})\cdot\widetilde{Q}^2 + \widetilde{E}(\widetilde{t})}{\widetilde{B}(\widetilde{t})} \tag{A1}$$

Next, the following coefficients of each order are calculated for each dimensionless time interval $\Delta\widetilde{t}$ [33].

$$k_1 = \dot{\widetilde{Q}}(\widetilde{t}_i, \widetilde{Q}_i) = -\frac{\widetilde{C}(\widetilde{t}_i)\cdot\widetilde{Q}_i + \widetilde{D}(\widetilde{t}_i)\cdot\widetilde{Q}_i^2 + \widetilde{E}(\widetilde{t}_i)}{\widetilde{B}(\widetilde{t}_i)} \tag{A2}$$

$$\begin{gathered} k_2 = \dot{\widetilde{Q}}\left(\widetilde{t}_i + \Delta\widetilde{t}/2, \widetilde{Q}_i + \Delta\widetilde{t}\cdot k_1/2\right) \\ = -\frac{\widetilde{C}(\widetilde{t}_i + \Delta\widetilde{t}/2)\cdot\left(\widetilde{Q}_i + \Delta\widetilde{t}\cdot k_1/2\right) + \widetilde{D}(\widetilde{t}_i + \Delta\widetilde{t}/2)\cdot\left(\widetilde{Q}_i + \Delta\widetilde{t}\cdot k_1/2\right)^2 + \widetilde{E}(\widetilde{t}_i + \Delta\widetilde{t}/2)}{\widetilde{B}(\widetilde{t}_i + \Delta\widetilde{t}/2)} \end{gathered} \tag{A3}$$

$$\begin{gathered} k_3 = \dot{\widetilde{Q}}\left(\widetilde{t}_i + \Delta\widetilde{t}/2, \widetilde{Q}_i + \Delta\widetilde{t}\cdot k_2/2\right) \\ = -\frac{\widetilde{C}(\widetilde{t}_i + \Delta\widetilde{t}/2)\cdot\left(\widetilde{Q}_i + \Delta\widetilde{t}\cdot k_2/2\right) + \widetilde{D}(\widetilde{t}_i + \Delta\widetilde{t}/2)\cdot\left(\widetilde{Q}_i + \Delta\widetilde{t}\cdot k_2/2\right)^2 + \widetilde{E}(\widetilde{t}_i + \Delta\widetilde{t}/2)}{\widetilde{B}(\widetilde{t}_i + \Delta\widetilde{t}/2)} \end{gathered} \tag{A4}$$

$$\begin{gathered} k_4 = \dot{\widetilde{Q}}\left(\widetilde{t}_i + \Delta\widetilde{t}, \widetilde{Q}_i + \Delta\widetilde{t}\cdot k_3\right) \\ = -\frac{\widetilde{C}(\widetilde{t}_i + \Delta\widetilde{t})\cdot\left(\widetilde{Q}_i + \Delta\widetilde{t}\cdot k_3\right) + \widetilde{D}(\widetilde{t}_i + \Delta\widetilde{t})\cdot\left(\widetilde{Q}_i + \Delta\widetilde{t}\cdot k_3\right)^2 + \widetilde{E}(\widetilde{t}_i + \Delta\widetilde{t})}{\widetilde{B}(\widetilde{t}_i + \Delta\widetilde{t})} \end{gathered} \tag{A5}$$

Each of the $k_i$ values above represents the time slope of the unknown function $\widetilde{Q}$ at dimensionless time $\widetilde{t}$.

Finally, the next value of $\widetilde{Q}$ in the time sequence is found using the following Equation (A6), which represents a weighted average of the $k_i$ slopes:

$$\begin{gathered} \frac{\widetilde{Q}_{i+1} - \widetilde{Q}_i}{\Delta\widetilde{t}} = \frac{1}{6}(k_1 + 2k_2 + 2k_3 + k_4) \rightarrow \\ \widetilde{Q}_{i+1} = \widetilde{Q}_i + \left[\frac{1}{6}(k_1 + 2k_2 + 2k_3 + k_4)\right]\cdot\Delta\widetilde{t} \end{gathered} \tag{A6}$$

## References


1. Wang, Y.-N.; Fu, L.-M. Micropumps and Biomedical Applications—A Review. *Microelectron. Eng.* **2018**, *195*, 121–138. [CrossRef]
2. Manopoulos, C.G.; Mathioulakis, D.S.; Tsangaris, S.G. One-Dimensional Model of Valveless Pumping in a Closed Loop and a Numerical Solution. *Phys. Fluids* **2006**, *18*, 017106. [CrossRef]
3. Thomann, H. A Simple Pumping Mechanism in a Valveless Tube. *Z. Für Angew. Math. Phys. ZAMP* **1978**, *29*, 169–177. [CrossRef]
4. Jung, E.; Peskin, C.S. Two-Dimensional Simulations of Valveless Pumping Using the Immersed Boundary Method. *SIAM J. Sci. Comput.* **2001**, *23*, 19–45. [CrossRef]
5. Ottesen, J.T. Valveless Pumping in a Fluid-Filled Closed Elastic Tube-System: One-Dimensional Theory with Experimental Validation. *J. Math. Biol.* **2003**, *46*, 309–332. [CrossRef]
6. Hickerson, A.I.; Rinderknecht, D.; Gharib, M. Experimental Study of the Behavior of a Valveless Impedance Pump. *Exp. Fluids* **2005**, *38*, 534–540. [CrossRef]
7. Avrahami, I.; Gharib, M. Computational Studies of Resonance Wave Pumping in Compliant Tubes. *J. Fluid Mech.* **2008**, *608*, 139–160. [CrossRef]
8. Timmermann, S.; Ottesen, J.T. Novel Characteristics of Valveless Pumping. *Phys. Fluids* **2009**, *21*, 053601. [CrossRef]
9. Shin, S.J.; Chang, C.B.; Sung, H.J. Simulation of a Valveless Pump with an Elastic Tube. *Int. J. Heat Fluid Flow* **2012**, *38*, 13–23. [CrossRef]
10. Lee, W.; Lim, S.; Jung, E. Dynamical Motion Driven by Periodic Forcing on an Open Elastic Tube in Fluid. *Commun. Comput. Phys.* **2012**, *12*, 494–514. [CrossRef]
11. Kozlovsky, P.; Rosenfeld, M.; Jaffa, A.J.; Elad, D. Dimensionless Analysis of Valveless Pumping in a Thick-Wall Elastic Tube: Application to the Tubular Embryonic Heart. *J. Biomech.* **2015**, *48*, 1652–1661. [CrossRef] [PubMed]
12. Loumes, L.; Avrahami, I.; Gharib, M. Resonant Pumping in a Multilayer Impedance Pump. *Phys. Fluids* **2008**, *20*, 023103. [CrossRef]

13. Yan, Q.; Yin, Y.; Sun, W.; Fu, J. Advances in Valveless Piezoelectric Pumps. *Appl. Sci.* **2021**, *11*, 7061. [CrossRef]
14. Hou, Y.; He, L.; Hu, D.; Zhang, L.; Yu, B.; Cheng, G. Recent Trends in Structures and Applications of Valveless Piezoelectric Pump—A Review. *J. Micromech. Microeng.* **2022**, *32*, 053002. [CrossRef]
15. Laser, D.J.; Santiago, J.G. A Review of Micropumps. *J. Micromech. Microeng.* **2004**, *14*, R35–R64. [CrossRef]
16. Manopoulos, C.; Mathioulakis, D. Valveless Pumping with an Unsteady Stenosis in an Open Tank Configuration. *Fluids* **2024**, *9*, 141. [CrossRef]
17. Parissis, H.; Graham, V.; Lampridis, S.; Lau, M.; Hooks, G.; Mhandu, P.C. IABP: History-Evolution-Pathophysiology-Indications: What We Need to Know. *J. Cardiothorac. Surg.* **2016**, *11*, 122. [CrossRef]
18. Abdolrazaghi, M.; Navidbakhsh, M.; Hassani, K. Mathematical Modelling of Intra-Aortic Balloon Pump. *Comput. Methods Biomech. Biomed. Engin.* **2010**, *13*, 567–576. [CrossRef]
19. Ferrari, G.; Khir, A.W.; Fresiello, L.; Di Molfetta, A.; Kozarski, M. Hybrid Model Analysis of Intra-Aortic Balloon Pump Performance as a Function of Ventricular and Circulatory Parameters. *Artif. Organs* **2011**, *35*, 902–911. [CrossRef]
20. De Lazzari, C.; De Lazzari, B.; Iacovoni, A.; Marconi, S.; Papa, S.; Capoccia, M.; Badagliacca, R.; Vizza, C.D. Intra-Aortic Balloon Counterpulsation Timing: A New Numerical Model for Programming and Training in the Clinical Environment. *Comput. Methods Programs Biomed.* **2020**, *194*, 105537. [CrossRef]
21. Schaaf, B.W.; Abbrecht, P.H. Digital Computer Simulation of Human Systemic Arterial Pulse Wave Transmission: A Nonlinear Model. *J. Biomech.* **1972**, *5*, 345–364. [CrossRef] [PubMed]
22. Benedict, R.P.; Carlucci, N.A.; Swetz, S.D. Flow Losses in Abrupt Enlargements and Contractions. *J. Eng. Power* **1966**, *88*, 73–81. [CrossRef]
23. Manopoulos, C. Experimental and Theoretical Determination of "Peristaltic Blood Pumps". Master's Thesis, National Technical University of Athens, Athens, Greece, 1999. Available online: https://dspace.lib.ntua.gr/xmlui/handle/123456789/49369?locale-attribute=en (accessed on 1 November 2019).
24. Miller, D.S. *Internal Flow Systems*; BHRA fluid engineering series; BHRA Fluid Engineering: Cranfield, UK, 1978.
25. Zamir, M. *The Physics of Coronary Blood Flow*; Biological and medical physics, biomedical engineering; Springer: New York, NY, USA, 2005.
26. Tao, R.; Ng, T.; Su, Y.; Li, Z. A Microfluidic Rectifier for Newtonian Fluids Using Asymmetric Converging–Diverging Microchannels. *Phys. Fluids* **2020**, *32*, 052010. [CrossRef]
27. Mehboudi, A.; Yeom, J. A Passive Stokes Flow Rectifier for Newtonian Fluids. *Sci. Rep.* **2021**, *11*, 10182. [CrossRef] [PubMed]
28. Hall, J.E.; Hall, J.E.; Guyton, A.C. *Guyton and Hall Textbook of Medical Physiology*, 12th ed.; Saunders: Philadelphia, PA, USA; Elsevier: Philadelphia, PA, USA, 2011.
29. Parissis, H.; Soo, A.; Leotsinidis, M.; Dougenis, D. A Statistical Model That Predicts the Length from the Left Subclavian Artery to the Celiac Axis; towards Accurate Intra-Aortic Balloon Sizing. *J. Cardiothorac. Surg.* **2011**, *6*, 95. [CrossRef] [PubMed]
30. Krishna, M.; Zacharowski, K. Principles of Intra-Aortic Balloon Pump Counterpulsation. *Contin. Educ. Anaesth. Crit. Care Pain* **2009**, *9*, 24–28. [CrossRef]
31. Zoghbi, W. Recommendations for Evaluation of the Severity of Native Valvular Regurgitation with Two-Dimensional and Doppler Echocardiography. *J. Am. Soc. Echocardiogr.* **2003**, *16*, 777–802. [CrossRef]
32. Papaioannou, T.G.; Mathioulakis, D.S.; Nanas, J.N.; Tsangaris, S.G.; Stamatelopoulos, S.F.; Moulopoulos, S.D. Arterial Compliance Is a Main Variable Determining the Effectiveness of Intra-Aortic Balloon Counterpulsation: Quantitative Data from an in Vitro Study. *Med. Eng. Phys.* **2002**, *24*, 279–284. [CrossRef]
33. Chapra, S.C.; Canale, R.P. *Numerical Methods for Engineers*, 7th ed.; McGraw-Hill Education: New York, NY, USA, 2015.

*vibration*

MDPI

*Article*

# Tunable High-Static-Low-Dynamic Stiffness Isolator under Harmonic and Seismic Loads

**Giovanni Iarriccio [1,*], Antonio Zippo [1], Fatemeh Eskandary-Malayery [2], Sinniah Ilanko [2], Yusuke Mochida [2], Brian Mace [3] and Francesco Pellicano [1]**

[1] Department of Engineering "Enzo Ferrari", University of Modena and Reggio Emilia, Via Vivarelli 10, 41125 Modena, Italy; antonio.zippo@unimore.it (A.Z.); francesco.pellicano@unimore.it (F.P.)

[2] School of Engineering, University of Waikato, Hamilton 3216, New Zealand; fe14@students.waikato.ac.nz (F.E.-M.); ilanko@waikato.ac.nz (S.I.); yusuke@waikato.ac.nz (Y.M.)

[3] Department of Mechanical and Mechatronics Engineering, University of Auckland, Auckland 1030, New Zealand; b.mace@auckland.ac.nz

* Correspondence: giovanni.iarriccio@unimore.it

**Abstract:** High-Static-Low-Dynamic Stiffness (HSLDS) mechanisms exploit nonlinear kinematics to improve the effectiveness of isolators, preserving controlled static deflections while maintaining low natural frequencies. Although extensively studied under harmonic base excitation, there are still few applications considering real seismic signals and little experimental evidence of real-world performance. This study experimentally demonstrates the beneficial effects of HSLDS isolators over linear ones in reducing the vibrations transmitted to the suspended mass under near-fault earthquakes. A tripod mechanism isolator is presented, and a lumped parameter model is formulated considering a piecewise nonlinear–linear stiffness, with dissipation taken into account through viscous and dry friction forces. Experimental shake table tests are conducted considering harmonic base motion to evaluate the isolator transmissibility in the vertical direction. Excellent agreement is observed when comparing the model to the experimental measurements. Finally, the behavior of the isolator is investigated under earthquake inputs, and results are presented using vertical acceleration time histories and spectra, demonstrating the vibration reduction provided by the nonlinear isolator.

**Keywords:** high-static-low-dynamic stiffness; vibration isolators; experiments; transmissibility; earthquakes

## 1. Introduction

Each year, over a million seismic events and the resulting large number of fatalities pose a challenge to the human population [1]. The development of new building technologies and early warning and prevention systems can counter the disastrous effects of earthquakes. In this framework, exploiting the inherent nonlinearities of seismic isolation systems represents a possible approach to developing high-performance devices for preventing buildings from undergoing disastrous damage during seismic events.

Vibration isolation systems can be classified as active or passive based on the presence or absence of integrated actuators for real-time operation. However, during seismic events, sudden power supply outages may occur, making active vibration control systems unreliable. Passive vibration isolators, functioning without external power or control, provide a cost-effective, robust, and reliable means of isolating buildings from ground motion [2].

Concerning the transmissibility of a linear suspension under harmonic base excitation, a standard mass-spring system allows for vibration isolation (i.e., absolute transmissibility amplitude $T < 1$) when the forcing frequency exceeds $\sqrt{2km^{-1}}$, where $k$ and $m$ denote the stiffness and mass of the system, respectively [3]. Thus, a low stiffness is desired to increase the working frequency range, but it results in larger static deflection, which may be impractical in some applications. To circumvent this issue, quasi-zero stiffness

(QZS) mechanisms, also known as zero spring rate mechanisms (ZSRMs) or High-Static-Low-Dynamic Stiffness mechanisms (HSLDS), can extend the working frequency range by leveraging kinematic nonlinearities. While effective, their performance is sensitive to mistuning [4], wherein any deviation in geometry or static loading from the structure (or payload) can alter the dynamic characteristics of the system, and it is worthwhile to note that the use of nonlinear isolators, which rely on the negative stiffness concept, is limited to applications where the mass of the payload is small, as these systems require a force preload of the same order of the payload weight [5].

In the analysis presented herein, only the mitigation of ground vibrations in the vertical directions is addressed. Studies in which HSLDS systems have been used to mitigate horizontal vibrations can be found in [6,7].

The HSLDS isolator considered here is based on an oscillating mass supported by a vertical linear spring in parallel with one or more horizontal linear springs. In dynamic conditions, the horizontal springs contribute to the equivalent dynamic vertical stiffness for small amplitude motions by reducing the stiffness and introducing hardening nonlinearities. The former effect increases the isolator working range, while the latter leads to jump-up/jump-down phenomena close to the saddle-node bifurcation points and reduces the isolation region since the resonance peak moves toward higher frequencies [8].

One of the early works on the subject was presented by Molyneux [9], where the arrangement of inclined springs was proposed to reduce the equivalent stiffness of a system. Many potential applications were suggested for road vehicle suspensions, torsional couplings, and vibration sensors. Woodard et al. [10] focused on the transition from linear to nonlinear behavior of a ZSRM mechanism. Considering the system around the equilibrium position, the authors demonstrated that increasing the cable preload decreases overall vertical stiffness, reducing the system's natural frequencies. Platus et al. [7] presented a series of isolators with multiple degrees of freedom capable of suppressing both vertical and horizontal oscillations. Comprehensive theoretical studies on the static analysis and nonlinear dynamics of QZS mechanisms are provided in Refs. [11–15], where force-displacement expressions and nonlinear equations of motion are derived, solved, and discussed for various configurations, parameters, and different forcing functions.

It is well established that an HSLDS system can be frequently modeled by the Duffing equation, where damping terms are added to simulate different dissipation mechanisms. Cheng et al. [16,17] and Shahraeeni et al. [18] considered a QZS system with linear viscous dampers. The resulting nonlinear damping has been approximated by terms like $\dot{x}x^n$, i.e., velocity times the $n$-th power of the displacement, in agreement with the lumped-mass system consisting of two angled dashpots considered in Ref. [19]. Den Hartog [20] and Ravindra and Mallik [21] addressed the role of dry friction on the dynamics of oscillators under base excitation, considering linear or nonlinear cubic stiffness terms. Donmez et al. [22] proposed the introduction of a dry friction element inside the isolator to reduce the transmissibility for frequencies around the resonance region. Liu and Yu [23] presented a QZS system with an oscillating mass supported by a vertical spring and surrounded by four lateral springs. Despite poor agreement between analytical and experimental results, attributed to a stiffness increment induced by the bending stiffness of the lateral springs not being considered in the theoretical model, experiments have shown a transmissibility reduction compared with the linear oscillator, confirming the effectiveness of the QZS system and the need to go beyond the classical model to predict QZS isolator performance.

The number of studies on QZS/HSLDS systems is increasing, reflecting growing interest in the topic [24]. Further theoretical and experimental studies can be found in the literature, proposing active stiffness control strategies [25], nonlinear absorbers [26], non-conventional springs [27], piecewise restoring forces [28], magnets [29,30], bio-inspired isolators [31], and X-shaped mechanisms [32,33] to avoid instabilities and expand the QZS working range.

In a recent study [34], theoretical and experimental results for a small-scale two-dimensional HSLDS isolator subjected to amplitude-scaled actual vertical earthquake signals were presented, demonstrating the isolator's effectiveness in reducing acceleration transmission under near-fault earthquakes. Based on these results, the present study investigated the behavior of a tunable HSLDS isolator for sensitive equipment under vertical base excitation. The design of a three-dimensional mechanism is proposed, and the fundamental components of the prototype under analysis are described. A theoretical model is formulated, incorporating piecewise restoring forces and Coulomb dry friction. The experimental analysis aims to evaluate the ground motion reduction achievable using such a device. To conduct the experiments, a high-power electrodynamic shaking table was utilized. The isolator's transmissibility was evaluated through harmonic base excitation tests, and a direct comparison with a linear suspension was made under replicated near-fault earthquakes, demonstrating the beneficial effects of the HSLDS isolator.

## 2. Description of the Real Isolator and Theoretical Dynamic Model

Figure 1a,b show the isolator, which consists of a tripod mechanism. A vertically translating payload of mass $m$ rests on a vertical spring of stiffness $k_v$; three oscillating rods of length $L$ connect the payload to three pushrods through spherical joints. Each pushrod compresses a horizontal spring of stiffness $k_h$. The rods are angularly spaced at 120 degrees from each other, and the spring preload can be adjusted by varying displacements $\delta_v$ and $\delta_h$ for the vertical and horizontal springs, respectively.

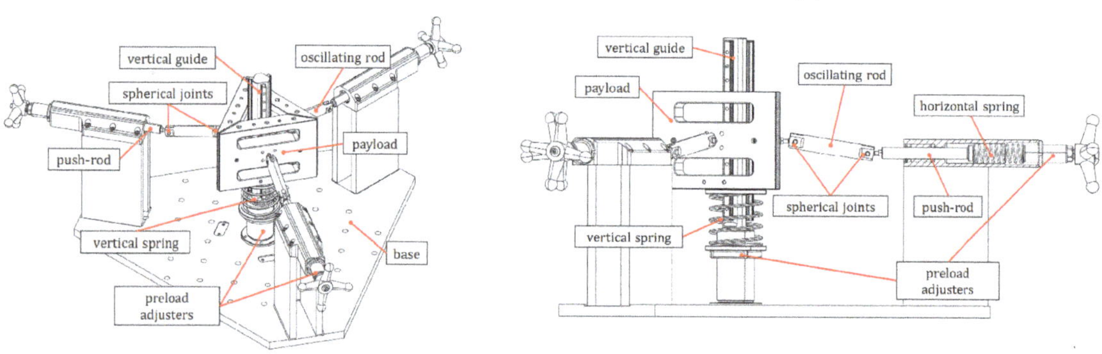


**Figure 1.** Tripod isolator with HSLDS characteristics: (**a**) isometric and (**b**) lateral views.

Figure 2 shows the lumped parameter model employed to determine the static characteristics of the isolator, where the cumulative effect of the three horizontal springs functioning in parallel is modeled as a single horizontal spring with equivalent stiffness $3 \times k_h$, $y$ denotes the payload displacement, $\theta$ is the actual angle between the rod and the horizontal line, and $F$ is the generic external static force.

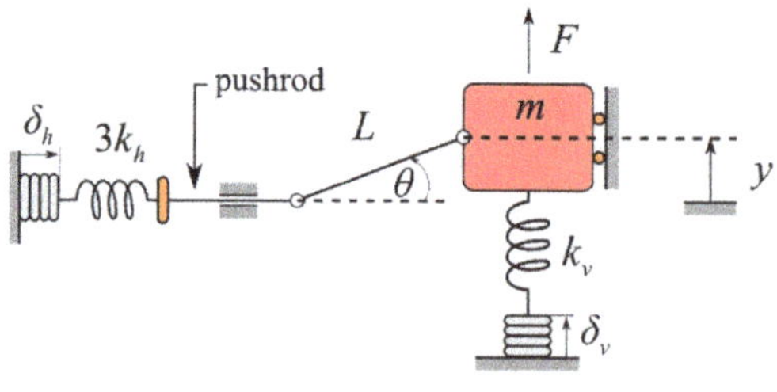


**Figure 2.** Lumped parameter model, static analysis.

The static behavior of the nonlinear isolator is governed by the equation,

$$F - F_s(y) = 0 \tag{1}$$

where $F_s(y)$ is the equivalent restoring force in the vertical direction. The horizontal springs are assumed to be purely compressive springs, i.e., they are unable to handle tractive forces. When the relative displacement $y$ reaches a critical value $y_{cr}$, the pushrod detaches from the spring, directly influencing the restoring force, which can be expressed as a piecewise function. Thus, by introducing the following nondimensional parameters

$$\begin{aligned} \hat{y} &= \frac{y}{L}, & r &= \frac{k_h}{k_v}, \\ \hat{\delta}_v &= \frac{\delta_v}{L}, & \hat{\delta}_h &= \frac{\delta_h}{L}. \end{aligned} \tag{2}$$

the nondimensional restoring force of the mechanism can be formulated as,

$$\frac{F_s(y)}{k_v\,L} = \hat{F}_s(\hat{y}) = \begin{cases} \hat{y} - \hat{y}_0 - \hat{\delta}_v - 3\,r\left[\hat{\delta}_h - \left(\sqrt{1-\hat{y}_0^2} - \sqrt{1-\hat{y}^2}\right)\right]\frac{\hat{y}}{\sqrt{1-\hat{y}^2}} & |\hat{y}| \leq |\hat{y}_{cr}| \\ \hat{y} - \hat{y}_0 - \hat{\delta}_v & |\hat{y}| > |\hat{y}_{cr}| \end{cases} \tag{3}$$

where $\hat{y}(0) = \hat{y}_0$ is the initial position of the payload. The critical displacement $\hat{y}_{cr}$, which is dependent on $\hat{\delta}_h$, is found by considering the null deflection of the horizontal spring,

$$\hat{\delta}_h - \sqrt{1-\hat{y}_0^2} + \sqrt{1-\hat{y}^2} = 0 \rightarrow \hat{y}_{cr} = \sqrt{1 - \left(\sqrt{1-\hat{y}_0^2} - \hat{\delta}_h\right)^2} \tag{4}$$

It is important to note that $\hat{F}_s(\hat{y})$ is defined without considering the detachment of the payload from the vertical spring, as this condition is not relevant for describing the experimental observation shown in the results section.

Figure 3 shows how the preload of the horizontal springs affects the restoring force $\hat{F}_s(\hat{y})$: the system can be tuned to function as an HSLDS mechanism for small $\hat{\delta}_h$ values or a QZS system for large $\hat{\delta}_h$ values. Specifically, by differentiating Equation (3) with respect to the displacement $\hat{y}$, the equivalent vertical stiffness reads,

$$\hat{k}_s(\hat{y}) = \frac{d}{d\hat{y}}\hat{F}_s = \begin{cases} 1 - 3\,r\left[\frac{\hat{\delta}_h - \sqrt{1-\hat{y}_0^2} + \sqrt[3]{(1-\hat{y}^2)^2}}{\sqrt[3]{(1-\hat{y}^2)^2}}\right] & |\hat{y}| \leq |\hat{y}_{cr}| \\ 1 & |\hat{y}| > |\hat{y}_{cr}| \end{cases} \tag{5}$$

Assuming $\hat{y}_0 = 0$, the minimum stiffness of the mechanism is obtained by evaluating $\hat{k}_s(\hat{y})$ for $\hat{y} = 0$, giving

$$\hat{k}_{s,min} = \hat{k}_s(0) = 1 - 3\,r\,\hat{\delta}_h \tag{6}$$

From Equation (6), depending on the value of $\hat{\delta}_h$, it is possible to observe that, in the neighborhood of the static equilibrium configuration, the system behaves in the following three ways:

- if $\hat{k}_{s,min} < 0$, the system presents a negative stiffness nonlinear spring;
- if $\hat{k}_{s,min} = 0$, the isolator works as a QZS mechanism;
- if $0 < \hat{k}_{s,min} < 1$, the isolator works as an HSLDS mechanism.

In order to study the dynamics of the isolator under base motion, the model depicted in Figure 4 is considered. The coordinate $x$ represents the absolute displacement of the payload, $x_b$ is the base displacement, and $z$ denotes the relative displacement of the payload.

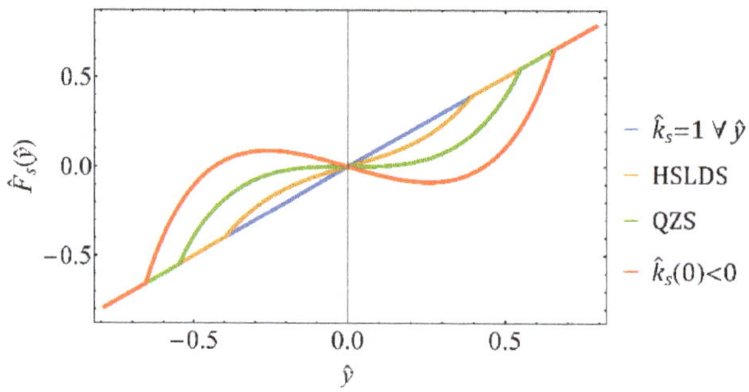


**Figure 3.** Non-smooth restoring force $\hat{F}_s(\hat{y})$ for different horizontal spring preloads (obtained assuming $\hat{y} = \hat{\delta}_v = 0$).

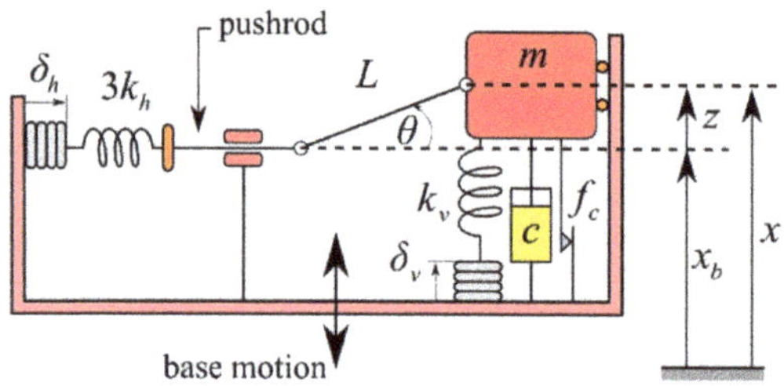


**Figure 4.** Lumped parameter model, dynamic analysis of the isolator under base motion.

Assuming that the system, under the action of the weight, is at rest in its static equilibrium position with the oscillating rods in a horizontal position, $z(0) = 0$, the restoring force $\hat{F}_s(\hat{z})$ is given by,

$$\hat{F}_s(\hat{z}) = \begin{cases} \hat{z} - \hat{\delta}_v - 3\,r\left(\hat{\delta}_h - 1 + \sqrt{1-\hat{z}^2}\right)\frac{\hat{z}}{\sqrt{1-\hat{z}^2}} & |\hat{z}| \le |\hat{z}_{cr}| \\ \hat{z} - \hat{\delta}_v & |z| > |\hat{z}_{cr}| \end{cases} \tag{7}$$

where $\hat{z} = z/L$ is the nondimensional relative displacement of the payload. The critical value $\hat{z}_{cr}$ is determined as,

$$\hat{z}_{cr} = \sqrt{1 - (1 - \hat{\delta}_h)^2} \tag{8}$$

Following the approach adopted in Ref. [34] to account for different dissipation mechanisms, a dissipation function $\hat{F}_d(z, \hat{z}')$ is considered in the model and is assumed to be a combination of viscous and dry friction forces, i.e.,

$$\hat{F}_d(z, \hat{z}') = 2\,\zeta \hat{z}' + sign(\hat{z}')\,\hat{f}_c \tag{9}$$

where $\zeta$ and $\hat{f}_c = f_c(k_v\,L)^{-1}$ denote the viscous damping ratio and the normalized Coulomb friction force, respectively.

The governing equation of the 1-DoF lumped parameter system can now be written as,

$$\hat{z}'' + \hat{F}_d(z, \hat{z}') + \hat{F}_s(\hat{z}) + \hat{F}_g = -\hat{x}_b'' \tag{10}$$

where $(\cdot)'$ denotes derivatives with respect to the normalized time $\tau = \omega_n t$, $\omega_n = \sqrt{k_v/m}$ is the natural frequency of the linear oscillator, and $\hat{F}_g = mg(k_v\,L)^{-1}$ is the normalized weight.

In Equation (7), to satisfy the condition for static equilibrium, the vertical spring preload must be $\hat{\delta}_v = \hat{F}_g$, and by defining the auxiliary restoring force $\widetilde{F}_s(z) = \hat{F}_s(z) + \hat{F}_g$, the equation of motion for the tripod, Equation (10), can be rearranged to a simpler form,

$$\hat{z}'' + \hat{F}_d(z, \hat{z}') + \widetilde{F}_s(z) = -\hat{x}_b'' \tag{11}$$

Equation (11) is computationally stiff and potentially expensive to solve numerically due to the presence of discontinuous functions. Thus, a smoothing approach is used to simplify the analysis [35]: the piecewise restoring function $\hat{F}_s(z)$ is substituted by a continuous Sigmoid-like function $\gamma(\beta, \hat{z})$, while the dry friction term is approximated by the hyperbolic tangent $\sigma(\alpha, \hat{z}')$, where

$$\begin{aligned} \gamma(\beta, \hat{z}) &= \tfrac{1}{1+e^{-\beta(\hat{z}+\hat{z}_{cr})}} - \tfrac{1}{1+e^{-\beta(\hat{z}-\hat{z}_{cr})}} \\ \sigma(\alpha, \hat{z}') &= \tfrac{e^{\alpha\hat{z}'} - e^{-\alpha\hat{z}'}}{e^{\alpha\hat{z}'} + e^{-\alpha\hat{z}'}} \end{aligned} \tag{12}$$

where $\alpha$ and $\beta$ are parameters for controlling the smoothing functions.

Figure 5 shows the smoothing functions (12) for different values of $\alpha$ and $\beta$; see that for smoothing parameters of about 1000, both functions well approximate the step behavior while preserving the continuity conditions.

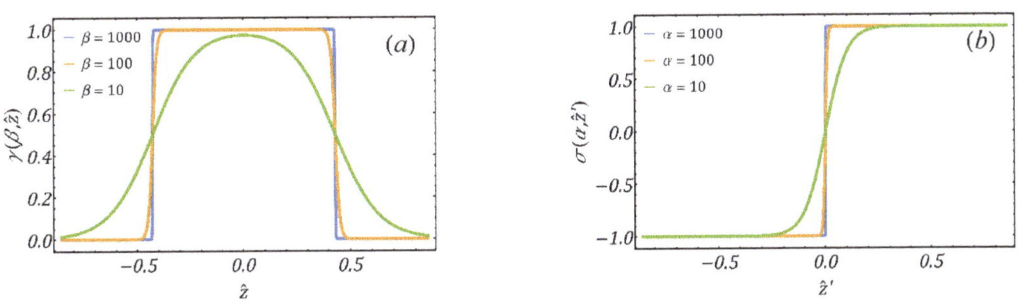


**Figure 5.** Smoothing functions: (**a**) sigmoid-based function and (**b**) hyperbolic tangent.

At this point, with the use of Equation (12), the equation of motion of the isolator can be rewritten in the state space as

$$\begin{cases} \hat{z}' = u \\ u' = -2\,\zeta u - \sigma(\alpha, u)\,\hat{f}_c - \hat{z} + 3\,r\left(\hat{\delta}_h - 1 + \sqrt{1-\hat{z}^2}\right)\,\frac{\hat{z}}{\sqrt{1-\hat{z}^2}}\gamma(\beta, \hat{z}) - \hat{x}_b'' \end{cases} \tag{13}$$

where $(\hat{z},\ u)$ are the state variables. Equation (13) is suitable for standard numerical integration. To this end, the Matlab solver ode23s for stiff ordinary differential equations [36] was used, assuming $\alpha = \beta = 1000$.

### 3. Experimental Setup and Shake Table Test

Experiments were conducted on a tripod isolator with payload mass $m = 20$ kg, oscillating rod length $L = 0.155$ m, vertical spring stiffness $k_v = 1848$ Nm$^{-1}$, and horizontal spring stiffness $k_h = 3747$ Nm$^{-1}$.

Figure 6a displays a schematic of the experimental setup: two piezoelectric accelerometers were used to measure the base and payload vibrations in the vertical direction; the seismic base motion was applied through a Dongling ET-40-370 shaker (40 kN peak force, 100 g maximum acceleration, 500 kg max static payload) controlled by a closed-loop algorithm, and signals were acquired using the LMS Scadas hardware with Testlab software.

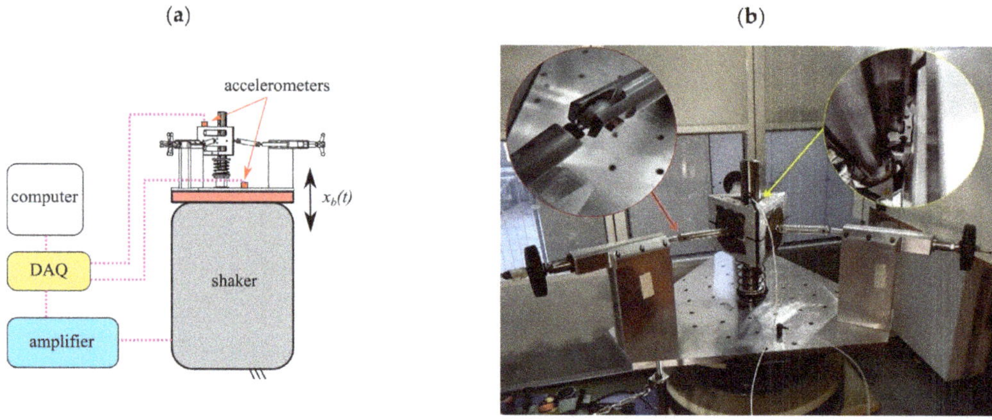


**Figure 6.** Shaking table tests: (**a**) schematic of the experimental setup, and (**b**) HSDLS mounted on the shaker.

Using the base and payload measurements, the absolute acceleration transmissibility was evaluated through the following expression,

$$T = \left| \frac{\text{RMS}(\ddot{x})}{\text{RMS}(\ddot{x}_b)} \right| \tag{14}$$

where RMS stands for root mean square.

### 3.1. Harmonic Excitation Tests

Stepped sine tests were carried out to evaluate the transmissibility $T$ of the isolator. Using the Siemens SCADAS Mobile system and the Siemens Testlab control software (Sine Control module), the shaker base motion was controlled in terms of vertical acceleration amplitude $|\ddot{x}_b|$ and forcing frequency $f_\Omega$, as shown in Figure 7. To assess the isolator's nonlinear characteristics, $f_\Omega$ was swept in the range of 0.5–10 Hz, both forward and backward. A variable frequency step $\Delta f_\Omega$ was chosen to span the resonance region with suitable frequency resolution, and for each frequency step, 120 periods of excitation were recorded with a sampling frequency of 400 Hz.

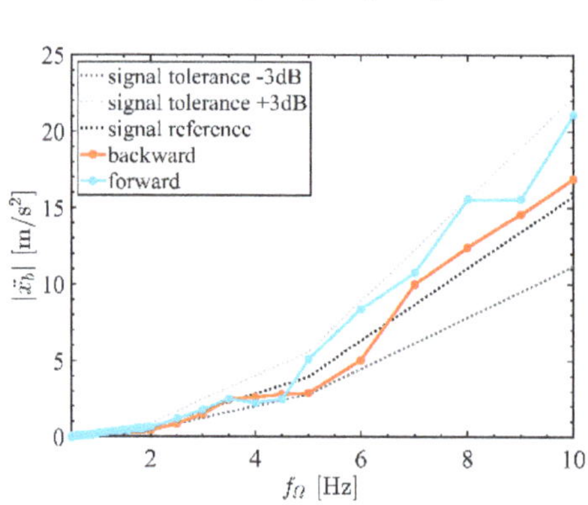


**Figure 7.** Harmonic excitation test. Shaker base acceleration amplitude $|\ddot{x}_b|$ at each frequency step for both backward and forward frequency sweep.

Controlling the shaking table at very low frequencies was demanding and critical due to the strict limitations on the maximum allowed displacement of the table (i.e., 51 mm, peak-to-peak). To address this limitation and ensure satisfactory control, the reference acceleration amplitude of the shaker base (black dotted line in Figure 7) was linearly varied at each step, with details about its slope provided in Table 1.

**Table 1.** Harmonic excitation test. Frequency step resolution and acceleration slope variations for each frequency range.

| | | | | |
|---|---|---|---|---|
| Frequency range $f_\Omega$ [Hz] | 0.5–1.05 | 1.05–2 | 2–5 | 5–10 |
| Frequency step $\Delta f_\Omega$ [Hz] | 0.050 | 0.025 | 0.5 | 1 |
| Slope of reference acceleration [m s$^{-2}$ Hz$^{-1}$] | 0.39 | 0.39 | 1.10 | 2.37 |

The investigated test cases are detailed in Table 2 and, more specifically, the following observations can be made:

- Each test is defined by a specific horizontal spring preload $\delta_h$;
- Before each experimental test, the isolator's static equilibrium position, characterized by having horizontal rods, $z(0) = 0$, was imposed by adjusting the vertical spring preload $\delta_v$. According to Equation (5), this angle was selected to provide the minimum stiffness around the static equilibrium position;
- To study the behavior of the isolator under HSLDS conditions for all the investigated cases, the displacement $\delta_h$ was kept below the QZS threshold given by Equation (6), $L(3\,r)^{-1} = 25$ mm;
- Cases where $\delta_h > 12$ mm were excluded because payload oscillations with amplitudes exceeding the maximum allowed stroke were observed, causing the premature end of the test;
- The natural frequency of the system can be obtained by the linearization of the restoring force, Equation (7), using the Taylor series. Considering Case 1 as a reference, a reduction in natural frequency of 22% and 27% was achieved for Case 2 and Case 3, respectively.

**Table 2.** Parameters of the HSLDS isolator for the three investigated cases.

| Parameter | Case 1 | Case 2 | Case 3 |
|---|---|---|---|
| Horizontal spring preload $\delta_h$ [mm] | 0 | 10 | 12 |
| Total preload $3\,k_h\delta_h$ [N] | 0 | 112.4 | 134.9 |
| Natural frequency [Hz] | 1.53 | 1.19 | 1.11 |

Figure 8a shows the transmissibility for a null preload. In this case, only the linear vertical spring contributes to the restoring force. In fact, the isolator behaves like a linear oscillator, except for the presence of dry friction, resulting in stiction at low frequencies when inertia is insufficient to overcome the friction force, as shown in the detailed view. By increasing the preload on the horizontal springs to 112.4 N, Figure 8b, a hardening nonlinearity is introduced. The tangent stiffness around the equilibrium position is reduced, leading to a lower natural frequency of the system. Additionally, the preload affects stick-slip thresholds: the greater the preload, the less relevant the role of dry friction becomes. Considering Figure 8c, where $3k_h\delta_h = 134.9$ N, the stiction phenomenon disappears, and at low frequencies, the transmissibility curves obtained in the case of forward and backward sweeps are perfectly overlapped, except in the multiple-valued jump up/down region. In all the investigated cases, the transmissibility obtained using the lumped parameter model shows excellent agreement with the experimental measurements and is able to capture the stiffness variation due to the loss of contact between the pushrods and the horizontal springs. Regarding the hysteretic phenomenon at low frequencies, the dry friction model considered in Equation (9) is not able to fit the experimental results for a narrow region in the case of null preload.

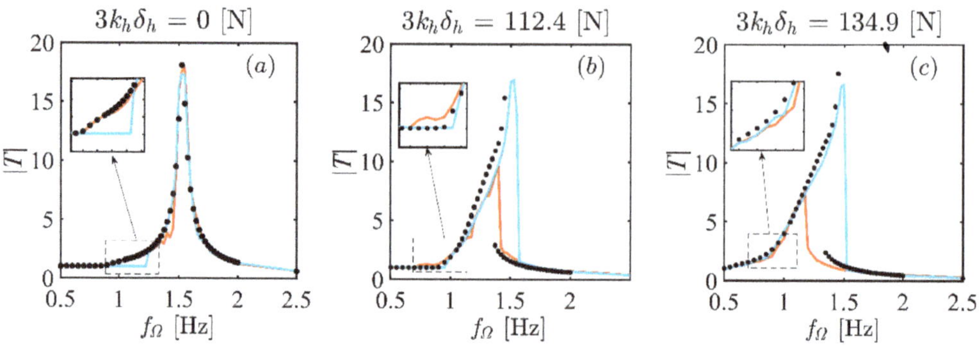


**Figure 8.** Magnitude of the transmissibility for different values of the horizontal springs preload: (**a**) 0 N, (**b**) 112.4 N, and (**c**) 134.9 N. Detailed views show the evolution of the stiction phenomena among the test cases. (— forward and — backward sweep, ●●● numerical solution).

Values of the damping ratio and the dry friction force adopted in the numerical model to fit the experimental data are given in Table 3. Variations in the damping coefficient must be sought in the different lubrication conditions between different tests and in the limitations of the linear viscous model considered herein. In fact, the motion of the horizontal links can lead to nonlinear dissipation terms [18]. On the other hand, the decrease in dry friction force is attributed to the load acting perpendicular to the linear guides. These guides are preloaded; consequently, as the payload oscillates, the forces exerted by the horizontal springs have a direct impact on the overall load normal to the linear ball bearings.

**Table 3.** Viscous damping ratio and dry friction force values obtained from the fits of the experimental transmissibility curves.

| Parameter | Case 1 | Case 2 | Case 3 |
|---|---|---|---|
| Viscous damping ratio $\zeta$ | 0.015 | 0.008 | 0.020 |
| Dry friction force $f_c$ [N] | 3.44 | 3.44 | 0.86 |

### 3.2. Earthquake Experimental Results

The experimental response of the tripod isolator to simulated earthquakes is presented in this section. The near-fault earthquake signals were obtained from the PEER database [37], and the experiments were conducted by controlling the shaker base using the Siemens SCADAS Mobile system and the Siemens Testlab control software (Single Axis Waveform Replication module) with the built-in online adaptive closed-loop control strategy. Two different test sets were carried out:

(i) Amplitude-scaled tests: the acceleration amplitude of the input signals was reduced to cope with the shaker limits (i.e., shaker base overtravel);
(ii) Time-scaled tests: the input signals were compressed in time to shift the earthquake energy content to higher frequencies in order to account for the limited suspended mass.

Focusing on the amplitude scale tests, time histories and spectra of the test acceleration signals are given in Figure 9a–f. Earthquake vibrations carry energy content at low frequencies; therefore, to reduce the noise components and prevent signal phase shifting, the acquired data were filtered using a fifth-order Butterworth low-pass digital filter with a cut-off frequency of 30 Hz.

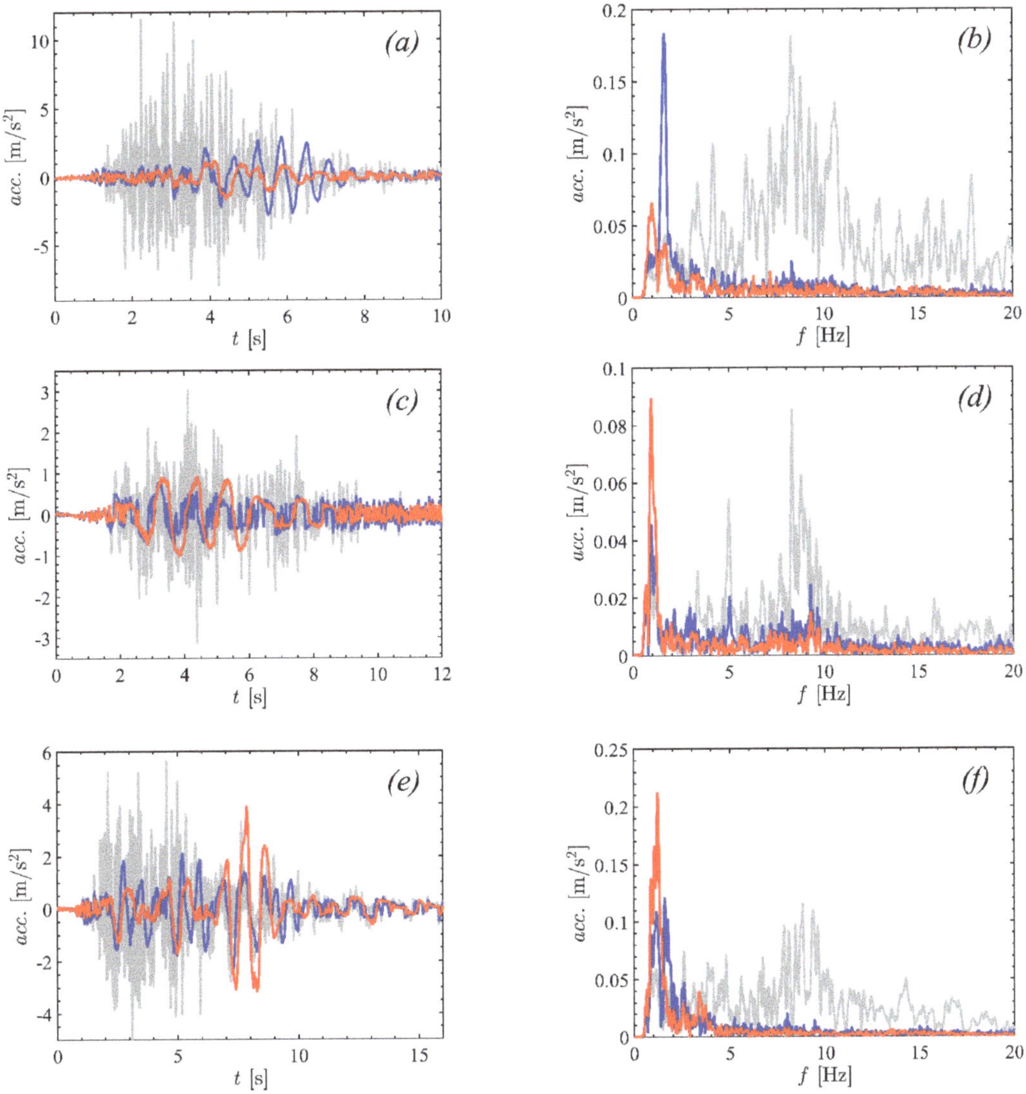

**Figure 9.** Amplitude-scaled tests. Acceleration time histories and spectra: (**a**,**b**) Christchurch, 70%; (**c**,**d**) Bam, 30%; and (**e**,**f**) Northridge, 90%. (— ground acceleration, — linear spring, and — HSLDS isolator).

Considering the Christchurch earthquake with an amplitude scaled by 70%, Figure 9a,b, the ground motion was particularly strong between [1–7 s], with a broadband spectrum governed by a peak at 8.24 Hz. Both the linear and the nonlinear (HSLDS Case 3, $3k_h\delta_h$ = 134.9 N) isolators were able to mitigate ground vibrations, Figure 9a, yet the HSLDS eliminated the resonance-like phenomenon exhibited by the linear system between [3.7–7.7 s], characterized by a harmonic component of the earthquake at 1.64 Hz, Figure 9b, hence close to the natural frequency of the system, which is 1.53 Hz according to Table 2. Conversely, for the Bam earthquake, with an amplitude scaling factor of 30%, the resonance of the HSLDS occurred in the range [2.5–6 s], Figure 9c. This behavior is attributed to the narrow frequency content of the ground motion, Figure 9d, localized at three frequencies:

0.95 Hz, 4.99 Hz, and 8.27 Hz. Despite the relatively small amplitude of the harmonic (0.036 m/s$^2$) at 0.95 Hz compared with the others, it was close enough to the HSDLS isolator resonance frequency (1.11 Hz) to induce strong payload oscillations. In Figure 9e, the linear and the HSLDS systems were compared under the Northridge earthquake, with an amplitude scaling factor of 90%. In this case, the HSLDS isolator reduction was higher than the linear system except between [6.5–9.3 s], where a sudden large amplitude oscillation of the HSDLS occurred. From the analysis of the frequency spectrum, Figure 9f, Northridge ground motion showed a double peak at 0.86 Hz and 1.15 Hz, causing the resonance of the HSDLS isolator. However, the frequency peaks were far enough from the resonance frequency of the linear system.

The results of the present section are summarized in the bar chart in Figure 10, where the peak acceleration and the RMS value provide a clear view of the performance of the linear and HSLDS isolators. Both systems were able to reduce ground vibrations in all the investigated cases, with the linear spring offering a higher reduction than the HSDLS isolator. However, for the Bam and Northridge cases, the HSDSL performance was conditioned by the activation of resonance-like phenomena.

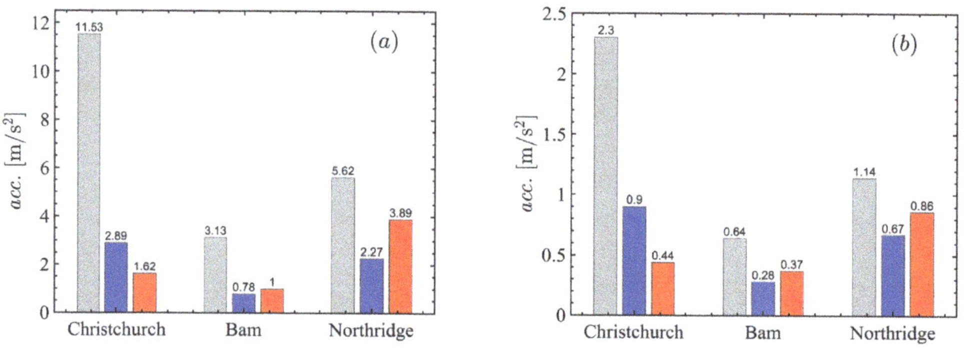


**Figure 10.** Amplitude-scaled test. Vertical acceleration: (**a**) maximum amplitude and (**b**) RMS. (— ground acceleration, — linear spring, and — HSLDS isolator).

Even though seismic isolators for actual buildings are designed to support large payloads and lead to very low resonance frequencies, the isolator under investigation was downsized for testing on a shake table having a limited payload of 300 kg. The limited value of the oscillating mass prevented the achievement of resonance frequencies below 1 Hz. Therefore, in order to perform realistic experiments on the downsized system, the original earthquake signals, having spectral energy at frequencies of about 1 Hz, were rescaled with respect to the time variable as $t_{scaled} = \xi t$, where $\xi > 1$ is the time-scaling factor. This scaling moves the signal spectral energy toward a higher frequency range that is more suitable for the downsized system. To determine a reasonable value for $\xi$, the following assumptions were considered: (i) small civil structures typically have resonance frequencies $f_R$ of about 3–10 Hz [38–40], and (ii) the linear system in question had a resonance frequency near $f_S$ = 1.5 Hz. Consequently, the time histories were scaled by a factor $\varepsilon = f_R f_S^{-1}$ = 2.

Time histories and spectra of the time-scaled earthquake tests are presented in Figure 11a–f. It is important to note that, in these cases, the amplitude of the ground acceleration signals remained unaltered (i.e., 100%). Similar to the amplitude-scaled earthquake tests, a fifth-order Butterworth low-pass digital filter was used to reduce data noise. A larger cut-off frequency (100 Hz) was considered to account for the effect of time scaling, broadening the input signal energy content.

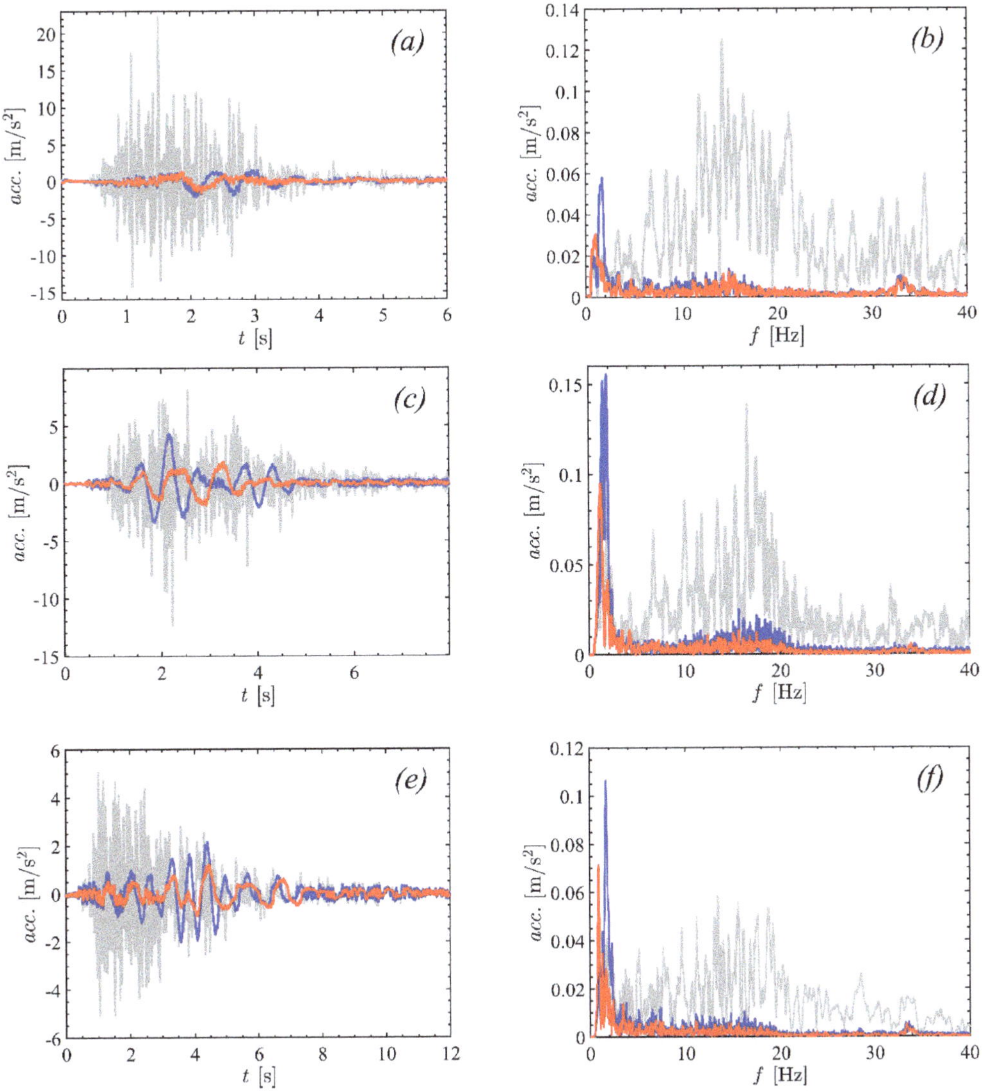

**Figure 11.** Time-scaled test. Acceleration time histories and spectra: (**a**,**b**) Christchurch, (**c**,**d**) Bam, and (**e**,**f**) Northridge. (— ground acceleration, — linear spring, and — HSLDS isolator).

In Figure 11a, both isolators behave similarly under the Christchurch earthquake, offering effective ground vibration reduction. Focusing on the acceleration RMS, an 81% reduction was achieved using the linear spring, while the HSDLS isolator demonstrated an 88% reduction. According to the spectrum, Figure 11b, the frequency content of the time-scaled Christchurch earthquake shifts to a higher frequency band, with a spectrum peak at 14.3 Hz, which is approximately one order of magnitude higher than the resonance frequencies of both the isolator configurations; thus, both the systems were working in the isolation regime.

Figure 11c shows the isolators' response to the Bam earthquake. The beneficial effects of the isolator are evident when the stiffness nonlinearity is retained. The payload's maximum acceleration amplitude measured with the HSLDS isolator (1.88 m/s$^2$) is lower than the value observed for the payload carried by the linear spring (4.29 $m/s^2$). Moreover, in the latter case, periodic payload oscillations were activated from Bam low-frequency components at 1.43 Hz and 1.88 Hz, see Figure 11d.

The experimental response acceleration time histories and spectra considering the Northridge earthquake ground motion are shown in Figure 11e,f, respectively. Despite the time-scaling procedure, the input signal has a broad energy distribution up to 20 Hz, and the low-frequency components induced an increasing payload response between [0–4.4 s]. Yet, the HSLDS isolator showed a higher isolation degree than the simple linear spring, reducing the ground motion by 88% RMS, while an 81% RMS reduction was observed in the linear spring case.

By shifting the frequency content of the tested earthquakes to a higher frequency, the time-scaling procedure allowed both isolators to work under optimal conditions. From the comparison, it is evident that the HSLDS isolator consistently performed better than a linear system for all the tested signals, as shown in the bar charts given in Figure 12.

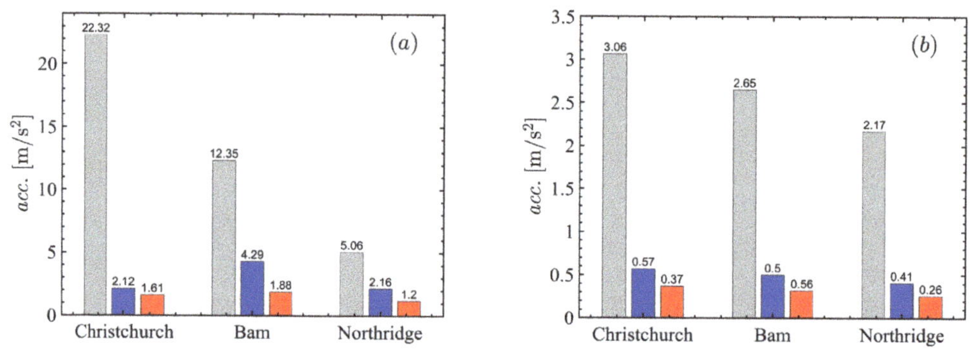


**Figure 12.** Time-scaled test. Vertical acceleration: (**a**) maximum amplitude and (**b**) RMS. (— ground acceleration, — linear spring, and — HSLDS isolator).

## 4. Concluding Remarks

In this study, the behavior of a nonlinear isolator under vertical base motion has been investigated. A three-dimensional tripod mechanism with tunable HSLDS characteristics was considered. A theoretical formulation incorporating a piecewise restoring force to approximate the behavior of compression springs was developed. Harmonic base motion experimental tests were conducted on a shaking table to assess the acceleration transmissibility, where different values of horizontal springs preload were used to tune the HSLDS characteristics. Additionally, ground vibration mitigation was evaluated under real-life seismic inputs. The concluding remarks of the work are summarized as follows:

The tunability of the proposed mechanism allows for adaptation to a wide range of loading conditions, making it a versatile solution for various applications.

(1) The numerically determined and experimental transmissibility exhibited noticeable agreement, underlining the importance of considering a piecewise nonlinear–linear restoring force to replicate the experimental observation;
(2) Harmonic excitation tests showed the limited capability of the Coulomb model to predict stiction phenomena, pointing out the need for refined friction and hysteresis models [41,42];
(3) Near-fault excitation tests revealed the beneficial effects of the HSLDS isolator in mitigating transmitted ground vibration to the payload. Results showed that a nonlinear isolator attained a higher vibration reduction than a linear spring isolator in four out of six investigated earthquakes;
(4) Experimental results align with existing literature [34], reinforcing the concept of the HSLDS mechanism as an effective means to prevent damage to sensitive objects;

(5) Insights into passive isolation systems are provided, revealing their susceptibility to resonance and emphasizing the importance of careful tuning to meet the safety requirements of suspended payloads.

**Author Contributions:** Conceptualization, S.I. and B.M.; methodology, all authors have contributed to the research methodology; software, G.I.; validation, G.I.; formal analysis, G.I., A.Z. and F.P.; investigation, G.I., A.Z. and F.P.; resources, F.P.; data curation, G.I. and F.P.; writing—original draft preparation, G.I., A.Z. and F.P.; writing—review and editing, F.E.-M., Y.M., S.I. and B.M.; visualization, G.I.; supervision, S.I. and F.P.; project administration, S.I.; funding acquisition, S.I. All authors have read and agreed to the published version of the manuscript.

**Funding:** The authors acknowledge the financial support provided by the Ministry of Business and Innovation and Employment (New Zealand) through the Smart Ideas scheme (Project No: UOWX1801, 'Omnidirectional earthquake isolation system').

**Data Availability Statement:** Data are available on request.

**Acknowledgments:** The authors acknowledge Kēpa Morgan (Pou Hautū, Mahi Maioro Professionals Ltd.) and Alan Park (CEO, Robinson Seismic Ltd.) for their suggestions regarding the experimental model.

**Conflicts of Interest:** The authors declare no conflicts of interest.

## References


1. Doocy, S.; Daniels, A.; Packer, C.; Dick, A.; Kirsch, T.D. The human impact of earthquakes: A historical review of events 1980-2009 and systematic literature review. *PLoS Curr.* **2013**, *5*, ecurrents.dis.67bd14fe457f1db0b5433a8ee20fb833. [CrossRef] [PubMed]
2. Parulekar, Y.M.; Reddy, G.R. Passive response control systems for seismic response reduction: A state-of-the-art review. *Int. J. Struct. Stab. Dyn.* **2009**, *9*, 151–177. [CrossRef]
3. Inman, D.J. *Engineering Vibration*, 4th ed.; Pearson International Education (Prentice Hall): Upper Saddle River, NJ, USA, 2013.
4. Abolfathi, A.; Brennan, M.J.; Waters, T.P.; Tang, B. On the effects of mistuning a force-excited system containing a quasi-zero-stiffness vibration isolator. *J. Vib. Acoust.* **2015**, *137*, 044502. [CrossRef]
5. Sarlis, A.A.; Pasala, D.T.R.; Constantinou, M.C.; Reinhorn, A.M.; Nagarajaiah, S.; Taylor, D.P. Negative Stiffness Device for Seismic Protection of Structures. *J. Struct. Eng.* **2013**, *139*, 1124–1133. [CrossRef]
6. Sarlis, A.A.; Pasala, D.T.R.; Constantinou, M.C.; Reinhorn, A.M.; Nagarajaiah, S.; Taylor, D.P. Negative Stiffness Device for Seismic Protection of Structures: Shake Table Testing of a Seismically Isolated Structure. *J. Struct. Eng.* **2016**, *142*, 04016005. [CrossRef]
7. David, L.P. Negative-stiffness-mechanism vibration isolation systems. In *Vibration Control in Microelectronics, Optics, and Metrology*; SPIE: Bellingham, WA, USA, 1992; Volume 1619. [CrossRef]
8. Gatti, G.; Brennan, M.J.; Tang, B. Some diverse examples of exploiting the beneficial effects of geometric stiffness nonlinearity. *Mech. Syst. Signal Process.* **2019**, *125*, 4–20. [CrossRef]
9. Molyneaux, W.G. Supports for vibration isolation. In *ARC/CP-322*; Aeronautical Research Council: London, UK, 1957.
10. Woodard, S.E.; Housner, J.M. Nonlinear behavior of a passive zero-spring-rate suspension system. *J. Guid. Control Dyn.* **1991**, *14*, 84–89. [CrossRef]
11. Carrella, A.; Brennan, M.J.; Waters, T.P. Static analysis of a passive vibration isolator with quasi-zero-stiffness characteristic. *J. Sound. Vib.* **2007**, *301*, 678–689. [CrossRef]
12. Carrella, A.; Brennan, M.J.; Waters, T.P. Optimization of a quasi-zero-stiffness isolator. *J. Mech. Sci. Technol.* **2007**, *21*, 946–949. [CrossRef]
13. Kovacic, I.; Brennan, M.J.; Waters, T.P. A study of a nonlinear vibration isolator with a quasi-zero stiffness characteristic. *J. Sound. Vib.* **2008**, *315*, 700–711. [CrossRef]
14. Brennan, M.J.; Kovacic, I.; Carrella, A.; Waters, T.P. On the jump-up and jump-down frequencies of the Duffing oscillator. *J. Sound. Vib.* **2008**, *318*, 1250–1261. [CrossRef]
15. Carrella, A.; Brennan, M.J.; Waters, T.P.; Lopes, V. Force and displacement transmissibility of a nonlinear isolator with high-static-low-dynamic-stiffness. *Int. J. Mech. Sci.* **2012**, *55*, 22–29. [CrossRef]
16. Cheng, C.; Li, S.; Wang, Y.; Jiang, X. Force and displacement transmissibility of a quasi-zero stiffness vibration isolator with geometric nonlinear damping. *Nonlinear Dyn.* **2017**, *87*, 2267–2279. [CrossRef]
17. Cheng, C.; Ma, R.; Hu, Y. Beneficial performance of a quasi-zero stiffness vibration isolator with generalized geometric nonlinear damping. *Noise Vib. Worldw.* **2021**, *52*, 59–71. [CrossRef]
18. Shahraeeni, M.; Sorokin, V.; Mace, B.; Ilanko, S. Effect of damping nonlinearity on the dynamics and performance of a quasi-zero-stiffness vibration isolator. *J. Sound. Vib.* **2022**, *526*, 116822. [CrossRef]
19. Jeong, B.; Cho, H.; Yu, M.F.; Vakakis, A.F.; McFarland, D.M.; Bergman, L.A. Modeling and measurement of geometrically nonlinear damping in a microcantilever-nanotube system. *ACS Nano* **2013**, *7*, 8547–8553. [CrossRef]

20. Den Hartog, J.P. Forced Vibration With Combined Viscous and Coulomb Damping. *Transcations Am. Soc. Mech. Eng.* **1930**, *9*, 801–817.
21. Ravindra, B.; Mallik, A.K. Hard duffing-type vibration isolator with combined Coulomb and viscous damping. *Int. J. Non Linear Mech.* **1993**, *28*, 427–440. [CrossRef]
22. Donmez, A.; Cigeroglu, E.; Ozgen, G.O. An improved quasi-zero stiffness vibration isolation system utilizing dry friction damping. *Nonlinear Dyn.* **2020**, *101*, 107–121. [CrossRef]
23. Liu, C.; Yu, K. Design and experimental study of a quasi-zero-stiffness vibration isolator incorporating transverse groove springs. *Arch. Civ. Mech. Eng.* **2020**, *20*, 67. [CrossRef]
24. Liu, C.; Zhang, W.; Yu, K.; Liu, T.; Zheng, Y. Quasi-zero-stiffness vibration isolation: Designs, improvements and applications. *Eng. Struct.* **2024**, *301*, 117282. [CrossRef]
25. Barbieri, M.; Ilanko, S.; Pellicano, F. Active vibration control of seismic excitation. *Nonlinear Dyn.* **2018**, *93*, 41–52. [CrossRef]
26. Samani, F.S.; Pellicano, F. Vibration reduction on beams subjected to moving loads using linear and nonlinear dynamic absorbers. *J. Sound. Vib.* **2009**, *325*, 742–754. [CrossRef]
27. Lan, C.C.; Yang, S.A.; Wu, Y.S. Design and experiment of a compact quasi-zero-stiffness isolator capable of a wide range of loads. *J. Sound. Vib.* **2014**, *333*, 4843–4858. [CrossRef]
28. Cheng, C.; Li, S.; Wang, Y.; Jiang, X. On the analysis of a piecewise nonlinear-linear vibration isolator with high-static-low-dynamic-stiffness under base excitation. *J. Vibroengineering* **2015**, *17*, 3453–3470.
29. Carrella, A.; Brennan, M.J.; Waters, T.P.; Shin, K. On the design of a high-static-low-dynamic stiffness isolator using linear mechanical springs and magnets. *J. Sound. Vib.* **2008**, *315*, 712–720. [CrossRef]
30. Dong, G.; Zhang, X.; Xie, S.; Yan, B.; Luo, Y. Simulated and experimental studies on a high-static-low-dynamic stiffness isolator using magnetic negative stiffness spring. *Mech. Syst. Signal Process.* **2017**, *86*, 188–203. [CrossRef]
31. Yan, G.; Zou, H.X.; Wang, S.; Zhao, L.C.; Wu, Z.Y.; Zhang, W.M. Bio-inspired vibration isolation: Methodology and design. *Appl. Mech. Rev.* **2021**, *73*, 1–21. [CrossRef]
32. Bian, J.; Jing, X. Analysis and design of a novel and compact X-structured vibration isolation mount (X-Mount) with wider quasi-zero-stiffness range. *Nonlinear Dyn.* **2020**, *101*, 2195–2222. [CrossRef]
33. Chong, X.; Wu, Z.; Li, F. Vibration isolation properties of the nonlinear X-combined structure with a high-static and low-dynamic stiffness: Theory and experiment. *Mech. Syst. Signal Process.* **2022**, *179*, 109352. [CrossRef]
34. Eskandary-Malayery, F.; Ilanko, S.; Mace, B.; Mochida, Y.; Pellicano, F. Experimental and numerical investigation of a vertical vibration isolator for seismic applications. *Nonlinear Dyn.* **2022**, *109*, 303–322. [CrossRef]
35. Kim, T.C.; Rook, T.E.; Singh, R. Effect of smoothening functions on the frequency response of an oscillator with clearance non-linearity. *J. Sound. Vib.* **2003**, *263*, 665–678. [CrossRef]
36. Shampine, L.F.; Reichelt, M.W. The MATLAB Ode Suite. *SIAM J. Sci. Comput.* **1997**, *18*, 1–22. [CrossRef]
37. *Peer Ground Motion Database*; Pacific Earthquake Engineering Research Center (PEER): Berkeley, CA, USA.
38. Eurocode 8. *Design of Structures for Earthquake Resistance—Part 3: Assessment and Retrofitting of Buildings*; Committee for Standardization: Brussels, Belgium, 2005; EN 1998-1.
39. Benedetti, D.; Carydis, P.; Pezzoli, P. Shaking table tests on 24 simple masonry buildings. *Earthq. Eng. Struct. Dyn.* **1998**, *27*, 67–90. [CrossRef]
40. Ceroni, F.; Sica, S.; Rosaria Pecce, M.; Garofano, A. Evaluation of the natural vibration frequencies of a historical masonry building accounting for SSI. *Soil. Dyn. Earthq. Eng.* **2014**, *64*, 95–101. [CrossRef]
41. Vaiana, N.; Rosati, L. Classification and unified phenomenological modeling of complex uniaxial rate-independent hysteretic responses. *Mech. Syst. Signal Process.* **2023**, *182*, 109539. [CrossRef]
42. Vaiana, N.; Rosati, L. Analytical and differential reformulations of the Vaiana–Rosati model for complex rate-independent mechanical hysteresis phenomena. *Mech. Syst. Signal Process.* **2023**, *199*, 110448. [CrossRef]

*vibration*

*Article*

# A Method for Applying the Use of a Smart 4 Controller for the Assessment of Drill String Bottom-Part Vibrations and Shock Loads

**Serhii Landar [1,2], Andrii Velychkovych [3,*], Liubomyr Ropyak [4] and Andriy Andrusyak [3]**

[1] Department of Oil and Gas Engineering and Technology, National University "Yuri Kondratyuk Poltava Polytechnic", 24 Pershotravneva Str., 36011 Poltava, Ukraine; s.landar@ungs-drilling.com.ua
[2] Ukrnaftagazservis Ltd., 3 Shevchenko Str., 36011 Poltava, Ukraine
[3] Department of Construction and Civil Engineering, Ivano-Frankivsk National Technical University of Oil and Gas, 15 Karpatska Str., 76019 Ivano-Frankivsk, Ukraine; andriiandrusiak@gmail.com
[4] Department of Computerized Mechanical Engineering, Ivano-Frankivsk National Technical University of Oil and Gas, 15 Karpatska Str., 76019 Ivano-Frankivsk, Ukraine; l_ropjak@ukr.net
* Correspondence: a_velychkovych@ukr.net

**Abstract:** Optimization of drilling processes for oil and gas and geothermal wells requires the effective use of mechanical energy for the destruction of rocks. When constructing a well, an important indicator of the drilling stage is the mechanical speed. Therefore, when performing drilling operations, operators usually use blade bits of an aggressive design and often use forced drilling modes. Drill bits under forced operation modes generate a wide range of vibrations in the drilling tools; in turn, a drill string, being a long-dimensional deformable body, causes the development, amplification, and interconnection of vibrations of different types. Vibration loads reduce the technical and economic indicators of drilling, with destructive effects on drill string elements, and cause complications and emergencies. The authors initiated the creation of an informational and analytical database on emergency situations that occurred as a result of excessive vibrations of the drill string during the construction of deep wells in the deposits of the Dnipro–Donetsk Basin. For the first time, the suitability and effectiveness of using the Smart 4 controller ("Innova Power Solutions", Calgary, Canada) for monitoring the vibration load of the drilling tool was tested in industrial conditions, while the controller was used as a separate element in the drill string. A special module was developed for the reliable installation of the Smart 4 controller, with a power battery in the layout of the lower part of the drill string. During the testing of the proposed device for measuring vibrations in the process of drilling an inclined well, verification of the registered data was carried out with the help of a high-cost telemetry system. The implementation of the proposed innovation will allow each operator to assess the significance of the impact of vibrations and shocks on the production process and, if necessary, adjust the drilling modes or apply vibration protection devices. In addition, service departments that operate and repair drilling equipment will be able to obtain an evidence base for resolving warranty disputes or claims.

**Keywords:** drill string vibrations; shock loads; sensor; bit; positive displacement motor; drilling tool

## 1. Introduction

In the process of drilling oil and gas and geothermal wells, the drill bit, which destroys rocks, can generate a wide range of drilling tool vibrations [1–3]. When using a drill string to carry out the drilling process, phenomena that are characteristic of classic elastic rods often occur in its sections. In particular, the string can experience local loss of stability; it can deform, acquiring a spiral shape; accumulate and then release the potential energy of elastic deformation; or carry out longitudinal, transverse and torsional oscillations [4–6]. In general, the drill string is a long-dimensional deformable body that, under certain conditions, can cause the development, amplification, and interrelationship of vibrations of different classes [7–9]. When drilling a well and drilling under a conductor, operators can very often visually observe a significant vibration of the lead pipe. As the depth of the well increases, the drilling tool near the hole continues to experience dynamic loads but without visible manifestations of vibration at the daylight surface.

Intense vibrations significantly reduce the technical and economic indicators of drilling; they destructively affect the elements of drill strings, bits, downhole engines, increase the specific energy consumption of the drilling process, and cause off-duty and emergency situations, which ultimately lead to an increase in financial costs for drilling each meter of the well [10–12]. A trivial solution to the problem is to reduce the rotation speed of the drilling tool and reduce the axial load on the bit. Such actions usually make it possible to reduce vibration loads to the permissible limit, but at the same time, the mechanical speed of drilling is lost [13,14]. Another, more effective way to solve the problem of the harmful effects of vibrations in drilling is based on the use of special vibration protection devices [15–17]. Among them, the most popular are drilling shock absorbers [18–21], dynamic single- and multi-mass vibration dampers [22,23], and specialized dampers with a high bearing capacity [24–27]. Similar anti-vibration devices, which prevent the occurrence of excessive dynamic loads, are often used in oil fields during the extraction of hydrocarbons by deep pumps [28,29]. For the effective use of these means of vibration protection for the drill string, high-quality flushing of the well should be ensured [30,31], and the surfaces of the steel parts of the shock absorbers, which are actively in contact with the abrasive, should be protected with functional coatings [32–35]. Also, issues related to early failure, loss of tightness, and the inadvertent self-loosening of threaded connections as a result of exposure to high vibrations should receive special attention [36–38].

The motivation for conducting this study was a number of real industrial complications and emergencies, the solution of which the authors of the article took a direct part in. When constructing the well, the main indicator of the drilling stage is mechanical speed, because the time for a complete cycle and the cost of constructing a well depend significantly on it. Today, during drilling operations in the deposits of the Dnipro–Donetsk Basin (Ukraine), in order to obtain high technical and economic indicators, drilling operators use blade bits of an aggressive design and apply forced drilling modes. As field studies have shown, uncontrolled cutting depths and stratified rocks with sharply changing characteristics very often lead to instant stops or jamming of such reaches. The most characteristic data on the occurrence of emergencies during drilling in the deposits of the Dnipro–Donetsk Basin were selected and are presented in Table 1. The presented information is taken from the "Accident Investigation Acts" compiled by a group of drilling and service companies. Some industrial cases noted in this table will be discussed and analyzed in more detail in the next section of the article. The main mistakes that led to complications in the drilling process were the incorrect selection of the layout of the drill string bottomhole assembly (BHA), the selection of an irrational drilling mode, and ignoring the use of anti-vibration devices. Drill string failures occurred due to overloading, excessive vibrations, or load fluctuations. Vibrations and cyclic loads mainly led to fatigue failure and loosening of the drill bit, while overloading the equipment caused loss of element stability, failure due to high contact stresses, or the phenomenon of material flow and rupture.

According to the data in Table 1, it is possible to distinguish the age of the rocks and the diameter of the bit, which caused complications in the drilling operations. The largest share of emergencies fell within the Triassic system (T) and the upper stage of the Carboniferous system (C3), which represents 62.5% and 18.75% of all the cases considered in the table. On the other hand, when drilling in the Cretaceous system (K), Jurassic system (J) and Serpukhovian stage of the Carboniferous system (C1s2), emergencies rarely occurred. Regarding the diameter of the bits, 50% of the abnormal cases happened with bits with a diameter of 444.5 mm, and 31.25% with bits with a diameter of 295.3 mm. To solve the described problem, an initiative has been proposed, which consists in conducting field research on the causes of emergencies caused by the BHA component integrity violation.

The main goal of this initiative is to collect industrial data on the nature and magnitude of downhole vibrations and to provide recommendations on optimizing drilling modes and the correct choice of BHA layout.

**Table 1.** Cases of emergencies in the Dnipro–Donetsk Basin fields.

| No | Drilling Depth, m | Stratigraphy | Place of Tool Damage | Type of Drill Bit | Diameter of the Bit, mm | Diameter of the Drilling Tool, mm | Axial Load on the Bit, kN | Rotational Frequency of the Upper Power Drive, rpm |
|---|---|---|---|---|---|---|---|---|
| 1 | 1009 | K | Adapter | GT55Cs | 393.70 | 241.30 | 120 | 60 |
| 2 | 1420 | J | PDM * | X616S | 444.50 | 241.30 | 70 | 30 |
| 3 | 2656 | T | PDM | GS1915T | 444.50 | 241.30 | 80 | 55 |
| 4 | 2293 | T | PDM | SI 516 HBPX | 295.30 | 203.2 | 140 | 40 |
| 5 | 2686 | T | PDM | VS616 | 444.50 | 241.30 | 40–80 | 20–40 |
| 6 | 2067 | T | PDM | SDI519MHBPX | 393.70 | 241.30 | 160 | 30–40 |
| 7 | 2286 | T | PDM | D516M HPX | 295.30 | 203.2 | 50–80 | 90 |
| 8 | 1695 | T | Shock absorber | TCTi + CR | 444.50 | 241.30 | 60–80 | 120 |
| 9 | 1688 | T | PDM | MSi619LHEBPX | 444.50 | 241.30 | 60–110 | 20 |
| 10 | 1752 | T | PDM | 8D31516HBPX | 295.30 | 203.2 | 100 | 50 |
| 11 | 1977 | T | PDM | U519S | 444.50 | 241.30 | 100 | 40 |
| 12 | 2372 | T | PDM | VS616DGHXU | 444.50 | 241.30 | 160 | 55 |
| 13 | 2558 | C3 | PDM | U519S | 311.15 | 203.2 | 120 | 40 |
| 14 | 2971 | C3 | PDM | XR + C | 295.30 | 203.2 | 90 | 40 |
| 15 | 2679 | C3 | PDM | CDi619VHBPX | 444.50 | 241.30 | 60–80 | 45–50 |
| 16 | 4375 | C1s2 | PDM | Z516S | 295.30 | 203.2 | 60 | 50 |

* PDM—positive displacement motor.

Machines, mechanisms, and devices that are equipped with various types of sensors and sensors are quite widely used in various fields of science and technology [39–41]. A number of critical infrastructure elements, such as oil and gas and energy facilities [42–44], as well as a number of critically important programs related to the use of a smart network, the Internet of Things, etc., cannot function properly without the participation of modern sensor components [45,46]. Today, the level of available technologies allows pothole vibrations to be measured using independent pothole sensors [47–49]. According to the principle of operation, such devices can be divided into two types: those that record data into the memory system and those that transmit a signal to the daylight surface in real time [50–53].

Recent studies have extensively investigated the dynamics of drill string vibrations and shock loads, emphasizing the critical role of sensor technology in this area [54]. Researchers have explored various types of sensors, including accelerometers, strain gauges, and fiber optic sensors, to monitor the behavior of drill strings in real time. For instance, several works [55,56] highlight the effectiveness of accelerometers in detecting vibration patterns that indicate potential failures in drilling operations. The development of wireless sensor networks allows for the real-time transmission of data from remote drilling sites, facilitating immediate analysis and decision-making. Research studies [57,58] have demonstrated how integrating wireless sensors with advanced telemetry systems can optimize drilling parameters by providing continuous feedback on operational conditions. Additionally, innovations in miniaturization and energy harvesting have made it feasible to deploy sensors in challenging environments, ensuring that data collection is both comprehensive and reliable [59]. The practical applications of sensor technology in drilling operations are numerous. Case studies, such as those conducted by the study in [60], illustrate how the real-time monitoring of vibrations and shock loads can lead to proactive maintenance strategies, reducing the risk of equipment failure. Machine learning algorithms and signal processing methods are increasingly employed to analyze complex datasets generated by sensors. Research by [61–63] highlights the potential of machine learning models to predict drill string failures based on vibration patterns, allowing for timely interventions.

The most advanced sensors transmit data during drilling to the daylight surface for a prompt response to changes in the drill string dynamic operating mode [64,65]. Real-time data transmission can be carried out by an electromagnetic channel (in this case, with a depth limitation) or a hydraulic communication channel (in this case, the quality of the drilling fluid, pumping group, and hydraulic system of the drilling rig has a significant impact on the transmission of the signal). For the customer of drilling works, the use of this type of equipment has one drawback—it increases the overall cost of the service. Therefore, in the oil and gas market of Ukraine, customers mostly refuse to use the described complex equipment.

Since we are trying to introduce technology and make it available to every customer in the Ukrainian market, we focus on the sensor that records the data and saves them in the memory system. Such data, obtained from the device's memory, can be compared with the events that occurred during drilling and with the performance of the equipment to increase the efficiency of drilling and to prevent complications. In the process of choosing an autonomous module, our attention was drawn to the Smart 4 technology presented by Innova Power Solutions. The main business of the company is the production of high-temperature lithium batteries for telemetry systems and sensors of various types. Starting in 2008, such batteries were equipped with Smart 2 and Smart 3 battery health monitors ("Innova Power Solutions", Calgary, Canada), which recorded temperature, residual current, and voltage. Since 2015, the technology of the controller has been improved to Smart 4, which is equipped with a three-axis accelerometer for recording shocks and vibrations [66–68]. The latest version of the controller includes gyroscopes that can measure the rotation frequency of the punching tool and capture the stick–slip phenomenon. According to the authors, this is the most interesting technology for implementation on the Ukrainian market, as it is the most affordable, and in terms of accuracy and frequency of data registration, it is not inferior to much more expensive analogs. However, the Smart 4 controller technology was developed to monitor the condition of lithium batteries and needs to be adapted in order to be used to register the dynamic mode of operation of the bottom of the drill string.

Our research aims to achieve the following:

- Analyze the typical examples of complications and emergencies that occurred during drilling operations in the deposits of the Dnipro–Donetsk Basin that are related to the dynamic operation mode of the of the drill string bottom part;
- Develop a module for using the selected sensor with a three-axis accelerometer and a gyroscope for recording shocks and vibrations at the lower part of the drill string and implement industrial approval of the proposed device.

Concluding the introductory section, we would like to note that the implementation of the proposed technology would provide an opportunity for each operator to assess the severity of the impact of vibrations and shocks when drilling deep oil and gas and geothermal wells and to compare the obtained data with critical events that occur during drilling. Service centers and service departments engaged in the operation and repair of drilling equipment will be able to obtain an evidence base for resolving warranty disputes or claims.

## 2. Materials and Methods

### *2.1. Case Reports of the Vibration Impact on the BHA Elements That Caused Complications or Emergencies*

In an attempt to increase the mechanical speed of drilling and minimize equipment costs, customers who rent drilling tools and PDMs from service centers often violate the provided equipment operation recommendations (i.e., they do not ensure that they use the proper BHA equipment, they exceed the recommended rotation frequency of the tool and axial load on the bit, they ignore the application anti-vibration devices, etc.). Such actions lead to the emergence of complications and emergencies in the drilling process, as well as guarantee disputes if there is equipment damage. Therefore, service centers are interested in the development and implementation of an inexpensive controller for recording the real drilling modes in which the drill string is operated.

#### 2.1.1. Effect of Transverse Vibrations

A PDC GT55Cs bit manufactured by the Halliburton company (Duncan, Oklahoma, USA) was used when drilling rocks of the Cretaceous system at one of the deposits in the Dnipro–Donetsk Basin. A shock absorber and a hydraulic pump were included in the BHA scheme. To isolate the PDM spindle section from axial vibrations, the customer decided to install the shock absorber below the motor, but no compromise was reached regarding the selection of the number and diameter of the support-centering elements. The following BHA scheme was used: 15 ½ Bit GT55Cs (0.48 m) + Crossover sub (0.38 m) + 9 ½ Shock Tool (4.3 m) + Crossover sub (0.28 m) + 9 ½ Drilling motor (10.55 m) + Crossover sub (0.35 m) + 15 ½ String Stabilizer (1.35 m) + 9 Drill Collar (17.65 m) + 15 ½ String Stabilizer (1.35 m) + 8 Drill Collar (24.07 m) + Crossover sub (0.45 m) + 8 Jar (5.39 m) + Crossover sub (0.44 m) + 8 Drill Collar (51.08 m) + Crossover sub (0.33 m) + 5 Drill Pipe (to surface).

During drilling at a depth of 1009 m, the pressure in the well flushing system decreased, and the mechanical speed of drilling decreased to zero. After lifting the tool, a fracture of the threaded nipple of the adapter between the PDM and the at-bit shock absorber was discovered (Figure 1a).

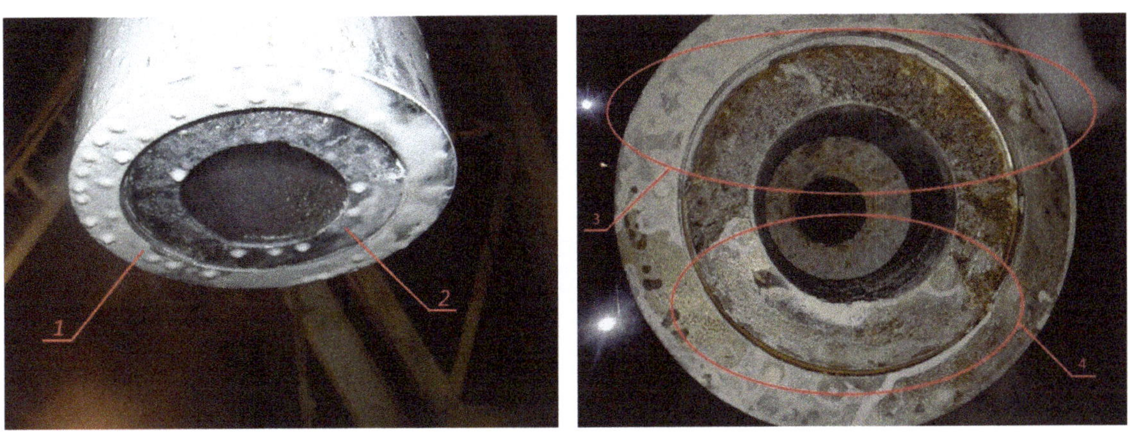


(**a**) (**b**)

**Figure 1.** The case of the destruction of the threaded nipple of the adapter between the PDM and the at-bit shock absorber: (**a**) View of the destroyed element of the drill string immediately after lifting from the well; (**b**) A detailed inspection of the place of damage to the tool; 1—the thrust end of the threaded coupling; 2—part of the adapter's nipple; 3—zone of fracture, 4—zone of crack formation and propagation.

Attention should be paid to the peculiar nature of the adapter nipple fracture (Figure 1b). Here, two separate zones of material destruction can be distinguished: zone 4 is a flat, smooth, polished surface where a fatigue crack has formed and grown; and zone 3 is a coarse-grained surface, which is a zone of sudden fracture.

The main reasons of the described complication are the insufficient stabilization of the BHA bottom part, partially the lack of a body stabilizer on the PDM and too-high rigidity at the part of the layout above the motor, since full-size calibrators were used. In the applied BHA, the distance from the bit to the lower calibrator was 15.5 m, which is essentially a pendulum scheme. The lack of a centering element on the PDM housing led to the formation of excessive deflections on the shock absorber and the PDM, which were subjected to cyclic loads of alternating signs during drilling. The use of two full-size calibrators above the motor limited the area of propagation of transverse vibrations, so the vibrations were concentrated in the area of the punching tools. The adapter between the shock absorber and the PDM was the most stressed, in which a fatigue crack formed under the action of transverse vibrations, which ultimately led to its destruction.

Analysis of the dynamics of the considered BHA was carried out using the specialized Halliburton (Duncan, OK, USA) software product, Wellplan-Software (version 5000.14.0) [69]. Figure 2a shows the dependence of the maximum bending moment in the drill string on the speed of the tool rotation (here, drilling was carried out in a combined way: a maximum of 99 rpm was provided by the PDM, and the rotor was used to increase the speed). Note that the z-axis is directed along the axis of the drill string, and the x- and y-axes are the main central axes of the cross-section of the string. One can see on this graph that at a rotation speed of around 162 rpm, there was a sharp disturbance of the bending moment, which reached more than 45 kNm. At the same time, Figure 2b illustrates that at a safe drilling speed, the maximum bending moment in any of the sections of the column did not exceed 10 kNm (the maximum value was registered in the area of the threaded adapter between the hydraulic cylinder and the shock absorber). In fact, the forced mode of drilling led to the occurrence of an excessive bending moment, which ultimately caused the emergency.

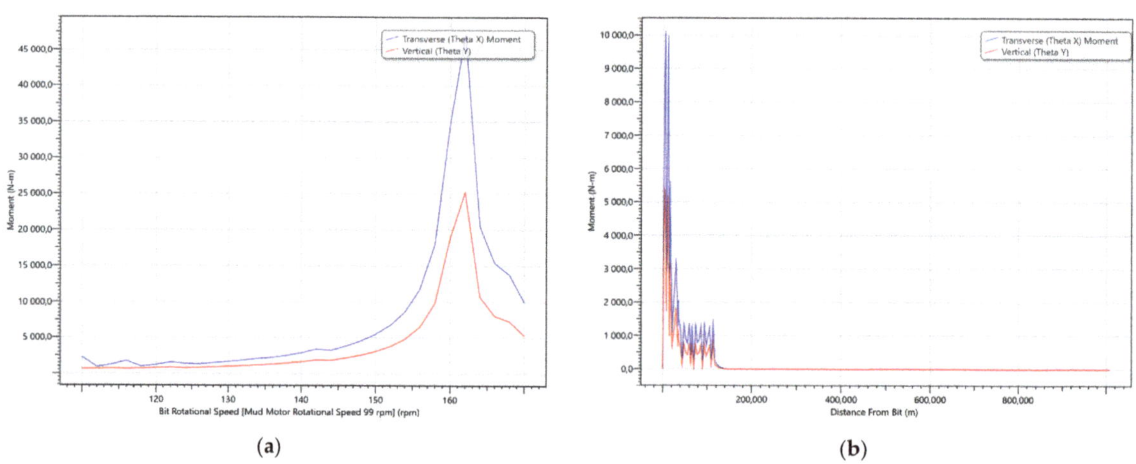


**Figure 2.** Bending moments in the drill string: (**a**) Dependence of the maximum bending moment on the rotation speed of the drill string; (**b**) Distribution of the maximum bending moment along the length of the column at a rotation speed of 153 rpm.

With regard to the considered complication, we note that a similar BHA, in which a drilling shock absorber was installed under the PDM while, at the same time, the motor body was equipped with an undersized centerer and calibrators were also used with a diameter smaller than the nominal diameter of the bit, successfully worked in the deposits of the Dnipro–Donetsk Basin.

### 2.1.2. Effect of Axial Vibrations

When drilling rocks of the Cretaceous and Jurassic systems in the range of 315–1208 m (for a technical column with a diameter of 340 mm), a BHA with a PDM and an over-bit shock absorber and a Smith SDGH roller bit was used (the shock absorber was installed above the engine). Actually, for vibration isolation of the drill string from axial vibrations, which are formed during the drilling of heterogeneous rocks of the upper sections, the shock absorber was included in the composition of the BHA. The following arrangement of the BHA was used: 17 ½ Bit (0.55 m)+ 9 ½ Drilling Motor (10.5 m) + 17 3/8 String Stabilizer (2.43 m)+ 9 ½ Drill Collar (8.2 m) + 9 ½ Shock Tool (3.75 m) + 9 ½ Drill Collar (9.22 m) + Crossover sub (0.73 m) + 15 3/8 String Stabilizer (0.96 m) + 9 Drill Collar (18.28 m) + 15 3/8 String Stabilizer (0.94 m) + 9 Drill Collar (23.27 m) + 15 ¼ String Stabilizer (0.96 m) + Crossover sub (0.45 m) + 8 Drill Collar (29.73 m) + Crossover sub (0.6 m) + 8 Drilling Jar (5.41 m) + Crossover sub (0.43 m) + 8 Drill Collar (37.77 m) + Crossover sub (0.43 m) + 5 ½ Drill Pipe (to surface).

The use of the shock absorber allowed the customer to create an axial load on the bit of up to 18 tons to obtain the maximum drilling speed without axial vibrations of the drill string being visible at the day surface. After reaching the design depth of the casing column, the drilling tool was lifted, and excessive axial backlash was found—40 mm on the spindle section of the gas turbine engine (with an allowable value of no more than 2 mm). When disassembling the PDM, the destruction of the thrust-bearing separators and mechanical damage to the internal elements of the spindle section of the PDM were found (Figure 3).

(**a**) (**b**)

**Figure 3.** Damage to thrust-bearing separators and internal elements of the spindle section of the screw engine: (**a**) Damaged elements of the downhole motor; (**b**) Inspection of the spindle part of the downhole motor; 1 and 2—upper and lower bearing brackets; 3 and 4—bearing rollers, 5—PDM bearing housing.

Such damage is caused by the excessive impact of axial vibrations on the bearing support of the engine and the high contact stresses that occur during the dynamic contact interaction between the support elements. Since the shock absorber was installed above the PDM, the thrust bearings of the spindle section turned out to be unprotected from vibration loads. In general, during the drilling of the Cretaceous, Jurassic, and Triassic systems with ball bits in the oil and gas fields of the Dnipro–Donetsk Basin, significant axial oscillations were observed at the wellhead when an axial load of more than 10 tons was created. In the case under consideration, an axial load of up to 18 tons was used on the bit; at the same time, there were no vibrations of the drill string and the oil system at the day surface, and no danger was visually observed (the shock absorber was activated above the PDM). Using the specialized software package "Halliburton: Wellplan-Software (version 5000.14.0)" [69], the stresses that arose in the BHA elements were analyzed. The graphs (Figure 4a) show the dependence of the maximum stresses in the drill string on the rotation speed. Figure 4b illustrates the distribution of stresses along the length of the column at the maximum rotation frequency of 144 rpm. It should be noted that the stresses on the graphs did not exceed the limit values; however, vibration loads at a high speed of rotation of the tool led to high dynamic contact stresses and caused damage to the PDM thrust bearing.

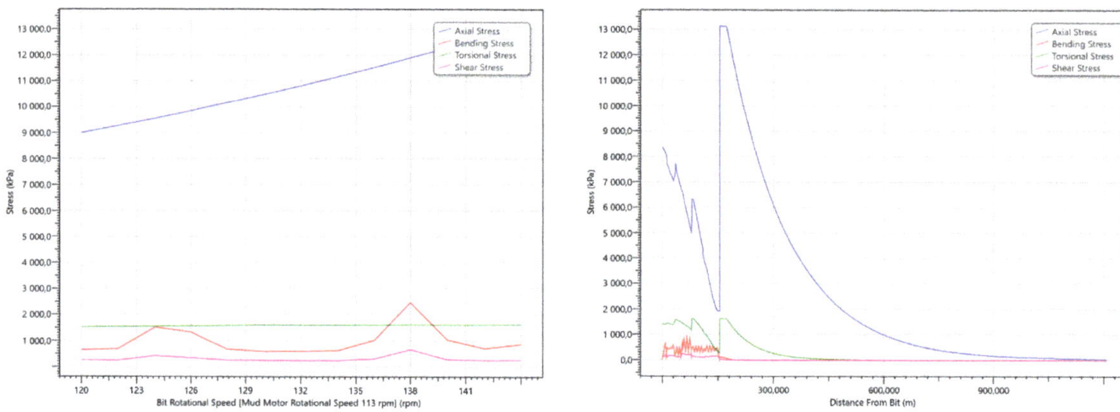


(a) (b)

**Figure 4.** Stresses in the elements of the drill string: (**a**) Dependence of the maximum stresses on the rotation speed of the drill string; (**b**) Stress distribution along the length of the column at a rotation speed of 144 rpm.

For the considered type of BHA, it is recommended to use a low-speed PDM for the possibility of reducing the rotation frequency of the bit from 145 rpm to 110 rpm. In order to ensure maximum drilling efficiency, to protect against the vibrations of the internal combustion engine, and to increase the duration of the trip, it is necessary to install a drilling shock absorber directly above the bit.

2.1.3. Effect of the Whirling Movement of Drilling Tools

A PDC Smith MSi619LHEBPX bit (Ackerman drilling services, Houston, TX, USA) and a 241 mm diameter rotary screw motor were used to drill the Triassic system in the interval of 1273–1688 m in one of the wells in the Dnipro–Donetsk Basin. A thick layer of sandy-clay soils with the inclusion of conglomerates and rocks prone to erosion and shedding represented the geological section in this interval. The following BHA was used for drilling: 17 ½ Drill Bit (0.63 m) + 17 ½ Dog Sub (0.60 m) + 9 ½ Drilling motor (10.20 m) + 17 ½ String Stabilizer (1.69 m) + 10 Drill Collar (8.18 m) + Crossover sub (0.44 m) + 9 ½ Shock Tool (3.29 m) + Crossover sub (0.4 m) + 10 Drill Collar (8.19 m) + 17 ½ String Stabilizer (1.69 m) + 10 Drill Collar (16.48 m) + 16 3/8 String Stabilizer (1.69 m) + 10 Drill Collar (8.30 m) + 9 Drill Collar (9.40 m) + Crossover sub (0.40 m) + 8 Drill Collar (18.82 m) + 8 Jar (9.7 m) + 8 Drill Collar (37.53 m) + Crossover sub (0.43 m) + 5 HWDP (56.30 m) + Crossover sub (0.39 m) + 5 ½ Drill Pipe (to surface).

When the bottomhole was at a depth of 1688 m, a cyclical change (increase–decrease) of the torque began in the range from 6 to 22 kNm with a constant axial load on the bit.

This process was accompanied by a rapid increase in the pressure of the washing liquid in the injection line in the range from 1.5 to 3.5 MPa. Every time the torque increased, the operator performed a smooth separation of the bit from the surface of the hole and started drilling again. During the next restoration of circulation after the bit was detached from the hole, a decrease in the idle pressure of the PDM by 1.5 MPa was recorded. A decision was made to raise the BHA for revision by the PDM. During the revision of the tool, a break in the spindle part of the PDM, destruction of the arms on the lower edges of the calibrator blades, and a partial loss of the diameter of the lower calibrator were found (Figure 5).

(a)

(b)

**Figure 5.** Breakage of the spindle part of the PDM due to the whirling movement of the drilling tool: (**a**) Damaged calibrator blades; (**b**) Detecting a break in the spindle part of a downhole screw motor when lifting a tool; 1—the zone of destruction of the lower edge of the calibrator blade; 2—slivers of the hard alloy armament of the stabilizer; 3—place of breakage of the PDM housing, 4—PDM housing.

Overshots with spiral/collet grabs and 350 and 380 mm diameter feed funnels were used to extract the remaining part of the BHA from the well. The lifting process and analysis of the condition of the emergency (broken) layout provided accurate indications that the actual diameter of the wellbore had exceeded the nominal one by 80%. Taking into account the above, it was concluded that during the drilling of the well, there was a whirling vibration of the BSC, which led to a significant increase in stress in the threaded elements of the PDM, uneven wear of the bit armature and calibrators, and, ultimately, to an emergency situation.

Modeling of the behavior of the considered BHA (Figure 6) indicates a surge of bending stresses and transverse displacements of the drilling tool during the forced drilling mode (the rotation frequency of about 130 rpm is very high for the existing engineering and geological conditions). According to the authors, the formation of vortex oscillations in the considered case is primarily caused by the high cavernousness of the wellbore, as well as non-compliance with the recommended drilling mode. To prevent similar situations when drilling unstable deposits of the Triassic system at this deposit, it is recommended to use a rotary BHA, as well as to include downhole shock absorbers in the drill string for damping longitudinal vibrations (when using rotary bits) and torsional vibrations (when using blade bits).

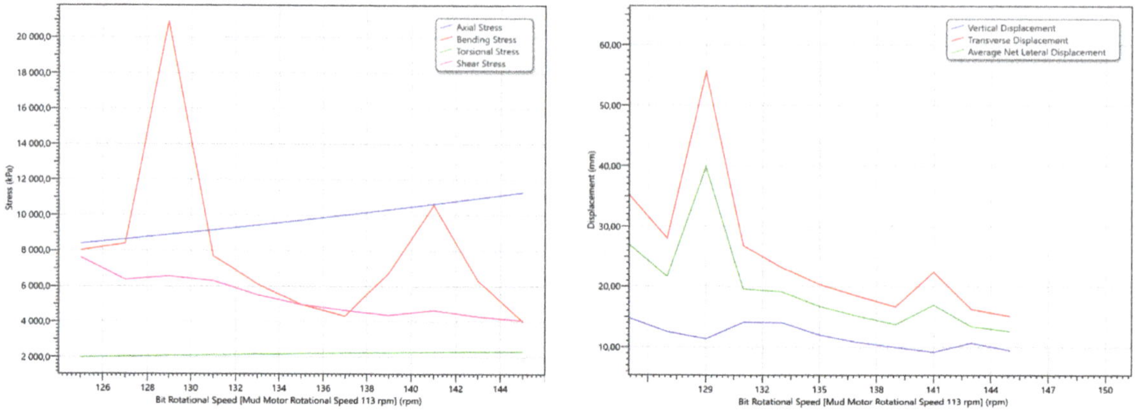


(a) (b)

**Figure 6.** The maximal stresses and displacements in the bottom elements of the drill string: (**a**) Dependence of the maximum stresses on the rotation speed of the drill string; (**b**) Dependence of the maximum movements on the rotation speed of the drill string.

2.1.4. Effect of Torsional Vibrations

During the drilling of a deep well in one of the fields of the Dnipro–Donetsk Basin, a 241 mm diameter drill bit and a Varel VS616DGHXU PDC drill (Varel Energy Solutions, Mexico) bit were used. The deepening took place in the Triassic system in the interval of 2189–2372 m. The rocks were mainly represented by red viscous clays. The following BHA was used: 17 ½ Bit (0.46 m) + 9 ½ Drilling Motor with 17 ¼ sleeve stabilizer (10.63 m) + 17 ¼ String Stabilizer (2.5 m) + 9 ½ Drill Collar (18.15 m) + 17 ¼ String Stabilizer (2.49 m) + 9 ½ Drill Collar (26.89 m) + Crossover sub (0.78 m) + 8 ½ Drill Collar (54.8 m) + 8 Jar (9.6 m) + 8 ¼ Drill Collar (26.98 m) + Crossover sub (0.67 m) + 6 ¾ Drill Collar (26.91 m) + 5 HWDP (55.05 m).

During the trip, the mechanical speed of drilling and the torque were constantly changing, which obviously indicates the layering of rocks. The operator tried to adjust the drilling mode according to the actual situation. At the drilling interval of 2365–2372 m, cyclic pressure increases in the range of 30.0–40.0 MPa began to occur, which were accompanied by stops of the upper power drive and further acceleration of the rotation of the drilling tool clockwise (these are convincing signs of torsional vibrations). With the next increase in pressure in the injection line, the operator performed the traditional procedure of reducing the supply of washing liquid and performing a smooth separation of the bit from the hole in order to prevent the destructive effect of vibration loads. After restoring the circulation of the washing liquid, an increase in the engine idle pressure from 22.5 to 24.5 MPa was noted, with constant performance of the drilling pumps. During further attempts to resume drilling, the absence of a differential drop in the PDM was recorded with an axial load on the bit of up to 8 tons. A decision was made to raise the drilling tool to determine the causes of the complication.

During a visual inspection of the gas turbine engine after lifting the drilling tool, a fracture of the body stabilizer at the place of the transition of the blades into the body as well as a fracture of the threaded nipple of the upper adapter of the angle regulator were detected (Figure 7).

The conclusion of the analysis of the emergency is that the damage was caused by distortion of the stick–slip torsional vibrations. Usually, when the reactive moment is released, the stabilizers are unscrewed, since the rotation of the drill string occurs in the opposite direction to the direction of the thread. In this situation, after the bit was jammed, there was a sharp start of the PDM with high angular acceleration, which led to the attachment of the threaded connection of the stabilizer and, as a result, to the violation of its integrity. According to the authors, the fracture of the threaded connection of the upper adapter of the angle regulator occurred simultaneously with the fracture of the stabilizer under the action of the same dynamic loads. The estimate of the relative angle of twisting of the BHA when a dynamic torque occurs was simulated in the "Halliburton (Duncan, OK, USA): Wellplan-Software (version 5000.14.0)" environment [69] and is presented in Figure 8. In the lower part of the drill string, we observed a disturbance in the angle of rotation of the cross-sections of the string around its longitudinal axis. In particular, the cross-sections of the PDM were turned more than a full revolution relative to the conventionally stationary cross-section of the column.

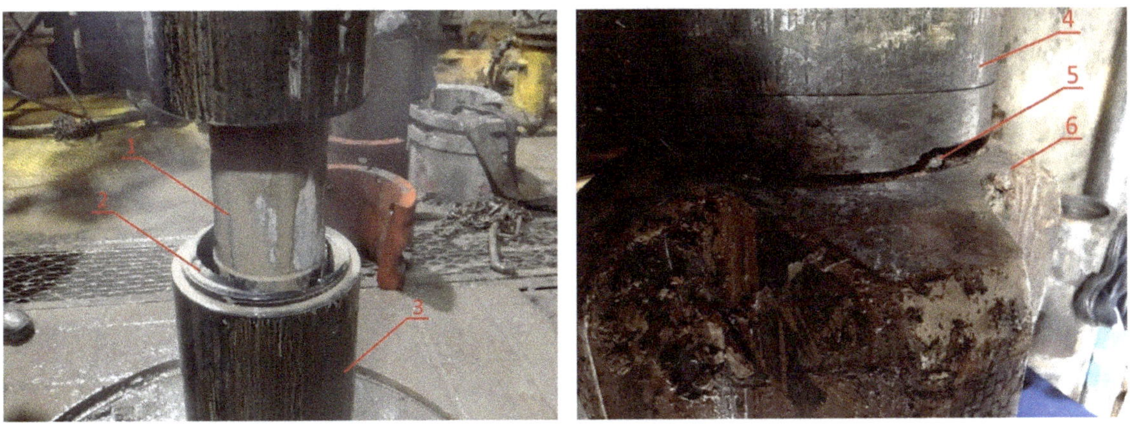


(**a**) (**b**)

**Figure 7.** A visual inspection of the damage to the PDM caused by the influence of torsional vibrations: (**a**) Lifting a damaged downhole screw motor; (**b**) Destroyed stabilizer housing; 1—the rotor of the power pair; 2—the place of the fracture of the threaded nipple of the angle regulator adapter; 3, 4—PDM housing; 5—the location of the housing stabilizer fracture; 6—stabilizer blades.

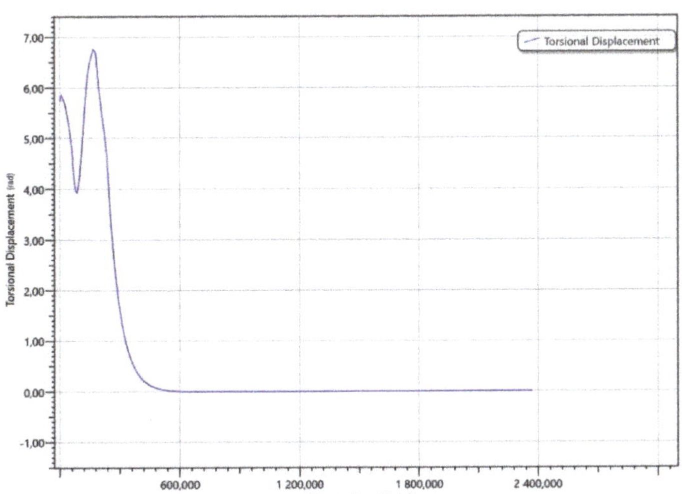


**Figure 8.** Twisting angles of cross-sections of the lower part of the drill string during complications.

### 2.2. A Device for Recording Bottomhole Vibrations

The main idea underlying the principle of operation of the sensor for recording bottomhole vibrations is to convert mechanical vibrations into electrical signals, which will then be processed and interpreted. Oscillations are detected using accelerometers and gyroscopes. When the sensor detects vibrations, it generates an electrical signal whose frequency and amplitude depend on the vibration parameters. In other words, the vibration sensor acts as an interface between the mechanical vibrations and the electronic data processing system, which allows useful information to be obtained about the dynamic mode of operation of the drill string.

A compact sensor with a three-axis accelerometer and a gyroscope (Figure 9) will be mounted inside a lithium thionyl chloride battery. The sensor has a very low power consumption and an extended memory capacity, which allows for uninterrupted logging of pitting data over 800–1000 h of operation.

**Figure 9.** Smart 4 controller for recording the vibrations of the bottom of the drill string (a coin is shown next to the controller for comparative size assessment).

The software product WatsonTM ("Innova Power Solutions", Canada) (Figure 10) is used to read the data, in the interface of which it is possible to change the scale of the curves and graphically visualize various parameters, including the temperature, rotation frequency, vibrations, and shocks along three mutually perpendicular axes [67]. The advantage of this interface is that digital data can also be obtained in a format convenient for further processing and interpretation; for example, they are available not only in the form of diagrams but also in the form of tables with a set of numerical data. The device has its own digital clock and records the event with reference to time. When connecting the controller to the computer, time synchronization occurs, which allows one to get a clearer picture of the pitting data. The device allows one to record the mean square value of the vibrations, as well as longitudinal and transverse vibrations along three axes, and can work at temperatures up to 180 °C. The unit of measurement of vibrations is g, and the frequency of data registration is within 10 s.

The sensor allows you to obtain the value of the amplitude of oscillations, which reflects the maximum deviation or acceleration of the drilling tool. This is an important indicator for determining the intensity of vibrations and their impact on drilling and equipment. It is also possible to analyze the oscillation frequency, which can indicate the source of the vibration. In addition, the sensor allows you to determine whether the oscillations are regular or irregular. Regular fluctuations may indicate problems with the drilling equipment, instability of the drilling process, or other technical problems, while irregular fluctuations may be caused by external factors such as geological features, such as rock stratification or heterogeneous inclusions. The data obtained by the sensor can be compared with the drilling modes, which are recorded 24/7 by the geological and technical research station. Furthermore, these data can be additionally superimposed onto the logs of the geophysical studies of the well to obtain a clear understanding of the influence of all possible factors on the dynamics of the layout of the bottom of the drill string. So, for example, by having a gamma log, one can easily monitor the change in the type of rocks, and profilometry gives an understanding of the actual diameter of the wellbore relative to the nominal one in one or another interval, which has a direct impact on the occurrence of fluctuations during drilling.

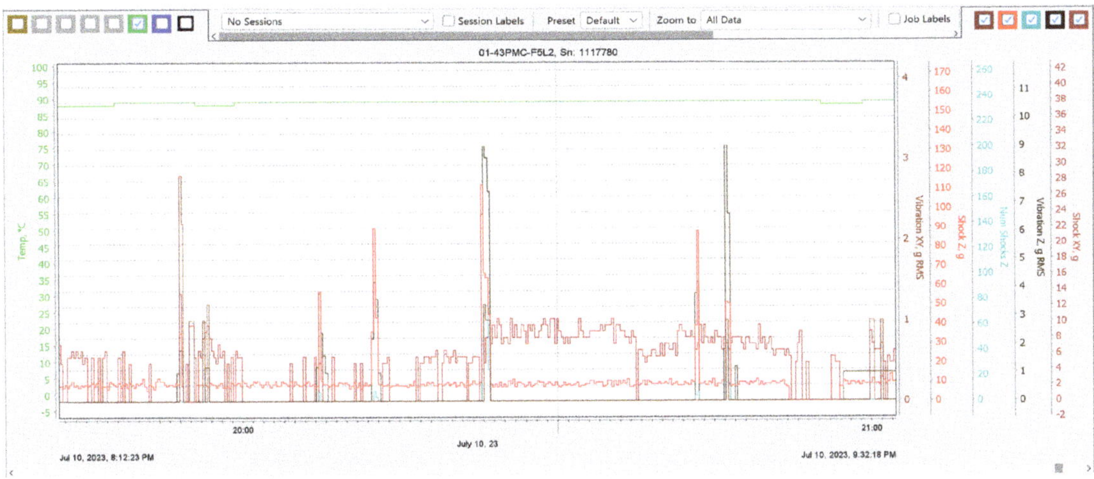


**Figure 10.** Software product interface for reading data from the controller memory.

### *2.3. The Design of the Module for Placing the Vibration Sensor in the Drill String Bottom Part*

An important problem that needs a rational solution is the way of placing the sensor to record vibrations and the power battery in the lower part of the drill string. Extremely limited diameter space and extreme operating conditions significantly complicate this task. Depending on the specific drilling conditions and the requirements for the downhole vibration recording system, its placement can be performed in different ways. Sometimes, sensors are built directly into the diametrical pocket of the drill string wall, or sensors, together with power elements, can be placed in special external modules that are attached to the outer surface of the drill string and have protective casings or shells [1,70]. The study proposes to place the recording device in a specially made module, which is mounted inside the lower part of the drill string. Creating such a module requires attention to design, materials, and the fastening method. The module is specially designed to accommodate the necessary sensor, the lithium thionyl chloride battery, and electronic communications. The module material is resistant to corrosion and abrasive materials, and the module is sealed to protect the electronics from moisture, dirt, and other harmful influences. The module is made as compact and ergonomic as possible to reduce its impact on the drill string diameter and to leave more space for the drilling fluid to flow, and the external shape facilitates the smooth movement of the drilling fluid around the module.

So, for the practical implementation of the ideas outlined above, as well as to avoid the need to use high-cost telemetry systems, a compact module is proposed that includes a Smart 4 vibration sensor and an Innova Power Solutions lithium thionyl chloride battery (Figure 11). A feature of the Smart 4 sensor is its low power consumption, which makes it possible to use a power battery that is three times smaller than the power batteries for traditional telemetry systems.

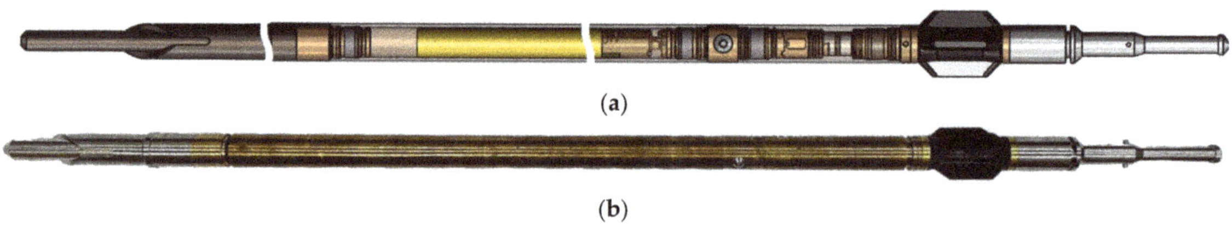


**Figure 11.** Compact module for installing a vibration sensor and power battery in the BHA: (**a**) Scheme of the module; (**b**) A sample of the assembled module.

Let us take a detailed look at the design and layout of the module from left to right. The guide tip (Figure 12a) is intended for sealing the module and for its landing in the guide sleeve, which is installed in the orienting adapter of the drill string. A two-pin adapter with sealing rings (Figure 12b) is required to connect the guide tip to the main body. The main body of the module (Figure 12c) contains a power element with a Smart 4 controller and a shock-absorbing spring in its inner cavity. The power battery is connected to the housing, which has a ventilation valve, through a support with a built-in cable, which plays the role of an MDM 9/Kintek adapter (Figure 12d). The Kintek connector is fixed with a special fastener and closed with a short upper case. The fishing neck (Figure 12f) is designed to seal the internal cavity of the module, and if necessary, it provides the possibility of lifting the module on a geophysical cable.

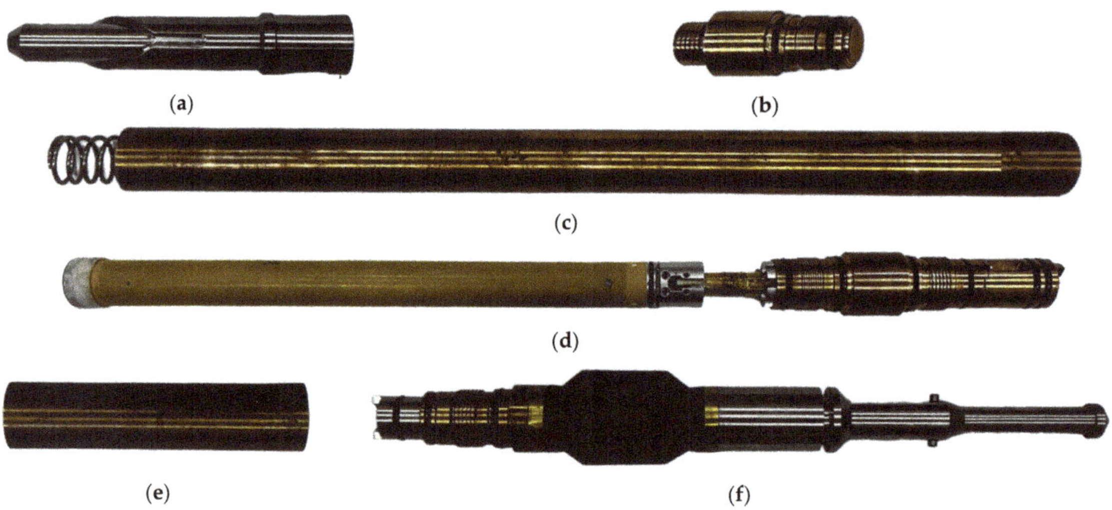


**Figure 12.** Layout diagram of the separate components used in the model assembly: (**a**) Guide tip; (**b**) Two-pin adapter; (**c**) The main body of the module with a shock-absorbing spring; (**d**) Power battery with vibration sensor and battery support; (**e**) The upper body of the contact group; (**f**) fishing neck.

After the module is assembled, it should be placed in the housing elements for its final installation at the bottom of the drill string. For this, it was decided to use a landing adapter (Figure 13a), into which the guide sleeve is installed, and the lower end of the module (guide tip) enters the sleeve itself. The sleeve is fixed with screws so that it does not turn and cannot move in the axial direction (Figure 13b). There is a special wedge in the sleeve, and a groove in the guide tip of the module. The wedge enters the groove, and the module remains in the fixed position (Figure 13c). The part of the module that protrudes beyond the landing adapter is hidden in a specially made pipe or in an element of a weighted drill pipe using an internal rubber centerer.

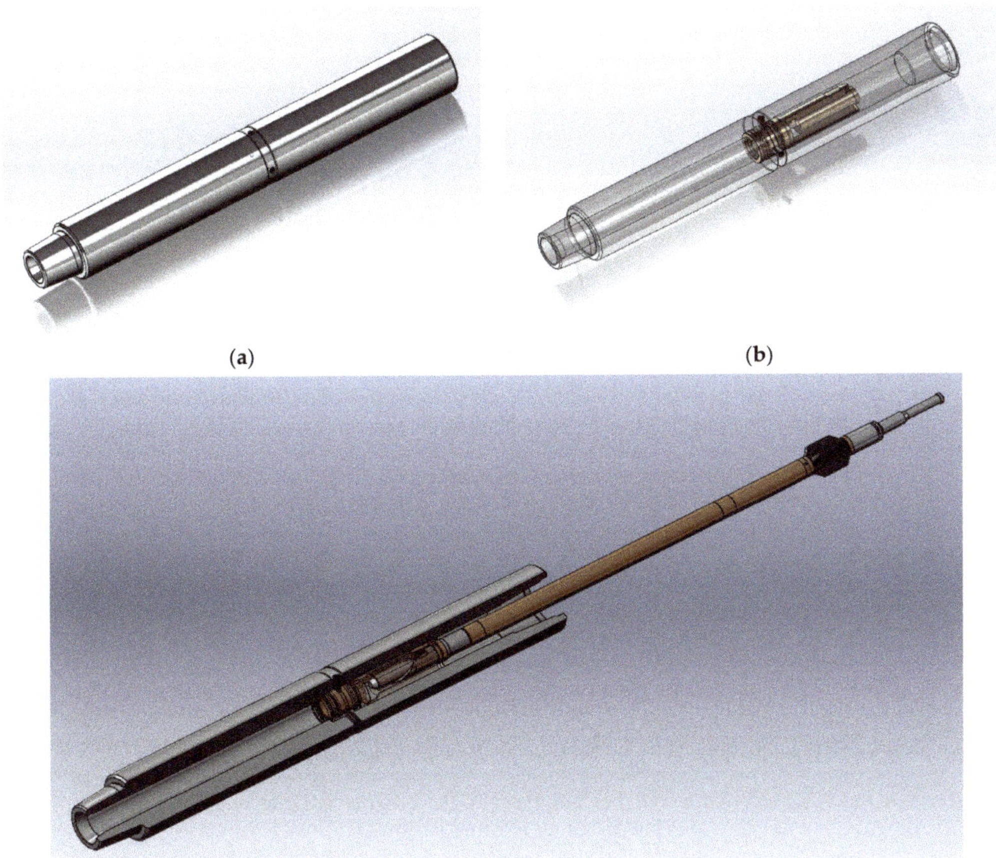

(a) (b) (c)

**Figure 13.** Assembly diagram of the assembled module in the landing adapter: (**a**) General view of the landing adapter; (**b**) Guide sleeve installed in the landing adapter and fixed with screws; (**c**) Fixation of the module in the guide sleeve.

Since magnetic interference does not affect the sensors of the Smart 4 controller, the body elements can be made of alloy steel, the characteristics of which correspond to industry standards, for example 45XHMA (AISI 4140, or AISI 4145). The landing gear, together with the module, will always be supplied by the service center. If the upper part of the module will be hidden in the weighted drill pipe (it is provided by the drilling contractor), then the final installation will take place on the drilling rig. If the upper part of the module is to be hidden in a special nozzle, then the device will be completely assembled at the production base of the service center. To obtain the recorded information, it will be enough to unscrew the upper fishing neck of the module when lifting the drill string, connect one end of the cable with a special connector to the contact group, and connect the other end of the cable to the laptop via the USB port. The described strategy will make it possible to effectively integrate the module with the vibration sensor into the drill string and ensure the reliable collection of data on the dynamic mode of operation of the drilling tool.

In order to expand the scope of application of the proposed device and to use it at different diameters of the wellbore, the housing part will be manufactured in diameters that correspond to the standard sizes of the weighted drill pipes and other elements of the BHA; namely, 241.1 mm, 228.6 mm, 203.2 mm, 171.4 mm and 127 mm. This will make it possible to measure vibrations when drilling with bits with a diameter of 444.5 mm (393.5 mm), 311.15 mm (295.3 mm), 215.9 mm, and 152.4 mm (165.1 mm), which covers all available diameters of wells in the deposits of the Dnipro–Donetsk Basin.

## 3. Results and Analysis

### 3.1. Industrial Approval of the Proposed Device for Registering Vibrations—Trip No.1

The Smart 4 controller ("Innova Power Solutions", Canada) with a lithium thionyl chloride power supply battery was installed in a compact module (Figure 11b) and mounted in the body parts at the production base of the service center. Such a device for assessing vibration and shock loads was installed in the lower part of the drill string.

Consider the history of the first trip. The process of construction of an inclined well was underway. After the completion of the next procedure of lowering and lifting operations to a depth of 2210 m, the drilling tool was planted, and the drilling process was resumed. A PDC Ulterra SPL616 (Ulterra Corporate, Calgary, Canada) drill bit was used during the voyage, and drilling took place in rocks of the Middle Carboniferous Bashkir stage. The following BHA was used: 12 ¼ Bit (0.37 m) + 8 Drilling Motor with a 12-sleeve stab and 1.5° bend (8.95 m) + 12 String Stabilizer (1.79 m) + 8 Pony Collar (1.84 m) + 8 UBHO sub (1.05 m) + 8 ¼ NMDC with MWD inside (7.04 m) + 8 ¼ NMDC (8.24 m) + Crossover Sub (0.71 m) + 8 Drill Collar (45.52 m) + Crossover Sub (0.76 m) + 8 Jar (8.69 m) + Crossover Sub (0.75 m) + 8 Drill Collar (27.91 m) + Crossover Sub (1.07 m) + 5 HWDP (27.56 m) + 6 5/8 Friction Reduction Tool (6.24 m) +5 HWDP (28.08 m) + Drill Pipe (to surface).

Starting from a depth of 3200 m, the stabilization section of the directional well was drilled. Initially, the well was deepened in the normal mode without any signs of complications. During drilling at the interval of 3387–3435 m, there was an uncontrolled drop in the zenith angle from 22.5 to 19.8 degrees, although slides were constantly made to set the zenith angle in the direction L10°–R10° (i.e., the deflector was directed toward the zenith angle). After this event, it was decided to lift the drilling tool to inspect the BHA. During the inspection of the bottom of the drill string assembly elements, the stabilizer was found to be loosened, the axial backlash in the BHA elements had increased from 0.5 mm to 30 mm, and the radial backlash was up to 5 mm (Figure 14).

In order to carry out a detailed study of the origin and causes of the detected damage, information was read from the memory of the pitting vibration recording device (the reading was carried out in the field). The diagram of all parameters of the measured oscillations and shocks is presented in Figure 15. Of course, the absolute value of acceleration is measured in the SI system in $m/s^2$. However, on the accelerogram, the unit of measurement is the acceleration of free fall ($1 m/s^2 \approx 0.102$ g).

Below, we detail the complex accelerogram and consider each component registered by the vibration sensor separately. Shocks during drilling are instantaneous transfers of energy from one body to another, which are caused by the impact of the bit, BHA, or drill string against the walls or downhole. Figure 16a shows the change of lateral shocks depending on time, including two directions: the x direction and the y direction (here, the x- and y-axes are the main central axes of the cross-section of the drill string). Figure 16b shows the change in axial shocks (in the direction along the $z$-axis) as a function of time. It should be noted that the severity of consequences of a shock depends on three parameters: the magnitude of the shock, the duration or length of time of the shock, and the frequency or number of shocks.

(**a**) (**b**)

**Figure 14.** Revision of the BHA after detection of signs of complications: (**a**) A PDM with an unscrewed body stabilizer; (**b**) The output shaft protrusion from the PDM housing exceeds the allowable value by three times (the cause was the occurrence of axial backlash due to the destruction of thrust-bearing separators).

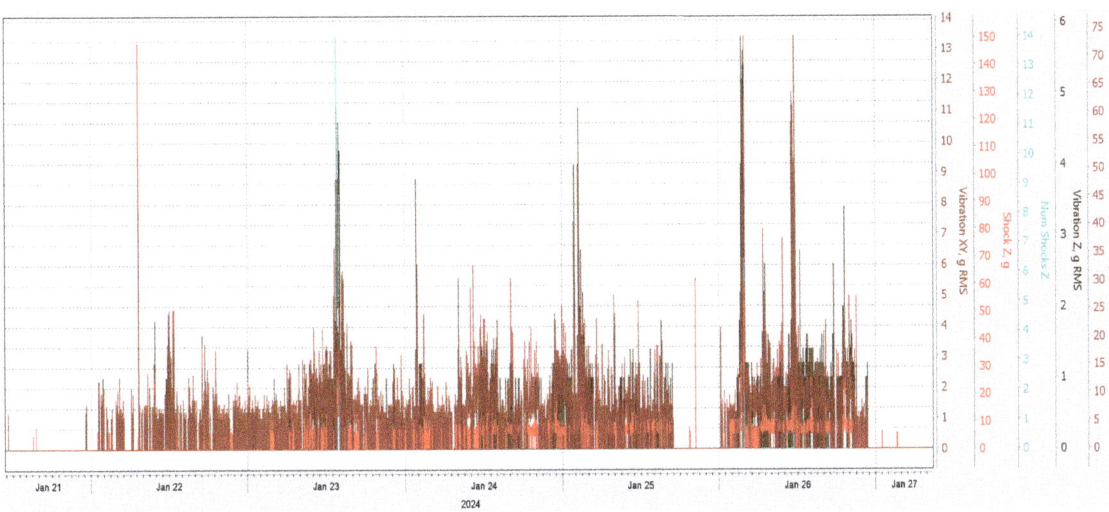


**Figure 15.** Complex accelerogram read in the field after raising the drilling tool.

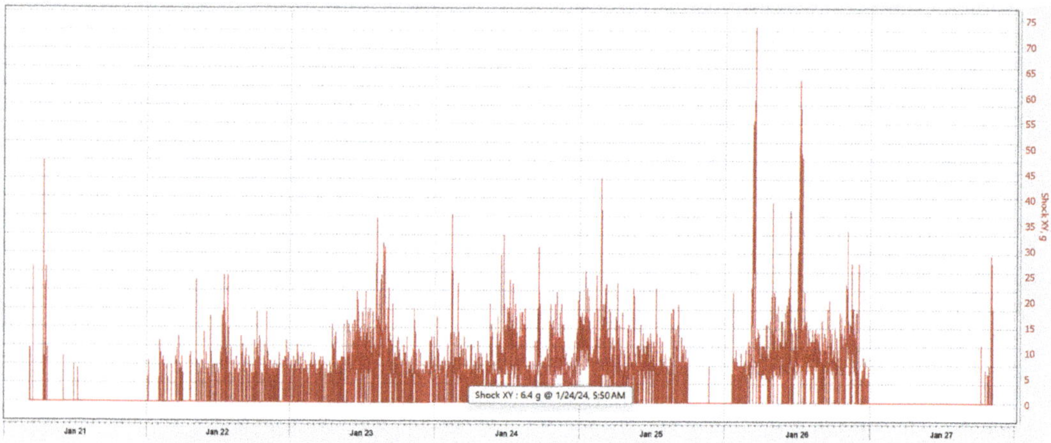

(a)

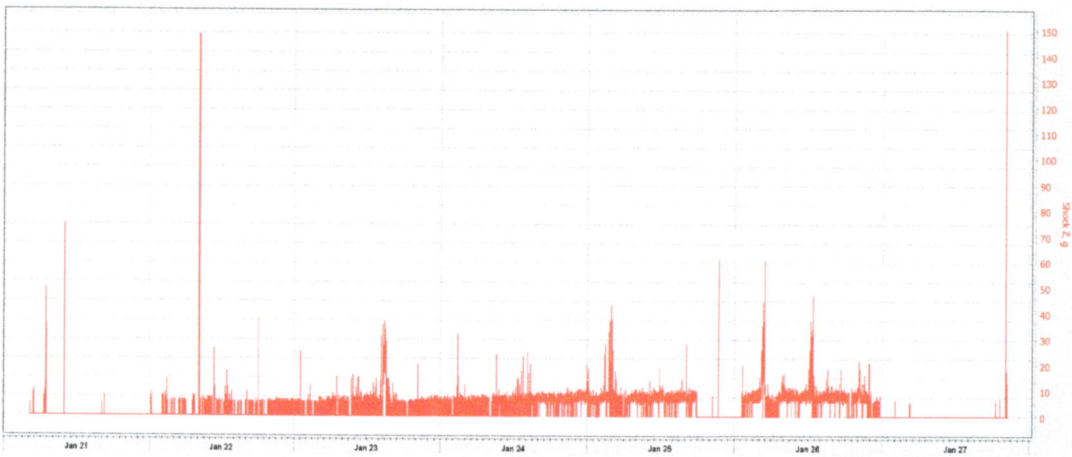

(b)

**Figure 16.** Registration of shocks in three mutually perpendicular directions during the first trip: (**a**) Change in the shocks in lateral directions; (**b**) Changes in the shocks in the axial direction.

For effective assessment of the dynamic regime of the drill string, vibration measurements (Figure 17) are carried out in the RMS (root mean square) system, which means the measurement of the root mean square value of the amplitude of the vibration signal. In the context of drilling tool vibration measurement, RMS indicates the average energy of the vibration signal as perceived by the mechanical system. This allows one to estimate the level of the vibration amplitude, taking into account both the maximum and minimum values of the signal while providing a convenient way to compare vibration signals. In the vibration diagram (Figure 17a), the RMS values of accelerations in the lateral direction (in the direction of the x- and y-axes) are determined on the basis of data that are monitored for ten seconds from the moment of registration of the previous value. The diagram of vibrations in the axial direction (in the direction of the z-axis) shown in Figure 17b is the root mean square value of accelerations in units of g determined from data recorded within ten seconds of the moment of fixation of the previous value in the diagram.

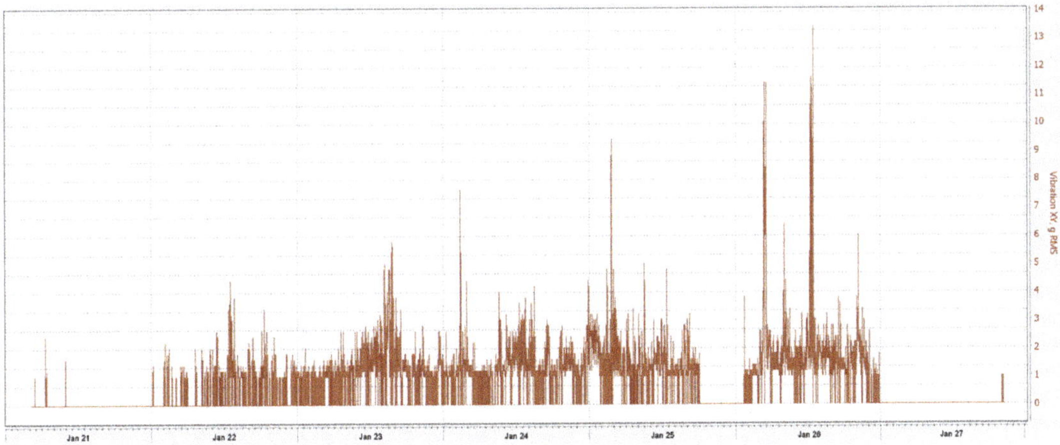


**(a)**

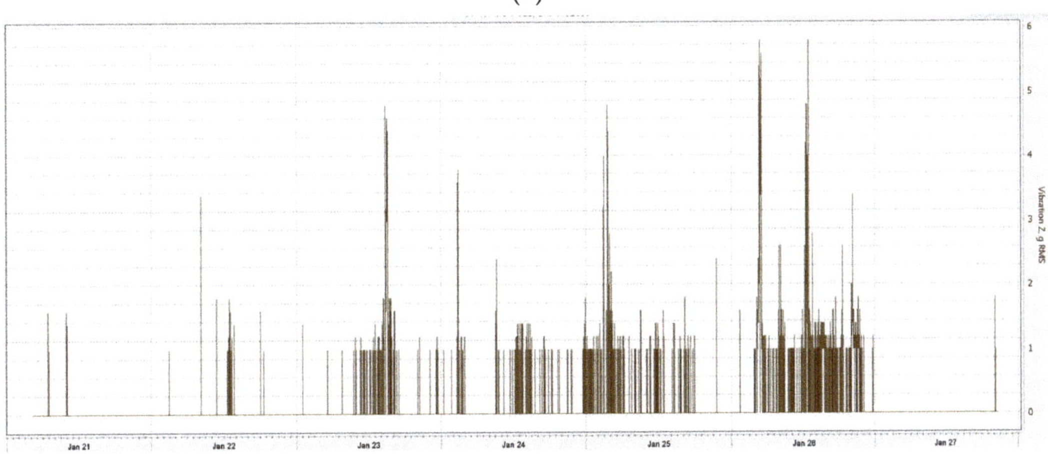


**(b)**

**Figure 17.** Registration of vibrations in three mutually perpendicular directions during the first trip: (**a**) Vibrations in the lateral directions; (**b**) Vibrations in the axial direction.

Figure 18 shows the dependence of "Num Shocks Z" on time. It should be noted here that the measurement takes place in the "CPS Mach (AMx)" mode. Let us reveal each component of the methodology of this measurement: CPS is an abbreviation of "cycles per second", which, in our case, means the number of repetitions of shocks per second; mach (AMx) means "maximum amplitude", which indicates the largest magnitude of shocks detected during measurement. Thus, the dependence in Figure 18 displays the number of shocks in the axial direction, which is measured by the "CPS mach (AMx)" method, and the vibration sensor registers shocks during drilling with an accuracy in relation to the number of oscillations per second, taking into account the maximum amplitude of the magnitude of these oscillations. For example (see Figure 18), a dangerous event was registered on January 23, in which 14 shocks of maximum amplitude in the axial direction occurred within one second.

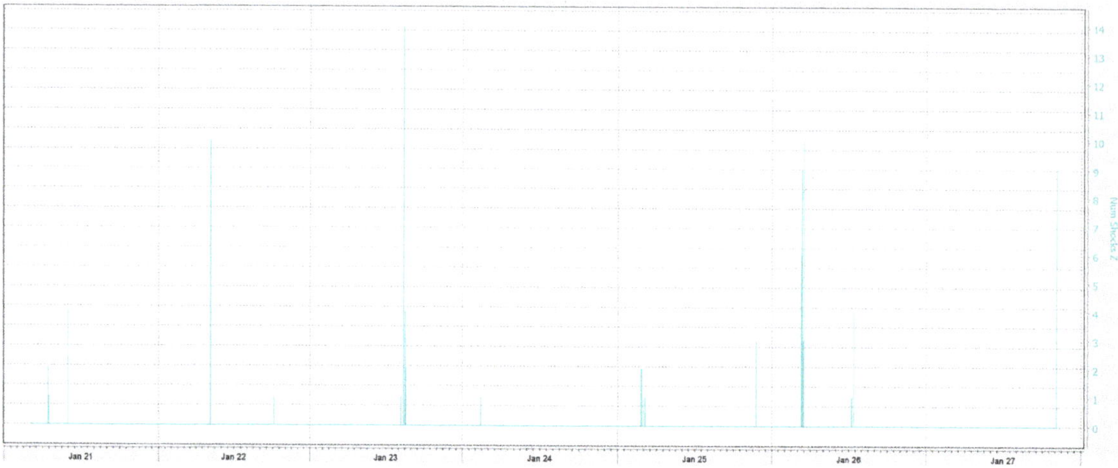


**Figure 18.** The number of amplitude shocks in the axial direction during one second (registered during the first trip).

The proposed system is also able to record the value of the number of revolutions of the drilling tool on the wellbore and to register torsional vibrations and evaluate the stick–slip phenomenon. Figure 19 clearly shows the peak values of the acceleration of the drill string during constant rotation of the upper drive system of the drilling rig. It is interesting that at the beginning of the trip, a negative value of the rotation frequency was recorded, that is, the drilling tool periodically moved in the opposite direction in the area of the hole. Analyzing the chronology and timing of the events, we can draw an unambiguous conclusion that the unscrewing of the casing stabilizer of the gas turbine engine took place on the first day of operation of the drilling motor in the well (that is, during the drilling of the interval of 2210–3200 m). Further drilling was accompanied by a significant level of transverse vibrations, which destroyed the thrust-bearing separators. At the end of the flight, an increase in the amplitude and value of axial vibrations and shocks was noted. First of all, this was caused by excessive backlash on the output shaft of the PDM, which was formed after the bearings were damaged. It also explains the unsatisfactory intensity of recruitment and the drop in the zenith angle, which was the reason for the lifting of the drilling tool.

First of all, the occurrence of vibrations is influenced by the rock-destructive tool, namely its type, design and aggressiveness. Also, the diameter of the well and placement of support-centering elements have a significant influence. That is, vibrations are primarily generated by those elements of the drill string that are in direct contact with the hole and the walls of the well. Increasing the depth of the well generally results in greater vibrations due to the longer drill string. A higher rotation speed can cause stronger vibrations (although there are exceptions). Excessive axial load can cause unstable behavior of the drill string and increase vibrations. High flushing fluid pressures and flow rates can affect vibrations. Rock properties, layering, and the presence of defects and inclusions in the rock, etc., have an effect on vibrations.

Accordingly, changing the parameters to reduce vibrations usually means reducing the rotation speed, optimizing the axial load, adjusting the flow rate and pressure of the flushing fluid, supplementing the BHA, using dampers, drilling shock absorbers and stabilizers, and choosing the optimal profile of the drill string.

However, effective control of drill string vibrations requires a comprehensive approach that combines the simultaneous analysis of a number of drilling parameters.

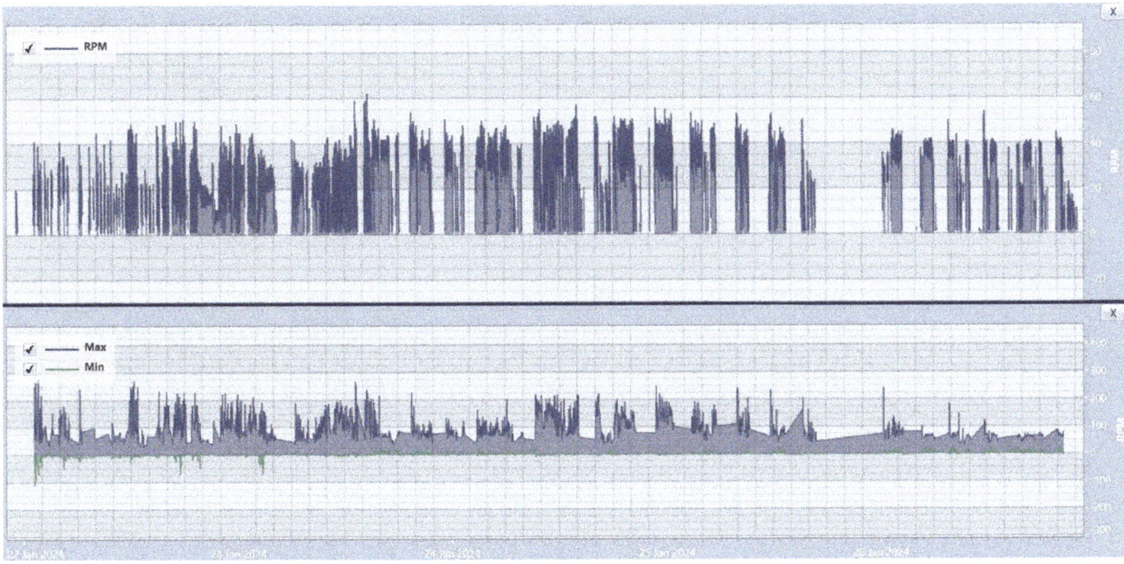


**Figure 19.** Registration of torsional vibrations in the lower part of the drill string.

### *3.2. Changing the Drilling Mode—Trip No. 2*

In the previous paragraph, we demonstrated the main capabilities of the controller for recording vibrations at the lower part of the drill string. In the case under consideration, there was a complication of the drilling process, so technological and structural corrections will be made in the next voyage. In general, even in the case of a problem-free trip, the recorded data should be compared with the regulations and recommendations for the safety and efficiency of drilling. If significant deviations from the standards are found, this may require correction of the drilling process or the installation of additional equipment, such as, for example, the drill shock absorber.

Therefore, after conducting a detailed analysis of the nature and causes of damage to the drilling equipment, the customer was offered the option of carrying out clear monitoring of vibrations in real time using a high-cost telemetry system for a prompt response in the case of dangerous values of vibration loads being registered. In order to prevent the occurrence of drilling complications, it was decided to use the technique of changing the drilling modes; namely, adjusting the rotation frequency of the drill string and varying the axial load. To continue the drilling of the stabilization section of the curvature parameters, the following BHA was assembled in the same well: 12 ¼ Bit (0.37 m) + 8 Drilling motor with a 12-sleeve stabilizer and 1.5° bend (9.06 m) + 11 ¾ String Stabilizer (1.78 m) + 8 Pony Collar (1.84 m) + 8 UBHO (1.05 m) + 8 NMDC with MWD inside (7.04 m) + 8 NMDC (8.24 m) + Crossover sub (0.71 m) + 8 Drill Collar (45.52 m) + Crossover sub (0.76 m) + 8 Drilling Jar (8.69 m) + Crossover Sub (0.75 m) + 8 Drill Collar (27.91 m) + Crossover Sub (1.07 m) + 5 HWDP (28.08 m) + 6 5/8 Friction Reduction Tool (6.37 m) + 5 HWDP (27.34 m) + Drill Pipe (to surface).

Drilling took place in the Bashkir layer, while a PDC Ulterra CF616 bit (Ulterra Corporate, Calgary, Canada) was used. After reaching a depth of 3549 m, inclined directional drilling was started. During the deepening of the well, constant slides were made to maintain its trajectory; between the slides, the drilling process took place in a combined way. Timely changes in drilling modes had a positive effect and made it possible to achieve the desired final result. When the well bore reached the depth of 4066 m, they started raising the drill string for the purpose of core selection. No damage was detected during the inspection of the drilling tool and the hydraulic cylinder (the axial play on the hydraulic cylinder was 1 mm). After lifting the drill string, data were extracted from the memory of the telemetry system and from the Smart 4 controller.

To assess vibrations during the second trip, we chose to measure vibrations in the RMS mode (Figure 20)—this is the determination of the root mean square value of the amplitude of the vibration signal during a certain time interval. Measurement in this mode provides a comprehensive picture of the average level of energy transmitted through vibrations during a certain period of time. This can be useful for assessing the overall impact of vibrations on the drill tool or for comparing vibration levels before and after vibration reduction measures are applied.

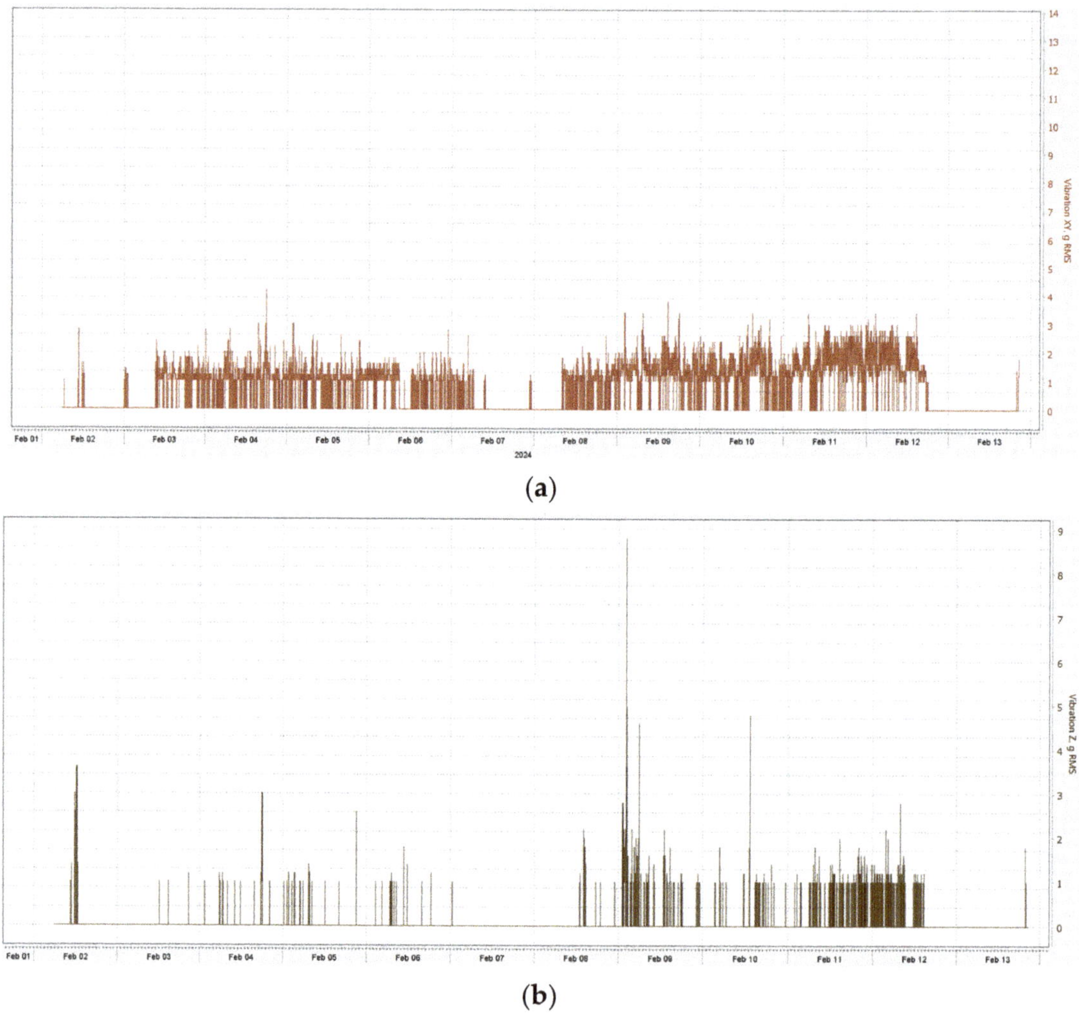


**Figure 20.** Registration of vibrations in three mutually perpendicular directions during the second trip: (**a**) Vibrations in the lateral directions; (**b**) Vibrations in the axial direction.

Comparing the diagrams shown in Figures 17 and 20, it can be concluded that reducing the intensity of vibrations makes it possible to increase the duration of the flight and to keep the out-of-hole equipment intact.

To assess the sensitivity of the proposed Smart 4 controller, a comparison of the data registered with the controller was made with the information obtained from the telemetry system. We graphically compared the absolute values of the vibration accelerations, which were registered without the use of additional mathematical algorithms (Figure 21). Lateral shocks measured in g units during the first trip were taken into consideration (this choice was made because during this flight, most of the events registered in the transverse plane). In the diagram (Figure 21), the LShk parameters are obtained from the Smart 4 controller (marked in orange), and the Shock XY data is obtained using the telemetry system (marked in blue).

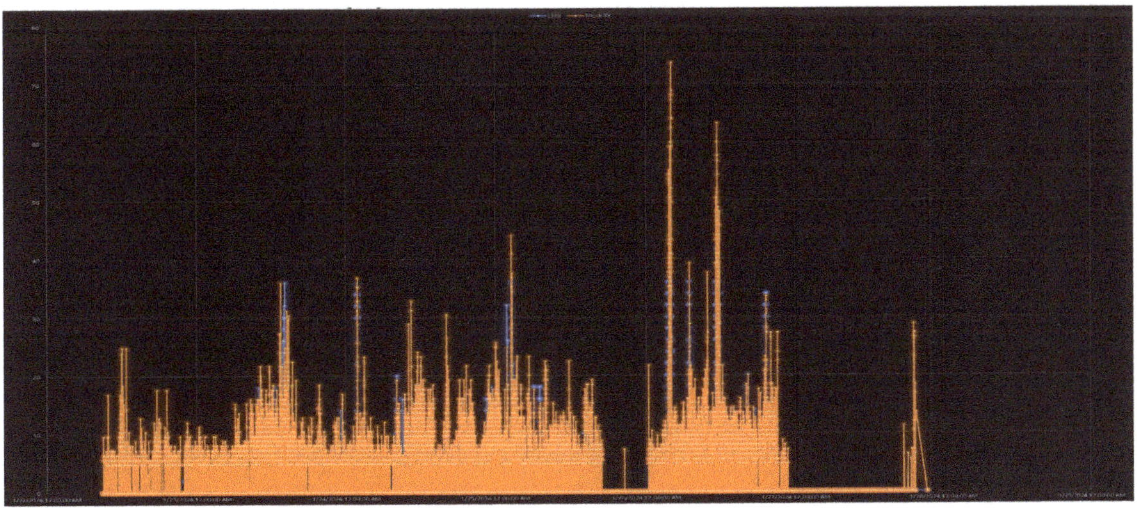

**Figure 21.** Comparison of the value of shocks in the lateral directions registered during the first trip using different devices: data from the Smart 4 controller are marked in orange; data from the telemetry system are marked in blue.

Graphical visualization shows a good convergence of data obtained by the different measuring devices. It should be understood that this is a pilot study, which first involved checking the design of a module for installing a vibration sensor in the wellbore area and checking the suitability of using a low-cost Smart 4 controller for recording the vibrations of the bottom of the drill string. In the future, it is important to compare the accelerogram data with other drilling parameters, such as the drilling speed, well pressure, well depth, gamma logging, and profilometry results. This will make it possible to better understand the reasons for the increase in vibrations of the tool and, based on the interpretation of accelerograms, to draw conclusions about the influence of vibrations on the drilling process, the efficiency of the drilling equipment, and the safety and quality of the drilling operations. If high-intensity vibrations are detected, it is important to develop measures to reduce them; in particular, to install drilling shock absorbers or other vibration protection devices, to adjust the operating mode of drilling equipment, and to supplement the BHA.

The use of live measurements of drill string vibrations can be useful for improving and refining existing analytical and numerical models of the dynamic processes that occur during the drilling of deep oil and gas wells [71–73]. Comparing the results of field measurements with the predictions of analytical and numerical models makes it possible to assess the accuracy of the models and to make the necessary corrections to improve their compliance in real conditions [74–76]. At the same time, field measurements of vibrations can be used to identify unknown model parameters using the inverse problem method or the system identification method, which will increase the reliability of modeling. Comparison of the simulation results with field measurements makes it possible to check the adequacy of the accepted assumptions and hypotheses laid down in the basis of the models. In addition, the analysis of natural vibration data can reveal new regularities and phenomena that are not taken into account in existing models [77,78]. This can be an impetus for the development of more advanced analytical and numerical models of the dynamic mode of drilling. Thus, the comprehensive use of on-site vibration measurements of the drill string provides an opportunity to increase the adequacy of analytical and numerical models of the dynamic behavior of the drilling tool.

## 4. Conclusions

Drill string vibration during the oil and gas deep-well construction process is one of the reasons for the high cost of drilling, and its harmful effect reduces the efficiency of drilling and can lead to complications or emergencies. The study carried out a detailed analysis of real case reports of the destructive impact of vibrations that occurred during drilling operations in the fields of the Dnipro–Donetsk Basin. Among the subjective reasons that led to an increase in the intensity of vibrations of the drilling tool, the following are highlighted: the use of aggressively designed drill bits, the use of forced drilling modes, errors in the assembly of the BHA, and ignoring the need to use anti-vibration devices. The main objective reasons for the increase in vibrations during drilling were rock stratification and the phenomenon of contact interaction of the drilling tool with the wall and the hole.

1. In this study, the search for an inexpensive method of recording downhole vibrations and shocks was initiated in order to adjust the dynamic mode of operation of the drill string.
2. To measure the vibrations of the drill string, it is proposed to use the Smart 4 controller, which is equipped with a three-axis accelerometer and a gyroscope, paired with a lithium thionyl chloride battery.
3. A compact module is designed for the reliable placement of the controller in the lower part of the drill string.
4. Industrial testing of the proposed device for measuring vibrations during drilling of an inclined well was carried out, and the registered data were verified using a high-cost telemetry system.

The feature of the proposed controller is its low power consumption and sufficient memory volume, which allows for uninterrupted recording of pothole data over 1000 h of operation, while the accuracy of the data recording is not inferior to much more expensive analogs. The developed module is easily assembled and disassembled, and with its help, the controller is placed in the center of gravity of the cross-section of the drill string, and the design of the module allows for quick unloading of the registered data in field conditions.

The implementation of the proposed method will allow operators to assess the impact of vibrations and shocks when drilling deep wells and to compare the recorded downhole data with critical events that occurred during drilling. At the same time, service departments engaged in the rental, operation, and repair of drilling equipment will be able to obtain an evidence base for resolving warranty disputes. The next stage of research will be devoted to the development of a methodology for conducting two-way accelerogram assessments with the aim of providing recommendations for adjusting the drilling regimes or for the additional completion of the BSC.

**Author Contributions:** Conceptualization, S.L. and A.V.; methodology, S.L. and A.V.; software, S.L. and A.V.; validation, S.L., L.R. and A.A.; formal analysis, L.R. and A.A.; investigation, S.L.; resources, A.V.; data curation, L.R. and A.A.; writing—original draft preparation, S.L., A.V., L.R. and A.A.; writing—review and editing, L.R. and A.A.; visualization, S.L.; supervision, S.L.; project administration, S.L., L.R. and A.A.; funding acquisition, A.V. All authors have read and agreed to the published version of the manuscript.

**Funding:** This research received no targeted external funding.

**Data Availability Statement:** Data are contained within the article.

**Conflicts of Interest:** Serhii Landar was employed by the Ukrnaftagazservis Ltd. The remaining authors declare that the research was conducted in the absence of any commercial or financial relationships that could be construed as a potential conflict of interest.

## References


1. Li, Y.; Wang, J.; Shan, Y.; Wang, C.; Hu, Y. Measurement and Analysis of Downhole Drill String Vibration Signal. *Appl. Sci.* **2021**, *11*, 11484. [CrossRef]
2. Riane, R.; Doghmane, M.Z.; Kidouche, M.; Djezzar, S. Observer-Based H∞ Controller Design for High Frequency Stick-Slip Vibrations Mitigation in Drill-String of Rotary Drilling Systems. *Vibration* **2022**, *5*, 264–289. [CrossRef]
3. Sharma, A.; Abid, K.; Srivastava, S.; Velasquez, A.; Teodoriu, C. A review of torsional vibration mitigation techniques using active control and machine learning strategies. *Petroleum* **2023**. [CrossRef]
4. Shlyun, N.V.; Gulyayev, V.I. Buckling of a drill-string in two-sectional bore-holes. *Int. J. Mech. Sci.* **2020**, *172*, 105427. [CrossRef]
5. Bembenek, M.; Grydzhuk, Y.; Gajdzik, B.; Ropyak, L.; Pashechko, M.; Slabyi, O.; Al-Tanakchi, A.; Pryhorovska, T. An Analytical–Numerical Model for Determining "Drill String–Wellbore" Frictional Interaction Forces. *Energies* **2024**, *17*, 301. [CrossRef]
6. Liu, W.; Yang, F.; Zhu, X.; Chen, X. Stick-slip vibration behaviors of BHA and its control method in highly-deviated wells. *Alex. Eng. J.* **2022**, *61*, 9757–9767. [CrossRef]
7. Kamgue Lenwoue, A.R.; Deng, J.; Feng, Y.; Li, H.; Oloruntoba, A.; Songwe Selabi, N.B.; Marembo, M.; Sun, Y. Numerical Investigation of the Influence of the Drill String Vibration Cyclic Loads on the Development of the Wellbore Natural Fracture. *Energies* **2021**, *14*, 2015. [CrossRef]
8. Shatskyi, I.; Makoviichuk, M.; Vaskovkyi, M. Transversal Straining of Pressurized Pipeline Caused by Vibration of Damaged Foundation. *Springer Proc. Math. Stat.* **2024**, *453*, 501–508. [CrossRef]
9. Kulke, V.; Thunich, P.; Schiefer, F.; Ostermeyer, G.-P. A Method for the Design and Optimization of Nonlinear Tuned Damping Concepts to Mitigate Self-Excited Drill String Vibrations Using Multiple Scales Lindstedt-Poincaré. *Appl. Sci.* **2021**, *11*, 1559. [CrossRef]
10. Velichkovich, A.S.; Dalyak, T.M. Assessment of stressed state and performance characteristics of jacketed spring with a cut for drill shock absorber. *Chem. Pet. Eng.* **2015**, *51*, 188–193. [CrossRef]
11. Kessai, I.; Benammar, S.; Doghmane, M.Z.; Tee, K.F. Drill Bit Deformations in Rotary Drilling Systems under Large-Amplitude Stick-Slip Vibrations. *Appl. Sci.* **2020**, *10*, 6523. [CrossRef]
12. Ejike, C.; Obuobi, I.F.; Avinu, S.; Abid, K.; Teodoriu, C. Investigation and Analysis of Influential Parameters in Bottomhole Stick–Slip Calculation during Vertical Drilling Operations. *Energies* **2024**, *17*, 622. [CrossRef]
13. Cheng, Z.; Zhang, L.; Hao, Z.; Ding, X.; Liu, Z.; Li, T. A New Bottom-Hole Assembly Design Method to Maintain Verticality and Reduce Lateral Vibration. *Processes* **2024**, *12*, 95. [CrossRef]
14. Chudyk, I.; Sudakova, D.; Dreus, A.; Pavlychenko, A.; Sudakov, A. Determination of the thermal state of a block gravel filter during its transportation along the borehole. *Min. Miner. Depos.* **2023**, *17*, 75–82. [CrossRef]
15. Liu, X.; Zhang, H.; Niu, Y.; Lin, G.; Liu, W.; Li, Y. Experiment on the influence of downhole drill string absorption & hydraulic supercharging device on bottom hole WOB fluctuation. *Energy Rep.* **2023**, *9*, 2372–2378. [CrossRef]
16. Velichkovich, A.S.; Popadyuk, I.I.; Shopa, V.M. Experimental study of shell flexible component for drilling vibration damping devices. *Chem. Petrol. Eng.* **2011**, *46*, 518–524. [CrossRef]
17. Riane, R.; Doghmane, M.Z.; Kidouche, M.; Tee, K.F.; Djezzar, S. Stick-Slip Vibration Suppression in Drill String Using Observer-Based LQG Controller. *Sensors* **2022**, *22*, 5979. [CrossRef]
18. Svitlytskyi, V.; Iagodovskyi, S.; Bilenko, N. Effect of vibration dampers on the dynamic state of a drill string. *Technol. Audit Prod. Reserves* **2023**, *5*, 32–36. [CrossRef]
19. Landar, S.; Velychkovych, A.; Mykhailiuk, V. Numerical and analytical models of the mechanism of torque and axial load transmission in a shock absorber for drilling oil, gas and geothermal wells. *Eng. Solid Mech.* **2024**, *12*, 207–220. [CrossRef]
20. Velichkovich, A.S. Design features of shell springs for drilling dampers. *Chem. Petrol. Eng.* **2007**, *43*, 458–461. [CrossRef]
21. Xu, Y.; Zhang, H.; Guan, Z. Dynamic Characteristics of Downhole Bit Load and Analysis of Conversion Efficiency of Drill String Vibration Energy. *Energies* **2021**, *14*, 229. [CrossRef]
22. Aarsnes, U.J.F.; van de Wouw, N. Effect of shock subs on self-excited vibrations in drilling systems. *J. Petrol. Sci. Eng.* **2019**, *181*, 106217. [CrossRef]
23. Velichkovich, A.; Velichkovich, S. Vibration-impact damper for controlling the dynamic drillstring conditions. *Chem. Petrol. Eng.* **2001**, *37*, 213–215. [CrossRef]
24. Dutkiewicz, M.; Velychkovych, A.; Shatskyi, I.; Shopa, V. Efficient Model of the Interaction of Elastomeric Filler with an Open Shell and a Chrome-Plated Shaft in a Dry Friction Damper. *Materials* **2022**, *15*, 4671. [CrossRef] [PubMed]
25. Shats'kyi, I.P.; Shopa, V.M.; Velychkovych, A.S. Development of full-strength elastic element section with open shell. *Strength Mater.* **2021**, *53*, 277–282. [CrossRef]
26. Saleh, M.K.A.; Nejatpour, M.; Yagci Acar, H.; Lazoglu, I. A new magnetorheological damper for chatter stability of boring tools. *J. Mater. Process. Technol.* **2021**, *289*, 116931. [CrossRef]
27. Machado, M.R.; Dutkiewicz, M.; Colherinhas, G.B. Metamaterial-based vibration control for offshore wind turbines operating under multiple hazard excitation forces. *Renew. Energy* **2024**, *223*, 120056. [CrossRef]

28. Ma, B.; Dong, S. Coupling Simulation of Longitudinal Vibration of Rod String and Multi-Phase Pipe Flow in Wellbore and Research on Downhole Energy Efficiency. *Energies* **2023**, *16*, 4988. [CrossRef]
29. Velichkovich, A.S. Shock Absorber for Oil-Well Sucker-Rod Pumping Unit. *Chem. Petrol. Eng.* **2005**, *41*, 544–546. [CrossRef]
30. Chudyk, I.I.; Dudych, I.F.; Sudakova, D.A.; Voloshyn, Y.D.; Bogoslavets, V.V. Influence of drilling mud pulsations on well cleanout efficiency. *Nauk. Visnyk Natsionalnoho Hirnychoho Universytetu* **2023**, *5*, 48–53. [CrossRef]
31. Panevnik, D.A.; Velichkovich, A.S. Assessment of the stressed state of the casing of the above-bit hydroelevator. *Neftyanoe Khozyaystvo Oil Ind.* **2017**, *1*, 70–73.
32. Prysyazhnyuk, P.; Molenda, M.; Romanyshyn, T.; Ropyak, L.; Romanyshyn, L.; Vytvytskyi, V. Development of a hardbanding material for drill pipes based on high-manganese steel reinforced with complex carbides. *Acta Montan. Slovaca* **2022**, *27*, 685–696.
33. Prysyazhnyuk, P.; Ivanov, O.; Matvienkiv, O.; Marynenko, S.; Korol, O.; Koval, I. Impact and abrasion wear resistance of the hardfacings based on high-manganese steel reinforced with multicomponent carbides of Ti-Nb-Mo-V-C system. *Procedia Struct. Integr.* **2022**, *36*, 130–136. [CrossRef]
34. Ropyak, L.; Shihab, T.; Velychkovych, A.; Bilinskyi, V.; Malinin, V.; Romaniv, M. Optimization of Plasma Electrolytic Oxidation Technological Parameters of Deformed Aluminum Alloy D16T in Flowing Electrolyte. *Ceramics* **2023**, *6*, 146–167. [CrossRef]
35. Shatskyi, I.; Makoviichuk, M.; Ropyak, L.; Velychkovych, A. Analytical Model of Deformation of a Functionally Graded Ceramic Coating under Local Load. *Ceramics* **2023**, *6*, 1879–1893. [CrossRef]
36. Onysko, O.; Kopei, V.; Barz, C.; Kusyi, Y.; Baskutis, S.; Bembenek, M.; Dašić, P.; Panchuk, V. Analytical Model of Tapered Thread Made by Turning from Different Machinability Workpieces. *Machines* **2024**, *12*, 313. [CrossRef]
37. Pryhorovska, T.O.; Ropyak, L. Machining Error Influnce on Stress State of Conical Thread Joint Details. In Proceedings of the 8th International Conference on Advanced Optoelectronics and Lasers (CAOL 2019), Sozopol, Bulgaria, 6–8 September 2019; pp. 493–497. [CrossRef]
38. Kopei, V.; Onysko, O.; Kusyi, Y.; Vriukalo, V.; Lukan, T. Investigation of the Influence of tapered Thread Pitch Deviation on the Drill-String Tool-Joint Fatigue Life. In *Lecture Notes in Networks and Systems, New Technologies, Development and Application V. NT 2022*; Karabegović, I., Kovačević, A., Mandžuka, S., Eds.; Springer: Cham, Switzerland, 2023; Volume 472, pp. 144–154. [CrossRef]
39. Hossain, N.; Rimon, I.H.; Mimona, M.A.; Mobarak, H.; Ghosh, J.; Islam, A.; Al Mahmud, Z. Prospects and challenges of sensor materials: A comprehensive review. *e-Prime* **2024**, *7*, 100496. [CrossRef]
40. Chen, Q.; Li, J. Study on the Mechanism and Suppression of Harmonic Vibration of AMB-Rotor System. *Vibration* **2024**, *7*, 83–97. [CrossRef]
41. Bembenek, M.; Makoviichuk, M.; Shatskyi, I.; Ropyak, L.; Pritula, I.; Gryn, L.; Belyakovskyi, V. Optical and Mechanical Properties of Layered Infrared Interference Filters. *Sensors* **2022**, *22*, 8105. [CrossRef]
42. Wright, R.F.; Lu, P.; Devkota, J.; Lu, F.; Ziomek-Moroz, M.; Ohodnicki, P.R., Jr. Corrosion Sensors for Structural Health Monitoring of Oil and Natural Gas Infrastructure: A Review. *Sensors* **2019**, *19*, 3964. [CrossRef]
43. Shatskyi, I.P.; Makoviichuk, M.V.; Shcherbii, A.B. Influence of Flexible Coating on the Limit Equilibrium of a Spherical Shell with Meridional Crack. *Mater. Sci.* **2020**, *55*, 484–491. [CrossRef]
44. Dutkiewicz, M.; Velychkovych, A.; Andrusyak, A.; Petryk, I.; Kychma, A. Analytical Model of Interaction of an Oil Pipeline with a Support of an Overpass Built in a Mountainous Area. *Energies* **2023**, *16*, 4464. [CrossRef]
45. Tsao, Y.-C.; Vu, T.-L. Determining an optimal sensor system for smart buildings with uncertain energy supply and demand. *J. Build. Eng.* **2023**, *71*, 106532. [CrossRef]
46. Liang, H.; Chen, H.; Zou, J.; Bai, J. Technical research on realizing remote intelligent diagnosis of petroleum drilling loss circulation under smart city strategy. *Future Gener. Comput. Syst.* **2021**, *125*, 91–99. [CrossRef]
47. Shan, Y.; Xue, Q.; Wang, J.; Li, Y.; Wang, C. Analysis of the Influence of Downhole Drill String Vibration on Wellbore Stability. *Machines* **2023**, *11*, 762. [CrossRef]
48. Li, Y.; Xue, Q.; Wang, J.; Wang, C.; Shan, Y. Pattern recognition of stick-slip vibration in combined signals of DrillString vibration. *Measurement* **2022**, *204*, 112034. [CrossRef]
49. Saadat, S.M.; Prakasan, H.; Poothia, T.; Pandey, G. A comprehensive study on vibration control and evaluation of drill string during drilling operation. *AIP Conf. Proc.* **2023**, *2855*, 040009. [CrossRef]
50. Dong, G.; Chen, P. A Review of the Evaluation, Control, and Application Technologies for Drill String Vibrations and Shocks in Oil and Gas Well. *Shock Vib.* **2016**, *2016*, 7418635. [CrossRef]
51. Maitra, E.K.; Al Dushaishi, M.F.; Sugiura, J.; Jones, S. Experimental Visualization of Downhole Drilling Vibration Using Industrial Drilling Dynamic Recorder. In Proceedings of the International Petroleum Technology Conference 2024, Dhahran, Saudi Arabia, February 2024. Paper Number: IPTC-23494-MS. [CrossRef]
52. Liu, J.; Huang, H.; Zhou, Q.; Wu, C. Self-Powered Downhole Drilling Tools Vibration Sensor Based on Triboelectric Nanogenerator. *IEEE Sens. J.* **2022**, *22*, 2250–2258. [CrossRef]
53. Mustapha, S.; Lu, Y.; Ng, C.-T.; Malinowski, P. Sensor Networks for Structures Health Monitoring: Placement, Implementations, and Challenges—A Review. *Vibration* **2021**, *4*, 551–585. [CrossRef]
54. Tian, J.; Wei, L.; Yang, L. Research and experimental analysis of drill string dynamics characteristics and stick-slip reduction mechanism. *J. Mech. Sci. Technol.* **2020**, *34*, 977–986. [CrossRef]
55. Qu, J.; Xue, Q.; Wang, J.; Sun, J.; Lu, J.; Zhang, H.; Sun, F.; Xinghua, T. From Design to Experimental Validation of the Fully Mechanical Gravity Tool Face Measurement-While-Drilling System. *SPE J.* **2024**, *29*, 2971–2987. [CrossRef]

56. Hassan, I.U.; Panduru, K.; Walsh, J. An In-Depth Study of Vibration Sensors for Condition Monitoring. *Sensors* **2024**, *24*, 740. [CrossRef] [PubMed]
57. Zhang, J.; Liang, H.; Chen, Z. The technology of intelligent recognition for drilling formation based on neural network with conjugate gradient optimization and remote wireless transmission. *Comput. Commun.* **2020**, *156*, 35–45. [CrossRef]
58. Liu, J.; Pan, G.; Wu, C.; Feng, Y. Research on Hybrid Vibration Sensor for Measuring Downhole Drilling Tool Vibrational Frequencies. *Appl. Sci.* **2024**, *14*, 5014. [CrossRef]
59. Wang, R.; Ren, J.; Ding, W.; Liu, M.; Pan, G.; Wu, C. Research on Vibration Accumulation Self-Powered Downhole Sensor Based on Triboelectric Nanogenerators. *Micromachines* **2024**, *15*, 548. [CrossRef] [PubMed]
60. Liu, Z.; Lei, K.; Song, J.; Li, L.; Li, T. A Designed Calibration Approach for the Measurement-While-Drilling Instrument. *Appl. Sci.* **2023**, *13*, 61. [CrossRef]
61. Wang, H.; Huang, H.; Wu, C.; Liu, J. A Ring-Shaped Curved Deformable Self-Powered Vibration Sensor Applied in Drilling Conditions. *Energies* **2022**, *15*, 8268. [CrossRef]
62. Yu, Y.; Liu, Q.; Han, B.S.; Zhou, W. Application of Convolutional Neural Network to Defect Diagnosis of Drill Bits. *Appl. Sci.* **2022**, *12*, 10799. [CrossRef]
63. Senjoba, L.; Sasaki, J.; Kosugi, Y.; Toriya, H.; Hisada, M.; Kawamura, Y. One-Dimensional Convolutional Neural Network for Drill Bit Failure Detection in Rotary Percussion Drilling. *Mining* **2021**, *1*, 297–314. [CrossRef]
64. Benmir, N.; Akroum, H.; Doghman, M.Z.; Kidouche, V. A Survey Study of Modeling, Analysis, and Control Drill String under Torsional, Axial, and Lateral Vibrations. In Proceedings of the Second International Conference on Energy Transition and Security (ICETS), Adrar, Algeria, 12–14 December 2023; pp. 1–11. [CrossRef]
65. Lyu, F.; Wang, Y.; Mei, Y.; Li, F. A High-Frequency Measurement Method of Downhole Vibration Signal Based on Compressed Sensing Technology and Its Application in Drilling Tool Failure Analysis. *IEEE Access* **2023**, *11*, 129650–129659. [CrossRef]
66. Zhang, L.; Ning, Z.; Peng, H.; Mu, Z.; Sun, C. Effects of Vibration on the Electrical Performance of Lithium-Ion Cells Based on Mathematical Statistics. *Appl. Sci.* **2017**, *7*, 802. [CrossRef]
67. Innova Power Solutions. Smart-4 Technolology. Available online: https://www.innova-power.com/smart-4-technology.html (accessed on 29 March 2024).
68. Hua, X.; Thomas, A. Effect of dynamic loads and vibrations on lithium-ion batteries. Journal of Low Frequency Noise. *Vib. Act. Control* **2021**, *40*, 1927–1934. [CrossRef]
69. Halliburton. Well Construction Suite. *Wellplan-Software*. Available online: https://www.halliburton.com/en/software/decisionspace-365-enterprise/decisionspace-365-well-construction/well-construction-suite/wellplan-software (accessed on 29 March 2024).
70. Deng, K.; Hu, W.; Ge, L.; Hu, Z.; Yang, Q.; Xiao, X. Study of Downhole Lateral Force Measurement Modelling and Devices in Petroleum Exploration. *Energies* **2022**, *15*, 5724. [CrossRef]
71. Liu, W.; Ni, H.; Wang, Y.; Guo, Y.; Gao, Y.; He, P. Dynamic modeling and load transfer prediction of drill-string axial vibration in horizontal well drilling. *Tribol. Int.* **2022**, *177*, 107986. [CrossRef]
72. Shatskii, I.P.; Perepichka, V.V. Shock-wave propagation in an elastic rod with a viscoplastic external resistance. *J. Appl. Mech. Tech. Phys.* **2013**, *54*, 1016–1020. [CrossRef]
73. Hai, W.; He, Y.; Xue, Q. Research on the Influence of Deep-Water Drilling Risers on Drillstring Motion Trajectory and Vibration Characteristics. *Machines* **2024**, *12*, 112. [CrossRef]
74. Moysyshyn, V.M.; Lyskanych, M.V.; Borysevych, L.V.; Vytyaz, O.Y.; Voznyi, I.I. Experimental Estimation of Design and Drilling Regime Option Influence on Drilling Tool Dynamics. *Metallofiz. I Noveishie Tekhnologii* **2021**, *43*, 689–712. [CrossRef]
75. Wang, R.; Liu, X.; Song, G.; Zhou, S. Non-Linear Dynamic Analysis of Drill String System with Fluid-Structure Interaction. *Appl. Sci.* **2021**, *11*, 9047. [CrossRef]
76. Shatskyi, I.; Perepichka, V. Problem of dynamics of an elastic rod with decreasing function of elastic-plastic external resistance. In *Dynamical Systems in Applications, Proceedings of the DSTA 2017, Lodz, Poland, 11–14 December 2017*; Awrejcewicz, J., Ed.; Springer: Cham, Switzerland, 2018; Volume 249, pp. 335–342. [CrossRef]
77. Ichaoui, M.; Schiefer, F.; Ostermeyer, G.-P. A Novel Mesoscopic Drill Bit Model for Deep Drilling Applications. *Modelling* **2023**, *4*, 296–322. [CrossRef]
78. Yan, B.; Tian, J.; Meng, X.; Zhang, Z. Stick–Slip Vibration Characteristics Study of the Drill String Based on PID Controller. *Energies* **2023**, *16*, 7902. [CrossRef]

*vibration*

MDPI

*Article*

# The Reduced-Order Modeling Approach for a Double-Damper Concept: A Comparison with a Single Damper for Comfort Analysis

**Behzad Hamedi [1,*], Sudarshan Shrikanthan [2] and Saied Taheri [2,†]**

1 Department of Engineering Mechanics, Virginia Tech, Blacksburg, VA 24061, USA
2 Department of Mechanical Engineering, Virginia Tech, Blacksburg, VA 24061, USA; sudarshan007@vt.edu (S.S.); staheri@vt.edu (S.T.)
* Correspondence: behzadh@vt.edu
† Professor and Director of Center for Tire Research (CenTiRe).

**Abstract:** This paper explores the modeling and simulation of an innovative double-damper suspension system, evaluating its effectiveness through different test scenarios. The double damper integrates two individual dampers into a unified assembly. The modeling process involves representing the damper as two distinct dampers and a body block, accounting for the additional degree of freedom introduced by combining the two dampers. Simulink/MATLAB is employed for modeling the pressure, discharge, and force equations of the damper. A simplified quarter-car model is designed to conduct simulations for different road profiles, evaluating the efficacy of this double-damper model. The reduced-order modeling approach, suitable for complex systems like dampers, is utilized. Dedicated mathematical models are utilized to examine both single- and double-damper configurations, with the resulting non-linear equations solved using Newton's iterative method. The equations derived for the single damper provide the basis for modeling the double-damper system. In this model, two separate dampers, each possessing similar properties, are simulated and considered to be rigidly linked at their connection point. Consequently, it is assumed that a portion of the force and velocity experienced by the lower damper is transmitted to the upper damper, and vice versa. Simulation results demonstrate that the innovative double-damper design outperforms a single passive damper in attenuating the oscillations of both the sprung and unsprung masses. Moreover, this innovative concept offers increased adaptability to balance between ride comfort and road holding, a feature previously limited to passive suspension systems.

**Keywords:** double damper; single damper; modeling; vehicle; simulation; ride comfort; road holding; sprung and unsprung mass

## 1. Introduction

Shock absorbers play a pivotal role in enhancing the ride quality and handling characteristics of vehicles by efficiently dissipating the energy generated during the displacement of suspension springs. Traditionally, shock absorbers are engineered to minimize damping force during compression, thereby reducing the impact of bumps, while enhancing damping force during rebound to dissipate stored energy within the suspension system. This hydraulic process involves the transfer of hydraulic oil between chambers via a network of valves, with damping accomplished through the resistance encountered by the oil as it flows through these valves. Early research in the field, such as Lang's seminal work in 1977 [1], focused on understanding the dynamics of automotive dampers, particularly at high stroking frequencies. Lang's pioneering research introduced one of the earliest parametric models for twin-tube automotive dampers, laying the foundation for modeling of dampers performance.

Kim [2] conducted a detailed analysis of a twin-tube damper with an emphasis on its integration into a vehicle's suspension system in 1993. This model accounted for chamber compliance and fluid compressibility, resulting in a differential equation for the chamber pressures, which was solved using the Runge–Kutta method. Discharge coefficients, determined through experimental methods, were incorporated into the model. By integrating damping characteristics into a quarter-car model, he numerically calculated the frequency response of the sprung mass and the tire deflection.

In 1995, Cafferty, Worden, and Tomlinson [3] introduced an alternative method for characterizing automotive shock absorbers using random excitation tests, building on the restoring force surface (RFS) approach from previous harmonic testing. Despite minor stochastic distortions in the force surfaces, their study demonstrated that random excitation is a viable method for identifying the dynamic properties of shock absorbers, highlighting its advantages and limitations compared to harmonic testing.

Reybrouck [4] developed one of the initial detailed parametric models for a monotube damper in 1994. He identified flow restriction forces through empirical formulas incorporating leak restriction, port restriction, and corrections for spring stiffness. After determining the individual internal forces, an additional empirical formula was utilized to compute the total damping force, and the model could also estimate pressure drops across specific flow restrictions. While the correction factors were grounded in physical principles, their precise values were obtained experimentally. The model exhibited strong correlation with experimental data within the 0.5 to 30 Hz frequency range, provided that hysteresis effects were minimal. However, the implementation of the model is complex due to the numerous correction factors necessary for accurate results. The model did not extensively discuss the causes of hysteresis beyond its frequency dependence. In 1997, Reybrouck refined his model to include a twin-tube damper, offering a more comprehensive explanation of hysteresis [5]. He found that hysteresis was affected not only by oil compressibility but also by the compressibility of gas bubbles transferred from the reserve chamber.

In 1997, Mollica and Youcef-Tuomi [6] demonstrated modeling of a monotube damper using the bond-graph method. Herr et al.'s research in 1999 at Tenneco automotive [7] presented an implicit method for simulating automotive hydraulic dampers or shock absorbers by combining computational fluid dynamics (CFD) with dynamic modeling techniques. Through CFD analysis, the study explores the flow characteristics of various damper components, including flow patterns, discharge coefficients, and pressure distribution. These findings are then integrated into a dynamic damper model to predict damping force accurately. Although the implicit and sophisticated model proposed by Herr et al. offers a comprehensive understanding of damper behavior with high accuracy, its complexity renders it less suitable for direct application in control algorithms. Despite its effectiveness in predicting damper performance, its intricacies may pose challenges for integration into real-time control systems or algorithmic frameworks. Talbott and Starkey's work in 2002 on gas-charged monotube racing dampers [8], and Rhoades' development of a parametric model in 2006 [9] focusing on damping forces for Formula SAE racecars have further advanced our understanding of shock absorber behavior through mathematical modeling.

Drawing insights from these foundational works on single-damper modeling, we aim to extend these principles to the modeling of double-damper systems. This extension is crucial for predicting the performance of double-damper configurations, which may offer enhanced performance characteristics.

This paper extends the research carried out at the CenTiRe lab of Virginia Tech, concentrating on an in-depth exploration of the double damper. The main question driving this study is whether the integration of two individual dampers into a unified assembly, known as the double damper, can offer superior performance compared to traditional single-damper setups. Specifically, the paper aims to address the following questions: Can the innovative double-damper design outperform a single passive damper in mitigating oscillations of both the sprung and unsprung masses? What are the potential benefits and challenges associated with double-damper systems? How can the modeling process of the double-damper architecture, encompassing components such as the gas chamber, bleed flow, piston valve flow, leakage flow, and semi-active flow, offer an understanding of its intricate dynamics? This study involves developing a mathematical model to describe damper behavior, as well as assessing the performance of various control algorithms using a shock dynamometer. However, this paper reviews only the modeling process, and the second part has been discussed in a separate article. Our modeling process involves representing the double damper as two individual dampers interconnected by a body block, accommodating the additional degree of freedom introduced by this type of arrangement. The novel double-damper technology, featuring semi-active solenoid valves, promises superior ride quality and enhanced car handling, showcasing notable advantages over single-damper systems. The simulation results, particularly those related to unsprung mass displacements, demonstrate the double damper's superior capability in mitigating high-frequency oscillations from road input. Moreover, this paper follows the reduced-order modeling approach, which is based on development of the analytical model for efficient assessments. The straightforward nature of the model facilitates the evaluation process, enabling us to gauge control algorithm effectiveness across diverse scenarios. While recognizing that the inherent complexities in certain conditions is challenging, this model can provide meaningful insights into control algorithm performance evaluation.

The concept of the double-damper configuration was first conceived at the CenTiRe Lab in June 2014. Yaswanth Siramdasu and Saied Taheri [10] introduced this concept alongside the development of a quarter-car model with two DoFs. This model was designed to simulate the responses of passive, semi-active, and active suspension systems, incorporating the novel idea of the double damper. Then, Sheetanshu Tyagi conducted further design improvement of the double damper at the same lab in 2016 [11]. This research concluded that the double damper could enhance the comfort and handling of the car, a capability not currently achievable with dampers available in the market. Moreover, Anish Gorantiwar and his team conducted additional research in 2020 [12], performing an experimental investigation to evaluate the efficacy of their newly developed semi-active double damper against a traditional single semi-active damper. The findings of this study demonstrated that the double semi-active damper outperformed commercial single dampers in terms of comfort at a comparable cost. This study extends earlier investigations, enriching the field of semi-active suspension technologies. Employing reduced-order modeling techniques helps simplify these complexities while retaining essential system functional performance. This simplification is particularly advantageous for conducting real-time simulations, designing control algorithms, and deepening overall system understanding. The simplified model provides a framework for optimizing control strategies more efficiently.

## 2. Research Gaps in Double-Damper Systems

Double-damper systems, despite their promise for applications such as isolation devices and structural vibration control, remain novel for use in vehicle suspension systems and present several unexplored areas. The complex dynamics of these systems, characterized by multi-physical interactions and non-linear behavior, lack reliable modeling methods that accurately capture their dynamic responses, transient behaviors, frequency-dependent damping properties, and stability considerations. Optimal design and control strategies also remain underdeveloped, highlighting the need for effective modeling methods. The reduced-order modeling approach is selected to explore the dynamic behavior of such systems, providing a foundation for optimization and the development of appropriate control algorithms. Further research is needed on experimental validation to address uncertainties in the predictive accuracy and practical applicability of these systems. Practical implementation challenges, including manufacturing complexity, cost-effectiveness, reliability, and integration with existing systems, are inadequately addressed. Additionally, there is a notable absence of comparative analyses of double-damper configurations, designs, and control strategies, which are crucial for understanding their relative advantages and limitations and for guiding future research directions.

## 3. Double-Damper Architecture

To initiate the analytical model, a comprehensive understanding of the shock absorber physics is crucial. The damper is meticulously engineered to enhance damping characteristics and seamlessly integrate with the vehicle's suspension system. Its core structure features a cylindrical tube housing the standard internal piston. Upon assembly, the tube is intelligently segmented into four distinct chambers: gas, rebound, bypass, and compression, each serving a specific purpose in optimizing the damper's functionality. Synthesizing these intricate details is essential in preparing for subsequent modeling stages to accurately capture the double damper's functional behavior.

The fundamental components of the double damper are depicted in Figure 1. Consisting of two smaller shocks joined together with an aluminum sleeve, its assembly involves precisely machining the eyelets of the shocks to facilitate secure assembly against each other, fastened in place with a bolt through each. Each individual damper's construction mirrors that of a single shock, essentially comprising two smaller shocks combined to match the size of a single unit. This design allows for effortless separation of the double damper into two individual dampers, thereby streamlining maintenance and servicing procedures. Additionally, solenoid valves are employed to provide input voltage to the controllers.

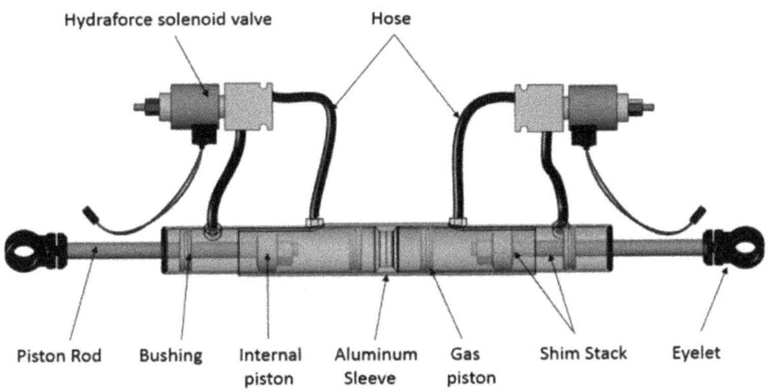


**Figure 1.** Double damper elements.

Figure 2 provides a comparison among a passive damper, double damper, and single semi-active damper, arranged from left to right:

At the top of the tube resides the gas chamber, strategically separated from the compression chamber by a floating piston. This piston serves the crucial role of isolating the nitrogen in the gas chamber from the oil in the compression chamber, preventing undesirable interactions between the two components. The compression chamber, situated between the floating piston and the main piston, houses synthetic mineral oil. On the opposite side of the piston, at the bottom end of the tube, is the rebound chamber, mirroring the compression chamber. Both chambers are meticulously filled with synthetic mineral oil to ensure optimal damping performance. The piston, intricately connected to the rod, extends through the rebound chamber and exits the bottom of the tube. To maintain the integrity of the damper's internal environment, a specially designed seal envelops the rod, preventing oil leakage, minimizing friction, and protecting against external contaminants. The external bypass valve, a critical component placed outside the damper, facilitates the connection between the two chambers. High-pressure hoses seamlessly link the SP10-24 valve to the shock body. During compression, fluid flows through two distinct paths: first through the internal piston, where damping is influenced by orifice area and shim stack stiffness, and second through the external bypass valve. The latter's damping effect is modulated by the orifice area controlled by the application of voltage. This process is reversed during the rebound stroke, ensuring consistent and controlled damping in both directions. The single semi-active damper is fabricated using these components, maintaining a size like that of the passive damper for equitable testing. This single damper serves as a benchmark for comparison with both the equivalent passive damper and the double-damper configuration. The double damper, a novel configuration featuring two smaller independent shocks held together by an aluminum sleeve, showcases a carefully engineered design. The shocks are seamlessly assembled against each other, held in place by a bolt passing through each, with additional support from an industrial hose clamp securing the shock body to the sleeve. This innovative design not only allows for easy maintenance and service but also enables the separation of the double damper into two individual dampers for independent testing, as demonstrated in subsequent sections.

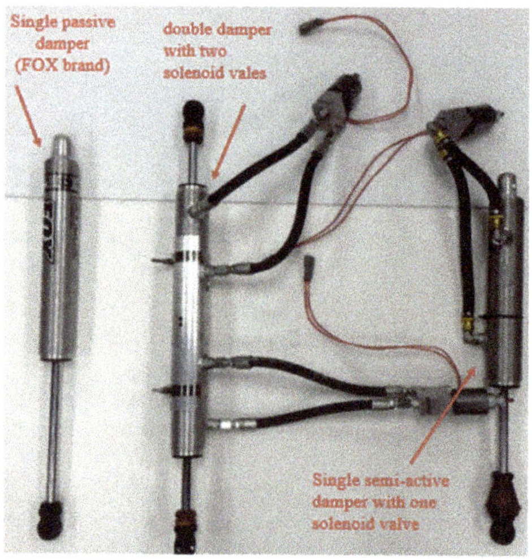


**Figure 2.** Dampers used from left to right: passive damper, double damper, single semi-active damper.

## 4. Damper Modeling

Talbott and Rhoades focused on developing mathematical models for high-performing dampers, incorporating various flow paths such as bleed orifice, piston leakage, and shim stack flows, along with considerations for the floating piston and shim stack stiffness characteristics. Their extensively validated models revealed significant insights: negligible compression side pressure variation compared to the rebound side due to the gas charge, effective modeling of valve shim stiffness using stacked thin circular plates, and the critical importance of bleed orifice settings at low speeds, as well as the dominance of shim stack stiffness at high speeds. Equations (1)–(24) presented in this section are based on their research, tailored for application in double-damper semi-active suspension concepts. Building on this foundation, Tyagi and Taheri further advanced this research by dissecting double dampers into individual components, formulating equations for each, and integrating both dampers into a unified model to study dynamic performance. These efforts laid the groundwork for constructing a mathematical model suitable for exploring various controller compatibilities in double-damper systems. The subsequent subsections will detail the specific equations and assumptions used in this modeling process, drawing on their significant contributions.

### *4.1. Dynamics of Fluid Flow*

Understanding the dynamics of fluid flow is essential for analyzing the behavior of such systems across various scenarios. The determination of the total flow rate across the piston necessitates the integration of four distinct flow paths, each playing a critical role in the overall hydraulic behavior. An assumption of the incompressibility of the hydraulic fluid within the damper lies at the core of this analysis, ensuring a consistent density throughout operational phases. Equation (1) serves as the cornerstone for illustrating the total flow rate (in$^3$/sec) during compression or rebound. Here, $Q_b$ represents the flow rate through the bleed orifice, responsible for regulating fluid flow during specific operational phases. $Q_p$ denotes the flow rate through the piston valve, governing the primary flow path within the damper assembly. $Q_{lp}$ accounts for flow rate of leakage past the piston seal, a factor crucial for maintaining operational integrity, and $Q_{sa}$ represents the flow rate through the semi-active valve, crucial for modulating damping characteristics.

$$Q = Q_b + Q_v + Q_{lp} + Q_{sa} \tag{1}$$

The total flow rate $Q$ corresponds to the volume of fluid displaced by the rod's insertion into a segment of the piston. Equation (2) defines the total flow rate ($Q'$) across a piston section due to rod insertion, with $A_c$ representing the area of the piston on the compression side and $A_p$ denoting the area of the piston on the rod side.

$$Q' = (A_c - A_p)\dot{x} \tag{2}$$

Term $\dot{x}$ represents the time derivative of the displacement $x$ with respect to time $t$. It signifies the velocity of the rod or piston within the damper assembly. In fluid dynamics, $\dot{x}$ often represents the rate of change of position, indicating how quickly the piston is moving either towards compression or rebound. By simplifying Equation (2) to account for the area change due to rod insertion, Equation (3) is derived:

$$Q = A_p\dot{x} \tag{3}$$

Further, Equation (4) represents the balance of flow terms, encapsulating the complex interplay between various hydraulic elements within the damper assembly:

$$A_p\dot{x} = Q_b + Q_v + Q_{lp} + Q_{sa} \tag{4}$$

To determine individual flow terms, a pressure difference ($\Delta p = p_c - p_r$) between compression and rebound chambers is considered. Bernoulli's Equation (5) contributes to the modeling of flow through an area ($A$), incorporating several factors influencing hydraulic performance.

$$Q = AC_d\sqrt{\frac{2\Delta p}{\rho}} \tag{5}$$

Here, $\rho$ is the density and $C_d$ embodies the steady-state discharge coefficient, a dimensionless parameter influenced by factors including the orifice diameter, Reynolds number, Cauchy number, and thickness-to-length ratio as described in Equation (6). This detailed understanding of fluid-flow dynamics forms the bedrock for precise modeling and optimization of hydraulic systems within the damper assembly. Lang showed that Bernoulli's equation could be used to model the unsteady flow through constant area passages in a damper:

$$C_d = f\left\{\frac{al}{v^2}, \frac{\mu}{\rho v l}, \frac{s}{l}\right\} \tag{6}$$

Lang also found that $C_d$ is constant, as evidenced by a strong correlation with experimental data. Various configurations were tested by adjustable bleed valve settings. The flow was measured experimentally and compared with the mathematical description defined in Equation (5) to determine the coefficient $C_d$. This equation models the flow through the bleed orifice, which is a component of the total flow. The total flow controls the damping force, as discussed in the other equations in the paper. The bleed flow is modeled as a separate block in Simulink.

### 4.2. Modeling of Bleed Flow

To characterize bleed flow, Bernoulli's equation provides a foundation for deriving the discharge through the bleed orifice of the piston, as depicted in Equation (7):

$$Q_b = A_b C_d \sqrt{\frac{2(p_c - p_r)}{\rho}} \tag{7}$$

Here, $A_b$ represents the area of the bleed orifice, determined experimentally by measuring the actual area of the damper's bleed orifice, which can vary. The value of $C_d$ is established through a comparison between simulations and experimental results.

### 4.3. Modeling of Piston Valve Flow

To facilitate modeling, the flow through the piston valve is divided into two components: the flow through the orifice and the portion of flow that interacts with the shim before exiting. This flow path involves two pressure drops: first, as the fluid exits the compression chamber and enters the piston orifice, denoted as $\Delta p_{po}$; second, as the fluid traverses the shim stack within the piston, denoted as $\Delta p_{valve}$. The flow rate through the piston orifice follows the same formulation as that of the flow through the bleed orifice, as described by Bernoulli's Equation (8).

$$Q = A C_d \sqrt{\frac{2\Delta p}{\rho}} \tag{8}$$

The principle of conservation of mass dictates that the flow through the valves must equal the flow through the piston orifice. This relationship is expressed by Equation (9):

$$Q_v = A_v C_d \sqrt{\frac{2\Delta p_{valve}}{\rho}} \tag{9}$$

The fluid flow exiting the orifice undergoes a 90-degree turn upon encountering the shim stack. This introduces complexity in modeling the $A_v$ term in Equation (9). The effective area of this flow path is equivalent to the area of a cylinder, where the circumference represents the shim's perimeter and the cylinder's height corresponds to the deflection of the shim stack, denoted as $y$:

$$A_v = \alpha \pi D_v y \tag{10}$$

Equation (10) presents the effective area $A_v$ of the flow path through the shim stack. The term $\pi D_v$ denotes the circumference of the largest shim in the stack, reflecting the total perimeter available for fluid flow. Additionally, the coefficient $\alpha$ in the equation serves as a correction factor, derived from empirical studies such as Talbott's research [8] on dampers featuring three compression and three rebound holes. For this scenario, $\alpha$ is assumed to be 0.5, indicating that half of the available perimeter is effectively utilized for flow passage. Upon substituting Equation (10) into (9), the complete expression for the flow rate through the piston valve, $Q_v$ is derived. This substitution enables us to determine how the effective area, influenced by both shim stack characteristics and the correction factor, contributes to the overall fluid-flow dynamics within the damper assembly:

$$Q_v = (\alpha \pi D_v y) A_v C_d \sqrt{\frac{2\Delta p_{valve}}{\rho}} \tag{11}$$

The subsequent step involves determining the deflection of the shim stack, denoted as $y$. This critical parameter is achieved by conducting a force balance on the valve assembly.

Following this, the discussion will proceed to determine the stiffness of the shim stack, represented by $k$. Summing forces in the $y$ direction leads to the following equation:

$$ky = \Delta p_{valve} A_v + F_m - F_{sp} \tag{12}$$

where $A_v$ signifies the area subjected to the valve pressure force, encompassing the effective area of fluid interaction with the shim stack. Meanwhile, $F_{sp}$ denotes the preload spring force, which, in this scenario, is assumed to be negligible due to the piston design of the damper. The momentum force, $F_m$, is determined based on the conservation-of-momentum principle within the valve assembly. This momentum force arises from the change in fluid direction within the valve, and its calculation is crucial for understanding the dynamic behavior of the system. Equation (13) describes the momentum force $F_m$ within the valve assembly, crucial for understanding the dynamic behavior of the fluid flow:

$$F_m = \rho v_{y,in} Q_{in} - \rho v_{y,out} Q_{out} \tag{13}$$

This force is determined based on the principle of conservation of momentum, which accounts for the change in fluid velocity within the valve. Here, $v_{y,out}$ represents the fluid velocity in the $y$ direction after passing through the valve, which is presumed to be zero due to the 90-degree shift in flow velocity. Consequently, the term $v_{y,in}$ is expressed in terms of the flow rate $Q_{in}$ and the effective area $A_o$. This adjustment ensures that the momentum-force equation accurately reflects the fluid dynamics within the valve assembly, resulting in:

$$v_{y,out} = 0 \tag{14}$$

and

$$\rho v_{y,in} = \frac{Q_{in}}{A_o} \tag{15}$$

Combining Equations (14) and (15) into Equation (13), we get

$$F_m = \rho \frac{Q_v^2}{A_o} \tag{16}$$

In accordance with Lang's research, a correction factor was incorporated into the momentum force equation to accommodate the non-zero velocity of the flow in the y direction, validated by comparing actual and predicted momentum forces. The momentum force coefficient, $C_f$, was determined to be 0.3. Implementing these adjustments to Equations (12) and (16) yields:

$$ky = \Delta p_{valve} A_v + \rho \frac{Q_v^2}{A_o} C_f - F_{sp} \tag{17}$$

### 4.4. Modeling of Leakage Flow

In the analysis of hydraulic systems, particularly in the context of dampers, accounting for oil leakage is crucial for an accurate representation of system behavior. The leakage flow path primarily addresses the oil seepage occurring between the piston seal and the cylinder walls. Lang (1997) [1] described this phenomenon by modeling it as laminar flow between parallel plates. This simplification is justified by the small distance (<0.004) between the cylinder seal and the walls, which is significantly smaller than the flow length. Here, the flow length corresponds to the height of the piston (b) inside the damper, while $D_p$ represents the diameter of the piston. To characterize the leakage flow mathematically, the Navier–Stokes equation—a fundamental equation in fluid mechanics—was employed. This equation describes the motion of viscous fluid substances and is particularly apt for modeling laminar flow regimes. The leakage flow rate, $Q_{lp}$, through the gap between the piston seal and the cylinder walls can be expressed as:

$$Q_{lp} = \left( \frac{(p_c - p_r)b^3}{12\mu l} + \dot{x}\frac{b}{2} \right) \pi D_p \tag{18}$$

where $p_c$ denotes the pressure at the cylinder side of the seal, $p_r$ represents the pressure at the rod side of the seal, $b$ signifies the height of the piston inside the damper, $\mu$ represents the dynamic viscosity of the fluid, $\dot{x}$ indicates the velocity of the piston, and $D_p$ represents the diameter of the piston.

Equation (18) provides a quantitative expression for the rate of leakage flow, considering the pressure differential across the seal, the geometry of the seal interface, and the velocity of the piston. By incorporating these factors, the model can effectively predict the extent of oil leakage, allowing for informed design decisions and optimization strategies to mitigate potential performance issues associated with excessive leakage.

### *4.5. Modeling of Semi-Active Flow*

The semi-active valve consists of a circular indent that allows for an adjustable change in area from 0 to 0.2 in$^2$. By altering the voltage applied to the solenoid, the fluid flow through the valve can be adjusted. This valve is conceptualized as a circular aperture with a controllable orifice area. The circular hole area is changed in Simulink to vary the area of the valve by changing the voltage supplied depending on the type of the control algorithm. The follow rate through the semi-active valve is depicted in Equation (19):

$$Q_{sa} = A_{sa} C_d \sqrt{\frac{2(p_c - p_r)}{\rho}} \tag{19}$$

In many commercial electro-hydraulic dampers, fast-acting solenoid valves are commonly employed. These valves are integrated in various configurations by different inventors. Some designs feature an externally mounted bi-directional solenoid valve, activated by applying voltage. This valve is typically attached to the outer tube of the damper in a twin-tube setup. A customized solenoid rod, resting against spring-loaded shims, facilitates fluid flow when voltage is applied, causing the solenoid to move against the spring. Other dampers utilize a valve assembly where fluid, after passing the solenoid, traverses multiple narrow passages to enhance damping [13].

### *4.6. Modeling of Gas Chamber*

In a monotube damper, the gas chamber accommodates the increase in volume due to the rod's insertion. Assuming that the oil is incompressible, the gas pressure within the chamber becomes a dynamic variable as a function of the piston displacement. During short operational periods, the gas chamber's temperature is assumed constant, allowing the use of the ideal gas law to determine the final gas chamber pressure:

$$P_f = \frac{P_i}{V_f} V_i \tag{20}$$

With the fluid assumed incompressible, the accelerations of the piston and gas piston are related as:

$$\ddot{z} = \frac{A_{rod}}{A_{gp}} \ddot{x} \tag{21}$$

where $\ddot{z}$ represents the acceleration of the gas piston, $A_{rod}$ signifies the cross-sectional area of the piston rod, $A_{gp}$ denotes the cross-sectional area of the gas piston, and $\ddot{x}$ indicates the acceleration of the piston. Considering the initial gas pressure $p_{gi}$ and the gas pressure at time $t$ as $p_g$, Equation (22) can be refined to represent the dynamic evolution of gas pressure with respect to piston displacement:

$$p_g = p_{gi} \frac{A_{gp} L_{ch}}{A_{gp} L_{ch} - A_{rod} x} \tag{22}$$

In this equation, $L_{ch}$ denotes the length of the gas chamber, and $x$ represents the displacement of the piston. The gas piston's force balance yields the compression chamber pressure:

$$p_c = \frac{A_{rod} m_{gp}}{A_{gp}^2} \ddot{x} + p_{gi} \frac{A_{gp} L_{ch}}{A_{gp} L_{ch} - A_{rod} x} \tag{23}$$

Equation (23) elucidates that the compression chamber pressure remains unaffected by the piston's velocity, solely contingent upon the piston's acceleration and displacement and the initial gas pressure.

### 4.7. Modeling of Damper Force

Once the chamber pressures are discerned, the calculation of damper force assumes paramount importance. Formulating the force balance in the $x$ direction yields:

$$F + p_r A_r - p_c A_c - F_f = m_p \ddot{x} \tag{24}$$

In Equation (24), $F$ represents the damper shaft force, $p_r A_r$ and $p_c A_c$ denote the compression and rebound pressure forces respectively, and $F_f$ embodies the piston's frictional force. The frictional force $F_f$ is typically derived from dynamometer measurements conducted at extremely low speeds, where pressure forces are negligible. Acceleration $\ddot{x}$ serves as a known input for the lower damper, while its computation is requisite for the upper damper. Pressure forces are computed leveraging the equations elucidated preceding this section. The mass of the piston $m_p$, encompassing the piston, rod, needle valves, and spherical bearings, can be directly quantified through measurement. Maintaining the gas pressure at a constant 200 psi across all tests, the sole remaining unknown pertains to the damper force, thereby facilitating its determination through the established framework.

For further clarity, all the flow equations are essentially used to determine the force in Equation (24). The area terms are design parameters of the damper, while the pressure terms are calculated from the various flow equations. The dynamic behavior of the damper depends on the velocity and acceleration terms, which are inputs to the damper model.

## 5. Modeling Double Damper on Simulink

Modeling the double damper in Simulink involved creating separate models for both the single- and double-damper configurations. The behavior of these dampers is governed by non-linear differential equations, which required a robust numerical method for solving. To achieve this, Newton's iterative method was employed, a technique well-suited for handling non-linear systems. The process began with an initial guess for the displacement and velocity of the damper. With these initial values, the Jacobian matrix, which contains partial derivatives of the system's equations with respect to the variables, was calculated. This matrix is crucial as it helps linearize the non-linear equations around the current guess, making them easier to solve. Next, the non-linear functions representing the damper's dynamics were evaluated at the current guess. These functions describe how the damper's displacement and velocity change over time under various forces. Using the evaluated functions and the Jacobian matrix, an update step was computed. This step involves solving a linear system to find the change needed in the guess to move closer to the true solution. The new guess for displacement and velocity was obtained by adding this update step to the old guess. This iterative process of evaluating the functions, computing the Jacobian, finding the update step, and refining the guess was repeated until the changes became negligibly small, indicating that the solution had converged. To ensure the simulations were both precise and efficient, the step sizes used in the iterative process were dynamically adjusted. This dynamic adjustment was crucial in capturing the transient behaviors accurately without imposing an excessive computational burden. This approach to solving the non-linear equations was adapted from the techniques outlined in *Numerical Methods for Engineers and Scientists* by Hoffman. [14] This methodology provided a structured framework for addressing complex non-linear systems using iterative techniques, ensuring that the model represented the real-world dynamics of the double damper. By employing this method, the Simulink model effectively captured the complex interactions within the double damper, offering reliable predictions of its performance under various operating conditions.

In the modeling process, the single damper is modeled in Simulink based on the derived damper equations, comprising three main blocks: compression, rebound, and final values.

In fact, the double-damper system incorporates two blocks, one for each of the individual damper components, while the single-damper system uses just one of these blocks. In the rebound block, the values are essentially zero when the velocity is positive. Although this could be managed using an 'if' block or equivalent conditional logic, modeling it separately simplifies troubleshooting if the results do not align with expectations. The input block receives values derived from a quarter-car model. In this model, the road input is the sole input to the entire system, driving the dynamic response of the damper.

Subsequently, the final-values block consolidates results from individual compression and rebound blocks, plotting them against velocity to generate final plots for the single semi-active damper. Additionally, a quarter-car block is constructed to incorporate sprung and unsprung masses, along with tire stiffness and damping coefficients, as detailed in [15]. Although the quarter-car model, with assumed unsprung, $k_{tire}$ and $c_{tire}$ values, is not correlated with real-time data due to testing limitations to the shock-dynamometer testbed, it provided a foundational step for further design iterations.

The double damper is then modeled in Simulink as two separate damper blocks, with a third 'body' block connecting them. The body block facilitates transmission of force and velocity between the lower and upper dampers. Both dampers were modeled similarly, employing pressure and discharge equations discussed earlier. The free-body diagram of the double damper is illustrated in Figure 3. Additionally, another quarter-car model block is implemented to account for sprung and unsprung masses, acting as an excitation source for the lower damper. A portion of the damper force is transmitted to the body block, serving as input for the upper damper. The disparity in pressures within the chambers, coupled with varied sprung and unsprung accelerations, contributes to both the lower and upper forces.

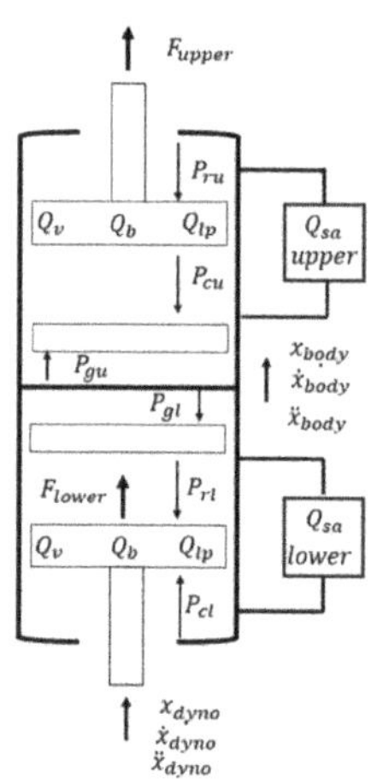


**Figure 3.** Free-body diagram of the double damper [11].

Parameters that are used for the damper modeling and functional simulation using Simulink have been presented in Table 1. While a variety of control algorithms have been explored and utilized to optimize the performance of the double-damper system, it is important to note that the specific details and results of these algorithm evaluations are not presented in this paper. Instead, the focus of this study lies on the modeling and simulation aspects of the double-damper system, as well as the comparison of its performance with a traditional single-damper setup.

**Table 1.** Parameters used for the damper modeling on Simulink.

| Parameter | Definition | Value | Parameter | Definition | Value |
|---|---|---|---|---|---|
| $\alpha$ (alpha) | Correction factor | 0.5 | $C_f$ | Momentum coefficient | 0.3 |
| $P_{gi}$ | Initial gas pressure | 24 psi | $k$ | Shim stack stiffness | 800 lbf/in |
| $A_{gp}$ | Area of gas piston | 1.5175 in$^2$ | $l$ | Length of piston leakage gap | 0.3 in |
| $L_g$ | Length of gas chamber | 0.8 in | $C_d$ | Dynamic discharge coefficient | 0.71 |
| $A_{rod}$ | Area of the rod | 0.1104 in$^2$ | $F_f$ | Seal Friction | 5 lbf |
| $m_{gp}$ | Mass of gas piston | 0.0273 lb. | $M_p$ | Mass of Piston/Rod assembly | 0.335 lb. |
| $D_p$ | Diameter of the piston | 1.13 in | $Q_b$ | flow through the bleed orifice | in$^3$/sec |
| $A_p$ | Area of the piston | 1.0029 in$^2$ | $Q_{lp}$ | leakage past the piston seal | in$^3$/sec |
| $A_r$, $A_c$ | Area of rebound/compression chamber | 0.829 in$^2$ | $p_c$, $p_r$ | compression and rebound chambers pressure | psi |
| $\rho$ | Fluid density of mineral oil | 0.033 lb/in$^3$ | $\Delta p_{valve}$ | pressure drop across the valve shim. | psi |
| $A_v$ | Area on which the valve pressure acts | 0.411 in$^2$ | $x_{dyno}$ | displacement input from the dyno | in |
| $\mu$ | Dynamic viscosity | $2.5 \times 10^{-5}$ lbs.sec/in$^2$ | $y$ | deflection of the shim stack | in |
| $Q_p$ | flow through the piston valve | in$^3$/sec | $m_{body}$ | body block mass | lb |
| $Q_{sa}$ | semi-active valve flow | in$^3$/sec | $x_{body}$ | displacement of body block | in |
| $F_{lower}$ $F_{upper}$ | Damping force by lower/upper damper | lbf | $A_{sa}$ | Area of semi-active orifice valve | in$^2$ |

The parameters without data are not design parameters and they are calculated through equations of motion from change of pressure.

## 6. Simulation Results

The simulation is performed to compare single- and double-damper models in terms of unsprung and sprung mass displacements and accelerations. Figure 4 illustrates the road input, characterized by three step inputs occurring at time intervals of 15, 24, and 64 s, thereby generating a sequence of bumps in the damper system's response. These step inputs were strategically positioned within the road input vector to emulate real-world scenarios. Analysis of the simulation results revealed a noticeable reduction in the unsprung mass displacement for the double-damper configuration compared to its single-damper counterpart. Figure 5 depicts the unsprung mass oscillations, illustrating that the single damper exhibits greater amplitude oscillations compared to the double damper. The observed lag in the double-damper response can be attributed to the delay in additional damping force being applied.

To further explore damper behavior under continuously fluctuating road conditions, a sinusoidal wave with an amplitude of 0.1 inches and a frequency of 25 Hz was introduced into the existing road input, as depicted in Figure 6. The simulation results, particularly regarding unsprung mass displacements, highlight the double damper's effectiveness in attenuating high-frequency sine oscillations in the road input compared to the single semi-active damper employing the same control logic, as evidenced in Figure 7. Furthermore, the double damper exhibits superior damping of sprung-mass oscillations compared to the single damper, albeit with a less pronounced difference. Notably, both dampers demonstrate nearly identical steady-state behavior, underscoring the efficacy of the double-damper configuration in damping oscillations across the system. Additionally, the double damper provides better damping for the oscillations of the sprung mass compared to the single damper. Although the difference is not as significant, both dampers show nearly identical behavior once they reach a steady state.

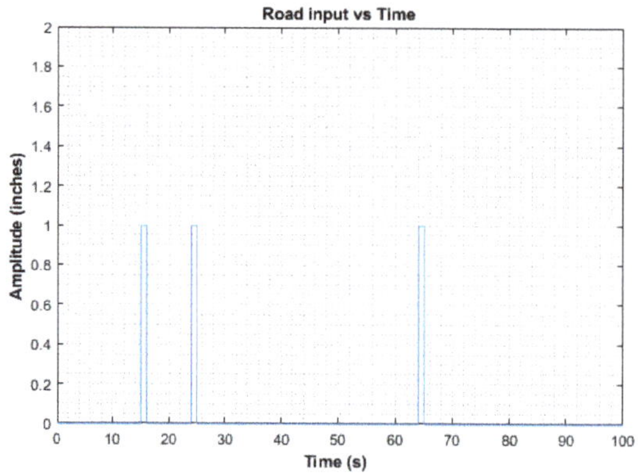


**Figure 4.** Road input with multiple impulse profile.

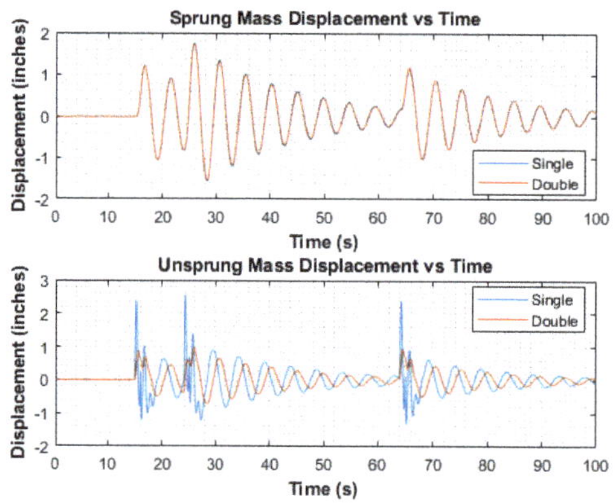


**Figure 5.** Comparisons between unsprung and sprung mass displacement for two dampers due to multiple impulse road profile.

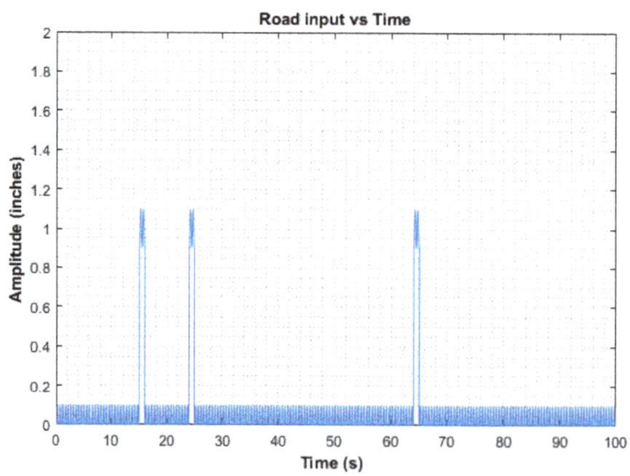


**Figure 6.** Road input including impulses and sinusoidal wave.

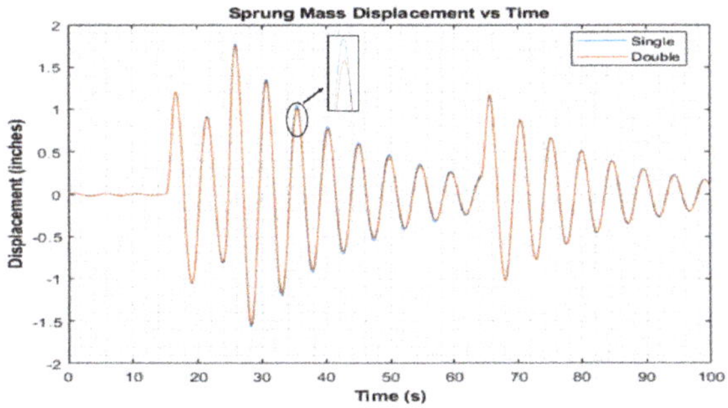


**Figure 7.** Comparison between sprung mass displacement of two damper concepts, single and double dampers.

However, due to certain constraints, it was not feasible to conduct the tests and simulations under identical conditions. The primary challenges we faced were equipment limitations and safety concerns. These constraints made it impractical to replicate the exact conditions for both numerical and experimental analyses. Despite these limitations, we carefully designed our study to provide meaningful comparisons between single- and double-damper systems. Both test and simulation environments, although not identical, were consistent within their respective categories, allowing for valid comparative analysis. Our goal was to highlight the relative performance improvements and differences between the two systems rather than an absolute validation under identical conditions. The mathematical model was employed for control algorithm adaptation, which is not in the scope of this paper.

## 7. Shock Dynamometer Testing Results

The experimental setup for damper testing depicted in Figure 8 provided a platform to evaluate the performance of the dampers. The shock dynamometer testing is conducted within the frequency range of 0.25 Hz to 3 Hz, effectively capturing a spectrum of vehicle dynamics including yaw, roll, and pitch dynamics [16]. Both the single and double dampers are subjected to testing at a suspension velocity of 2 in/sec on the shock dynamometer. This testbed machine facilitates systematic evaluation of the dampers' responses across a range of velocities, enabling detailed analysis of their force-vs.-velocity characteristics. As presented in reference [15], by subjecting the dampers to varying velocities, the testing captures their performance under dynamic conditions representative of real-world driving scenarios. The force-vs.-velocity graphs illustrate the force buildup characteristics with varying velocity.

These force-vs.-velocity graphs provide valuable insights into the behavior of the dampers under different loading conditions and velocities. These plots illustrate the force buildup characteristics exhibited by the dampers as the velocity varied, offering a comprehensive understanding of their damping capabilities across a wide range of operating conditions.

Figure 9 displays the force-vs.-velocity graph for both single and double dampers tested at different combinations of voltages, showcasing their behavior. The different combinations of damping achievable by supplying various voltages to the upper and lower damper, along with the adjustable behavior of the shim stack, allow for fine-tuning of the dynamic response of the damper. Specifically, varying the input voltage or current to a semi-active damper alters its performance characteristics. This is because semi-active dampers use electromagnetic or magnetorheological principles to modulate the damping force in real time based on the supplied electrical input. By adjusting the voltage, the viscosity of the damping fluid or the force exerted by the damper can be controlled, resulting in different damping rates. This ability to precisely control the damping force is crucial for optimizing the ride comfort and handling performance of the vehicle. Therefore, the choice of supplied voltages is determined based on the desired dynamic response and performance characteristics of the damper in various operating conditions. When voltage is applied, the solenoid moves upward against the spring, permitting fluid to pass through. Damping through the internal piston relies on orifice area and shim-stack stiffness, while flow through the bypass valve is regulated by orifice size controlled by voltage application. Two 12 V power supply was utilized to maintain the voltage, with the shock receiving input from 0 to 5 V for the single and 22 V for the double damper.

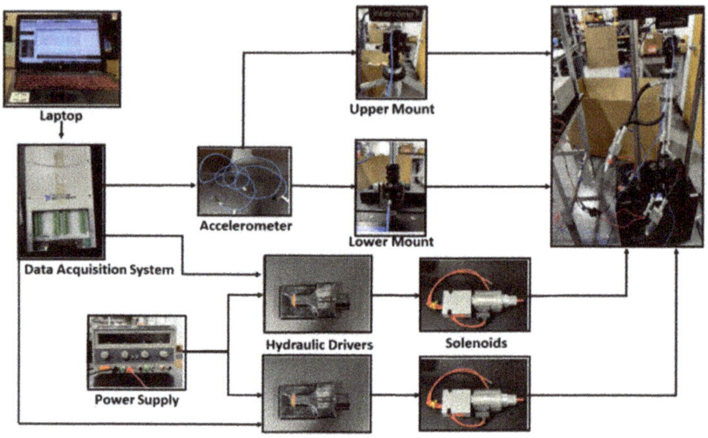


**Figure 8.** Damper experimental setup.

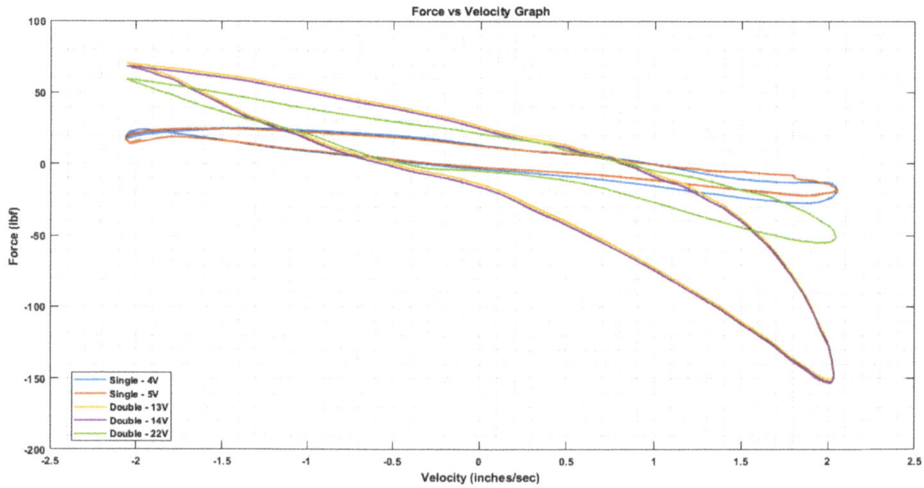


**Figure 9.** Force vs. velocity for the single and double damper tested with different combinations of voltages at 2 in/sec.

Fundamentally, rebound forces are notably higher in all cases due to asymmetrical shim stacks. Across all configurations, the double damper exhibits greater force in both compression and rebound compared to the single damper. Even the 22nd configuration of the double damper, though less stiff than the 13th and 14th configurations, produces more damping force than the single damper.

Furthermore, as illustrated in Figure 10, the double damper generates greater force than the single damper for similar suspension deflection. The higher rebound force compared to compression aligns with simulation results, indicating an improvement in ride quality with the double damper in the suspension system.

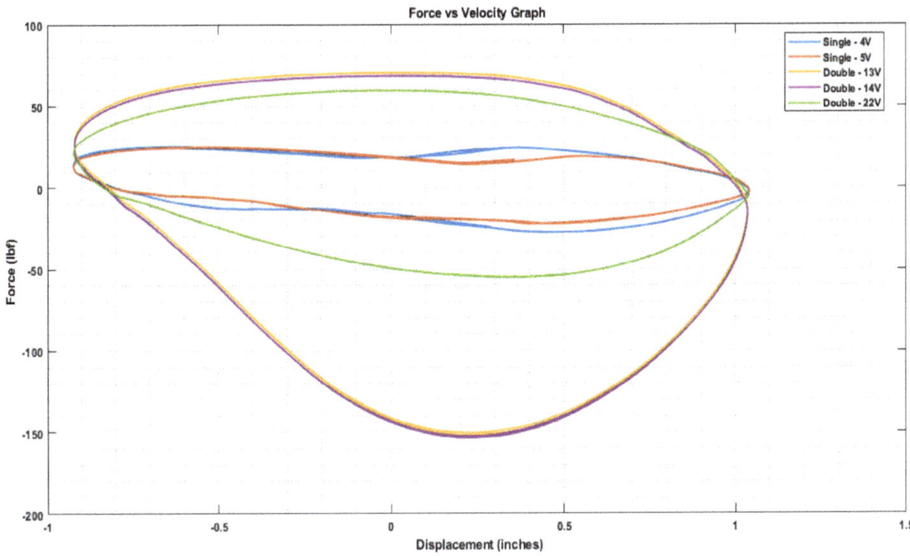


**Figure 10.** Force vs. displacement for the single and double damper tested with various voltage combinations at 2 in/sec.

Another observation from Figures 9 and 10 is that for the single damper, the damping forces decrease progressively with increasing voltage, ranging from fully closed (0 V) to fully open (5 V/1.2 A), resulting in varied rebound forces due to asymmetrical shim stacks. Conversely, the double damper, operating with identical upper and lower solenoid voltages, behaves passively in the fully closed scenario (0 V), with flow through the bleed orifice and shim stacks, leading to the highest compression and rebound forces. Beyond the 22 V configuration, the double damper exhibits a shift in behavior, transitioning from a digressive to a progressive nature, influenced by factors like increased frictional forces and constant flow through the semi-active orifices at low speeds. This adaptability in damper response holds potential for enhancing ride comfort in SUVs across diverse road conditions.

## 8. Conclusions

This paper presents the results of a study on the modeling, simulation, and performance evaluation of a novel double-damper system against a traditional single-damper configuration using Simulink/MATLAB. The modeling process involved representing the double damper as two individual dampers interconnected with a body block, thus accommodating the additional degree of freedom introduced by the integration of two dampers. Leveraging the reduced-order modeling (ROM) approach proved apt for capturing the complexities inherent in damper systems, enabling the application of equations to both passive and double-damper configurations.

Simulations were conducted across various road profiles employing a simplified quarter-car model. The results demonstrate the superior performance of the innovative double-damper design in mitigating oscillations of both the sprung and unsprung masses, compared to a single damper. Furthermore, the paper also focuses on the pros and cons associated with the double-damper system. Highlighted benefits of the double-damper system include its cost-effectiveness, simplicity, versatility across vehicle types, enhanced reliability, ease of integration, and its ability to strike a balanced combination of ride comfort and handling performance by having two independent damping characteristics. However, challenges such as higher length constraints for compact cars, friction dynamics with dual cylinders and pistons, and the need for extensive testing and market validation are duly acknowledged. Thorough modeling of the double-damper architecture, encompassing components such as the gas chamber, bleed flow, piston valve flow, leakage flow, and semi-active flow, provides invaluable insights into its intricate dynamics. Special attention is given to modeling the gas chamber, accounting for the incompressibility of oil and utilizing the ideal gas law to determine gas pressure. Simulation results showed significant improvements in unsprung mass displacement for the double damper compared to the single damper, especially under impulse and sinusoidal road inputs.

Shock dynamometer testing demonstrates the superior performance of the double damper, showcasing increased force in both compression and rebound compared to the single damper. The force-vs.-velocity and force-vs.-displacement graphs highlight the enhanced damping capabilities of the double damper. The progressive nature of the damper depends on factors such as frictional forces and the absence of a shim stack. This flexible characteristic, coupled with its progressive nature and ability to handle greater damping peak loads, makes the double damper suitable for SUVs, where ride comfort is crucial for both on-road and off-road conditions. These conclusions are supported by findings detailed in reference [15]. Additionally, an observed trend suggests that force varies predictably with voltage, with the lowest applied voltage resulting in the highest force output.

Importantly, this modeling approach lays a solid foundation for the analysis of different control algorithms. The straightforward nature of the model facilitates a streamlined evaluation process, enabling researchers to gauge control algorithm effectiveness across diverse scenarios. This study emphasizes the potential of the model to provide meaningful insights into control algorithm performance, making it a valuable tool in the iterative process of algorithm development and refinement. In conclusion, this study not only contributes valuable insights into the potential of the double-damper system for advanced vehicle suspension but also positions itself as an asset for the analysis and optimization of different control algorithms to achieve the best performance.

**Author Contributions:** Conceptualization, S.S. and S.T.; methodology, S.S.; software, S.S.; validation, S.S., S.T. and B.H.; formal analysis, S.S.; investigation, B.H.; resources, S.T.; data curation, S.S.; writing—original draft preparation, B.H.; writing—review and editing, B.H.; visualization, B.H.; supervision, S.T.; project administration and funding acquisition for prototyping and testing. All authors have read and agreed to the published version of the manuscript.

**Funding:** This research received no external funding.

**Data Availability Statement:** The data presented in this study including detail design of the double damper are available on request from S.T. The technical data is not publicly available due to confidentiality.

**Conflicts of Interest:** The authors declare no conflict of interest.

## References

1. Lang, H.H. AStudy of the Characteristics of Automotive Hydraulic Dampers at High Stroking Frequencies. Ph.D. Dissertation, The University of Michigan, Ann Arbor, MI, USA, 1977.
2. Kim, D. Analysis of Hydraulic Shock—Absorber Implementation on the Vehicle Suspension Systems. Master's Thesis, Seoul National University, Seoul, Republic of Korea, 1993.
3. Cafferty, S.; Worden, K.; Tomlinson, G. Characterization of Automotive Shock Absorbers Using Random Excitation. *Proc. Inst. Mech. Eng. Part D J. Automob. Eng.* **1995**, *209*, 239–248. [CrossRef]
4. Reybrouck, K.G. *A Non Linear Parametric Model of an Automotive Shock Absorber*; SAE Technical Paper Series 940869; SAE International: Warrendale, PA, USA, 1994.
5. Duym, S.W.; Steins, R.; Baron, G.V.; Reybrouck, K.G. *Physical Modeling of the Hysteretic Behaviour of Automotive Shock Absorbers*; SAE Technical Paper Series 970101; SAE International: Warrendale, PA, USA, 1997.
6. Mollica, R.; Youcef-Toumi, K. A Nonlinear Dynamic Model of a Monotube Shock Absorber. In Proceedings of the American Control Conference, Albuquerque, NM, USA, 6 June 1997; pp. 704–708.

7. Herr, F.; Mallin, T.; Lane, J.; Roth, S. *A Shock Absorber Model Using CFD Analysis and Easy5*; SAE Technical Paper 1999-01-1322; SAE International: Warrendale, PA, USA, 1999. [CrossRef]
8. Talbott, M.S.; Starkey, J. *A Mathematical Model of a Gas-Charged Mono-Tube Racing Damper*; SAE International: Warrendale, PA, USA, 2002. [CrossRef]
9. Rhoades, K.S. Development and Experimental Verification of a Parametric Model of an Automotive Damper. Master's Thesis, Texas A&M University, College Station, TX, USA, 2006.
10. Siramdasu, Y.; Taheri, S. A New Semi-Active Suspension System for Vehicle Applications. In Proceedings of the ASME 2017 Dynamic Systems and Control Conference, Tysons, VA, USA, 11–13 October 2017.
11. Tyagi, S. Development of a Semi-active Suspension System for Lightweight Automobiles. Thesis for Master of Science in Mechanical Engineering, Virginia Polytechnic Institute and State University, Blacksburg, VA, USA, 2016.
12. Gorantiwar, A.; Nalawade, R.; Nouri, A.; Taheri, S. Experimental Analysis of a Novel Double Damper System with Semi-Active Control. *J. Electron.* **2020**, *9*, 1518. [CrossRef]
13. Reybrouck, K.G.; Schurmans, R.J. Damper with Externally Mounted Semi-Active System. Google Patents US6321888B1, 27 November 2001.
14. Hoffman, J.D. *Numerical Methods for Engineers and Scientists*, 2nd ed.; Marcel Dekker, Inc.: New York, NY, USA, 2001.
15. Shrikanthan, S. Modeling, Development and Experimental Validation of Different Control Strategies for a Double-Damper Semi-Active Suspension System. Dissertation for the degree of Master of Science in Mechanical Engineering, Virginia Polytechnic Institute and State University, Blacksburg, VA, USA, 2024.
16. Bhanot, N. Artificial Neural Networks Based Modeling and Analysis of Semi-Active Damper System. Ph.D. Thesis, Virginia Polytechnic Institute and State University, Blacksburg, VA, USA, 2017.

*vibration*

MDPI

*Review*

# Experimental Linear and Nonlinear Vibration Methods for the Structural Health Monitoring (SHM) of Polymer-Matrix Composites (PMCs): A Literature Review

**Loan Dolbachian *, Walid Harizi and Zoheir Aboura**

Centre de Recherche Royallieu, Roberval (Mechanics Energy and Electricity), Université de Technologie de Compiègne, CEDEX CS 60 319, 60 203 Compiègne, France; walid.harizi@utc.fr (W.H.); zoheir.aboura@utc.fr (Z.A.)
* Correspondence: loan.dolbachian@utc.fr

**Abstract:** The goal of this article is to provide a review of the experimental techniques and procedures using vibration methods for the Structural Health Monitoring (SHM) of Polymer-Matrix Composites (PMCs). It aims to be a guide for any researchers to carry out vibration experiments. The linear methods are first introduced. But, as PMC is a complex material, these classic methods show some limits, such as low accuracy for small damages and a high environmental dependency. This is why the nonlinear methods are secondly studied, considering that the complexity of PMCs induces a nonlinear behavior of the structure after damage occurrence. The different damage mechanisms are well-explained in order to evaluate the potential of each vibration method to detect them.

**Keywords:** Polymer-Matrix Composites (PMC); Structural Health Monitoring (SHM); vibration analysis; linear/nonlinear methods

## 1. Introduction

Nowadays, the use of PMCs in industrial sectors, such as aeronautics and the automotive sector, is sharply increasing thanks to their high mechanical properties, low weight, high resistance to corrosion environments, etc. [1–10], permitting them to challenge classical materials like metals. Table 1 references some mechanical characteristics of three different types of PMCs in comparison with two classical materials, steel and aluminum. The key synthetic parameters used to assess the mechanical performance of a composite include the specific modulus (the ratio of elasticity modulus to density $E/\rho$) and the specific strength (the ratio of ultimate strength to density $R_e/\rho$). Typically, these properties are governed by the fibers. When comparing mechanical properties with mass considerations, the three PMC materials seem to offer more promise than the two metallic materials. In recent years, there has been a significant increase in the number of articles published about PMCs (Figure 1), with a surge starting around 2010. PMCs are made of fiber and matrix. The combination and variation of the fraction or orientation of fibers make it a complex, heterogeneous, and anisotropic material, which complicates its damage mechanisms. This is why measuring and understanding them as well as evaluating their impact on the structure are important to ensure the continuity of services. This is known as Structural Health Monitoring (SHM). The concept of SHM brought these four steps [11]: (i) detect damage in the structure; (ii) localize it; (iii) quantify its severity and qualify it; and (iv) give a conclusion regarding the service life of the structure. This review mainly explores the first two steps of SHM by focusing on the methods used and does not cover the prognostic part.

**Table 1.** Absolute and reduced characteristics of three PMCs and two metals [12].

| Material | Volume Mass $\rho$ (g/cm$^3$) | Young's Modulus E (GPa) | Strength $R_e$ (MPa) | Tenacity $K_{IC}$ (MPam$^{1/2}$) | $E/\rho$ | $E^{1/2}/\rho$ | $E^{1/3}/\rho$ | $R_e/\rho$ |
|---|---|---|---|---|---|---|---|---|
| Composite | | | | | | | | |
| Carbon-Fiber-Reinforced Polymer (58% uniaxial fiber in epoxy) | 1.5 | 189 | 1050 | 32–45 | 126 | 9 | 3.8 | 700 |
| Glass-Fiber-Reinforced Polymer (50% uniaxial fiber in epoxy) | 2.0 | 48 | 1240 | 42–60 | 24 | 3.5 | 1.8 | 620 |
| Kevlar-Fiber-Reinforced Polymer (60% uniaxial fiber in epoxy) | 1.4 | 79 | 1240 | - | 54 | 6.2 | 3.0 | 886 |
| Metal | | | | | | | | |
| High-resistance steel | 7.8 | 207 | 1000 | 100 | 27 | 1.8 | 0.76 | 128 |
| Aluminum alloy | 2.8 | 71 | 500 | 28 | 25 | 3.0 | 1.5 | 179 |

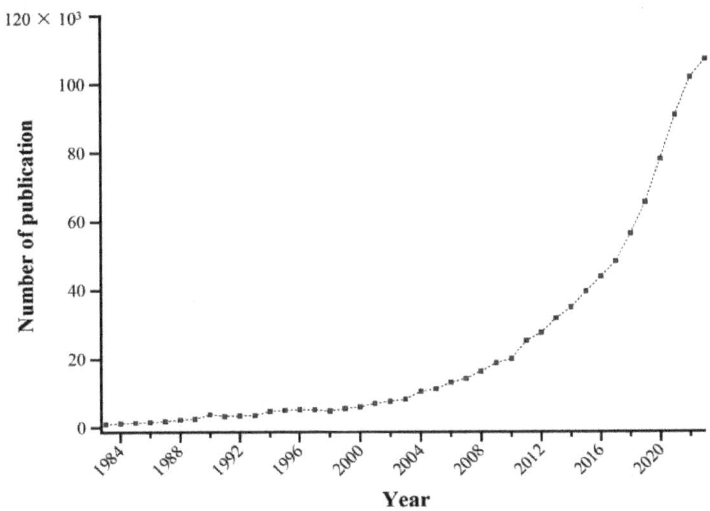


**Figure 1.** Number of articles published during the past 40 years about PMC. (source = https://app.dimensions.ai/discover/publication, accessed on 6 March 2024).

Many Non-Destructive Testing (NDT) techniques are available for SHM purposes [13], such as Ultrasonic Testing (UT) [14,15], Acoustic Emission (AE) [16,17], Digital Image Correlation (DIC) [18,19], X-ray tomography [20,21], Infrared thermography [22], Vibration analysis [23], and others. In the literature, many reviews have focused on the numerical aspects of vibration analysis and/or do not distinguish the type of material. Given the growing utilization of PMC materials and their complex behavior, it is necessary to understand how they vibrate and what influences the damage mechanisms have on the vibration.

The vibration of a structure provides a wealth of information regarding its physical properties, which motivates its use as a damage indicator [24–31]. Indeed, the vibration of a system follows the equation of motion of a forced-damped system:

$$m\ddot{x} + c\dot{x} + kx = F(t) \quad (1)$$

where $\ddot{x}$, $\dot{x}$ represent the second and first time-derivatives of the displacement $x$, $m$ is the mass, $c$ is the viscous coefficient, $k$ is the stiffness of the system, and $F(t)$ is an external force applied to the system as a function of time. The vibration parameters, such as natural frequency ($f_0$) or damping ratio ($\zeta$), can be calculated from this formula and clearly show that a change in the material characteristics due to damage will influence the vibration:

$$\begin{gathered} \ddot{x} + 2\zeta\dot{x} + {\omega_0}^2 x = \frac{F(t)}{m} \\ \zeta = \frac{c}{2\sqrt{mk}} \qquad\qquad f_0 = 2\pi\omega_0 = 2\pi\sqrt{\frac{k}{m}} \end{gathered} \tag{2}$$

The vibration methods for SHM are divided into two parts: the linear and the nonlinear methods. Linear vibration methods have been known and utilized for a long time, but innovations in their application still represent a current field of research. On the other hand, nonlinear methods have shown increasing interest from the 2010s to the present day. Given the anisotropy and the heterogeneous aspects of PMCs, it turns out that the detection and quantification of their damage mechanisms are more relevant using nonlinear vibration methods [32]. Considering that, it is worthwhile to undertake a review that highlights the greater effectiveness of the nonlinear methods over the linear ones for PMC material while still introducing both approaches. As it aims to be a guide to the different experimental vibration methods, the equipment, the devices used for each method, and the damages studied in the literature will be described in detail.

This article is divided as follows. The first part introduces the PMCs: their composition, the different kinds of architecture, and the principal damage mechanisms that can occur. The second part describes the methods and the equipment used in the literature. The third part is the main part of this review, and it introduces the experimental linear and nonlinear vibration analyses and the damage specificities.

## 2. Reminder about PMCs and Their Damage Mechanisms

A composite is a structural material that combines two different components. One is the matrix and the other is the reinforcement. There exist several types of composites depending on the used constituents: the Ceramic-Matrix Composites (CMCs), the Metal-Matrix Composites (MMCs), and the PMCs. As their application domains are different, this article only focuses on the PMC, which is a material of wide diffusion.

### 2.1. Composition and Structure of PMC

In PMC, the reinforcing part is made of continuous or short fibers and the matrix is made of polymers, such as thermoplastic or thermoset resins [6,33]. The volume or the orientation of the fibers, the number of layers, the different possible structures, and the different possible constituents expand the possibilities in the creation of PMC. A list of possible constituents of PMCs is given in Table 2.

**Table 2.** PMC composition.

| Reinforcement (Fiber) | Matrix (Polymer) | |
|---|---|---|
| | **Thermoplastics** | **Thermosets** |
| Carbon<br>Glass<br>Aramid (Kevlar)<br>Polypropylene<br>Hemp<br>Flax<br>etc. | Nylon<br>Polypropylene<br>Polycarbonate<br>Cellulose Acetate<br>Polystyrene<br>Polyethylene<br>Polyvinyl chloride<br>Acrylonitrile-butadiene-styrene<br>Polyether-ether-ketone<br>etc. | Phenolic<br>Polyimide<br>Polyurethane<br>Polyepoxide<br>Polyester<br>etc. |

Due to the polymer resin, which can dissipate energy through internal friction and molecular relaxation processes [34], a PMC exhibits a viscoelastic behavior that significantly influences vibration analysis. This viscoelasticity can be characterized through dynamic mechanical analysis, which involves the application of a small oscillatory stress to the material while measuring the strain resulting from this load. The stress σ and the strain ε can then be linked with a dynamic modulus G:

$$\sigma = G\,\varepsilon \qquad\qquad G = G' + iG'' \tag{3}$$

where $G'$ is the storage modulus representing the elastic behavior and $G''$ is the loss modulus representing the viscous behavior. The phase angle $\delta$ represents the combination of viscous and elastic behavior within the material and will be included between 0 and 90° ($\delta$ = 0° indicates purely elastic behavior (energy stored and restituted by the material) and $\delta$ = 90° indicates purely viscous behavior (energy dissipated in the form of heat)):

$$\tan\,\delta = \frac{G''}{G'} \tag{4}$$

The common structures made of PMC are the laminates and the sandwiches, both of which are shown in Figure 2. The laminated composite consists of the assembly of several plies with different fiber orientations, while the sandwich is made of two thin skins joined with an adhesive to a core; the skin is made of laminates and the core is made of a thick, low-strength material that can be foam material, such as polyurethane, honeycombs, and others. The layers of laminates can either be made of unidirectional, short, or woven fibers [35]. The woven fabric consists of the interlacing of warp and weft fibers with a specific pattern (plain, satin, twill, etc.) (Figure 3).

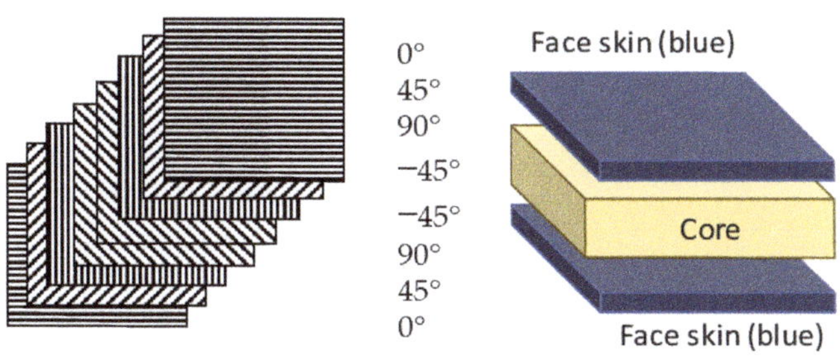


**Figure 2.** Laminated (**left**) and sandwich composite (**right**).

Recently, many 3D woven composites have been developed for particular applications (aerospace and defense domains), and some studies of their mechanical properties have been proposed [36].

### 2.2. Damage Mechanisms

In a PMC's service life, four main damage mechanisms can cause structure failure: matrix cracking, fiber breaking, delamination, and fiber–matrix interfacial debonding. They can be observed in Figure 4 coming from an impact. Many articles have used impact damage in their research, including, mainly, the BVID (Barely Visible Impact Damage) [37], which is a common way of introducing damage within PMC materials through low-velocity impact. The damage introduced through BVID could be a mix of the four main damage mechanisms while varying based on the architecture and composition of the PMC structure. In general, damage starts with very small matrix cracking (microcracks) and the three other mechanisms will be induced from this start, resulting in a combination of them. On a laboratory scale, these damages can be artificially generated: fiber-breakage can be made from fiber cutting before the manufacturing step, delamination can be introduced by inserting Teflon or foreign film between two plies, etc. In service, impact damage is common, and it results in the same damage mechanisms. The more forceful the impact, the more the structure is damaged. In reference [38], a clarification is made regarding the characteristic failure mechanisms of PMC. In Figure 5, we can observe the evolution of damages during a tensile test.

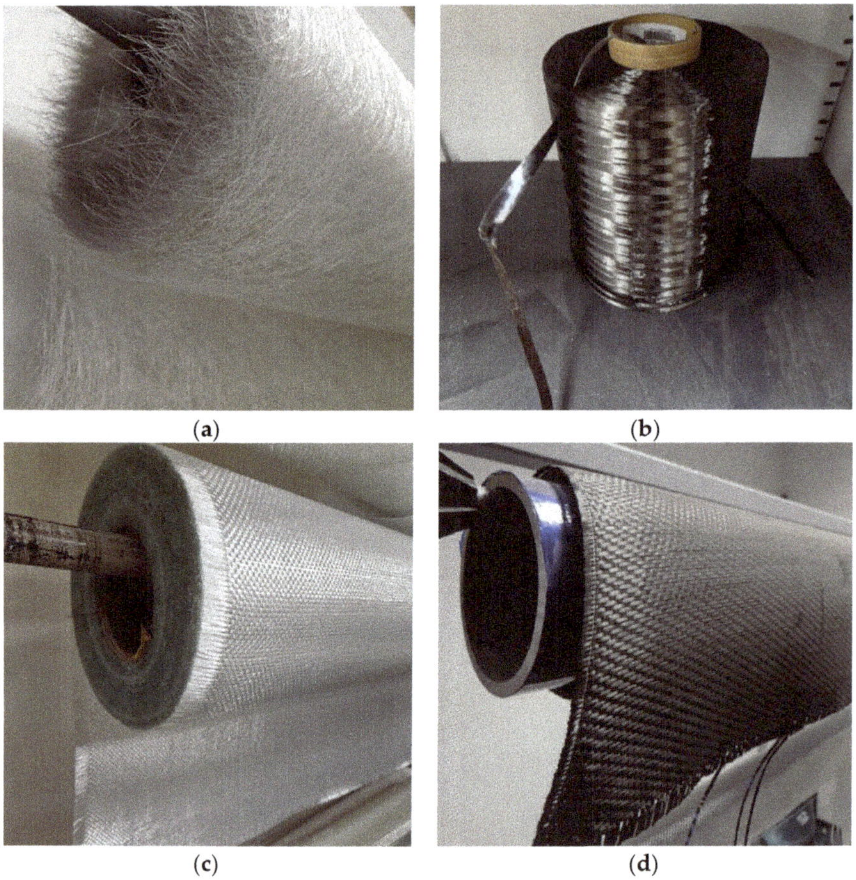

**Figure 3.** Different architectures of fibers: (**a**) mat short glass fiber, (**b**) continuous carbon fiber, (**c**) plain glass fiber, and (**d**) twill 2/1 carbon fiber.

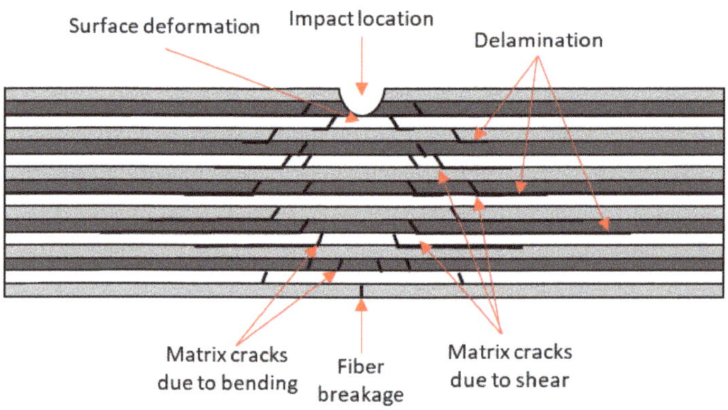


**Figure 4.** Impact damage mechanisms in composite laminate (adapted from [39]).

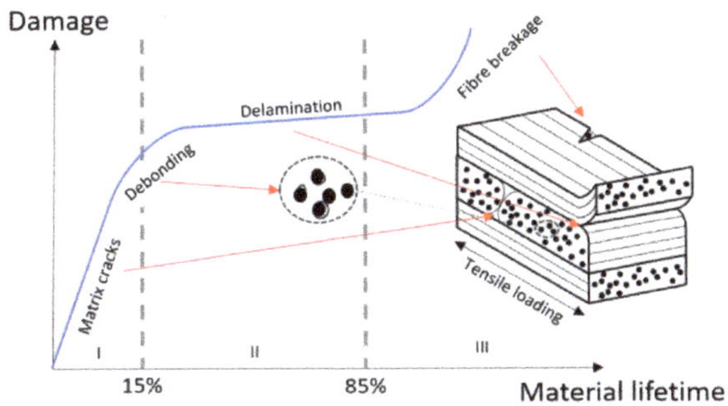


**Figure 5.** Evolution of damage in composite laminate (adapted from [40]).

### 2.2.1. Matrix Cracking

The first step in the damage process is the microscopic matrix cracks consisting of the breaking of the resin. They have two forms: transverse or longitudinal. The transverse ones propagate perpendicular to the direction of the mechanical load, whereas the longitudinal ones follow the direction of the loading. Therefore, the two forms depend on the type of applied mechanical stress (tensile, compression, shear or bending, etc.) and the orientation of the fibers within the composite material. With increasing stress, these cracks will progressively propagate and reach the fibers. At this step, the mechanical properties are slightly degraded, and it is difficult to detect these matrix cracks due to the low energy release that they produce. The damage process will either be stopped or will continue with the propagation of the cracks in the less energetic way, which corresponds to the interface between the fiber and the matrix. This will lead to fiber-matrix debonding.

### 2.2.2. Fiber-Matrix Interfacial Debonding

Fiber-matrix interfacial debonding usually happens when the cohesion of the fiber-matrix is lower than the shear strength of the matrix or when the adherence of the fiber-matrix is weak. The matrix crack induces debonding without fiber breaking. The amount of debonding will increase while the stress increases. This damage mechanism is more energetic than the matrix cracks, but it is still low and does not have a big influence on the mechanical properties. As the matrix cracks and the debonding will propagate, the interlaminar cracks will occur and the load will progressively increase on the fibers, resulting in delamination and the fibers breaking.

### 2.2.3. Delamination or Interlaminar Cracking

Delamination [41] concerns the laminated composites. It consists of an interlaminar cracking of the resin and leads to the separation between two successive plies. Delamination can be internal or near-surface. It can grow and extend its area, and then it will lead to the failure of the laminated composite. It can be considered one of the most severe damages for laminate composite materials, which explains why many referenced articles focus on detecting it.

### 2.2.4. Fiber Breaking

Fiber breaking is also one of the main reasons for PMC structure failure; when the stress reaches the fracture strength of the fiber, it will break. It usually happens when the stress direction is similar to the orientation of the fiber. The rupture of a fiber releases high energy and strongly degrades the mechanical properties of the structure. This is why most of the NDT methods are willing to detect such a violent damage mechanism. The final failure of the PMC material is a combination and an accumulation of these various damage mechanisms.

## 3. Equipment and Setup for Vibration Analysis

A focus on the modes of vibration (how the structure vibrates at specific frequencies) of the structure is given; this is called modal analysis. How to perform it is first introduced to have the basis to understand the linear and nonlinear techniques.

A brief and nice introduction to modal analysis is given in [42]. In a dynamic analysis, like vibration one, we have to introduce the system that excites the structure (the actuator) and all equipment necessary (the input equipment), the system that catches the information (the sensor) and the equipment that processes the output data and displays results (the output equipment), the system that holds the structure, and all additional setup and equipment.

### 3.1. Fixation Setup

To perform vibration analysis, there exist several ways of fixing the structure according to the boundary conditions we are looking for. The classical one is to suspend the structure, with elastic wires or threads, to obtain free–free boundary conditions. This will isolate the dynamic response of the structure. The other can be the cantilever configuration with the free–clamped condition (one edge clamped) or by putting the structure on absorbing foam. These three configurations can be seen in Figure 6. Some researchers also used the clamped–clamped configuration to perform their analysis. As the vibrations of bridges are often investigated, another method consists of laying the structure on several supports; it can be one support for each edge or even some support in the middle of the structure.

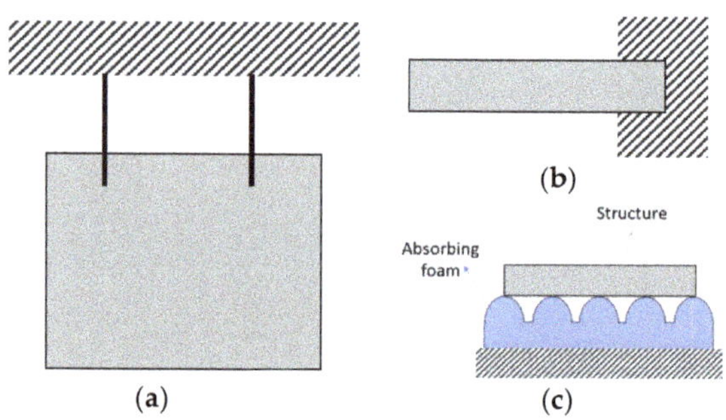


**Figure 6.** Schematic of different fixation setups: suspended with wire (**a**), clamped in cantilever configuration (**b**), and isolated with foam (**c**).

### 3.2. Actuators and Sensors

The classical modal analysis is performed with an impact hammer and accelerometers [43]. The hammer has a force sensor to record the input force. For repeatability purposes, it can be replaced by a shaker, which permits applying different kinds of excitation: white noise, sweep signals, etc. It needs a generator and an amplifier to provide an electrical signal. The hammer test is the easiest to perform, it will provide a global view of the vibration of the structure, and it is often used as a first test to obtain the natural frequencies and their damping ratios. Exciting the structure with white noise will give similar results. Once these parameters are obtained, a lot of researchers use different signals, such as harmonic signals, sweep-sine signals, etc., with the intention of focusing on specific modes or working on a large frequency band.

Different actuators and sensors were used in [44,45] to perform modal analyses and identify natural frequencies. The actuators were an impact hammer, a loudspeaker, a fan, and an in situ PVDF piezoelectric transducer (Polyvinylidene fluoride). PVDF and a loudspeaker were used with a white noise generator and piezo-amplifier (the output of the amplifier was 500 V to excite the PVDF), and the accelerometer and the PVDF received the output signal. These articles present a lot of possibilities to perform modal analysis. The setup necessary to excite with PVDF is given in Figure 7. The SQV 1/500 piezo amplifier is necessary to provide the 500 V at the PVDF.

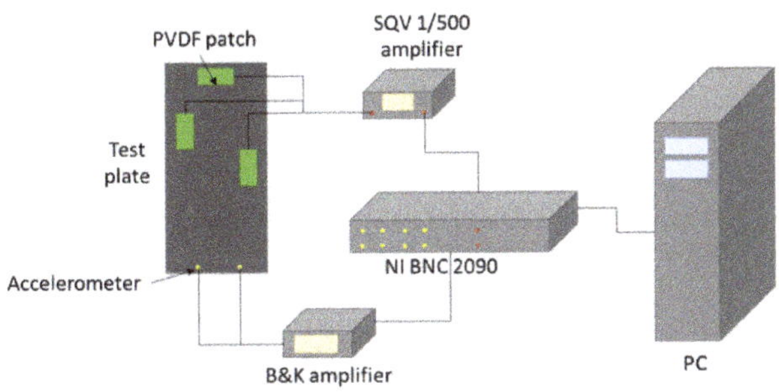


**Figure 7.** Setup vibration analysis with PVDF as actuator (adapted from [44]).

Piezoelectric transducers deform when they are receiving an electrical signal, which creates vibration, and inversely emit a voltage signal while stress is applied to them (this is known as the piezoelectric effects) [46,47]. There exist several piezoelectric transducers, such as PVDF, PZT—piezoceramic (Lead Zirconate Titanate), MFC (Macro Fiber Composite), and PFC (Piezoelectric Fiber Composite). They have all been used for modal analysis in the literature, where they are mostly surface-bonded, but researchers have investigated the possibility of the in situ transducers [13,48–51]. They can be used as actuators and sensors, but PZT needs a lower amplitude of excitation to excite the structure (20–100 V for PZT against 500 V for PVDF); this is why PZTs are often used as actuators and PVDFs as sensors. Table 3 presents a non-exhaustive list of equipment used for modal analysis.

**Table 3.** List of possible equipment for modal analysis.

| | Equipment | Additional Equipment | Details |
|---|---|---|---|
| Fixation | Foam<br>Wire<br>Clamping system<br>Fixed–fixed | " | " |
| Input | Impact hammer<br>Manual action<br>Fan | " | " |
| | Shaker<br>Loudspeaker | Generator—Amplifier | " |
| | PZT | Generator—<br>Piezo amplifier | 40–100 V out of the amplifier |
| | PVDF | | 500 V out of the amplifier |
| | MFC | Generator | " |
| Output | Accelerometer<br>Laser Vibrometer (LV)<br>Piezoelectric sensors | Analyzer and PC<br>Oscilloscope | " |

### 3.3. Data Processing

The response from the sensors usually goes into an analyzer that directly gives the needed information: Frequency Response Function (FRF) [52], Auto power function [53], Coherence, and Spectrum. There exist several analyzers, such as Simcenter Testlab (also called LMS Testlab from Siemens, Munich, Germany). Without an analyzer, data from the sensors have to be processed to obtain the information that we are looking for. There exist plenty of data processing techniques; the most famous is the Fast Fourier Transform (FFT) to obtain output in the frequency domain. Some tools, such as, for example, Matlab® software or Python libraries, also allow for extracting such information from the output signal [54]. Some new tools, such as BeagleBoneBlack (from Texas Instruments, Dallas, TX, USA) or Rasberry cards (from Raspberry Pi Foundation, Cambridge, England), propose low and small acquisition systems that permit easily implementing them on operational structures.

### *3.4. Sensor Placement*

In modal analysis, the placement of the sensors is very important to obtain the highest amplitude of the response possible or even to obtain a response. Apart from the LV, the accelerometer and the piezoelectric sensors need optimal placement. There exist several techniques and algorithms to optimize it [55–57]. The easiest way is to make a complete modal analysis of the structure experimentally with an LV or even numerically and to observe the mode shapes in order to see the displacement or the strain field in the structure. Indeed, the accelerometers catch the acceleration of the structure at a single point, which is directly linked to the velocity and then the displacement. On the other hand, piezoelectric transducers catch the stress applied to them, which is directly linked to the strain. Accelerometers and piezoelectric transducers do not give the optimal signal at the same position. An example is given in Figure 8; there is the displacement field of the first four modes of vibration of a free plate, and the accelerometers should not be placed in the blue area in order to record as much acceleration as possible. Obviously, for a free plate, the highest displacement will be on its edge (the red area). In Figure 9, we can see the optimal placement obtained for a set of accelerometers and PVDF transducers using the Effective Independence (EI) method based on the maximum kinetic energy [45].

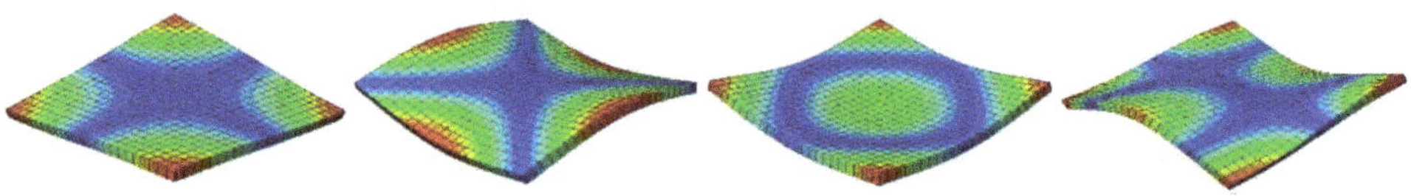

**Figure 8.** First four modes of vibration of a free–free plate (displacement is shown, blue means no displacement and red maximum displacement).

**Figure 9.** Optimal placement obtained from the Effective Independence (EI) for 3 accelerometers (**left**) and 3 PVDFs (**right**) (adapted from [45]).

## 4. Vibration Methods for SHM

### *4.1. Linear Vibration Methods*

#### 4.1.1. Natural Frequencies

The natural frequencies of a structure are the frequencies where the structure will tend to vibrate after applying an external force. Using these frequencies to excite the structure will introduce the effect of resonance, which creates a higher amplitude of vibration. They are directly linked to the mass and stiffness of the structure, as well as its boundary condition. This is what pushed many researchers to use them as damage indicators. Indeed, the damage will create a loss of stiffness, which induces a decrease in natural frequencies (Figure 10). Moreover, this technique is one of the cheapest and easiest to perform. In a review published in 1996 [58], it is indicated that the first one to use natural frequency loss as a damage indicator was Lifshitz [59] in 1969.

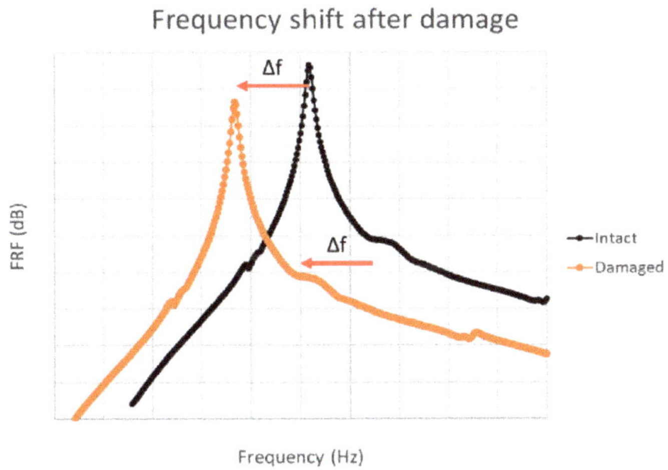


**Figure 10.** Natural frequency shift induced by damage.

In [60], the change of the first natural frequency in E-glass/epoxy-laminated composite plates with two different dimensions (13 × 6 cm$^2$ and 19 × 6 cm$^2$) and three stacking sequences ([90/90/90/90]°, [0/0/0/0]°, and [0/90/0/90]°) with a cantilever configuration (Figure 11) was investigated. Damage was introduced by cutting the inner fibers in two locations: in the middle and at the root of the structure (which plies is unknown). The fiber cutting dimension was 2.1 × 1 cm$^2$ and 3.1 × 1 cm$^2$. PZT transducers were used to excite the plate but also to catch the output signal. An oscillating voltage with different amplitudes was transmitted to the actuator (from 40 to 80 V) and the output was displayed on a sensing unit. The frequency was manually varied from 0 to 40 Hz until a sharp increase in the sensing voltage was observed. The frequency of this increase corresponded to the first natural frequency of the structure. The frequency decreased (1.5 to 4 Hz for the smaller plate and 0.3 to 1 Hz for the larger plate) when the structure was damaged for the [90/90/90/90] stacking sequence. However, this was not the case for the other stratifications due to the high stiffness in the [0/90/0/90] sequence.

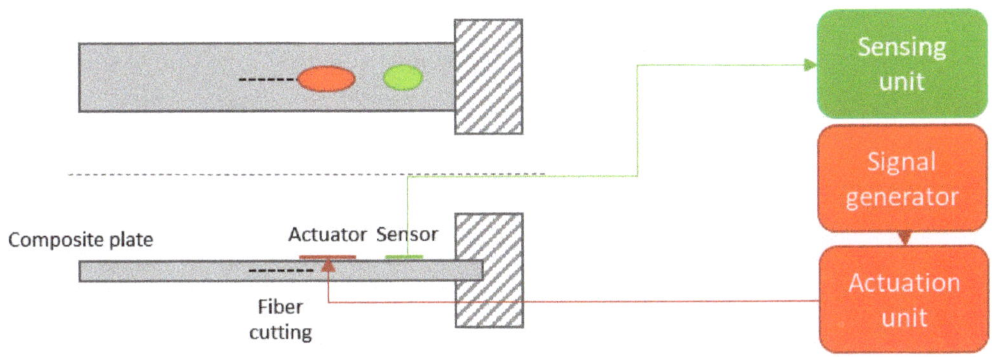


**Figure 11.** Schematic experimental cantilever setup: middle damage (adapted from [61]).

In [61,62], two kinds of inner piezoelectric transducers, MFC and PFC, were used in glass/epoxy laminate and sandwich beams in a cantilever configuration. Different severities of delamination (10/20/30/40/50/60% of the area of the structure) and crack (2/3/7/14/20/27/34% of the length of the structure) were created in laminates (300 × 32 × 21 mm$^3$) and only delamination between core and skin (10/30/45/60/75% of the area of the structure) in the sandwich material (300 × 32 × 14 mm$^3$). The excitation was introduced by displacing the tip of the beam by 0.5 inches, and the transducers caught the vibration of the structure. Here, the change in the first two natural frequencies was investigated. The frequencies decreased when the severity of damage increased; the decrease was almost linear and slow, which meant no clear distinction was possible before a large amount of damage (20% for the delamination, 15% for the crack of the laminates, and 60% for the sandwiches), which meant the small damage could not be detected. The loss of natural frequency was higher in the laminated composite.

In [63], PZTs were surface-bonded and used as actuators and sensors on a graphite/epoxy [90/+45/−45/0]° plate (250 × 50 × 1 mm$^3$) clamped at one end. The natural frequencies of six modes of vibration were used for several types of damage: a 6.4 mm diameter hole in the center, matrix cracks induced by fatigue testing and mallet hit, and delamination (8%) made of Teflon strip and cut. A signal generator was used to transmit the voltage to the PZT and an LV was used to compare with the results from the PZT. Clear distinctions regarding frequency loss for any type of damage from LV or PZT were not possible.

In [64,65], experimentation was conducted on a plate (305 × 244 × 2.16 mm$^3$) made of eight plies of carbon/epoxy stack in $[0]_s$. To isolate the dynamical response, the plate was suspended with wires. An impact hammer made the structure vibrate, and the signal from surface-bonded MFCs and accelerometers was caught; then, it was processed in a data acquisition LMS system and displayed on a PC analyzer interface, Testlab. Results were difficult to analyze and no real distinctions between undamaged and damaged structures (made from hole and impact—no information is given regarding the dimensions of damages) were observable. For this reason, the authors used different metrics that compared healthy and damaged signals (see FRF section).

The effects of damage on natural frequencies of the first three modes have been investigated on glass/epoxy laminate (200 × 30 × 8 mm$^3$) [66] and sandwiches (250 × 40 × 26 mm$^3$) [67,68] with PVC foam core and glass/epoxy skin on a cantilever configuration. The damages were debonding between the core and skin for the sandwich (from 20 to 180 mm of debonding by 20 mm step) and fatigue induced for the laminate. The structure was excited with a shaker and a power amplifier, and the vibration was sensed with an accelerometer. The excitation signal was a swept-sine around bending mode frequency with an amplitude of 50 mV. The natural frequency decreased with damage severity, and the higher mode seemed more influenced by this severity (Figure 12). In these references, the change in natural frequency was studied and investigated, including, especially, the loss factor and the nonlinear elastic and dissipative parameters. These magnitudes will be introduced later.

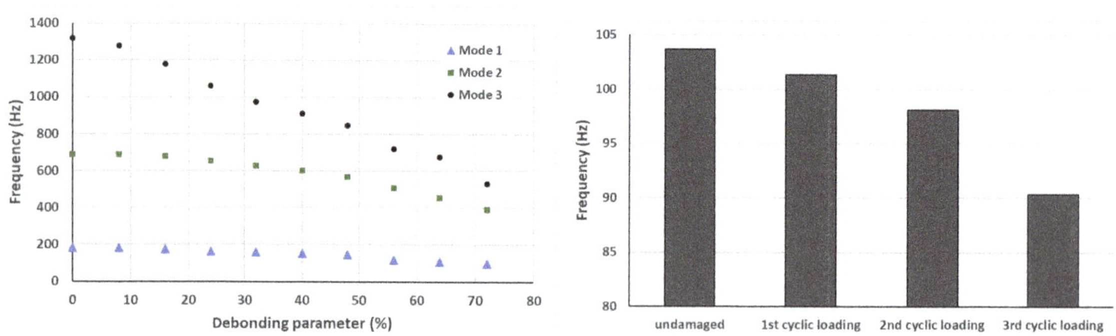


(a) (b)

**Figure 12.** Loss of natural frequency with increase in damage severity for sandwiches (**a**) (adapted from [68,69]) and laminate (**b**) (adapted from [66]).

Through these articles, detecting damages from the variation in natural frequencies seems efficient only when damage is already consequent. However, when damage is small (<20%), this technique showed some limits. Table 4 gives additional bibliographical references using this method. Nowadays, it is less and less used because of its high sensitivity to environmental effects, like temperature or boundary conditions.

**Table 4.** Other articles about the experimental use of natural frequency shift for SHM.

| Structure | Fixation | Input | Output | Damage | Description | Ref. |
|---|---|---|---|---|---|---|
| Graphite/epoxy-laminated plate | Fixed in V-blocks | | Accelerometers | Delamination (Fluorinated Ethylene Propylene (FEP) patches) | 4 first natural frequencies were used to observe that delamination has effects on the frequency loss, and 1/3 of the area of the specimen will lead to 20% of max defect | [69] |
| Carbon/cyanate-laminated plate | Cantilever configuration | Surface bonded bimorph PZT patch actuator QP25N (150 V) Agilient 33120A waveform generator CX Quickpack Amplifier | LV | Delamination (plastic film) | Natural frequencies are used to identify delamination in composite plates; a small loss in natural frequencies is observed, especially for higher modes (20% decrease) | [70] |
| Glass/epoxy-laminated beam | Cantilever configuration | Impact hammer G1195 PZT bonded patch (sine sweep signal) Amplifier | PVDF bonded patch Analyzer (HP35665A) | Delamination (Teflon patches) | Natural frequencies shift induced by delamination is used to feed an NN; the natural frequencies became lower when the structure was damaged | [71] |
| Carbon- and Kevlar/epoxy-laminated beam | Cantilever configuration | Impact hammer PZT | Accelerometer PZT | Delamination (FEP patches) | Natural frequencies decrease is observable for the damaged structure for both PZT and accelerometer sensors | [72] |
| Circular glass/epoxy-laminated plate | Suspended with 3 flexibles | Loudspeaker Oscillator and amplifier | Accelerometer Oscilloscope | Delamination (impact 45 J) | Natural frequencies tend to decrease when the delamination area gets bigger and bigger; for mode 1 and 6, the natural frequency increases after reaching a delamination area (probably induced by a local change in the geometry) | [73] |
| Glass/polyester-laminated beam | Suspended by flexible | Impact hammer | Accelerometer Analyzer | Crack | Natural frequencies tended to decrease when the crack size increases, and some modes seemed more influenced by this damage | [74] |

### 4.1.2. Mode Shapes

The mode shapes correspond to the displacement of a structure for a particular natural frequency (a particular mode). The simplest example is the vibration of a fixed string; for each mode, there will be nodes (the fixed part) and antinodes (the maximum displacement) (Figure 13).

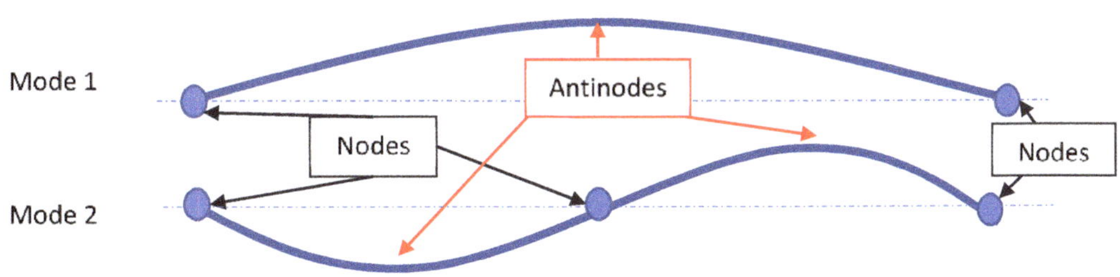


**Figure 13.** Two first modes of vibration of a fixed string.

The first thing is to obtain the mode shapes of the structure, so it is necessary to perform a modal analysis that takes the vibration measurements at several locations. This means having a network of sensors, moving excitation location and sensor position [42,75], or scanning the structure with an LV.

Several techniques exist using modal shapes to detect and localize damage; the most classical ones are the Mode Shape Displacement difference (MSD) or the use of assurance criteria, such as Modal Assurance Criterion (*MAC*) [76], and criteria derived from it, such as Co-Ordinate *MAC* (COMAC) and others [77]. Damage indexes derived from these criteria can also be used, such as the Mode Shape Damage Index (MSDI).

The *MAC* [78] is the main statistical tool regarding mode shapes. It compares two sets of modal vectors using this formula:

$$MAC(\{\varphi r\},\{\varphi s\}) = \frac{\left|\{\varphi_r\}^T\{\varphi_s\}\right|^2}{\left(\{\varphi_r\}^T\{\varphi_s\}\right)\left(\{\varphi_s\}^T\{\varphi_r\}\right)} \tag{5}$$

where $\varphi_s$ and $\varphi_r$ are the modal vectors. *MAC* will be close to one if there exists a linear relationship between two modes (if the vectors have a similar way of moving); otherwise, it will be close to zero (Figure 14). In other words, *MAC* can detect which modes will be affected by the damage. A *MAC* can have similar modal vectors even for an intact structure; this is why the reference of *MAC* from the intact structure is necessary to detect the variation. The *MAC* is also commonly used to see the coherence between experimental and numerical vibration analysis.

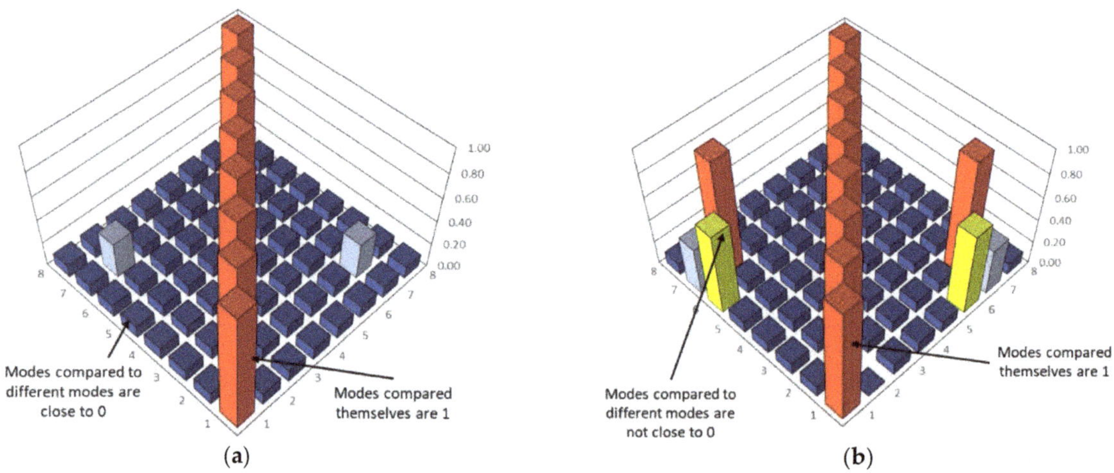


**Figure 14.** Typical MAC. No similar way of moving of the modes (**a**) and some modes with a similar way of moving (**b**).

In [79], the modal parameters of a hemp fiber/epoxy resin plate (150 × 150 × 5 mm$^3$) in a free–free condition were used to detect six damage scenarios: cut of a layer with different sizes and positions (20 × 20, 15 × 15, 10 × 10 mm$^2$ in the center and 20 × 10 mm$^2$ on the edge), cut of two layers (10 × 10 mm$^2$), and three layers (10 × 10 mm$^2$). The classical impact hammer/accelerometer combination was used to perform the modal analysis (Figure 15). A grid of 6 × 6 was created on the plate to acquire 36 data points through the roving hammer method [75]. The MAC and COMAC confusion matrixes between intact plates were created in order to have a reference of both matrixes without damage. Then, it was made between the intact and the six damaged plates, and both detected damage occurrence in the structure. Moreover, the COMAC has higher sensitivity to small damage than MAC.

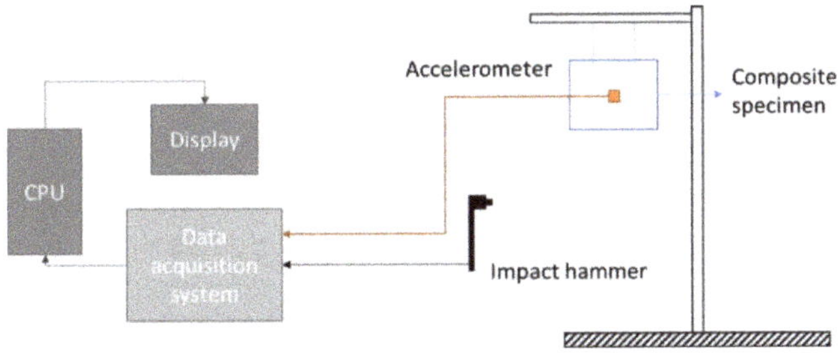


**Figure 15.** Schematic of the experimental setup (adapted from [79]).

In [80], COMAC was used to detect and localize damage in a helicopter composite blade made of glass-fiber-reinforced plastic with a honeycomb (Figure 16). The structure was suspended to obtain a free–free condition for modal analysis. The actuator was an electrodynamic shaker implemented with a burst random signal, and the sensors were accelerometers positioned on 55 localizations. "Fake" damage was introduced by adding an extra mass of 300 g at one location. The goal of the study was to localize this mass. The modal analysis was performed for both intact and damaged blades. The COMAC detected the presence of damage very close to the exact position.

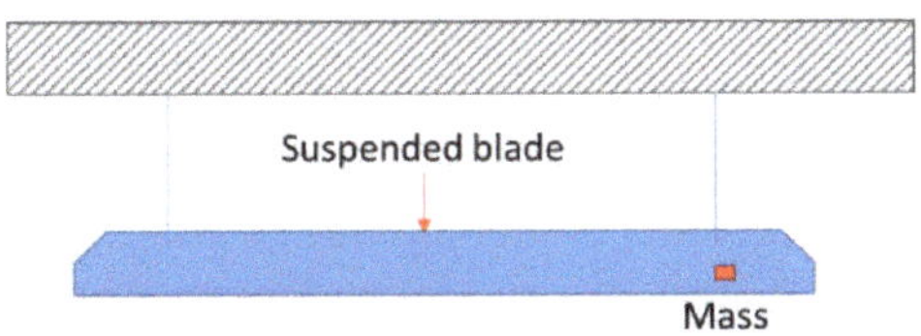


**Figure 16.** Helicopter blade made of composite suspended with wires (adapted from [80]).

Two derivations of MAC, PrMAC and WECoMAC, were used in [81] on composite plates made of carbon/epoxy. Damage was introduced from a drop-weight impact test, and then compression after the impact test was made in order to determine the compressive residual strength of the structure. The mode shapes were obtained from the roving hammer test at 25 different locations, and damage indicators were calculated at each damage step. PrMAC showed the highest sensitivity regarding damage detection.

To summarize, using changes in mode shape to detect and localize damage has been used by many researchers; however, there exist many techniques and damage indicators, and the classical ones are MAC and COMAC. In [77], a large list of such criteria derived from MAC is given. This technique shows good results in many studies; however, it needs a lot of sensors or heavy measurement processes to acquire the modal deformation accurately. There exist several vibration methods derived from the mode shapes, such as the mode shapes curvature and the modal strain energy, and this is why it is common to see articles using several and comparing them.

Table 5 summarizes other studies that have used the mode shapes methods for the SHM of PMC materials.

**Table 5.** Other articles about the experimental use of mode shapes for SHM.

| Method | Structure | Fixation | Input | Output | Damage | Description | Ref |
|---|---|---|---|---|---|---|---|
| MSC | T-stiffened panels of carbon/epoxy | Elastic bands | Shaker (sinusoidal wave (0–10 kHz)) | LV | Delamination (PTFE Film) Porosity | Change in mode shape displacement between damaged and intact structures was not the best indicator of damage presence | [82] |

**Table 5.** *Cont.*

| Method | Structure | Fixation | Input | Output | Damage | Description | Ref |
|---|---|---|---|---|---|---|---|
| MSC and mode shape slope change | Carbon/epoxy-laminated plate | Suspended with 2 cotton strings | Impulse hammer | Accelerometer | Surface and penetrated crack | The experimental results were not able to detect crack location using both methods (Mode Shape Displacement and Slope Change) | [83,84] |
| Gapped-Smoothing Method (GSM) | E-glass/epoxy-laminated beam | Cantilever configuration | PZT (Continuous-sweep sine 140 V) Power amplifier | LV (PSV 400 SLV) and PVDF | 3 delaminations (Teflon insertion), impact damage, and saw-cut | GSM is able to localize damage through LV or a network of PVDF | [85] |

4.1.3. Mode Shape Curvature

The mode shape curvature is the second derivative of the mode shapes. It can be calculated from the central difference method as [86,87]:

$$\varphi''_{ij} = \frac{\varphi_{i+1} - 2\varphi_{ij} + \varphi_{i-1}}{h^2} \quad (6)$$

where $\varphi_{ij}$ is the $j$th mode shape of the measurement point $i$ and $h$ is the distance between points $i+1$ and $i-1$.

The first step is usually to determine the difference in mode shape curvatures (named Mode Shape Curvature criterion (*MSC*)) between intact and damaged structures:

$$MSC = \Delta\varphi'' = \left|\varphi''_d - \varphi''_i\right| \quad (7)$$

The largest value will give the location of the damage (Figure 17). As for the mode shapes method, the main disadvantage is the number of sensors needed to obtain the most accurate results.

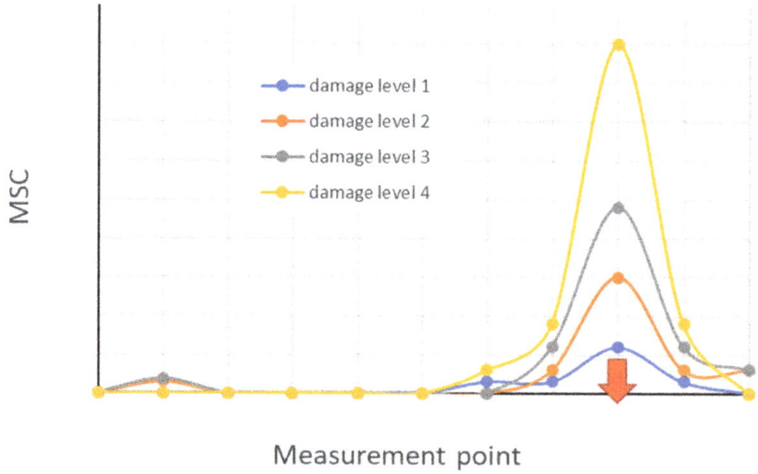


**Figure 17.** Example of MSC for a different level of damage—damage at the red arrow.

Damage indicators have been derived from the *MSC*. The classical step is to normalize the *MSC* (*NMSC*) and then the Normalized Curvature Damage Factor (*NCDF*) [88]:

$$NMSC = \left[1 + \frac{\Delta\varphi''}{max(\Delta\varphi'') - min(\Delta\varphi'')}\right]^2 \quad (8)$$

$$NCDF_i = \sum_{j=1}^{n} NMSC \tag{9}$$

There exist a lot of possible damage indexes, such as Modal Curvature Change Rate (MCI) and Curvature Mode Difference (CMD).

Modal curvatures were used in [89] in order to detect damage in glass/epoxy plates in a cantilever configuration. The modal analysis was performed with 10 × 5 localization for the PCB impact hammer, and a PCB accelerometer was fixed on the end of the plate. The damage was a crack made from a 1 mm thick blade positioned at 10% of the length of the structure close to the fixed end. NCDF detected well the damage position.

In [90,91], the modal shape curvature was tested on a carbon/epoxy beam ($241.3 \times 25.4 \times 1.76$ mm$^3$) in a cantilever configuration damaged from different sources: an impact, a saw-cut notch, and three different types of delamination (Figure 18). The modal analysis was performed using the PZT or impact hammer as actuators and 16 PVDF as sensors, all surface-bonded (Figure 19). Both NCDF and the DI were used to localize damage; these methods were efficient for three damage scenarios—the saw cut, the impact, and the delamination far from the fixed end—but the conclusion that more sensors are necessary for the two other damage scenarios was made (the two delaminations close to the fixed end).

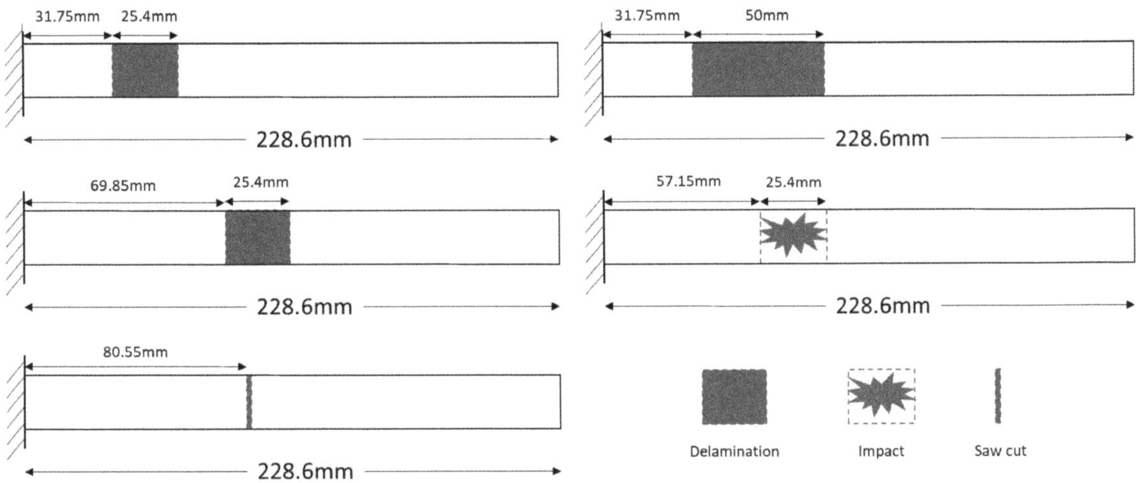


**Figure 18.** Different damage made on carbon/epoxy cantilever beam (adapted from [90,91]).

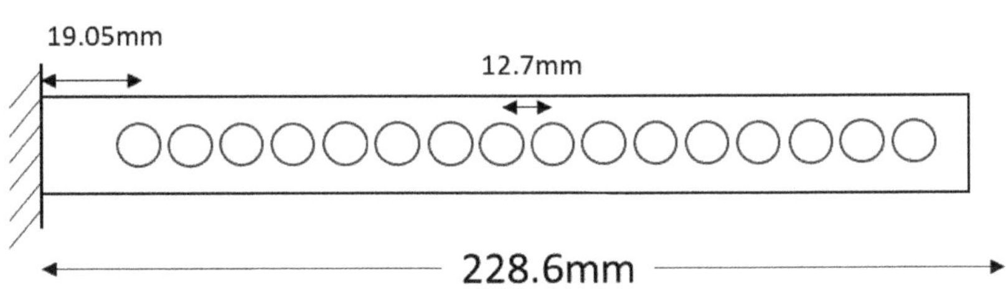


**Figure 19.** Schematic of the 16 PVDF sensor layout (adapted from [90,91]).

Table 6 summarizes other studies that have used the mode shapes curvatures methods for the SHM of PMC materials.

**Table 6.** Other articles about the experimental use of mode shapes curvature for SHM.

| Method | Structure | Fixation | Input | Output | Damage | Description | Ref. |
|---|---|---|---|---|---|---|---|
| MSC | T-stiffened panels of carbon/epoxy | Elastic bands | Shaker (sinusoidal wave (0–10 kHz)) | LV | Delamination (PTFE Film) Porosity | MSC showed that it can detect delamination accurately, but porosity was not detectable | [82] |
| MCI and CMD | Carbon/epoxy-laminated beam | Cantilever configuration | Modal hammer | Accelerometer | Delamination (Teflon patch) | CMD showed higher sensitivity than MCI to localized damage | [92] |
| MSC | Carbon/epoxy beams | Suspended on a frame | Shaker (periodic chirp) Impulse hammer | LV | Delamination and crack | The damage was identified efficiently by the MSC | [93] |

### 4.1.4. Modal Strain Energy

Modal Strain Energy (MSE) is the energy that is stored in the structure for each mode of vibration. According to [94,95], the MSE methods are more sensitive to SHM than the natural frequencies and mode shapes. Many articles use this method on offshore platform structures.

In [96], the ways of calculating MSE for a beam and a plate were given. A review of MSE-based methods for SHM was made in [97]. Four groups were introduced: Damage Indexes (DI), such as Stubbs' DI (SDI) or MSE Decomposition (MSED), Modal Strain Energy Chance (MSEC), Cross-Modal Strain energy (CMSE), and others. In that review, the methods, their calculations, and an experimental study using them were presented. Moreover, a lot of other methods derived from MSE can be found in the literature.

Several articles [84,98,99] used the MSE method on several composite structures; for example, in [98], on a carbon/epoxy plain-woven laminate (310 × 222 × 2.2 mm$^3$), the damage was a surface crack (16 × 1 mm$^2$) made from a knife located in the middle of the small dotted area in Figure 20. First, a tensile test was performed to determine the mechanical properties of the material. The modal analysis was performed with an impact hammer and an accelerometer fixed on the plate (Figure 20). The total ($U_k$) and sub-region (delimited by the grid—$U_{k,ij}$) strain energies induced by mode shapes for healthy and damaged plates were calculated from plate formulation. From these equations, the fractional energies ($F_{k,ij}$) for healthy and damaged (*) plates, and the damage index ($\beta_{ij}$) and its normalization ($Z_{ij}$) for each sub-region, were calculated as follows:

$$F_{k,ij} = \frac{U_{k,ij}}{U_k} \tag{10}$$

$$\beta_{ij} = \frac{\sum_{k=1}^{m} F^*_{k,ij}}{\sum_{k=1}^{m} F_{k,ij}}; \qquad Z_{ij} = \frac{\beta_{ij} - \overline{\beta_{ij}}}{\sigma_{ij}} \tag{11}$$

where $\overline{\beta_{ij}}$ and $\sigma_{ij}$ are the mean and standard deviation. This damage index was able to localize it accurately after using at least three modes of vibration by creating a map of DI on the (x,y) position.

The same helicopter blade [80,100] as used with the COMAC method previously discussed (Figure 16) has been used, and the "fake" damage was a mass adding. The modal analysis was performed with an impact hammer and shaker with a burst random signal; the sensors were 55 accelerometers. The method used to calculate strain energy was based on a Euler–Bernoulli beam and used the modal shape curvature. These calculations are similar to the previous one of the plates. The main goal is to calculate a damage index. The modal strain energy methods showed higher sensitivity than COMAC and were able to detect and localize damage, with a high value on an XY map.

Table 7 summarizes other studies that have used the modal strain energy methods for the SHM of PMC materials.

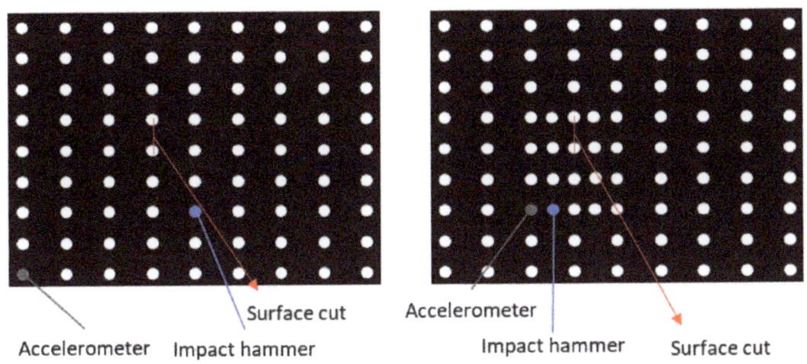


**Figure 20.** Woven laminate and the grid of measured points for global and local testing (adapted from [98]).

**Table 7.** Other articles about the experimental use of modal strain energy for SHM.

| Method | Structure | Fixation | Input | Output | Damage | Description | Ref. |
|---|---|---|---|---|---|---|---|
| MSEC | Carbon/epoxy-laminated plate | Suspended with 2 cotton strings | Impulse hammer | Accelerometer | Surface and penetrated crack | MSEC was able to locate both types of damage | [83] |
| MSE and Differential Quadrature (DQM) | Carbon/epoxy-laminated plate | Suspended with 2 cotton strings | Impulse hammer | Accelerometer | Surface and penetrated crack | For MSE, peaks were present around the location of damage, but some also occurred at other positions; this is why the DQM was implemented, and very good results were obtained with it | [84] |
| MSEC | E-glass/epoxy-laminated beam | Cantilever configuration | PZT (sweep sine 140 V) Power amplifier | LV (PSV 400 SLV) and PVDF | 3 delaminations (Teflon insertion), impact damage, and saw-cut | Strain energy detected well the location of damage, except for special cases with some doubt caused by other high value | [85] |
| MSECR | E-glass/epoxy skin and PVC foam for sandwich beams | Cantilever configuration | Electrodynamic shaker | Accelerometer | Debonding + remove face (Teflon sheet) | MSECR correctly located damage for single and multiple damage cases | [101] |

#### 4.1.5. Modal Flexibility

The modal flexibility method (MFM or MF) is another kind of damage detection technique that seems more sensitive than others and needs less experimental data. The modal flexibility is defined as a matrix:

$$[F] = \sum_{i=1}^{n} \frac{1}{f_i^2} \{X\}_i \{X\}_i^T \tag{12}$$

where $f_i$ is the $i$-th natural frequency and $\{X\}_i$ is the mass normalized mode shape vectors. What is interesting to observe using MFM is the flexibility difference between undamaged and damaged structures:

$$[\Delta F] = |[F_D] - [F_U]| \tag{13}$$

The highest value of variation will give the location of the damage (Figure 21).

In [102], the MF was also compared to other methods previously seen (Figure 22). In this experiment, the MF seemed the most accurate SHM method.

In [103], the MF and a derivate damage indicator (MDI) were used on a carbon fiber bi-directional corrugated sandwich panel with the dimensions 200 × 200 × 15 mm$^3$. A grid of 9 × 9 points was created on the surface in order to obtain the vibration at different locations and localize the damages. They created two debonding (1.2% of total area) and three core damage (0.3% of total volume of the core) cases, with single and multiple damage cases. The vibration analysis was performed by suspending the plate, and with two accelerometers and an impact hammer, an LMS signal analyzer stored all of the output and input data of the first four modes of vibration. Only the MF was not able to clearly distinguish the damage location, as several peaks with similar amplitude were observable. These peaks also occurred in the MDI, but the peak close to the damaged area had a higher value, which avoids the fake damage detection of the MF. They concluded that the new MDI is efficient in detecting single and multiple damages in composite structures.

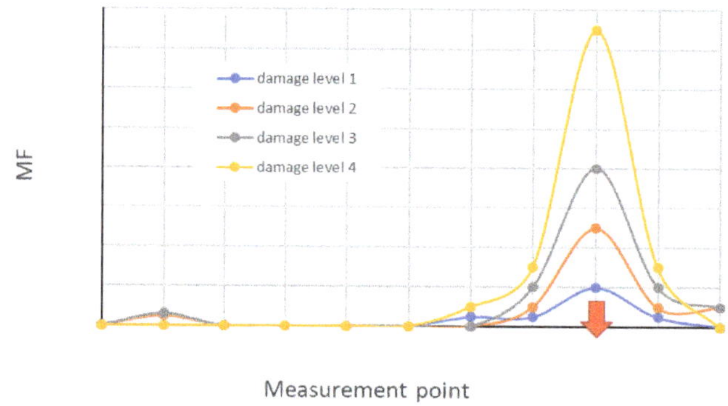


**Figure 21.** Example of flexibility change for different levels of damage—damage at red arrow.

| Case | COMAC | ECOMAC | MSC | MSE | MF |
|---|---|---|---|---|---|
| 1 | X | O | O | O | O |
| 2 | X | X | OO | O | O |
| 3 | X | X | OO | OO | OO |
| 4 | X | OO | OO | O | O |
| 5 | X | X | X | X | O |
| 6 | X | OO | O | O | O |
| 7 | X | OO | X | OO | OO |
| 8 | X | OO | OO | OO | O |

**O**: damage detected; **X**: damage not detected; **OO**: damage detected not accurately

**Figure 22.** Comparison of MF with mode shapes, curvature, and strain energy methods for locating the different damage scenarios (adapted from [102]).

#### 4.1.6. Damping

Damping corresponds to the energy of dissipation of material (as seen in Section 2.1). In [104], a review of the use of damping for SHM is provided. According to this article, the change in damping has higher sensitivity than natural frequencies or mode shapes to detect damage, but research is still in progress. The simplest analysis of damping is the difference in damping between intact and damaged structures; usually, it will be higher when the structure is damaged, and it shows higher sensibility than the change in natural frequency (Figure 23). Damage identification using damping can either be linear or nonlinear.

In modal analysis, the damping ratio is mostly calculated from the −3 dB method (or half-power bandwidth) (Figure 24) in the frequency domain, but time domain and time–frequency domain methods are also found in the literature. In the time domain, the most frequent method used is the logarithmic decrement [105], and in the time–frequency domain, the use of analyzing wavelets, such as the Gabor wavelet, is necessary [106]. There exist several quantities to define damping:

$$\eta = \frac{1}{Q} = 2\zeta = \frac{\%Cr}{50} = \tan\Phi = \frac{\Delta\omega_{3dB}}{\omega_0} \tag{14}$$

where $\eta$ is the loss factor, $Q$ is the damping of the quality factor, $\zeta$ is the damping ratio, %Cr is the percent of critical damping ($\%Cr = 100\% \times \zeta$), and $\Phi$ is the phase angle between cyclic stress and strain.

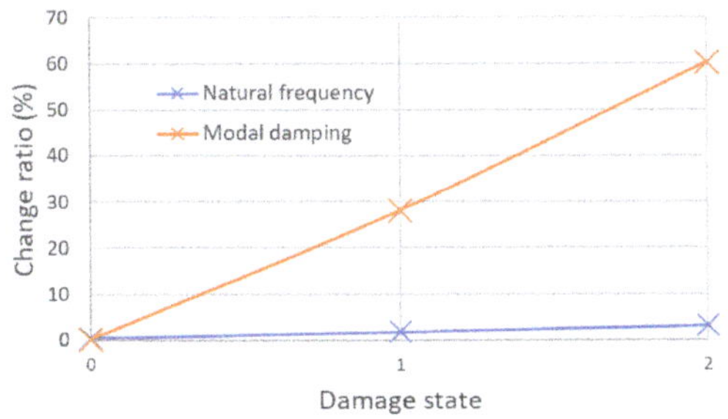


**Figure 23.** Difference of sensibility between damping and natural frequency methods (adapted from [104]).

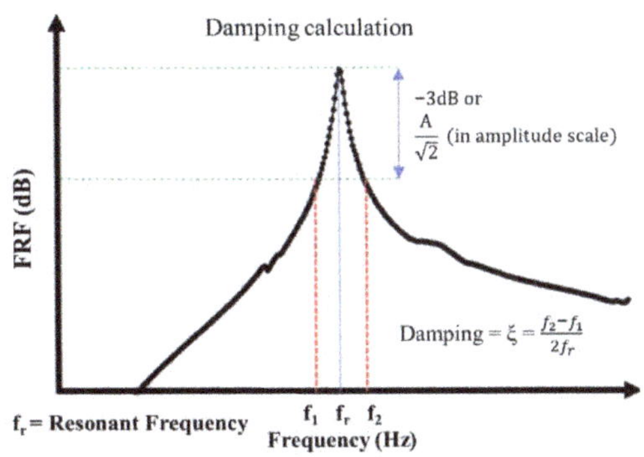


**Figure 24.** −3 dB method.

Damping to detect damage is also used with a Damage Index, and it is often mixed with other methods, such as modal shape, natural frequencies, modal strain energy, etc.

In [65], in addition to comparing natural frequency, a comparison of the difference in damping between damaged and intact carbon/epoxy plates was made in a free–free configuration from impact hammer excitation and two accelerometer sensors. Two damages were studied: one impact damage and one hole (of unknown diameter) both in the center of the plate. No specific pattern in the difference in damping ratio was detectable.

In [66–68], the effects of damage on loss factors were also investigated; on the laminated and sandwich composite with delamination and fiber cutting, it was shown that the loss factor increased with damage severity (Figure 25).

In [107,108], a method was developed using damping mixed with Plane Shape Function (*PSF*) to localize damage. The name is Damping Damage Indicator (*DaDI*). It was tested on a woven carbon/phenolic plate damaged from impacts and quasi-static pressure. The structure was suspended with two nylon strings to obtain a free–free configuration. The *DaDI* needs modal strain shapes and modal damping factors to localize damage. The excitation was made with an electromagnetic shaker and a multisite excitation force, and the sensor was LV at four different points. The *PSF* for mode n and *DaDI* at coordinate (*i,j*) was calculated as:

$$PSF_{ij}^{n} = \frac{\left|x\varepsilon_{ij}^{n}\right|}{max\left|x\varepsilon_{ij}^{n}\right|} + \frac{\left|y\varepsilon_{ij}^{n}\right|}{max\left|y\varepsilon_{ij}^{n}\right|} \qquad DaDI_{ij} = \frac{\sum_{n=1}^{N}\left(PSF_{ij}^{n}\ \Delta\eta_{n}^{D}\right)}{\sum_{n=1}^{N}\left(PSF_{ij}^{n}\right)} \tag{15}$$

where $x\varepsilon_{ij}^{n}$ and $y\varepsilon_{ij}^{n}$ are the strain in both directions and $\Delta\eta_n^D$ is the difference in damping ratio between damaged and intact structures. In the article, *DaDI* seems to determine the location of both damages well. Adding mass to the structure with transducers may decrease the accuracy of this method. In [109], similar experiments were conducted on carbon/epoxy plates. Several indexes were used, such as Multiparameter Damage Indicator (MuDI), which is created from two others, the *DaDI* and the frequency damage indicator (FreDI), to localize the damage. A weighting function was introduced to emphasize the role of FreDI or *DaDI* in the computation of MuDI; this weighting function showed the importance of the correct localization of damage.

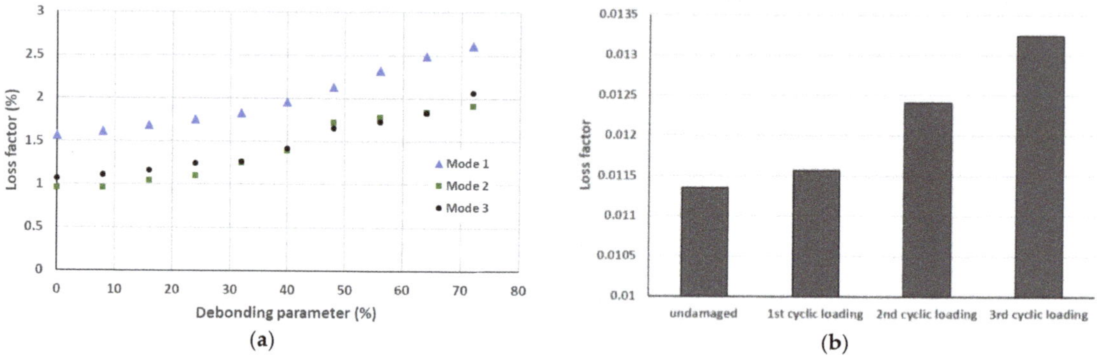


**Figure 25.** Loss factor increase with the increase in damage severity for the sandwich (**a**) (adapted from [68,69]) and laminate (**b**) (adapted from [66]).

4.1.7. Frequency Response Function Metrics

Comparing the FRF of intact and damaged structures can be a way to detect damage [110]. Several damage indicators based on FRF have been introduced in the literature [64], such as the Frequency Domain Assurance Criterion (FDAC) [111] or the Frequency Response Assurance Criterion (FRAC) [112].

In [64,65,113], the metrics introduced were tested on a carbon/epoxy plate suspended in a free–free configuration; the sensors were accelerometers and MFC, and the actuator was an impact hammer. The plates were damaged by a central hole and an impact (dimensions are unknown). The change in natural frequencies was not sufficient to detect damage, but the metrics gave interesting results, and it seemed that metrics using more than one sensor gave higher sensitivity.

4.1.8. Transmittance or Transmissibility Function (TF)

TF is a technique to determine the localization of the damage. It is built from the ratio of the vibration response of the structure at two separate locations and the location of the input force. It is often written as $T_{ij}^k$.

$$T_{ij}^{k} = \frac{x_i^k(\omega)}{x_j^k(\omega)} \tag{16}$$

where $x$ is the displacement at $i$ and $j$ location, where the force is applied on k. TF can also be calculated from the velocity of the structure instead of displacement. Then, detecting damage is performed with a damage indicator on a frequency range $[\omega_1;\ \omega_2]$, which compares the TF for healthy ($h$) and damaged ($d$) structures.

$$D_{ij}(\omega_1;\omega_2) = \frac{\int_{\omega_1}^{\omega_2} \left| T_{ij}^h(\omega) - T_{ij}^d(\omega) \right| d\omega}{\int_{\omega_1}^{\omega_2} \left| T_{ij}^h(\omega) \right| d\omega} \tag{17}$$

A review of TF is given in [114]. This technique is mostly used numerically, but with a network of sensors or an LV, it can be used experimentally. One issue with this method is the selection of the frequency range to obtain accurate results. As we can see in Figure 26, damage on paths 2–3 can be detected with the frequency range [0, 39] but not with the range [0, 300]. Usually, the upper value of the range is chosen as the first natural frequency of the structure. The position of the input force will also have a strong influence on the result.

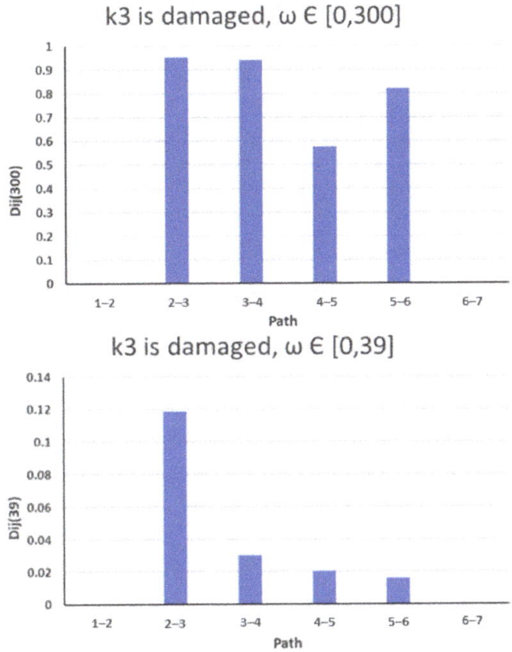


**Figure 26.** Damage indicator of TF for different frequency ranges, damage on path 2–3 (adapted from [114]).

To summarize, in the review, they said that the perfect localization of the damage is possible when the damage is located next to the input force localization or by choosing the correct frequency band, which means the global vibration of the structure has to be known.

In [115], TF was used on a cantilever fiberglass beam (112 × 1.18 × 0.635 cm$^3$) with a PZT actuator and accelerometers or PZT patches as sensors (Figure 27). The PZT actuator was excited with a random noise signal coming out of an amplifier. It was shown that using an impact hammer and shaker as actuators did not give nice repeatability of the experimentation, but the PZT actuator gave correct repeatability. Translational and curvature TF were investigated. The damage was created by bonding a steel plate (5.08 × 5.08 × 0.635 cm$^3$) to the beam between paths 1 and 2. The translational and curvature TF showed efficiency in localizing moderate damage. The curvature one seemed more accurate while the frequency range was higher (10–20 kHz). A very high-frequency range did not allow for detecting damage, and the curvature TF seemed more sensitive for small damage.

In [116], the modal analysis of a wind turbine blade made of fiberglass in a free–free boundary condition was performed. PZT patches were used with a periodic chirp signal at 200 V as an actuator and an LV as the sensor. The damage was introduced as "fake" damage by clamping a steel plate (203.2 × 152.4 × 19.05 mm$^3$) near one edge close to the path 11–15. The damage indicator of the TF indicated the correct location for the "fake" damage (Figure 28); this damage indicator should be near 0 for the healthy structure and healthy paths.

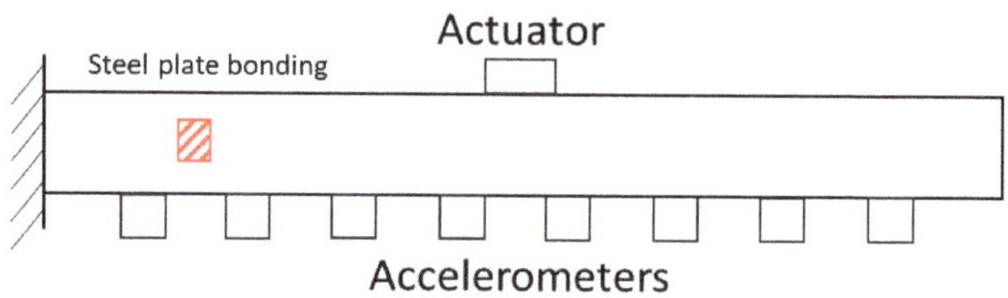


**Figure 27.** Experimental setup of the cantilever fiberglass beam with 8 accelerometers and one PZT actuator (adapted from [115]).

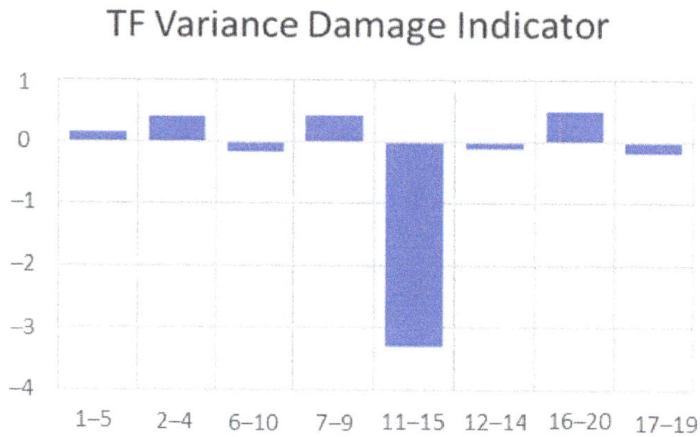


**Figure 28.** TF variance DI, damage between path 11–15 (adapted from [116]).

In [117], two models based on transmittance function were created and studied, the first one using only one transmittance model for the healthy state named "single transmittance model" (STM) and the second one using several transmittances models, the "multiple transmittance models" (MTM). Both models were based on a data-based response-only methodology using the autoregressive with exogenous excitation model (ARX). The study was conducted using 29 composite beams made of carbon fiber in unidirectional or woven fabrics. The damage was introduced in the middle of two beams with an impact hammer at 5 and 15 J. The modal analysis was performed in a cantilever configuration with a shaker as an actuator and two accelerometers. The STM scheme detected most of the damage, but it gave some false alarms and failed to detect some damage; on the other hand, the MTM scheme was able to detect all damage cases without false alarms.

4.1.9. Wavelet Transforms

In vibration analysis, the Wavelet Transform (WT) has been used a lot to process the vibration data coming from various sensors. The principle relies on dividing a time-series signal into a set of sub-signals (wavelets) and extracting data from these sub-signals. There exist different kinds of Wavelets, such as Daubechies and Symlet. A review of WT is given in [118].

In [119], small delamination (0.11%, 0.167% and 0.22% of the total surface) of glass/epoxy plates ($360 \times 250 \times 4$ mm$^3$) with piezoelectric transducers (one actuator and four sensors) was studied. The signals were decomposed for each sensor into 16 sub-signals with Daubechies Wavelet, and a damage indicator was introduced for intact and damaged structures based on the energy of these sub-signals and then by setting a threshold for this damage indicator. The conclusion was that using a threshold of 20% with WT helps to detect small delamination of an area superior to 0.12% of the total area of the structure. In Table 8, it is possible to see the damage indicator calculated for each sub-wavelet. If one of the sixteen sub-wavelet damage indicators are higher than the threshold, the authors considered the delamination detected (grey cells).

**Table 8.** DI results for 3 different delamination areas and 16 sub-wavelets with a threshold at 20% [120].

| | | Delamination Area of the Total Area | | |
|---|---|---|---|---|
| | | **0.11%** | **0.167%** | **0.22%** |
| Sub-Wavelets | 0 | 1.703 | 2.491 | 0.908 |
| | 1 | 8.053 | 6.078 | 21.32 |
| | 2 | 0.403 | 3.718 | −1.005 |
| | 3 | −0.634 | 2.197 | −1.818 |
| | 4 | −14.10 | −45.38 | −6.603 |
| | 5 | 12.83 | 5.221 | 36.42 |
| | 6 | 0.395 | 1.546 | 3.626 |
| | 7 | 6.811 | 5.449 | 17.76 |
| | 8 | 0.961 | 3.068 | −2.736 |
| | 9 | 3.575 | −0.045 | 8.691 |
| | 10 | 0.561 | 4.663 | −1.279 |
| | 11 | 0.165 | 1.226 | −2.389 |
| | 12 | −12.30 | −31.87 | −8.129 |
| | 13 | 3.143 | −2.751 | 17.04 |
| | 14 | 0.412 | 4.026 | −1.193 |
| | 15 | 0.602 | 3.390 | 0.225 |

### *4.2. Nonlinear Vibration Methods*

As the damage mechanisms in a composite structure are complexes, the tendency nowadays is to focus on nonlinear vibration methods, as they give the opportunity to increase accuracy and detect smaller damages. Most of these methods consider that the damage of the composite structure will induce nonlinearity, caused by some microcracks, delaminations between layers, or fiber breakages. As an indication, opening and closing consecutive phases of a matrix cracking can generate a nonlinear response.

This behavior has been investigated in many studies, and some nonlinear methods have emerged. Some of them cross the boundary between ultrasonic and vibration fields, but they are still introduced to see the possibilities and the potential setup needed. Moreover, as these methods are less used than linear ones, the number of experimental studies focused on PMC materials is smaller.

#### 4.2.1. Single Frequency Excitation (SFE) and Sub-/Super-Harmonics Generation

By definition, a harmonic is a multiple of a fundamental frequency (which is the harmonic n°1). Sub- and super-harmonics are ratios of this harmonic, the sub-harmonics are $\frac{f}{n}$, and super-harmonics are $n \times f$ (with $f$ the harmonic). The damage detection technique relies on the fact that by exciting the structure through a harmonic signal, the response will only show this harmonic for an intact structure, and will show super-harmonics (or higher harmonics) and sub-harmonics of smaller amplitude for a damaged structure (Figure 29).

Some authors highlighted the fact that better results can be obtained using a frequency of excitation corresponding to a ratio of a natural frequency, ½, 2, etc. Higher amplitude of sub- and super-harmonic can be observed. But a lot of them used ultrasonic frequency.

The nonlinearities can also be observed as distortions in the time-domain response (Figure 30). It can be velocity, acceleration, or voltage signal distortions. A phase portrait of velocity vs. displacement or acceleration vs. velocity graphs can be made. The phase portrait should not show any distortions for an intact structure (circle or ellipse), while it should show some for a damaged structure (Figure 31).

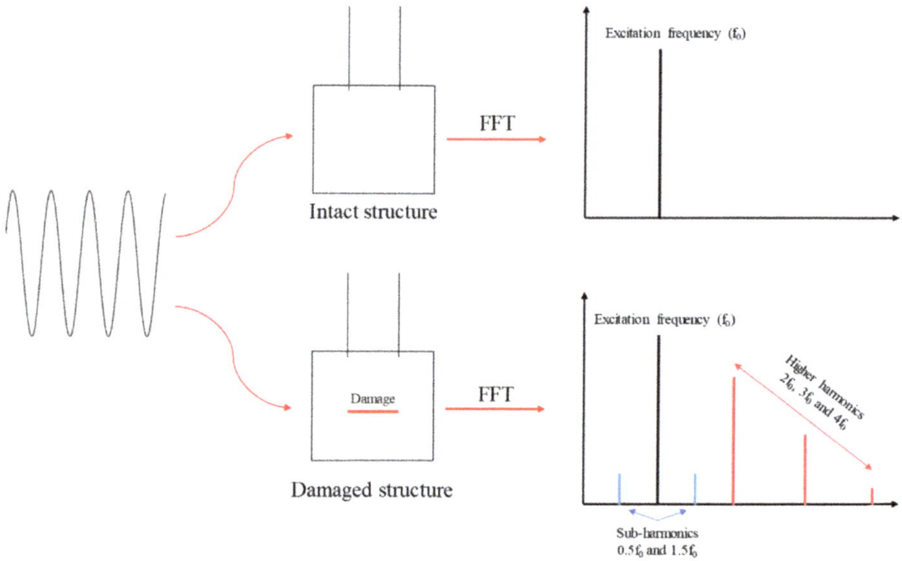


**Figure 29.** Schematic of the SFE method (adapted from [120]).

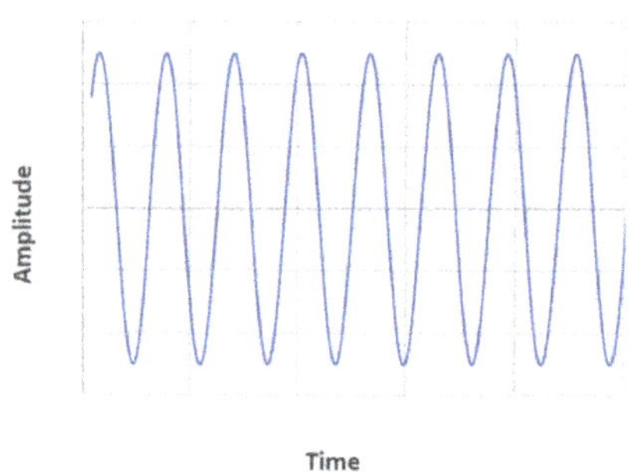


(**a**)

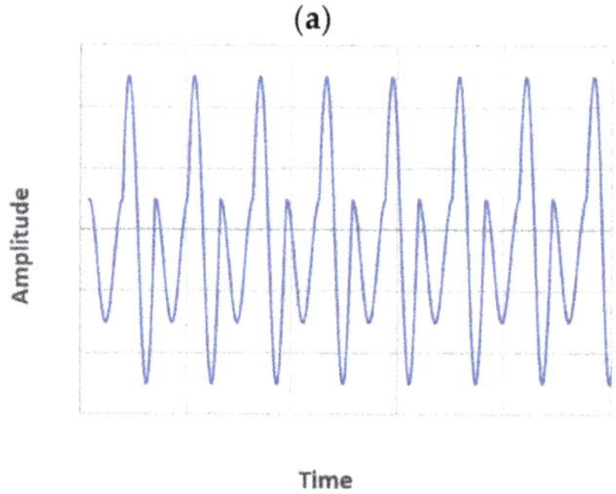


(**b**)

**Figure 30.** Time-domain signal of the voltage of the structure under a harmonic signal excitation—intact (**a**) and damaged (**b**) structures.

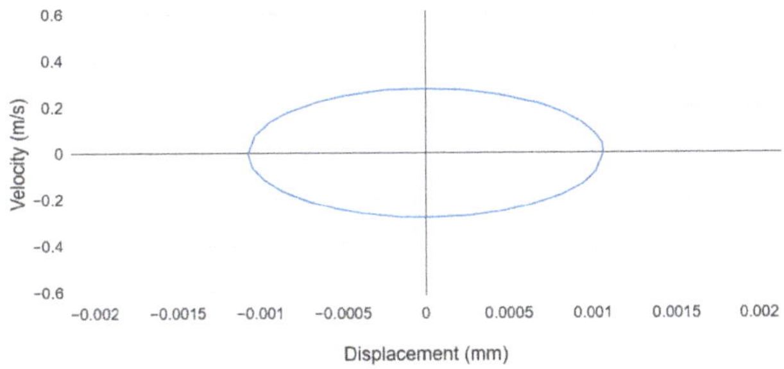


(**a**)

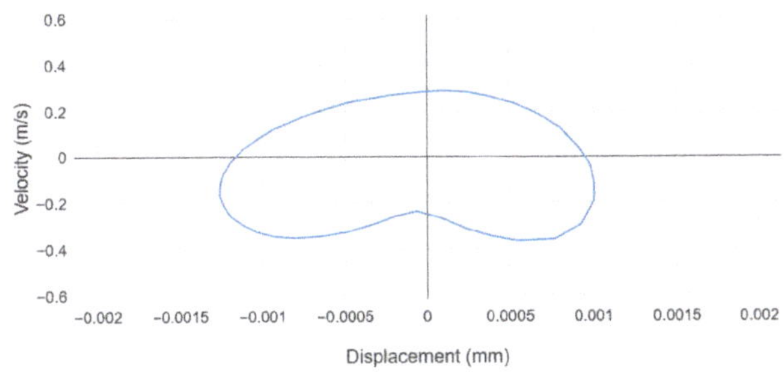


(**b**)

**Figure 31.** Phase portrait: velocity vs. displacement for intact (**a**) and damaged (**b**) structures (adapted from [121]).

In [120], three different methods to detect damages in carbon/epoxy laminate plates ($200 \times 200 \times 2.08$ mm$^3$) were investigated. The structure was suspended freely with thin wires and the damages were introduced by inserting Teflon sheets located under the fourth plies from the surface (Figure 32), which created delamination ($35 \times 35$, $40 \times 40$ and $45 \times 45$ mm$^2$). One of these methods was the SFE, and five piezoelectric transducers were used, one as an actuator and four as sensors. They were bonded onto the surface. From the actuator, the plate was excited with a sinusoidal wave of 120 V and with a frequency varying from 20 to 100 kHz. The additional equipment was a generator, a power amplifier, and an oscilloscope. Higher harmonics in the damage cases were observable. Some frequencies (20–40 kHz) have higher sensitivity, which means the choice of the best one is important. A damage indicator based on the amplitude of harmonics was defined to easily detect damage.

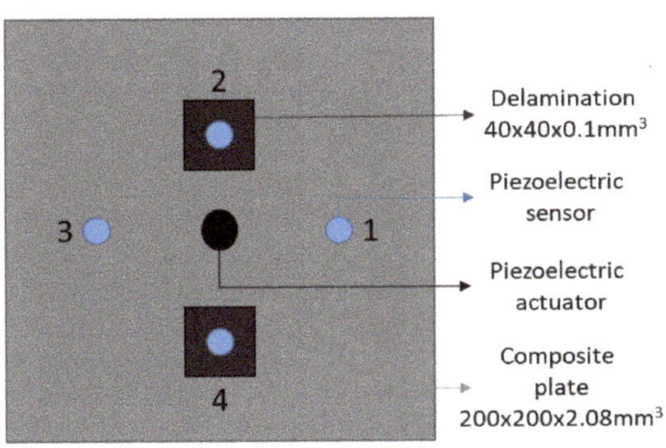


**Figure 32.** Composite plate with 5 PZT transducers and 2 delaminations (adapted from [120]).

In [122], the SFE methods were tested on a PMC plate embedded with an in situ PVDF sensor and a shaker actuator; an accelerometer sensor was also used to validate the effectiveness of the PVDF. Three natural frequencies were chosen for excitation purposes and two ratios of them (1/2 and 1/3), which gives six tested frequency. The excitation was a sinus with an amplitude of 1 V. Several plates with different damages (delamination ($35 \times 35$ mm$^2$), fiber cutting (70 mm), and holes (from 1.5 to 10 mm diameter)) were tested (Figure 33). The comparison of the amplitude of the sub- and super-harmonics was used as a damage indicator and showed sensitivity to detecting all of the introduced damages.

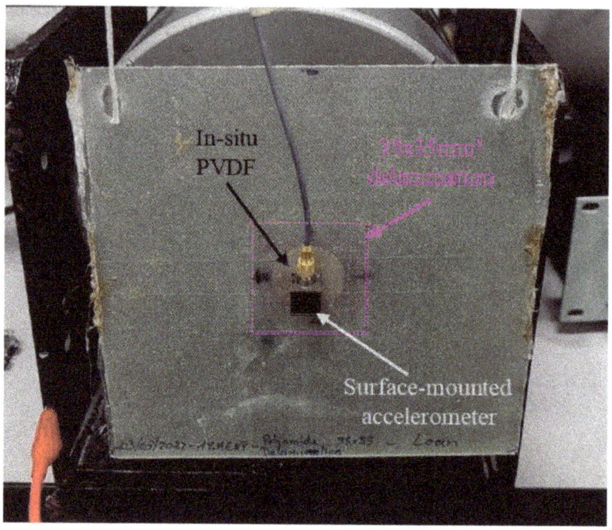


**Figure 33.** PMC plate used to perform the nonlinear vibration analysis. Delamination is introduced in the center of the plate. PVDF and accelerometer are used as sensors [122].

A similar technique was used in [123], named Second-harmonic generation, on a carbon/epoxy plate damaged from impact (at low (18 mm length) and high (32 mm length) energy) or 2 Teflon-FEP ($5 \times 5 \times 0.012$ or $10 \times 10 \times 0.012$ mm$^3$) insertion (Figure 34). The excitation was introduced from an in situ PZT with a sweep signal and a frequency band from 20 to 500 kHz. The goal was to determine which frequencies show the higher amplitudes of the output signal, and these frequencies became the input frequencies of the tests. Then, a signal with different amplitude from 60 to 100 V was sent, and vibration from an acoustic sensor on the surface was measured. A second harmonic generation was observable for the damaged plate. Similar experimentation with an LV was conducted in order to see if the chosen frequency was correct. In [124], a damage localization method was created based on the amplitude of the second harmonic generation and a network of surface-bonded PZTs. A nonlinear parameter is calculated as $\beta = \frac{A_{2fs}}{(A_{fs}^2)\, L_{ij}}$ ($A_{2fs}$ and $A_{fs}$ are the amplitudes of the second and the first harmonics; $L_{ij}$ is the distance of the path). For every path of the network (actuator/sensor path), the one with the highest value of this parameter becomes the 'damaged path', and a map can be created to visualize the expected localization. This method showed nice results to localize damage with an error from 4 mm to 22 mm.

Another technique quite similar to the SFE was used in [120] with the same equipment as the SFE method. Instead of using a harmonic signal, this one used a sweep signal excitation, and the higher-harmonics band was observed for a damaged structure (Figure 35).

Table 9 summarizes other studies that have used the SFE method for the SHM of PMC materials.

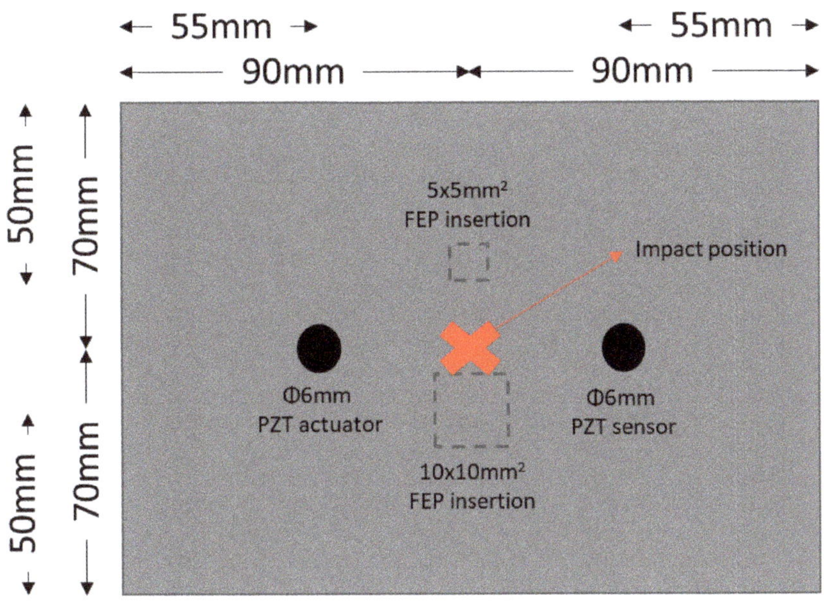


**Figure 34.** Specimen and damage specification (adapted from [123]).

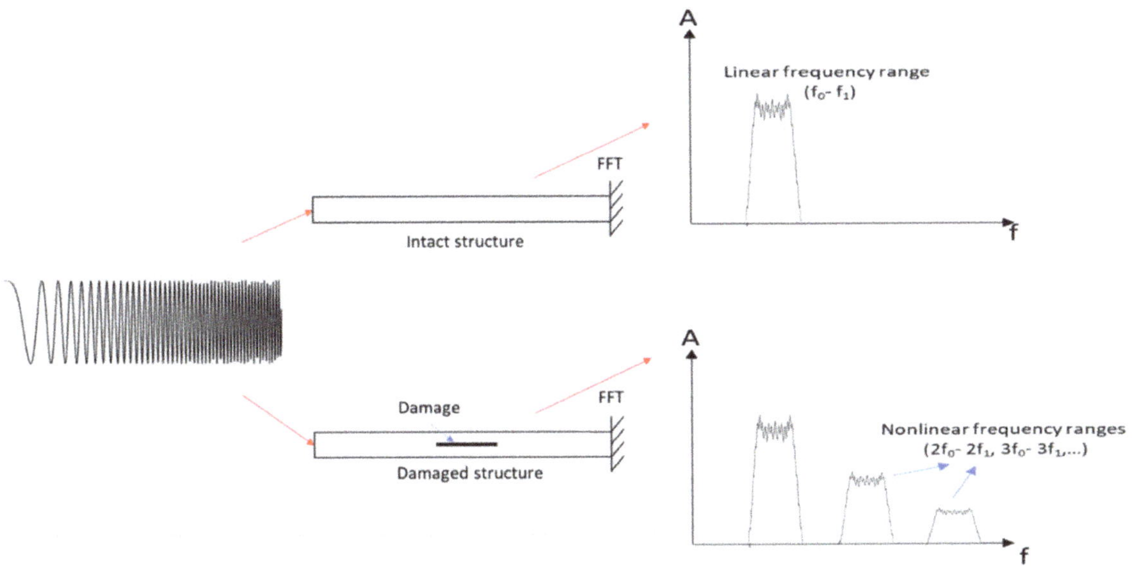


**Figure 35.** Schematic of the SWP method (adapted from [120]).

**Table 9.** Other articles about Single Frequency Excitation for SHM of PMC materials.

| Method | Structure | Fixation | Input | Output | Damage | Description | Ref. |
|---|---|---|---|---|---|---|---|
| Higher harmonics | Carbon/epoxy-laminated plates | - | Signal generator Power amplifier Periodic signal | Accelerometer | BVID | Higher harmonics are observable and their amplitudes get bigger with higher damage severity | [125] |

**Table 9.** *Cont.*

| Method | Structure | Fixation | Input | Output | Damage | Description | Ref. |
|---|---|---|---|---|---|---|---|
| Damage indicator based on the amplitude of higher harmonics | Carbon/epoxy-laminated beam | Suspended with nylon cords | PI PL055.31—piezo actuator Harmonic excitation—6 V—1580, 2860, and 3915 Hz | PZT Oscilloscope | Impact damage | The first thing observed is that the response of the sensor is a linear function of the excitation amplitude for the intact structure, and this is not the case for the damaged ones. The damage index for the first frequency clearly detected the damage presence, but it was not the case for the two other frequencies. This is why they used a damage index combining all of the frequencies | [126] |
| Sub- and super-harmonics Velocity and acceleration distortions Phase portraits | Skin-stiffener composite structure | Freely suspended by elastic wire | Electromechanical shaker Single tone harmonic excitation signal—4th bending mode frequency—at different amplitude of excitation | LV | Damaged from impact testing (delamination) | Super-harmonics were present for the damaged structure Velocity, acceleration and also phase portrait showed distortion in the time-domain for damaged structure. Link between the motion of the structure and the distortions has been made. Opening and closing phases did not give high distortions, but the contact phase gave these distortions | [127] |
| Scaling subtraction method (SSM) | Carbon-fiber-reinforced polymer | Freely suspended by elastic ropes | Piezoelectric ceramic | LV | Far-end delamination | SSM methods showed very good results to filter noise and to effectively localize the delamination | [128] |

4.2.2. Frequency Shifts from Different Excitation Amplitude

This technique consists of finding the natural frequencies of an intact and damaged structure for several amplitudes of excitation (for example, from 0.1 to 1 V every 0.1 V step). The frequency will be the same for an intact structure, while a loss of frequency will be observed for a damaged structure (Figure 36). The classical linear frequency shift can be observed between intact and damaged structures, but only the slope of decreasing frequency is used.

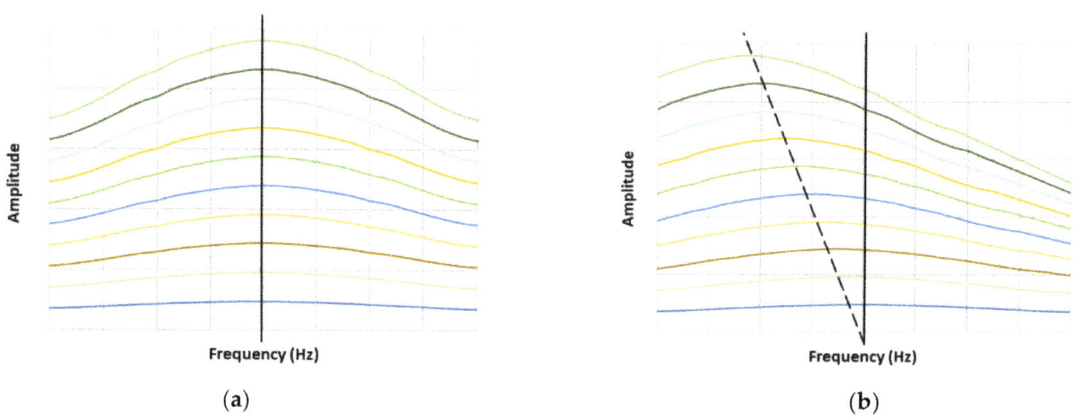


**Figure 36.** Frequency shift for several amplitudes of excitation for intact (**a**) and damaged (**b**) structure.

In [66–68], this method was used on the same equipment as previously seen. A small frequency shift was present for intact beams and a big one for damaged beams. By using the elastic nonlinear parameter:

$$\alpha_f = -\frac{1}{\varepsilon}\left(\frac{f - f_0}{f_0}\right) \quad (18)$$

($f$ and $f_0$ are the current and the initial frequencies, ε is the strain), a higher sensitivity was shown compared to the linear natural frequency change (Figure 37). It reduces linearly as a function of the excitation amplitude.

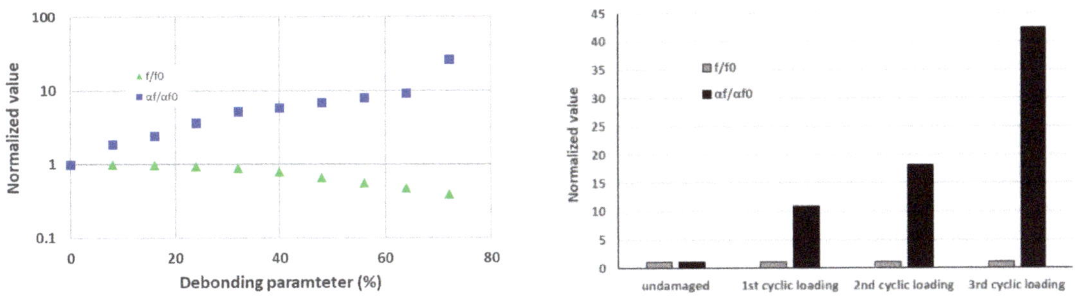


**Figure 37.** Nonlinear elastic parameter sensibility against the linear frequency loss parameter: debonding on sandwich (**a**) (adapted from [67]) and fatigue damage induced on laminate (**b**) (adapted from [66]).

Another name for this method is the single-mode nonlinear resonance ultrasound (NRUS). In [129], several low-velocity impact damages (10 J with different sizes of impactors) were detected in carbon/epoxy plates. The excitation was a sweep signal around the first bending natural frequency; several experimentations were conducted by varying the amplitude of excitation. To visualize the delamination area (between 2.9 and 3.9 cm$^2$), a C-scan was used, and then the delamination area function of the nonlinear parameter α showed that this parameter is highly sensitive to damage area (Figure 38). In [130], the limits of the techniques in the previous article were investigated. A clamped carbon/epoxy plate was used with a speaker and an accelerometer. To investigate the influence of the clamping system on this method, the clamping setup was two steel plates with a torque system. By modifying the torque value, a frequency shift that increased while the torque value increased was observable; this directly means that the presence of a frequency shift is not only caused by damage but also by boundary conditions. This phenomenon can give false alarms about damage detection or hide them. For this, special care has to be taken when using it.

In [131], the "fast dynamic" and "slow dynamic" were investigated on concrete and composite materials. The fast dynamic is the successive measurement of resonant curves with different amplitudes of excitation without time between two excitations to avoid the material relaxing. The slow dynamic starts with the highest excitation and is kept for several minutes to "condition" the material. Then, the excitation is stopped, allowing the material to relax. To follow this relaxation, a low-excitation amplitude is introduced into the material, which permits following the evolution of the resonant curve in time. In this thesis, it is shown that the phenomenon of frequency shifts from different excitation amplitudes is induced by the conditioning and relaxing phenomena of the material. The fast dynamic on PMC was studied by focusing around the sixth bending mode with a frequency of 16.685 kHz and an amplitude varying from 10 to 100 mV amplified with a 52 dB gain. PZT patches as the actuator and sensor were used, and low damages with a three-point bending test were gradually introduced in four steps. Then, with the same equipment, the slow dynamic was studied by using the 100 mV excitation. The conditioning process was first analyzed, and by letting the structure relax, the times of conditioning and relaxation were analyzed. By focusing on the time of relaxation, it was shown that this is a promising damage indicator for SHM; it seems that the damaged structure will have a lower natural frequency (linear method) but also that the damaged structure will relax to an initial frequency slower than an intact structure (Figure 39).

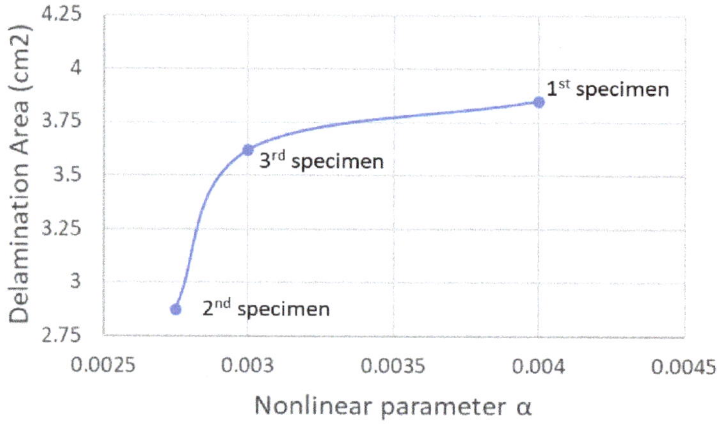


**Figure 38.** Delamination area function of the nonlinear parameter (adapted from [129]).

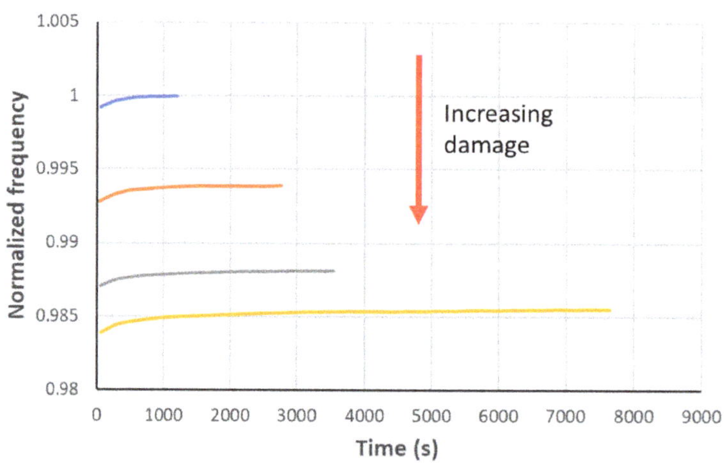


**Figure 39.** Time of relaxation of 4 damage levels of a PMC structure (adapted from [131]).

In [122], in addition to the SFE method, the NLR was also performed with the same configuration, damages, and equipment (Figure 33). Six natural frequencies were chosen for the excitation. The fiber cutting showed the biggest sensitivity in this method, with a big frequency shift for two tested modes, the delaminations showed a small sensitivity for one mode, and the holes showed sensitivity on five modes, with increasing frequency shift according to the increasing diameter of the hole.

Table 10 presents other studies using frequency shifts from different amplitudes of excitation methods for the SHM of the PMC materials.

**Table 10.** Other articles about frequency shifts from the different amplitudes of excitation methods for SHM.

| Structure | Fixation | Input | Output | Damage | Description | Ref |
|---|---|---|---|---|---|---|
| CFRP-laminated plate | Supported by foam | Waveform generator Amplifier Speaker—100 Hz sweep signal around resonance frequencies—from 1.05 to 1.3 V (0.05 V step) | Accelerometer | Pendulum impact damage (repeated) | An obvious frequency shift is observable with increasing excitation amplitude. The damage indicator used in this article increased with the increase in the damage. | [132] |
| Rectangular beam of fibre-cemented slate | Suspended with nylon wires | Function generator Speaker Excited at the lowest flexural resonance mode at different amplitude | Accelerometer Labview | Hydrothermal shocks | The micro-cracks, induced by the hydrothermal shocks, clearly exhibited nonlinear behavior in the vibration response of the structure, with a frequency shift increasing with the damage severity. | [133] |
| Steel–TiC composite beams Glass rods Polymer-based beams | | PZT or electromagnetic shaker Swept-sine signal—around bending mode frequencies—from 1 to 10 V | Accelerometer | Damage from tensile test | The classical nonlinear frequency shift gave nice results, but it was also combined with the harmonic-generation method, which showed higher sensitivity. | [134] |
| Flax/elium cross-ply and unidirectional composite | Clamped–Free | Impact hammer Shaker | LV Accelerometer | Damage from tensile test | Greater sensitivity obtained from the nonlinear resonance methods over the linear ones. | [135] |

#### 4.2.3. Vibro-Acoustic Modulation—VAM

The VAM methods may be beyond the scope of this article because some of the excitation frequencies used are in the ultrasonic field (ultrasound starts around 15~20 kHz). However, this is an interesting technique. It consists of exciting the structure with a pumping signal having a low frequency and applying a probing signal simultaneously with a high frequency; both excitations are harmonic signals. The nonlinearities caused by damage will induce super-harmonics of the pumping signal frequency (similar to the SFE method) and sub- and super-harmonics made of both "frequencies" modulation (also called sidebands) (Figure 40). This technique can have several names regarding the frequencies used: vibroacoustic modulation, Vibro-ultrasonic, etc. If pumping and probing frequencies are both very high, it will enter the ultrasonic field more than the vibration one. In most cases, the pumping frequency corresponds to the natural frequency of the structure, while the probing frequency is much higher (ultrasound). The reference [136] gives a small overview of this technique.

A lot of articles used the amplitude of this sideband as a damage indicator by comparing them with the sidebands of an intact structure (which should be very low). A similar technique is the Impact Modulation; this one uses an impact as the pumping excitation and an actuator as probing excitation. Wave distortions in the time domain (velocity, acceleration, voltage) can also be observed by performing VAM methods.

Some researchers name it the Nonlinear Wave Modulation Technique (NWMS). In [129], NWMS was used with two natural frequencies of carbon/epoxy plate (200 Hz and 2245 Hz). Damage in the plate was introduced through low-velocity impact, and the excitation of the structure was performed with a loudspeaker. Unfortunately, we do not have information about the sensing equipment (the accelerometer is the most commonly used). The results clearly showed the super-harmonics after the pumping frequency and the sidebands around the probing frequency for the damaged structure. When the damage severity is elevated, the sideband amplitude is higher.

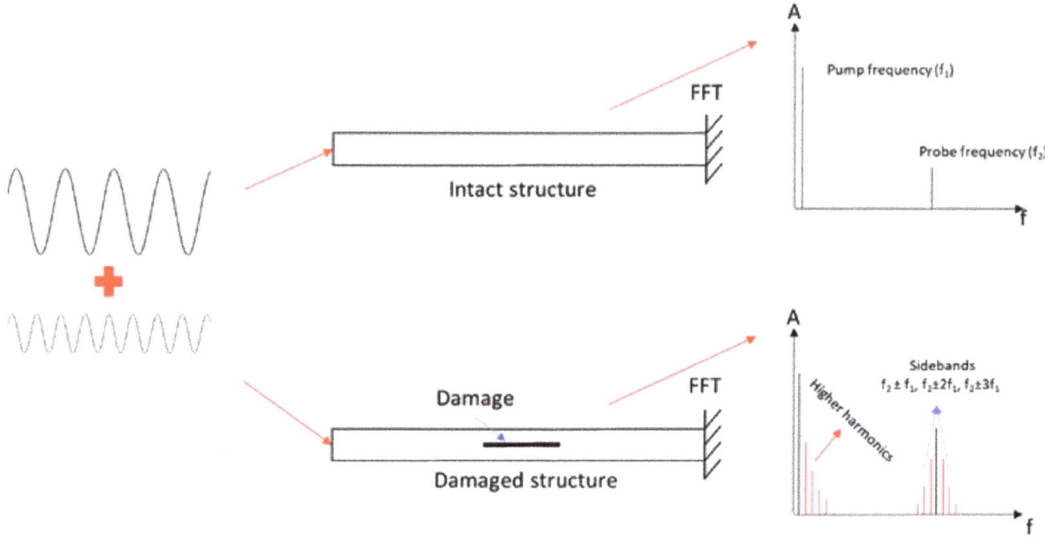


**Figure 40.** Schematic of the VAM method (adapted from [120]).

In [137], the VAM method was used with two in situ PZTs as actuators and one acoustic sensor on the surface of a carbon/epoxy plate. The setup was the same as with the SFE method. One PZT was excited with a harmonic signal at 18 kHz and another one at 192 kHz (these values were obtained after finding the best combination that generates higher modulation). Sidebands were observable only for the damaged plates, and they confirmed this with LV experimentation. In [130], the limits of VAM methods were also investigated. The same configuration as previously seen with the SFE method was used, with the carbon/epoxy plate clamped between two steel plates with different torque values. Higher harmonics and sidebands for undamaged structures were observable, and the higher the torque value, the higher the amplitude of harmonics and sidebands. This means that false damage detection can happen while performing this method, and we should be careful about the boundary condition.

As we have seen through these articles, the choice of the best combination of frequencies is important to increase the accuracy of the method. This is why in [138] this choice was investigated through several experimentations on carbon/epoxy plates with impact damage. First, a classical modal analysis with an electromagnetic shaker using white noise and LV was performed in order to choose the best natural frequencies for the VAM method. Secondly, the high-frequency sinusoidal signal with a piezoelectric transducer and the low-frequency signal with the electromagnetic shaker were sent to the structure. The two natural frequencies previously determined were used for the low-frequency excitation, while Local Defect Resonance (LDR—[139–141]) and a random high frequency were used for the probing signal. The results obtained from an LV showed that the combination with higher sensitivity is to use a natural frequency corresponding to a mode with an out-of-plane motion as a pumping signal and the LDR as a probing signal.

In [120], the vibration was introduced with PZT in a carbon/epoxy plate with delamination created from Teflon-FEP insertion. The pumping signal had a 5 kHz frequency, and the probing signal had a 100 kHz frequency. Then, the pumping and the probing frequency were changed sequentially. In order to evaluate the influence of each frequency, a damage index calculated from the amplitude of the sidebands was used.

In [142,143], this technique was investigated with surface-bonded PZTs, and especially its limitations. On a carbon/epoxy plate, with impact damage in a cantilever configuration, the clamping conditions affected the amplitude of sidebands for undamaged structures, as previously seen, but also for damaged structures. Similarly, with a lot of articles, the sidebands happened around the probing frequency for damaged cases but also for undamaged cases. Then, the observation that the amplitude of fundamental increases, while clamping force is increasing for undamaged plates, was made, and it is the reverse for the damaged ones. Otherwise, the amplitude of sidebands decreases while the clamping force increases for undamaged plates, and it is the reverse for damaged ones (Figure 41).

| When clamping force ↗ | Undamaged | Damaged (10J impact) |
| --- | --- | --- |
| Fundamental amplitude | ↗ | ↘ |
| Sidebands amplitude | ↘ | ↗ |

**Figure 41.** Effect of increasing clamping force on fundamental and sidebands amplitude for the undamaged and damaged plate: ↗ and ↘ mean increasing and decreasing (adapted from [142]).

In [144], the same structure as in [127,145], the skin-stiffener composite one, was used with the VAM method. This structure was damaged through impact testing, which mainly created delamination. A shaker was used to introduce the pumping signal with a frequency of 1455 Hz and a piezoelectric diaphragm was used to introduce the probing one with a frequency of 50 kHz. Then, the vibration was caught by an LV. By using a Hilbert Transform, instantaneous amplitude and frequency were obtained and used to plot them over time. Two damage indicators represented by the peak-to-peak value of the signal were used, and their distributions over the length of the structure showed that the instantaneous amplitude has high sensitivity in the area of the damage; this is not the case for instantaneous frequency.

In [146,147], both probing and pumping signals were sent to the same actuator, which is an interesting way of using this technique.

Table 11 presents other studies using the VAM method for the SHM purposes of PMC materials.

#### 4.2.4. Damping

According to [104], damping can also have nonlinear features. One is called the instantaneous damping coefficient, and the other is the nonlinear damping parameter.

The instantaneous damping coefficient is determined from Continuous Wavelet Transform (CWT) on the vibration response of the structure; it was created to show the variation of damping over time. The instantaneous amplitude and phase are extracted from the CWT to create the instantaneous damping coefficient $h_0$.

$$h_0(t) = -\frac{\dot{A}}{A} - \frac{\dot{\omega}}{2\omega} \tag{19}$$

where $A$ is the instantaneous amplitude and $\omega$ is the damped angular frequency, with their first derivative.

The nonlinear damping parameter $\alpha_Q$, also known as the dissipative nonlinear parameter, can be found using the same method as "frequency shift with different amplitude of excitation". Indeed, the dissipative parameter can be calculated from [66,67]:

$$\alpha_Q = \frac{1}{\varepsilon}\left(\frac{1}{Q} - \frac{1}{Q_0}\right) \tag{20}$$

where $Q$ is the loss quality factor (1/loss factor), $Q_0$ is the loss of quality factor for the first low amplitude, and $\varepsilon$ is the strain.

**Table 11.** Other articles about the VAM method for SHM.

| Structure | Fixation | Input | Frequencies | Output | Damage | Description | Ref. |
|---|---|---|---|---|---|---|---|
| Carbon/epoxy-laminated beam | Glued on an aluminum stud linked to the rod of the shaker | Pump:<br>• Generator (Agilent 3200 A)—Sinusoidal—0.5 to 2 V<br>• Shaker<br>Probe:<br>• Generator (TTi TG2000)—Sinus 20 V<br>• PZT | Pump: 155, 282 and 494 Hz<br><br>Probe: 6710 and 7180 Hz | Accelerometer | Delamination and fiber break (impact damage at 1.8 and 2.4 J) | VAM clearly detected damage occurrence in composite beam, but the accuracy of it depends on the choice of the pumping and probing frequencies.<br>The lowest pumping frequency showed higher performance. | [148] |
| Carbon-reinforced fiberglass wind turbine | On service | Pump:<br>• Rotational speed of the rotor<br>• Impact<br>• PCB Piezotronics disk<br>Probe:<br>• 9 V-battery and power amplifier—Sinusoidal<br>• MFC | Pump: 3.1, 6.2, 9.4, 400 Hz<br><br>Probe: 5–10 kHz | MFC | Crack | VAM can be used while the turbine is rotating, and it can detect the presence of damage. The sidebands for damage cases were higher than for the healthy structure, but it still showed high sidebands for the healthy structure. | [149] |
| CFRP | Fix-end | Pump:<br>• Surface-bonded piezoceramic<br>• Voltage amplifier WMA-300<br>Probe:<br>• Instron 8802L2741 | Pump: 3 Hz<br><br>Probe: 185–220 kHz | Surface-bonded piezoceramic | Hole | The pump wave for VAM was created from a fatigue mechanical loading, which is an original way. Phase modulation showed higher sensitivity than frequency modulation. | [150] |

In [66,67], the dissipative parameter was used to evaluate its sensibility to damage against the classical loss factor. The configuration was the same as for the elastic parameter on the sandwich and the laminate. A higher sensibility of this dissipative nonlinear parameter over the linear one was found (Figure 42). This behavior was also been investigated in [134], where the dissipative parameter showed a high sensitivity for damaged cases.

### 4.3. *Summary of the Vibration Methods and Decision Tree*

Many methods have been presented, and most of the researchers investigated several of them in their articles. For example, the natural frequencies shift is often combined with the damping ratio and the modes shape methods: modes shapes, modes shape curvature, modal strain energy, and modal flexibility (as in Figure 22). These three methods linked to modes shape need only to be calculated, but it would not be wise not to use them. In addition, for all of the methods, a general overview of the vibration of the structure over a large frequency band is needed. From this result, the natural frequencies and the damping ratio only have to be detected and calculated, and this is why they are always used as a first reference. In nonlinear methods, the sub- and super-harmonics generation induced by damage is also often investigated through SFE and VAM in the same articles. Indeed, the techniques are a bit different, but the phenomena observed are similar. These two techniques are both vibro-ultrasonic methods, where the frequencies used for excitation can either correspond to the modes of vibration or ultrasonic wave, but both showed that they can be used within the vibration field efficiently.

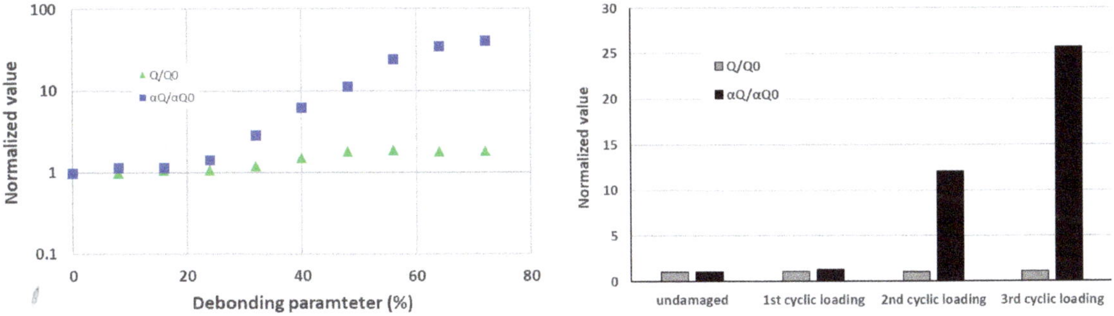


(a) (b)

**Figure 42.** Nonlinear dissipative parameter sensibility against the linear loss factor parameter: debonding on a sandwich (**a**) (adapted from [67]) and fatigue damage induced on the laminate (**b**) (adapted from [66]).

In Tables 12 and 13, a summary of the linear and nonlinear techniques is given, as well as the accuracy of each method for specific damages (these tables are built from the cited article). This review helped to build a decision tree about the SHM of PMC performing vibration analysis (Figure 43).

**Table 12.** Linear vibration methods for SHM (Detected Size represents the minimum percentage of the detected damaged area divided by the total area).

| Technique | What Is Needed? | How to? | Drawbacks | Damage | Detected Size |
|---|---|---|---|---|---|
| Natural frequencies shift | The vibration of the structure in the frequency domain for a large frequency range is needed to evaluate the natural frequencies. | Send a white noise or a sweep signal with a large frequency band. Compare the natural frequencies of intact and damaged structures. | Only big damages are detectable. It can be influenced by some external conditions: temperature, boundary conditions, etc. Some modes show a higher shift, which means several modes have to be tested. | Delamination | >20% |
| | | | | Crack | >20% |
| | | | | Impact | X |
| | | | | Fiber cutting | X |
| | | | | Debonding | >16% |
| Modes shape (MS) | The vibration of the structure in the frequency domain for a large frequency range and at several locations is needed to record the mode shapes. | With LV, a network of sensors, or a roving input method, displacement at several locations can be obtained. Several criteria comparing intact and damaged structures can be used: MAC, CoMAC, ECoMAC, PrMAC, etc. | A large number of sensors or heavy processes are needed. | Delamination | >10% |
| | | | | Crack | >10% |
| | | | | Impact | >10% |
| | | | | Layer cutting | >0.4% |
| Mode shape curvature | The modes shape. | Several criteria comparing intact and damaged structures can be used: MSC, NMSC, NCDF, etc. | | Crack | >27.7% |
| | | | | Surface cut | >7% |
| | | | | Debonding | >10% |
| Modal strain energy | The modes shape. | Several criteria comparing intact and damaged structures can be used: MSEC, MSED, CMSE, etc. | | Delamination | >11% |
| | | | | Surface cut | >11% |
| | | | | Impact | >11% |
| Modal flexibility | The modes shape. | Check the flexibility change. | | Debonding | >1.2% |
| | | | | Core | >0.3% |

**Table 12.** *Cont.*

| Technique | What Is Needed? | How to? | Drawbacks | Damage | Detected Size |
|---|---|---|---|---|---|
| Damping | The vibration of the structure is needed, and the damping can be obtained with several techniques, but the most classical one is the −3 dB method. | Send a white noise or a sweep signal with a large frequency band. Compare the damping parameter of intact and damaged structures. | Mainly influenced by operational factors and uncertainty in the damping characterization. | Delamination | >48% |
| | | | | Fatigue | 3rd cycle |
| | | | | Impact | BVID |
| FRF | Input and output are needed to build the FRF. | Some metrics can be used to observe differences in the FRF of intact and damaged structures for a large frequency range. | | Impact | |
| | | | | Hole | |
| TF | The displacements or velocities at several locations of the structure for the same input are needed. | A ratio of two displacements or velocities has to be made, and a damage indicator is calculated from this ratio. | A large number of sensors or heavy processes are needed. | Bonding steel plate | >20% |
| | | | | Impact | >5 J |
| WT | Vibration output in the time domain. | The general vibration of the structure is needed, and the time domain signal should be divided into several sub-signals to compare each sub-signal energy. | Extra processing is needed, and the correct choice of wavelets should be made. | Delamination | >0.12% |

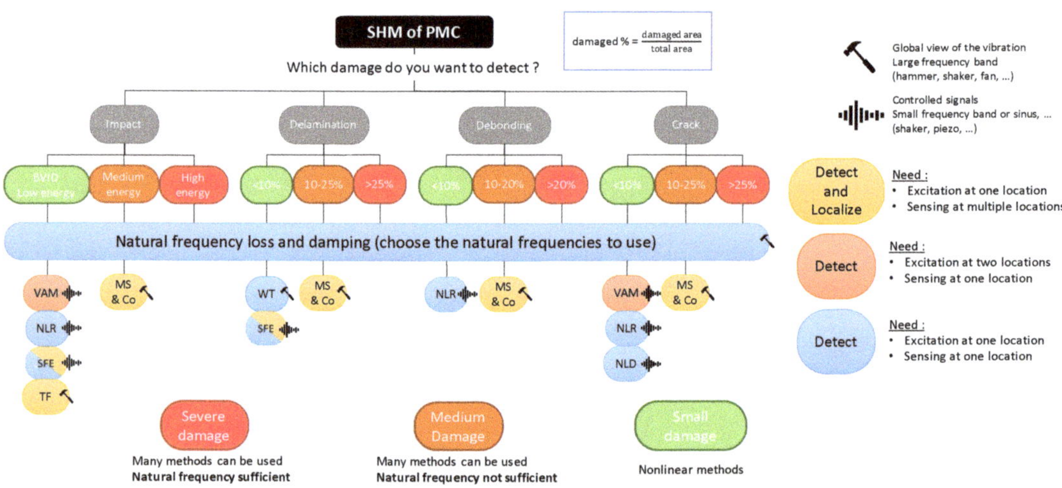


**Figure 43.** Decision tree of vibration methods for specific damage in PMC structure.

**Table 13.** Nonlinear vibration methods for SHM (Detected Size represents the minimum percentage of the detected damaged area divided by the total area).

| Technique | What Is Needed? | How to? | Drawbacks | Damage | Detected Size |
|---|---|---|---|---|---|
| SFE | Global knowledge of the vibration of the structure to choose the best frequency for the harmonic signal excitation. | Compare damaged and intact structures through distortions in time-domain and phase portraits, sub- and super-harmonics, and damage indicators. | Small peaks can be hidden in the noise in operational detection. | Impact | BVID (>7 J) |
| | | | | Delamination | >3% |
| Nonlinear resonance (NLR) and damping | Global knowledge of the vibration of the structure to choose the natural frequencies used for this technique with a harmonic or a sweep signal around these frequencies. | Compare the resonance curves obtained from different amplitudes of excitation for damaged and intact structures. Use nonlinear elastic and dissipative parameters as damage indicators. | A lot of tests have to be performed to obtain the resonance curves (10 in general) for 1 natural frequency. The frequency shift is also influenced by the boundary condition. | Debonding | >8% |
| | | | | Fatigue crack | 1st cycle |
| | | | | Impact | BVID (10 J and 2400 times 3 J) |

**Table 13.** *Cont.*

| Technique | What Is Needed? | How to? | Drawbacks | Damage | Detected Size |
|---|---|---|---|---|---|
| VAM | Global knowledge of the vibration of the structure to choose the best frequency of excitation for the pumping signal and, in some cases, also for probing one. | Compare the occurrence of super-harmonics and sidebands between damaged and intact structures. Damage indicators based on their amplitude are also used. | This method is influenced a lot by the boundary conditions, so special care should be made with VAM. | Impact | BVID (>1.8 J) |
| | | | | Crack | |
| Nonlinear Damping (NLD) | Time-frequency domain signal is needed in order to perform the CWT. | Extract the instantaneous amplitude and phase of the response to calculate the instantaneous damping. Compare the variation of it between intact and damaged structures. | Post-processing techniques (CWT) have to be used to obtain the instantaneous parameters, which are necessary for the damping calculation. | Delamination | >32% |
| | | | | Fatigue crack | 2nd cycle |

## 5. Conclusions

An important amount of techniques and the main methods of vibration analysis for SHM have been presented in this article, with a focus on PMC materials. The goal of this review was to study the techniques, used equipment, and configurations, as well as their effectiveness and, for some techniques, their drawbacks and limitations. We also reviewed the different damage mechanisms detected or not with these techniques, which allows for selecting the best method according to the studied damage.

Firstly, the classical linear methods have been introduced, even though they are considered insufficient to detect small and complex damage in the composite structures, and secondly, the nonlinear methods have shown interesting capacities in the last few years in detecting and locating damage with nonlinear behavior of the structure. The nonlinear methods are still in development, and nowadays there already exist three main techniques that can easily be used to perform SHM of PMC: the SFE, the VAM, and the nonlinear resonance.

This review highlighted the possibility of using vibration analysis for SHM of PMC. However, some limitations have been observed, such as relatively low sensitivity to early-stage damage, environmental effects, optimal sensor position, signal processing, etc. In the introduction, the concept of SHM has been presented, and from the cited articles, it is clear that the step of qualifying the damage mechanism is not yet well-developed in vibration analysis, whereas detection and localization are possible through several techniques. This indicates that further advancements are needed in this field, whether through the development of more optimized and/or internal sensors within the structure (smart materials), improved signal processing, or the development of new vibration analysis methods. A growing interest in Artificial Intelligence (AI) [151] has showed promising potential in the context of data processing in SHM and vibration analysis. This could become a significant asset for advancing the qualification aspect of SHM by enabling real-time data interpretation and adaptive damage detection strategies and thus providing a more robust and automated assessment of service life estimation. In addition, most of them investigated the presence of damage on laboratory-scale composite structures. This is important to take into consideration as the transposition to the operational scale still needs improvement and development.

To counteract these limitations for SHM diagnostics, current studies focus on combining the results of several vibration methods and also with other NDT results (such as AE, US, etc.). Indeed, this combination permits us to obtain a multi-sources and multi-physics response of the structure, thus enhancing the diagnostics. Indeed, a large range of NDT exists, which allows for combining them and obtaining different information about the damaged structure. Here, the AI algorithms have also demonstrated that they can become a major element in this combination in order to visualize similar patterns induced by the different types of damage through all of these techniques. This will help to avoid fake detection/localization and to increase accuracy in the SHM process, with the goal of qualifying the damage mechanisms efficiently.

**Author Contributions:** Conceptualization, W.H., Z.A. and L.D.; methodology, W.H., Z.A. and L.D.; formal analysis, W.H. and L.D.; investigation, W.H. and L.D.; data curation, L.D.; writing—original draft preparation, W.H. and L.D.; writing—review and editing, W.H., Z.A. and L.D.; supervision, W.H., Z.A. and L.D.; project administration, W.H. and Z.A.; funding acquisition, W.H. All authors have read and agreed to the published version of the manuscript.

**Funding:** The authors would like to thank the Ministère de l'Enseignement Supérieur et de la Recherche of France for funding this work as part of the doctoral thesis of Loan Dolbachian. This research was funded by Ministère de l'Enseignement Supérieur et de la Recherche of France.

**Data Availability Statement:** The raw/processed data required to reproduce these findings cannot be shared at this time as the data also form part of an ongoing study.

**Conflicts of Interest:** The authors declare no conflicts of interest.

## Glossary

| | |
|---|---|
| AE | Acoustic Emission |
| AI | Artificial Intelligence |
| BVID | Barely Visible Impact Damage |
| CFRP | Carbon-Fiber Reinforced Polymer |
| CMC | Ceramic-Matrix Composite |
| DaDI | Damping Damage Indicator |
| DI | Damage Indicator |
| DIC | Digital Image Correlation |
| FEP | Fluorinated Ethylene Propylene |
| FFT | Fast Fourier Transform |
| FRF | Frequency Response Function |
| GFRP | Glass-Fiber-Reinforced Polymer |
| LDR | Local Defect Resonance |
| LV | Laser Vibrometer |
| MAC | Modal Assurance Criterion |
| MF | Modal Flexibility |
| MFC | Macro-Fiber Composite |
| MMC | Metal-Matrix Composite |
| MSC | Modal Shape Curvature |
| MSD | Modal Shape Displacement |
| MSE | Modal Strain Energy |
| NDT | Non-Destructive Techniques |
| NWMS | Nonlinear Wave Modulation Technique |
| PFC | Piezo-Fiber Composite |
| PM | Process Monitoring |
| PMC | Polymer-Matrix Composite |
| PSF | Plane Shape Function |
| PVC | PolyVinyl Chloride |
| PVDF | Polyvinylidene Fluoride |
| PZT | Lead Zirconate Titanate |
| SFE | Single Frequency Excitation |
| SHM | Structural Health Monitoring |
| SWP | Sweep Signal Excitation |
| TF | Transmittance or Transmissibility Function |
| UT | Ultrasonic Testing |
| VAM | Vibro-Acoustic Modulation |

## References


1. Harizi, W.; Chaki, S.; Bourse, G.; Ourak, M. Mechanical Damage Characterization of Glass Fiber-Reinforced Polymer Laminates by Ultrasonic Maps. *Compos. B Eng.* **2015**, *70*, 131–137. [CrossRef]
2. Harizi, W.; Chaki, S.; Bourse, G.; Ourak, M. Mechanical Damage Assessment of Polymer-Matrix Composites Using Active Infrared Thermography. *Compos. B Eng.* **2014**, *66*, 204–209. [CrossRef]
3. Harizi, W.; Chaki, S.; Bourse, G.; Ourak, M. Mechanical Damage Assessment of Glass Fiber-Reinforced Polymer Composites Using Passive Infrared Thermography. *Compos. B Eng.* **2014**, *59*, 74–79. [CrossRef]
4. Saba, N.; Jawaid, M. A Review on Thermomechanical Properties of Polymers and Fibers Reinforced Polymer Composites. *J. Ind. Eng. Chem.* **2018**, *67*, 1–11. [CrossRef]
5. Al-Haik, M.S.; Hussaini, M.Y.; Garmestani, H. Prediction of Nonlinear Viscoelastic Behavior of Polymeric Composites Using an Artificial Neural Network. *Int. J. Plast.* **2006**, *22*, 1367–1392. [CrossRef]
6. Sajan, S.; Philip Selvaraj, D. A Review on Polymer Matrix Composite Materials and Their Applications. *Mater. Today Proc.* **2021**, *47*, 5493–5498. [CrossRef]
7. Jia, Z.; Li, T.; Chiang, F.P.; Wang, L. An Experimental Investigation of the Temperature Effect on the Mechanics of Carbon Fiber Reinforced Polymer Composites. *Compos. Sci. Technol.* **2018**, *154*, 53–63. [CrossRef]
8. Indra Reddy, M.; Vijaya Kumar Raju, P.; Bhargava, N.R.M.R. Experimental Investigation on the Mechanical and Thermal Properties of Sprouts Center Stem (Asian Palmyra) Fiber Reinforced Polymer Composites. *Mater. Today Proc.* **2018**, *5*, 7808–7817. [CrossRef]
9. Colombo, C.; Vergani, L.; Burman, M. Static and Fatigue Characterisation of New Basalt Fibre Reinforced Composites. *Compos. Struct.* **2012**, *94*, 1165–1174. [CrossRef]
10. Harizi, W.; Chaki, S.; Ourak, M.; Harizi, W.; Chaki, S.; Bourse, G.; Ourak, M. Characterization of the Damage Mechanisms in Polymer Composite Materials by Ultrasonic Waves, Acoustic Emission and Infrared Thermography. In Proceedings of the ECCM15, Venice, Italy, 24–28 June 2012.
11. Rytter, A. Vibrational Based Inspection of Civil Engineering Structures. *Fract. Dyn.* **1993**, *R9314*, 44.
12. Jones, D.R.H.; Ashby, M.F. *Engineering Materials 1—-An Introduction to Their Properties & Applications*; Butterworth-Heinemann: Oxford, UK, 1996.
13. Tuloup, C.; Harizi, W.; Aboura, Z.; Meyer, Y.; Khellil, K.; Lachat, R. On the Use of In-Situ Piezoelectric Sensors for the Manufacturing and Structural Health Monitoring of Polymer-Matrix Composites: A Literature Review. *Compos. Struct.* **2019**, *215*, 127–149. [CrossRef]
14. Bastianini, F.; di Tommaso, A.; Pascale, G. Ultrasonic Non-Destructive Assessment of Bonding Defects in Composite Structural Strengthenings. *Compos. Struct.* **2001**, *53*, 463–467. [CrossRef]
15. Aymerich, F.; Meili, S. Ultrasonic Evaluation of Matrix Damage in Impacted Composite Laminates. *Compos. B Eng.* **2000**, *31*, 1–6. [CrossRef]
16. Bourchak, M.; Farrow, I.R.; Bond, I.P.; Rowland, C.W.; Menan, F. Acoustic Emission Energy as a Fatigue Damage Parameter for CFRP Composites. *Int. J. Fatigue* **2007**, *29*, 457–470. [CrossRef]
17. Harizi, W.; Chaki, S.; Bourse, G.; Ourak, M. Damage Mechanisms Assessment of Glass Fiber-Reinforced Polymer (GFRP) Composites Using Multivariable Analysis Methods Applied to Acoustic Emission Data. *Compos. Struct.* **2022**, *289*, 115470. [CrossRef]
18. Hack, E.; Rastogi, P. *Optical Methods for Solid Mechanics*; Wiley: Hoboken, NJ, USA, 2012.
19. McCormick, N.; Lord, J. Digital Image Correlation. *Mater. Today* **2010**, *13*, 52–54. [CrossRef]
20. Schilling, P.J.; Karedla, B.P.R.; Tatiparthi, A.K.; Verges, M.A.; Herrington, P.D. X-ray Computed Microtomography of Internal Damage in Fiber Reinforced Polymer Matrix Composites. *Compos. Sci. Technol.* **2005**, *65*, 2071–2078. [CrossRef]
21. Wang, L.B.; Frost, J.D.; Voyiadjis, G.Z.; Harman, T.P. Quantification of Damage Parameters Using X-ray Tomography Images. *Mech. Mater.* **2003**, *35*, 777–790. [CrossRef]
22. Gaussorgues, G.S.C. *Infrared Thermography*; Springer: Berlin/Heidelberg, Germany, 1993.
23. Chaupal, P.; Rajendran, P. A Review on Recent Developments in Vibration-Based Damage Identification Methods for Laminated Composite Structures: 2010–2022. *Compos. Struct.* **2023**, *311*, 116809. [CrossRef]
24. Gomes, G.F.; Mendez, Y.A.D.; da Silva Lopes Alexandrino, P.; da Cunha, S.S.; Ancelotti, A.C. A Review of Vibration Based Inverse Methods for Damage Detection and Identification in Mechanical Structures Using Optimization Algorithms and ANN. *Arch. Comput. Methods Eng.* **2019**, *26*, 883–897. [CrossRef]
25. Bovsunovsky, A.; Surace, C. Non-Linearities in the Vibrations of Elastic Structures with a Closing Crack: A State of the Art Review. *Mech. Syst. Signal Process.* **2015**, *62*, 129–148. [CrossRef]
26. He, M.; Zhang, Z.; Ramakrishnan, K.R. Delamination Identification for FRP Composites with Emphasis on Frequency-Based Vibration Monitoring–A Review. *SDHM Struct. Durab. Health Monit.* **2018**, *12*, 213–256. [CrossRef]
27. Hou, R.; Xia, Y. Review on the New Development of Vibration-Based Damage Identification for Civil Engineering Structures: 2010–2019. *J. Sound. Vib.* **2021**, *491*, 115741. [CrossRef]
28. Montalvão, D.; Maia, N.M.M.; Ribeiro, A.M.R. A Review of Vibration-Based Structural Health Monitoring with Special Emphasis on Composite Materials. *Shock. Vib. Dig.* **2006**, *38*, 295–324. [CrossRef]

29. Sinou, J.-J. A Review of Damage Detection and Health Monitoring of Mechanical Systems from Changes in the Measurement of Linear and Non-Linear Vibrations. In *Mechanical Vibrations: Measurement, Effects and Control*; Nova Science: Hauppauge, NY, USA, 2013.
30. Worden, K.; Farrar, C.R.; Haywood, J.; Todd, M. A Review of Nonlinear Dynamics Applications to Structural Health Monitoring. *Struct. Control Health Monit.* **2008**, *15*, 540–567. [CrossRef]
31. Fan, W.; Qiao, P. Vibration-Based Damage Identification Methods: A Review and Comparative Study. *Struct. Health Monit.* **2011**, *10*, 83–111. [CrossRef]
32. Hassani, S.; Mousavi, M.; Gandomi, A.H. Structural Health Monitoring in Composite Structures: A Comprehensive Review. *Sensors* **2021**, *22*, 153. [CrossRef]
33. Akay, M. *An Introduction to Polymer-Matrix Composites*; Bookboon: London, UK, 2015; ISBN 9788740309805.
34. Tang, X.; Yan, X. A Review on the Damping Properties of Fiber Reinforced Polymer Composites. *J. Ind. Text.* **2020**, *49*, 693–721. [CrossRef]
35. Mohan Kumar, S.; Raghavendra Ravikiran, K.; Govindaraju, H.K. Development of E-Glass Woven Fabric/Polyester Resin Polymer Matrix Composite and Study of Mechanical Properties. *Mater. Today Proc.* **2018**, *5*, 13367–13374. [CrossRef]
36. Tableau, N.; Aboura, Z.; Khellil, K.; Laurin, F.; Schneider, J. Multiaxial Loading on a 3D Woven Carbon Fiber Reinforced Plastic Composite Using Tensile-Torsion Tests: Identification of the First Damage Envelope and Associated Damage Mechanisms. *Compos. Struct.* **2019**, *227*, 111305. [CrossRef]
37. Loi, G.; Marongiu, G.; Porcu, M.C.; Aymerich, F. Vibro-Acoustic Modulation with Broadband Pump Excitation for Efficient Impact Damage Detection in Composite Materials. *IOP Conf. Ser. Mater. Sci. Eng.* **2023**, *1275*, 012008. [CrossRef]
38. Talreja, R. PMC Failure Mechanisms. In *Comprehensive Composite Materials II*; Elsevier: Amsterdam, The Netherlands, 2017; pp. 107–117. [CrossRef]
39. Greenhalgh, E. Defects and Damage and Their Role in the Failure of Polymer Composites. In *Failure Analysis and Fractography of Polymer Composites*; Woodhead: Cambridge, UK, 2009; pp. 356–440. [CrossRef]
40. Duchene, P.; Chaki, S.; Ayadi, A.; Krawczak, P. A Review of Non-Destructive Techniques Used for Mechanical Damage Assessment in Polymer Composites. *J. Mater. Sci.* **2018**, *53*, 7915–7938. [CrossRef]
41. Bolotin, V. Mechanics of Delaminations in Laminate Composite Structures. In *Mechanics of Composite Materials*; CRC Press: Boca Raton, FL, USA, 2001; Volume 37.
42. Avitabile, P. Experimental modal analysis. *Sound Vib.* **2001**, *35*, 20–31.
43. Kranjc, T.; Slavič, J.; Boltežar, M. A Comparison of Strain and Classic Experimental Modal Analysis. *JVC J. Vib. Control.* **2016**, *22*, 371–381. [CrossRef]
44. Cugnoni, J.; Guzman, E.; Cugnoni, J.; Gmür, T. Use of PVDF Transducers as Actuators for Operational Modal Analysis on Composite Samples. In Proceedings of the ICCE21, Tenerife, Spain, 21–27 July 2013.
45. Guzman, E.; Cugnoni, J.; Gmur, T. Monitoring of Composite Structures Using a Network of Integrated PVDF Film Transducers. *Smart Mater. Struct.* **2015**, *24*, 055017. [CrossRef]
46. Xin, Y.; Sun, H.; Tian, H.; Guo, C.; Li, X.; Wang, S.; Wang, C. The Use of Polyvinylidene Fluoride (PVDF) Films as Sensors for Vibration Measurement: A Brief Review. *Ferroelectrics* **2016**, *502*, 28–42. [CrossRef]
47. Bregar, T.; Starc, B.; Čepon, G.; Boltežar, M. On the Use of PVDF Sensors for Experimental Modal Analysis. In *Proceedings of the Society for Experimental Mechanics Series*; Springer: Berlin/Heidelberg, Germany, 2021; pp. 279–281.
48. Tuloup, C.; Harizi, W.; Aboura, Z.; Meyer, Y.; Ade, B.; Khellil, K. Detection of the Key Steps during Liquid Resin Infusion Manufacturing of a Polymer-Matrix Composite Using an in-Situ Piezoelectric Sensor. *Mater. Today Commun.* **2020**, *24*, 101077. [CrossRef]
49. Janeliukstis, R.; Mironovs, D. Smart Composite Structures with Embedded Sensors for Load and Damage Monitoring–A Review. *Mech. Compos. Mater.* **2021**, *57*, 131–152. [CrossRef]
50. Paradies, R.; Ruge, M. In Situ Fabrication of Active Fibre Reinforced Structures with Integrated Piezoelectric Actuators. *Smart Mater. Struct.* **2000**, *9*, 220. [CrossRef]
51. Kovalovs, A.; Wesolowski, M.; Barkanov, E.; Gluhihs, S. Application of Macro-Fiber Composite (MFC) as a Piezoelectric Actuator. *Vibromech. J. Vibroeng.* **2009**, *11*, 105–112.
52. Siemens Company. What Is a Frequency Response Function (FRF)? Available online: https://community.sw.siemens.com/s/article/what-is-a-frequency-response-function-frf (accessed on 11 March 2022).
53. Siemens Company. Autopower Function. Available online: https://community.sw.siemens.com/s/article/the-autopower-function-demystified (accessed on 11 March 2022).
54. Baptiste Chomette MIMO Modal Parameters Identification in Frequency Domain–File Exchange–MATLAB Central. Available online: https://fr.mathworks.com/matlabcentral/fileexchange/82380-mimo-modal-parameters-identification-in-frequency-domain (accessed on 11 April 2022).
55. An, H.; Youn, B.D.; Kim, H.S. A Methodology for Sensor Number and Placement Optimization for Vibration-Based Damage Detection of Composite Structures under Model Uncertainty. *Compos. Struct.* **2022**, *279*, 114863. [CrossRef]
56. Mallardo, V.; Aliabadi, M.H. Optimal Sensor Placement for Structural, Damage and Impact Identification: A Review. *SDHM Struct. Durab. Health Monit.* **2014**, *9*, 287–323. [CrossRef]

57. Ferreira Gomes, G.; SimõesSim, S.; da Cunha, S., Jr.; da Silva Lopes Alexandrino, P.; Silva de Sousa, B.; Carlos Ancelotti Jr, A. Sensor Placement Optimization Applied to Laminated Composite Plates under Vibration. *Struct. Durab. Health Monit.* **2022**, *58*, 2099–2118. [CrossRef]
58. Doebling, S.W.; Farrar, C.R.; Prime, M.B.; Shevitz, D.W. Damage Identification and Health Monitoring of Structural and Mechanical Systems from Changes in Their Vibration Characteristics: A Literature Review. *Shock. Vib. Dig.* **1996**, *30*, 249299.
59. Lifshitz, J.M.; Rotem, A. Determination of Reinforcement Unbonding of Composites by a Vibration Technique. *J. Compos. Mater.* **2016**, *3*, 412–423. [CrossRef]
60. Kuriakose, V.M.; Sreehari, V.M. Experimental Investigation on the Enhancement of Vibration and Flutter Characteristics of Damaged Composite Plates Using Piezoelectric Patches. *Compos. Struct.* **2021**, *275*, 114518. [CrossRef]
61. Prasad Konka, H. Embedded Piezoelectric Fiber Composite Sensors for Applications in Composite Structures. Ph.D. Thesis, Louisiana State University and Agricultural and Mechanical College, Bâton-Rouge, LA, USA, 2011.
62. Konka, H.P.; Wahab, M.A.; Lian, K. On Mechanical Properties of Composite Sandwich Structures with Embedded Piezoelectric Fiber Composite Sensors. *J. Eng. Mater. Technol. Trans. ASME* **2012**, *134*, 011010. [CrossRef]
63. Kessler, S.S.; Spearing, S.M. In-Situ Sensor-Based Damage Detection of Composite Materials for Structural Health Monitoring. In Proceedings of the Collection of Technical Papers–AIAA/ASME/ASCE/AHS/ASC Structures, Structural Dynamics and Materials Conference, Denver, CO, USA, 22–25 April 2002; American Inst. Aeronautics and Astronautics Inc.: Reston, VA, USA, 2002; Volume 4, pp. 2706–2716.
64. De Medeiros, R.; Sartorato, M.; Tita, V.; De Medeiros, R.; Sartorato, M.; Vandepitte, D.; Tita, V. SHM of Composite Plates: Vibration Based Method by Using PZT Sensors. In Proceedings of the ISMA2014, Le Mans, France, 7–12 July 2014.
65. de Medeiros, R.; Sartorato, M.; Marques, F.D.; Tita, V.; De, R.; Murilo, M.; Flávio, S.; Marques, D.; Vandepitte, D. Vibration-Based Damage Identification Applied for Composite Plate: Experimental Analyses. In Proceedings of the COBEM 2013, Rio de Janeiro, Brazil, 3–7 November 2013.
66. Idriss, M.; El Mahi, A. Linear and Nonlinear Resonant Techniques for Characterizing Cyclic Fatigue Damage in Composite Laminate. *Compos. B Eng.* **2018**, *142*, 36–46. [CrossRef]
67. Idriss, M.; El Mahi, A.; El Guerjouma, R. Characterization of Sandwich Beams with Debonding by Linear and Nonlinear Vibration Method. *Compos. Struct.* **2015**, *120*, 200–207. [CrossRef]
68. El Mahi, A.; Idriss, M.; Assarar, M.; El Guerjouma, R.; Dazel, O. Effets de l'endommagement Sur Le Comportement Vibratoire Des Matériaux Sandwichs Endommagés. In Proceedings of the 19ème Congrès Français de Mécanique, Marseille, France, 24–28 August 2009.
69. Tracy, J.J.; Manager, S.; Pardoen, G.C. Effect of Delamination on the Natural Frequencies of Composite Laminates. *J. Compos. Mater.* **1989**, *23*, 1200–1251.
70. Ghoshal, A.; Chattopadhyay, A.; Schulz, M.J.; Thornburgh, R.; Waldron, K. Experimental Investigation of Damage Detection in Composite Material Structures Using a Laser Vibrometer and Piezoelectric Actuators. *J. Intell. Mater. Syst. Struct.* **2003**, *14*, 521–537. [CrossRef]
71. Okafor, A.C.; Chandrashekhara, K.; Jiang, Y.P. Delamination Prediction in Composite Beams with Built-in Piezoelectric Devices Using Modal Analysis and Neural Network. *Smart Mater. Struct.* **1996**, *5*, 338–347. [CrossRef]
72. Islam, A.S.; Craig, K.C. Damage Detection in Composite Structures Using Piezoelectric Materials. *Smart Mater. Struct.* **1994**, *3*, 318–328. [CrossRef]
73. Hou, J.P.; Jeronimidis, G. Vibration of Delaminated Thin Composite Plates. *Compos. Part A Appl. Sci. Manuf.* **1999**, *30*, 989–995. [CrossRef]
74. Lakhdar, M.; Mohammed, D.; Boudjemâa, L.; Rabiâ, A.; Bachir, M. Damages Detection in a Composite Structure by Vibration Analysis. *Energy Procedia* **2013**, *36*, 888–897. [CrossRef]
75. Siemens Company. Modal Tips: Roving Hammer versus Roving Accelerometer. Available online: https://community.sw.siemens.com/s/article/modal-tips-roving-hammer-versus-roving-accelerometer (accessed on 7 March 2022).
76. Pastor, M.; Binda, M.; Harčarik, T. Modal Assurance Criterion. *Procedia Eng.* **2012**, *48*, 543–548. [CrossRef]
77. Allemang, R.J. The Modal Assurance Criterion (MAC): Twenty Years of Use and Abuse. *Sound. Vib.* **2003**, *37*, 14–23.
78. Siemens Company. Modal Assurance Criterion (MAC). Available online: https://community.sw.siemens.com/s/article/modal-assurance-criterion-mac (accessed on 4 March 2022).
79. Patil, S.; Mallikarjuna Reddy, D. Damage Identification in Hemp Fiber (*Cannabis sativa*) Reinforced Composite Plates Using MAC and COMAC Correlation Methods: Experimental Study. *J. Nat. Fibers* **2020**, *19*, 1249–1264. [CrossRef]
80. dos Santos, F.L.M.; Peeters, B.; van der Auweraer, H.; Góes, L.C.S.; Desmet, W. Vibration-Based Damage Detection for a Composite Helicopter Main Rotor Blade. *Case Stud. Mech. Syst. Signal Process.* **2016**, *3*, 22–27. [CrossRef]
81. Pérez, M.A.; Gil, L.; Oller, S. Impact Damage Identification in Composite Laminates Using Vibration Testing. *Compos. Struct.* **2014**, *108*, 267–276. [CrossRef]
82. Herman, A.P.; Orifici, A.C.; Mouritz, A.P. Vibration Modal Analysis of Defects in Composite T-Stiffened Panels. *Compos. Struct.* **2013**, *104*, 34–42. [CrossRef]
83. Hu, H.; Wang, B.-T.; Su, J.-S. Application of Modal Analysis to Damage Detection in Composite Laminates. In Proceedings of the ESDA04, Manchester, UK, 19–22 July 2004.

84. Hu, H.; Wang, B.T.; Lee, C.H.; Su, J.S. Damage Detection of Surface Cracks in Composite Laminates Using Modal Analysis and Strain Energy Method. *Compos. Struct.* **2006**, *74*, 399–405. [CrossRef]
85. Qiao, P.; Lestari, W.; Shah, M.G.; Wang, J. Dynamics-Based Damage Detection of Composite Laminated Beams Using Contact and Noncontact Measurement Systems. *J. Compos. Mater.* **2007**, *41*, 1217–1252. [CrossRef]
86. Sampaio, R.P.C.; Maia, N.M.M.; Silva, J.M.M. Damage Detection Using the Frequency-Response-Function Curvature Method. *J. Sound. Vib.* **1999**, *226*, 1029–1042. [CrossRef]
87. Pandey, A.K.; Biswas, M.; Samman, M.M. Damage Detection from Changes in Curvature Mode Shapes. *J. Sound. Vib.* **1991**, *145*, 321–332. [CrossRef]
88. Radzieński, M.; Krawczuk, M.; Palacz, M. Improvement of Damage Detection Methods Based on Experimental Modal Parameters. *Mech. Syst. Signal Process.* **2011**, *25*, 2169–2190. [CrossRef]
89. Govindasamy, M.; Kamalakannan, G.; Kesavan, C.; Meenashisundaram, G.K. Damage Detection in Glass/Epoxy Laminated Composite Plates Using Modal Curvature for Structural Health Monitoring Applications. *J. Compos. Sci.* **2020**, *4*, 185. [CrossRef]
90. Lestari, W.; Qiao, P.; Hanagud, S. Curvature Mode Shape-Based Damage Assessment of Carbon/Epoxy Composite Beams. *J. Intell. Mater. Syst. Struct.* **2007**, *18*, 189–208. [CrossRef]
91. Hamey, C.S.; Lestari, W.; Qiao, P.; Song, G. Experimental Damage Identification of Carbon/Epoxy Composite Beams Using Curvature Mode Shapes. *Struct. Health Monit.* **2004**, *3*, 333–353. [CrossRef]
92. He, M.; Yang, T.; Du, Y. Nondestructive Identification of Composite Beams Damage Based on the Curvature Mode Difference. *Compos. Struct.* **2017**, *176*, 178–186. [CrossRef]
93. Mehdizadeh, M. Curvature Mode Shape Analyses of Damage in Structures. Ph.D. Thesis, School of Aerospace, Mechanical & Manufacturing Engineering College of Science, Engineering and Technology, Bundoora, Australia, 2009.
94. Sazonov, E.; Klinkhachorn, P. Optimal Spatial Sampling Interval for Damage Detection by Curvature or Strain Energy Mode Shapes. *J. Sound. Vib.* **2005**, *285*, 783–801. [CrossRef]
95. Cha, Y.J.; Buyukozturk, O. Structural Damage Detection Using Modal Strain Energy and Hybrid Multiobjective Optimization. *Comput. Aided Civ. Infrastruct. Eng.* **2015**, *30*, 347–358. [CrossRef]
96. Eraky, A.; Anwar, A.M.; Saad, A.; Abdo, A. Damage Detection of Flexural Structural Systems Using Damage Index Method—Experimental Approach. *Alex. Eng. J.* **2015**, *54*, 497–507. [CrossRef]
97. Wang, S.; Xu, M. Modal Strain Energy-Based Structural Damage Identification: A Review and Comparative Study. *Struct. Eng. Int.* **2019**, *29*, 234–248. [CrossRef]
98. Hu, H.; Wang, J. Damage Detection of a Woven Fabric Composite Laminate Using a Modal Strain Energy Method. *Eng. Struct.* **2009**, *31*, 1042–1055. [CrossRef]
99. Hu, H.; Lee, C.H.; Wu, C.B.; Lu, W.J. Detection of Matrix Cracks in Composite Laminates by Using the Modal Strain Energy Method. *Mech. Compos. Mater.* **2010**, *46*, 117–132. [CrossRef]
100. dos Santos, F.L.M.; Peeters, B.; van der Auweraer, H.; Góes, L.C.S. Modal Strain Energy Based Damage Detection Applied to a Full Scale Composite Helicopter Blade. *Key Eng. Mater.* **2013**, *569–570*, 457–464.
101. Kumar, M.; Shenoi, R.A.; Cox, S.J. Experimental Validation of Modal Strain Energies Based Damage Identification Method for a Composite Sandwich Beam. *Compos. Sci. Technol.* **2009**, *69*, 1635–1643. [CrossRef]
102. Meng, F.; Yu, J.; Alaluf, D.; Mokrani, B.; Preumont, A. Modal Analysis and Damage Detection for Suspension Bridges: A Numerical and Experimental Investigation. *Smart Struct. Syst.* **2019**, *23*, 15–29.
103. Yang, J.S.; da Liu, Z.; Schmidt, R.; Schröder, K.U.; Ma, L.; Wu, L.Z. Vibration-Based Damage Diagnosis of Composite Sandwich Panels with Bi-Directional Corrugated Lattice Cores. *Compos. Part. A Appl. Sci. Manuf.* **2020**, *131*, 105781. [CrossRef]
104. Cao, M.S.; Sha, G.G.; Gao, Y.F.; Ostachowicz, W. Structural Damage Identification Using Damping: A Compendium of Uses and Features. *Smart Mater. Struct.* **2017**, *26*, 043001. [CrossRef]
105. Inman, D.J. *Engineering Vibration*; Pearson Education Limited: London, UK, 1994; ISBN 978-0-13-287169-3.
106. Slavič, J.; Simonovski, I.; Boltežar, M. Damping Identification Using a Continuous Wavelet Transform: Application to Real Data. *J. Sound. Vib.* **2003**, *262*, 291–307. [CrossRef]
107. Montalvão, D.; Ribeiro, A.M.R.; Duarte-Silva, J. A Method for the Localization of Damage in a CFRP Plate Using Damping. *Mech. Syst. Signal Process.* **2009**, *23*, 1846–1854. [CrossRef]
108. Montalvão, D.; Ribeiro, A.M.R.; Duarte-Silva, J.A.B. Experimental Assessment of a Modal-Based Multi-Parameter Method for Locating Damage in Composite Laminates. *Exp. Mech.* **2011**, *51*, 1473–1488. [CrossRef]
109. Montalvão, D.; Karanatsis, D.; Ribeiro, A.M.; Arina, J.; Baxter, R. An Experimental Study on the Evolution of Modal Damping with Damage in Carbon Fiber Laminates. *J. Compos. Mater.* **2015**, *49*, 2403–2413. [CrossRef]
110. Kessler, S.S.; Spearing, S.M.; Atalla, M.J.; Cesnik, C.E.S.; Soutis, C. Damage Detection in Composite Materials Using Frequency Response Methods. *Compos. Part B* **2002**, *33*, 87–95. [CrossRef]
111. Sampaio, R.P.C.; Maia, N.M.M.; Silva, J.M.M. The Frequency Domain Assurance Criterion as a Tool for Damage Detection. *Key Eng. Mater.* **2003**, *245–246*, 69–76. [CrossRef]
112. Chen, A. Frequency Response Assurance Criterion and Applications to Model Correlation of Body Structures. *SAE Tech. Pap.* **2003**. [CrossRef]

113. de Medeiros, R. Development of a Criterion for Predicting Residual Strength of Composite Structures Damaged by Impact Loading. Ph.D. Thesis, Mechanical Engineering, University of Sao Paulo–USP Sao Carlos School of Engineering, Sao Paulo, Brazil, 2016.
114. Chesné, S.; Deraemaeker, A. Damage Localization Using Transmissibility Functions: A Critical Review. *Mech. Syst. Signal Process.* **2013**, *38*, 569–584. [CrossRef]
115. Zhang, H.; Schulz, M.J.; Ferguson, F.; Pai, P.F. Structural Health Monitoring Using Transmittance Functions. *Mech. Syst. Signal Process.* **1999**, *13*, 765–787. [CrossRef]
116. Sundaresan, M.; Ferguson, F.; Pai, P.; Sundaresan, M.J.; Schulz, M.J.; Hill, J.; Wheater, E.A.; Ferguson, F.; Pai, P.F. Damage Detection on a Wind Turbine Blade Section. In Proceedings of the International Society for Optical Engineering, San Jose, CA, USA, 26–28 January 1999.
117. Poulimenos, A.G.; Sakellariou, J.S. A Transmittance-Based Methodology for Damage Detection under Uncertainty: An Application to a Set of Composite Beams with Manufacturing Variability Subject to Impact Damage and Varying Operating Conditions. *Struct. Health Monit.* **2019**, *18*, 318–333. [CrossRef]
118. Liu, C.-L. A Tutorial of the Wavelet Transform. *NTUEE* **2010**, *21*, 22.
119. Yan, Y.J.; Yam, L.H. Detection of Delamination Damage in Composite Plates Using Energy Spectrum of Structural Dynamic Responses Decomposed by Wavelet Analysis. *Comput. Struct.* **2004**, *82*, 347–358. [CrossRef]
120. Singh, A.K. Detection of Defects in Composite Laminates with Vibro-Ultrasonic Methods. Ph.D. Thesis, National University of Singapore, Singapore, 2019.
121. Andreaus, U.; Baragatti, P. Experimental Damage Detection of Cracked Beams by Using Nonlinear Characteristics of Forced Response. *Mech. Syst. Signal Process.* **2012**, *31*, 382–404. [CrossRef]
122. Dolbachian, L.; Harizi, W.; Aboura, Z. Structural Health Monitoring (SHM) Study of Polymer Matrix Composite (PMC) Materials Using Nonlinear Vibration Methods Based on Embedded Piezoelectric Transducers. *Sensors* **2023**, *23*, 3677. [CrossRef]
123. Andreades, C.; Ciampa, F. CFRP Composites with Embedded PZT Transducers for Nonlinear Ultrasonic Inspection of Space Structures. In Proceedings of the EWSHM2018, Manchester, UK, 10–13 July 2018.
124. Andreades, C.; Malfense Fierro, G.P.; Meo, M. A Nonlinear Ultrasonic SHM Method for Impact Damage Localisation in Composite Panels Using a Sparse Array of Piezoelectric PZT Transducers. *Ultrasonics* **2020**, *108*, 106181. [CrossRef]
125. Polimeno, U.; Meo, M.; Almond, D.P. Smart Nonlinear Acoustic Based Structural Health Monitoring System. *Adv. Sci. Technol.* **2008**, *56*, 426–434. [CrossRef]
126. Loi, G.; Porcu, M.C.; Aymerich, F. Impact Damage Detection in Composite Beams by Analysis of Non-Linearity under Pulse Excitation. *J. Compos. Sci.* **2021**, *5*, 39. [CrossRef]
127. Ooijevaar, T.H.; Rogge, M.D.; Loendersloot, R.; Warnet, L.L.; Akkerman, R.; Tinga, T. Nonlinear Dynamic Behavior of an Impact Damaged Composite Skin-Stiffener Structure. *J. Sound. Vib.* **2015**, *353*, 243–258. [CrossRef]
128. Wei, L.; Chen, J. Far-End Delamination Characterization of CFRP Laminates Using Nonlinear Ultrasonic Spectra of Scaling Subtracted Vibration Response. *Compos. Commun.* **2024**, *45*, 101811. [CrossRef]
129. Meo, M.; Polimeno, U.; Zumpano, G. Detecting Damage in Composite Material Using Nonlinear Elastic Wave Spectroscopy Methods. *Appl. Compos. Mater.* **2008**, *15*, 115–126. [CrossRef]
130. Polimeno, U.; Meo, M. Understanding the Effect of Boundary Conditions on Damage Identification Process When Using Nonlinear Elastic Wave Spectroscopy Methods. *Int. J. Non Linear Mech.* **2008**, *43*, 187. [CrossRef]
131. Bentahar, M. Nonlinear Acoustics: Application to Ultrasonic Characterisation of Damage in Heterogeneous Materials and Remaining Life Determination. Ph.D. Thesis, INSA Lyon, Lyon, France, 2005.
132. Wei, Q.; Zhu, L.; Zhu, J.; Zhuo, L.; Hao, W.; Xie, W. Characterization of Impact Fatigue Damage in CFRP Composites Using Nonlinear Acoustic Resonance Method. *Compos. Struct.* **2020**, *253*, 112804. [CrossRef]
133. van den Abeele, K.E.A.; Sutin, A.; Carmeliet, J.; Johnson, P.A. Micro-Damage Diagnostics Using Nonlinear Elastic Wave Spectroscopy (NEWS). *NDT E Int.* **2001**, *34*, 239–248. [CrossRef]
134. Novak, A.; Bentahar, M.; Tournat, V.; el Guerjouma, R.; Simon, L. Nonlinear Acoustic Characterization of Micro-Damaged Materials through Higher Harmonic Resonance Analysis. *NDT E Int.* **2012**, *45*, 1–8. [CrossRef]
135. Haggui, M.; Jendli, Z.; El Mahi, A.; Akrout, A.; Haddar, M. Damage Analysis in Flax/Elium Composite Using Linear and Nonlinear Resonance Techniques. *J. Vib. Eng. Technol.* **2023**. [CrossRef]
136. Pieczonka, L.; Klepka, A.; Martowicz, A.; Staszewski, W.J. Nonlinear Vibroacoustic Wave Modulations for Structural Damage Detection: An Overview. *Opt. Eng.* **2015**, *55*, 011005. [CrossRef]
137. Andreades, C.; Malfense Fierro, G.P.; Meo, M.; Ciampa, F. Nonlinear Ultrasonic Inspection of Smart Carbon Fibre Reinforced Plastic Composites with Embedded Piezoelectric Lead Zirconate Titanate Transducers for Space Applications. *J. Intell. Mater. Syst. Struct.* **2019**, *30*, 2995–3007. [CrossRef]
138. Klepka, A.; Pieczonka, L.; Staszewski, W.J.; Aymerich, F. Impact Damage Detection in Laminated Composites by Non-Linear Vibro-Acoustic Wave Modulations. *Compos. B Eng.* **2014**, *65*, 99–108. [CrossRef]
139. Adebahr, W.; Solodov, I.; Rahammer, M.; Gulnizkij, N.; Kreutzbruck, M. Local Defect Resonance for Sensitive Non-Destructive Testing. In Proceedings of the AIP Conference Proceedings, Liptovsky Jan, Slovakia, 9–11 April 2014; American Institute of Physics Inc.: Melville, NY, USA, 10 February 2016; Volume 1706.

140. Solodov, I.; Bai, J.; Bekgulyan, S.; Busse, G. A Local Defect Resonance to Enhance Acoustic Wave-Defect Interaction in Ultrasonic Nondestructive Evaluation. *Appl. Phys. Lett.* **2011**, *99*, 211911. [CrossRef]
141. Solodov, I.; Bai, J.; Bekgulyan, S.; Busse, G. A Local Defect Resonance to Enhance Wave-Defect Interaction in Nonlinear Spectroscopy and Ultrasonic Thermography. In Proceedings of the World Conference on Nondestructive Testing, Durban, South Africa, 16–20 April 2012.
142. Aymerich, F.; Staszewski, W.J. Impact Damage Detection in Composite Laminates Using Nonlinear Acoustics. *Compos. Part A Appl. Sci. Manuf.* **2010**, *41*, 1084–1092. [CrossRef]
143. Aymerich, F.; Staszewski, W.J. Experimental Study of Impact-Damage Detection in Composite Laminates Using a Cross-Modulation Vibro-Acoustic Technique. *Struct. Health Monit.* **2010**, *9*, 541–553. [CrossRef]
144. Ooijevaar, T.H.; Loendersloot, R.; Rogge, M.D.; Akkerman, R.; Tinga, T. Vibro-Acoustic Modulation Based Damage Identification in a Composite Skin-Stiffener Structure. *Struct. Health Monit.* **2016**, *15*, 458–472. [CrossRef]
145. Ooijevaar, T.H.; Loendersloot, R.; Warnet, L.L.; de Boer, A.; Akkerman, R. Vibration Based Structural Health Monitoring of a Composite T-Beam. *Compos. Struct.* **2010**, *92*, 2007–2015. [CrossRef]
146. Chen, B.Y.; Soh, S.K.; Lee, H.P.; Tay, T.E.; Tan, V.B.C. A Vibro-Acoustic Modulation Method for the Detection of Delamination and Kissing Bond in Composites. *J. Compos. Mater.* **2016**, *50*, 3089–3104. [CrossRef]
147. Prevorovsky, Z.; Krofta, J.; Kober, J.; Chlada, M.; Kirchner, A. Non-Linear Ultrasonic Spectroscopy of 3D Printed Metallic Samples. *Insight Non-Destr. Test. Cond. Monit.* **2019**, *61*, 157–161. [CrossRef]
148. Loi, G.; Uras, N.; Porcu, M.C.; Aymerich, F. Damage Detection in Composite Materials by Flexural Dynamic Excitation and Accelerometer-Based Acquisition. *IOP Conf. Ser. Mater. Sci. Eng.* **2022**, *1214*, 012007. [CrossRef]
149. Kim, S.; Adams, D.E.; Sohn, H.; Rodriguez-Rivera, G.; Myrent, N.; Bond, R.; Vitek, J.; Carr, S.; Grama, A.; Meyer, J.J. Crack Detection Technique for Operating Wind Turbine Blades Using Vibro-Acoustic Modulation. *Struct. Health Monit.* **2014**, *13*, 660–670. [CrossRef]
150. Willmann, E.; Boll, B.; Mikaelyan, G.; Wittich, H.; Meißner, R.H.; Fiedler, B. Vibro-Acoustic Modulation Based Measurements in CFRP Laminates for Damage Detection in Open-Hole Structures. *Compos. Commun.* **2023**, *42*, 101659. [CrossRef]
151. Ruiz, D.V.; de Bragança, C.S.C.; Poncetti, B.L.; Bittencourt, T.N.; Futai, M.M. Vibration-Based Structural Damage Detection Strategy Using FRFs and Machine Learning Classifiers. *Structures* **2024**, *59*, 105753. [CrossRef]

MDPI AG
Grosspeteranlage 5
4052 Basel
Switzerland
Tel.: +41 61 683 77 34

*Vibration* Editorial Office
E-mail: vibration@mdpi.com
www.mdpi.com/journal/vibration

www.ingramcontent.com/pod-product-compliance
Lightning Source LLC
Chambersburg PA
CBHW061612220926
51850CB00005B/486